Grasses

of COLORADO

"... all flesh is grass ..."
Isaiah 40:6

Grasses

of COLORADO

Robert B. Shaw

UNIVERSITY PRESS OF COLORADO

© 2008 by the University Press of Colorado

Published by the University Press of Colorado
5589 Arapahoe Avenue, Suite 206C
Boulder, Colorado 80303

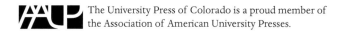 The University Press of Colorado is a proud member of
the Association of American University Presses.

The University Press of Colorado is a cooperative publishing enterprise supported, in part, by Adams State College, Colorado State University, Fort Lewis College, Mesa State College, Metropolitan State College of Denver, University of Colorado, University of Northern Colorado, and Western State College of Colorado.

∞ The paper used in this publication meets the minimum requirements of the American National Standard for Information Sciences—Permanence of Paper for Printed Library Materials. ANSI Z39.48-1992

Library of Congress Cataloging-in-Publication Data

Shaw, Robert Blaine, 1949–
  Grasses of Colorado / Robert B. Shaw.
    p. cm.
  Includes bibliographical references and index.
  ISBN 978-0-87081-883-7 (hardcover : alk. paper) — ISBN 978-1-60732-139-2 (pbk. : alk. paper) 1. Grasses—Colorado. 2. Grasses—Colorado—Identification. I. Title.
  QK495.G74S545 2008
  584'.909788—dc22

                                                                    2007037115

Design by Daniel Pratt

To Ashley and Chris

CONTENTS

Contents

PREFACE

This systematic treatment of Colorado grasses is meant to assist students, botanists, ecologists, agronomists, range scientists, naturalists, and other interested individuals in identifying members of this most important and unique plant family. It is intended to include all native, introduced, naturalized, and adventive taxa occurring within Colorado that have been collected outside of cultivation. Ornamentals are not addressed in this treatment unless they have escaped and persist outside of cultivation. Some native grass species that are used as ornamentals (*Panicum virgatum, Schizachyrium scoparium, Sorghastrum nutans,* and similar species) obviously will be found in the book.

Floral and vegetative structures within the Poaceae (Grass Family) are some of the simplest and most uniform in the plant kingdom. Unfortunately, a unique vocabulary is associated with them, and they are diminutive in size. To assist with the terminology, a glossary of scientific terms is included, as well as a brief description of the structures that compose the typical grass plant. For further assistance with the terms, one can use H. D. Harrington's marvelous little book, *How to Identify Grasses and Grasslike Plants* (1977), where variations among structures that compose the grass plant are discussed and illustrated. It is very difficult, if not impossible, to correctly identify grasses without the use of a binocular dissecting microscope. Magnification as well as the use of both hands for dissection and securing the very "jumpy" floral structures is vital. Full or partial illustrations are provided for each species to help in initially verifying the identification. I suggest that any identification made from a flora or treatment be considered "tentative" until it can be verified by a comparison with a documented herbarium specimen.

The book is much more a compilation than a scientific endeavor. As is the case with most treatments of this sort, it builds upon previous works related to the locale and taxonomic group. All agrostologists are indebted to A. S. Hitchcock for his invaluable contribution, *Manual of the Grasses of the United States* (1935), and to A. M. Chase for the 1951 revision. Many of H. D. Harrington's grass keys and descriptions for *Grasses of Colorado* (1946) and *Manual of the Plants of Colorado* (1954) come from Hitchcock's earlier work. The often used generic key from F. W. Gould and R. B. Shaw's *Grass Systematics* (1983) is modified for use in this book, as are many of the generic descriptions. Information on diagnostic and key characters and species characteristics is available in Hitchcock (1951), Cronquist et al. (1977), *Flora of North America*, vols. 24 and 25 (Barkworth et al. 2003, 2007), Weber and Wittmann (2001a, 2001b), and Janet Wingate's helpful book, *Illustrated Key to the Grasses of Colorado* (1994). "In ours" is implied for all descriptions, or, in other words, the descriptions only include characteristics of the subfamilies, tribes, genera, and species found in Colorado. Habitat information is found in Harrington (1954), Weber and Wittmann (2001a, 2001b), and Wingate (1994). County distribution data are available in the University of Colorado–Specimen Database of Colorado Vascular Plants, Colorado State University–Herbarium Database, Weber and Wittmann (1992, 2001a, 2001b), and Lynn Rubright's excellent grass atlas (2000). Additional county distribution data were made available by Ronald L. Hartman and B. Ernie Nelson, Rocky Mountain Herbarium, University of Wyoming, Laramie; Janet Wingate, Denver Botanic Gardens; and Jennifer Ackerfield, Colorado State Herbarium, Colorado State University, Fort Collins. Stubbendieck, Hatch, and Landholt's (2003) *North American Wildland Plants* provides interesting comments about uses and forage value of the most common range grasses.

Much of the credit for this work goes to Nancy Hastings, who compiled many of the descriptions, edited keys, and made numerous editorial changes. Thanks to Tracy Wager for the original illustrations and drawing. I am indebted to Marta Lynn Laven and Marie Held for assistance with the distribution maps and editing and to Lisa Blay and Melanie Arnett for early assistance in compiling descriptions. Paul Block was a great asset in editing, indexing, and word processing.

Permission was gratefully received from the University of Nebraska Press for use of figures on the grass plant, spikelet and floret structure, inflorescence types, and ligule forms from *North American Wildland Plants* (Stubbendieck, Hatch, and Landholt 2003).

Grass identification is difficult enough, but it is even more so without illustrations. Permission was graciously given to use the fine and nicely detailed illustrations from University of Washington Press [*Vascular Plants of the Pacific Northwest Part 1* (Hitchcock et al. 1969)] and Utah State University [*Flora of North America*, vols. 24 and 25 (Barkworth et al. 2003, 2007)]. Illustrators and their identifying marks are listed below.

Figures 15–22 were composed of representative spikelets taken from illustrations and used with the permission of Utah State University.

Identifying marks for the different illustrators are:

UTAH STATE UNIVERSITY

Ƙ	Karen Klitz	ꓞP	Hana Pazdírková
SL	Sandy Long	cꓦ	Cindy Roché
Ƈ	Annaliese Miller	Ƀ	Andy Sudkamp
ᗷ	Linda Bea Miller	ᵥᵥ	Linda Ann Vorobik

VASCULAR PLANTS OF THE PACIFIC NORTHWEST
JRJ   JR Janish

NORTH AMERICAN WILDLAND PLANTS
Bellamy Parks Jansen

CENTER FOR ENVIRONMENTAL
MANAGEMENT OF MILITARY LANDS (CEMML)
T@   Tracy Wager

Funding was provided by the Center for Environmental Management of Military Lands and the generosity of Dr. Joyce Berry, then dean of the Warner College of Natural Resources, Colorado State University.

Also, thanks to my many grass taxonomy students who have shown me many of the mistakes or misunderstandings in the various keys and descriptions. My wish is that this book helps everyone correctly identify their grass specimens.

Grasses

of C O L O R A D O

INTRODUCTION

The Poaceae (Gramineae) has approximately 785 genera and 10,000 species (Watson and Dallwitz 1992). Based on genera, this places the grasses as the third-largest family of flowering plants after the Asteraceae (sunflowers) and Orchidaceae (orchids). Grasses are fifth in the number of species behind the Asteraceae, Fabaceae (legumes), Orchidaceae, and Rubiaceae (madders) (Good 1953). Grasslands occupy a third of the earth's land surface (Schantz 1954), and almost two-thirds is used for grazing. Grasses occur on every continent and within nearly every terrestrial ecosystem. Hartley (1954) estimates that there are more individual grass plants than all other flowering plants combined. Based on completeness of representation in all regions of the world and on their percentage of the world's vegetation, grasses far surpass all others (Gould and Shaw 1983). The major uses of grasses described next support the assumption that they are by far the most important agricultural and ecological members of the flowering plants (Figure 1).

FOOD FOR HUMAN CONSUMPTION

The cereal grains (barley, corn, millet, oats, rice, rye, sorghum, wheat) supply the bulk of food consumed by humans (Figure 1). Rice feeds more people than any other food product. Wheat cultivation covers more area than any other crop. No crop covers a wider geographic range than corn. The world's main source of sugar is sugarcane. Even the "woody" grasses are a source of nutrition (e.g., bamboo shoots).

1

Figure 1. Grasses are most important as a source of food for humans (a. wheat field, b. harvesting corn, c. most barley production in the United States is used in the manufacturing of beer), as forage for wild and domestic animals (d. native pasture, e. hay cut from improved grass pastures), and in soil conservation (f. native range improvement practice).

The major dietary substance found in grass caryopses is carbohydrates. These carbohydrates are stored in the endosperm, nutritive tissue used during seed germination and seedling establishment, which, along with the embryo, is the major part of the grass caryopsis. The seed containing the endosperm is what the human race desires. It is what makes flour, cornmeal, rice, oats, beer, sake, and some whiskeys (Figure 1). This is why all cereal species are annual plants. Annuals put most of their energy into reproduction (creating more seeds) rather than roots and vegetative structures, which will die at the end of the growing season. The seeds used to ensure propagation of the annual species are harvested and put to multiple uses by humans. Carbohydrates in the endosperm are the substance upon which most civilizations have developed and been maintained. It is a farmer's ability to feed many individuals, not just himself and his family, that allows others to pursue such activities as the arts, manufacturing, trade, education, bureaucracies, and, alas, war—all part of the fabric of human civilizations. The classic example of endosperm is the large, white "exploded" portion of a piece of popcorn. The soft, sweet portion of a partially

Table 1. Area Planted (in the case of hay, area harvested), Production, and Value of Major Grass Crops in Colorado.[1] Data are from Colorado Agricultural Statistics 2005 (on-line).

Crop	Area (× 1,000)		Production (× 1,000)	Total Value (× 1,000)
Barley	80 a	32 h	9,086 b	$25,895
Corn				
Grain	1,200 a	486 h	140400 b	$301,860
Silage	100 a	40 h	5,400 t	$54,450
Oats	75 a	30 h	1,100 b	$1,980
Proso millet	370 a	150 h	7,920 c	$21,384
Grass hay	750 a	304 h	1,125 t	$76,500
Sorghum				
Grain	280 a	113 h	5,400 b	$9,223
Silage	19 a	8 h	266 t	$5,187
Wheat				
Winter	2,300 a	931 h	45,900 b	$146,880
Summer	15 a	6 h	980 b	$3,087
Grand Total	5,189 a	2,100 h	–	$646,446

1. Area is presented as acres (a) and hectares (h), while production is bushels (b), tonnage (t), or hundredweight (c).

popped kernel, sometimes referred to as "old maids" or "duds," is the embryo. The golden covering of the endosperm and embryo, which more often than not gets stuck between one's teeth, is the seed coat.

Roughly 13 percent of Colorado's land area was planted in cereal crops or harvested for hay in 2004 (Table 1). Corn, wheat, and grass hay were the largest grass products based on dollar value. Wheat had nearly twice as much area planted as corn did, but corn's total dollar value was more than twice that of wheat (Table 1). Corn, in fact, accounted for 55 percent of total grass products in Colorado in 2004 ($356,310,000).

FORAGE FOR WILD AND DOMESTIC ANIMALS

The majority of large herbivores characteristic of the world's grasslands are dependent on grasses as a major portion of their diet. Man, in turn, depends upon wild and domestic herbivores as a major source of protein and other nutrients (meat, blood, milk). These animals are also significant as the source of leather (hides) and animal fiber (wool, mohair, and the like). Livestock and livestock products were valued at over $3.5 billion in Colorado in 2004 (Colorado Agricultural Statistics 2005, on-line).

In Western societies, large infrastructures have been developed to supply these animal products to an ever enlarging and demanding population. Corn and sorghum silage is harvested for livestock in feedlots (Figure 1). Also, much of the corn, millet, and sorghum grain is used as feed for cattle, swine, and poultry (Table 1). Table 2 shows the estimated number of domestic livestock in Colorado. Colorado's highest rankings among the fifty states in terms of livestock are for the number of sheep and lambs (5th), wool production (5th), lamb crop (9th), and total number of cattle (10th).

While it is easy to envision the role and importance of grasses to the large herds of herbivores that once occupied the grasslands of Africa, Eurasia, and North America, one often forgets about the importance of grasses to other wildlife. There are many graminivorous

Table 2. Colorado Livestock Numbers and Rankings among All States. Data are from Colorado Agricultural Statistics 2005 (on-line).

Commodity	Unit	Number (× 1,000)	Categories[1]	Rank
Cattle and calves	head	2,500	—	10
All cows	head	—	740	22
Beef cattle	head	—	639	18
Calf crop	head	—	740	19
Milk cows	head	—	102	23
Chickens	head	4,991	—	23
Layers	head	—	3,960	24
All hogs and pigs	head	800	—	15
Pig crop	head	—	2,385	—
All sheep and lambs	head	365	—	5
Lamb crop	head	—	165	9
Wool production	pounds	2,570	—	5

1. Categories do not sum to total number by commodity.

upland birds and small mammals. Also, grasses compose a significant portion of waterfowl diets. Mannagrasses (*Glyceria*), cutgrasses (*Leersia*), and wildrices (*Zizania, Zizaniopsis*) are of special importance to birds that utilize swamps, lakes, and marshes. Some migratory birds not only use wetlands dominated by grasses but also overwinter on grain fields and improved pastures (Gould and Shaw 1983).

SOIL CONSERVATION AND LAND IMPROVEMENTS

A majority of the world's most productive soils now used for grain production were developed under perennial grassland cover (Gould and Shaw 1983). Removal of this native, perennial grass cover by plowing or poor grazing management has led to both wind and water erosion, and a large amount of topsoil has been lost over the centuries (Figure 1). Reestablishment of a perennial grass cover is a common practice in soil conservation and range improvements. A "good" cover of grasses not only stabilizes the site, reducing erosion, but helps in rebuilding depleted soils. Numerous state and federal agencies have developed over the past century to assist landowners in reducing erosion and returning lands to a more productive and economically viable state. Specific industries have developed, focusing on improving the condition of rangelands and restoring other disturbed sites (e.g., land rehabilitation, mine reclamation).

Some grasses adapted to restricted ecological niches have been used for specific erosion control purposes. For example, vetiver grass (*Vetiveria*) is used extensively in southern Asia and Oceania for erosion control and terrace production (National Research Council 1993). Beachgrass (*Ammophila*), with its extensive rhizomes, has been used to stabilize sand dunes. The same is true for the native grass *Redfieldia,* used to stabilize "blowouts" in southeastern Colorado.

TURF AND ORNAMENTALS

Often overlooked but not insignificant is the use of grasses for turf and ornamentals (Figure 2). Grasses with rhizomes, stolons, or both (sod formers) have been used for centuries to

Figure 2. Grasses are also used for turf and ornamentals. Ornamental grasses commonly used in Colorado include (a) *Saccharum ravennae*, (b) *Miscanthus sinensis*, (c) inflorescences of *M. sinensis*, (d) *Calamagrostis* x *acutifolia* "Karl Foerster," (e) *Pennisetum alopecuroides*, and (f) *Helictotrichon sempervirens*.

produce a dense, thick cover that will resist use (primarily foot traffic) and be aesthetically pleasing (generally of uniform color and density). Most common uses of turf grasses are for lawns, parks, highway right-of-ways, golf courses, and athletic fields of all sorts (Gould and Shaw 1983). Maintenance of turf supports a multi-billion-dollar industry, supplying seed, fertilizer, herbicides, insecticides, specialized machinery, sprinklers, hoses, and similar materials, as well as lawn care services. In Colorado the dominant turf grasses are familiar to most: bluegrasses (*Poa*), fescues (*Festuca*), bentgrasses (*Agrostis*), and ryegrasses (*Lolium*). Many of the turf grasses have escaped cultivation and occur as permanent members of our grass flora, and they are found in this book.

Recently, grasses have become much more popular as ornamentals. Bamboos (*Bambusa, Dendrocalamus, Phyllostachys,* and other genera) have traditionally been cultivated, but now many herbaceous grass species are being incorporated into human landscapes (Figure 2). This popularity is based on the relatively cost-effective propagation of herbaceous grasses, their ease of maintenance, adaptability to a wide range of soil textures and nutritive levels, and versatility. Their uses can vary from ground cover to single specimen plants (*Erianthus* [*Saccharum*], *Miscanthus*), mass planting (*Calamagrostis, Festuca, Pennisetum,* and other genera), and erosion control. Both exotic and native species (e.g., *Panicum, Andropogon, Sorghastrum, Chondrosum, Bouteloua*) are used. Colorado native taxa used for ornamentals will be found in this book, but the ornamentals that have not escaped cultivation are not included.

OTHER USES

I have heard that more structures are composed of bamboo than of stone, brick, and wood combined. While this may be an exaggeration, it points out the importance of bamboo to millions of people living in tropical Central and South America, Africa, Asia, and Oceania. Not only is bamboo used in construction, but it can be a source of nutrition (e.g., bamboo shoots) as well as an occasional source of grain in times of famine (Gould and Shaw 1983). The nonfood uses of bamboo are endless, but here are a few: poles and posts for building construction, fence posts, bridges, boat masts, ladders, cages, flooring; shafts, spears, bows and arrows, fishing poles; handles for tools, whips, knives; furniture; window shades; woven articles such as mats, roofs, baskets; rope and cordage; water conduits and drainpipes; musical instruments; toys; cooking utensils; and a number of miscellaneous items such as chopsticks, pipe stems, sieves, writing paper, facial tissue, and cigarette papers (Gould and Shaw 1983; Chapman 1996).

Lemongrass (*Cymbopogon*) is used as a flavoring in cooking and, along with vetiver grass (*Vetiveria*), is harvested for essential oils (National Research Council 1993). Other aromatic grasses (*Hierochloë, Anthoxanthum*) that contain coumarin are used for perfumes, hair tonics, and flavoring vodka.

HARMFUL GRASSES

Grasses are fairly benign, doing little harm to man or beast compared to the benefits they afford. That said, some fungi that use grasses as a host have caused serious problems throughout history, and the alkaloids produced by the fungus *Claviceps purpura* L. have caused many deaths when ingested in significant quantities. "Ergot" is the term most commonly used for these endophytic fungal diseases. Ergot outbreaks caused hundred of thousands of deaths in the Middle Ages after contaminated grain (rye, for the most part) was ingested (Lorenz 1979). "St. Anthony's fires" is the term often used for spontaneous hysteria caused by ergot poisoning. I recall hearing that the hallucinations and hysteria during the Salem witch episode in 1692 may have been caused by ergot poisoning. Infected kodo millet (*Paspalum scrobiculatum* L.), a staple food of many in northern India, has caused intoxication and poisoning. The fungus *Aspergillus tamaril* Kita, growing on these plants, produces cyclo-piazonic acid, which is toxic to man (Krishnamachari and Bhat 1978). Poisoning by ergot is very rare in modern times because of high agricultural grain inspection requirements. Surprisingly, the alkaloids produced by these fungi have shown the potential for medicinal use (Chapman 1996).

Along with ergot infestation causing fungal secondary compound toxicity, Allred (2005) lists these harmful compounds produced by grasses that can cause poisoning: coumarin, cyanide, nitrates, and oxalates. He also lists other harmful effects of grasses, including dermatitis, hay fever, asthma, and photosensitivity. Often forgotten are the harmful impacts caused by mechanical injury. It is generally the awns or stiff bristles that penetrate soft tissues of the eyes, nose, throat, and underbelly that cause the problem (Allred 2005).

Some feel the accumulation of secondary compounds, mutualism with fungi, and mechanical defenses are adaptations to combat herbivory (Chaplick and Clay 1988; Vicari and Bazely 1993).

PHYSIOGRAPHY AND ECOREGIONS OF COLORADO

Colorado is the eighth largest of the fifty states and covers 104,100 sq mi (269,618 sq km). The highest point in the state is Mount Elbert, at 14,433 feet (4,400 m) above sea level. The lowest point is where the Arkansas River exits the state on the eastern border, at 3,350 ft (1,021 m) above sea level. This makes Colorado the highest U.S. state, with an average elevation of 6,800 ft (2,073 m) (Colorado Geography–NETSTATE, on-line).

Perhaps it is this spectacular variation in elevation that produces the variety of habitats that support such a large diversity of grasses. As a comparison, Texas, considered to have one of the most diverse grass floras of any state (523 species), has over 2.5 times the land area of Colorado (Gould 1975). This current book reports 335 species from Colorado. Dividing the total land area by the number of species provides an estimate of area per species that can be used as a measure of diversity. Colorado averages about 310 sq mi (805 sq km) per species, while Texas has 500 sq mi (1,295 sq km) per species. Thus, based on land area, the grass flora in Colorado is approximately 1.5 times more diverse than that in Texas.

The review that follows of the variation in physiography and ecoregions of Colorado might help explain the richness of grass species found within the state.

PHYSIOGRAPHY

Geomorphologists have divided Colorado into five more or less homogeneous landforms:

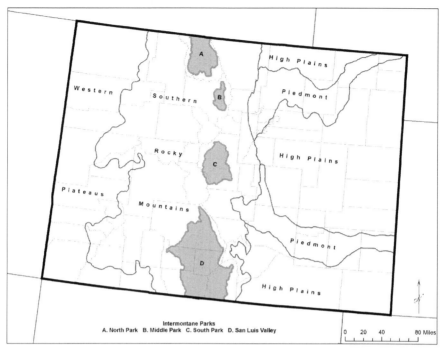

Figure 3. The five major landforms (High Plains, Piedmont, Southern Rocky Mountains, Intermountain Parks, and Western Plateaus) found in Colorado (Griffiths and Rubright 1983).

High Plains, Piedmont, Southern Rocky Mountains, Intermountain Parks, and Western Plateaus (Griffiths and Rubright 1983; Weber and Wittmann 2001a, 2001b). Figure 3 shows that the High Plains and Piedmont form the eastern part of the state, the Southern Rocky Mountains and Intermountain Parks constitute the central region, and the Western Plateaus are confined to the western part of Colorado. Characteristics of each major landform are presented next and represent a brief summary of a much more detailed account of Colorado geology and physiography in the fine works by Chronic and Chronic (1972) and Griffiths and Rubright (1983). Weber and Wittmann (2001a, 2001b) also provide interesting notes on the complexity of Colorado landscapes and vegetation.

High Plains

The northern and central parts of the High Plains are a gently rolling upland with extensive areas of flat plain interspersed with eroded areas of broad rolling hills, all sloping gently to the east. The entire region is underlain by Jurassic-age (180–135 million years before present [YBP]) limestones, shales, and sandstones. Low bluffs occur along streams, and buttes may outline the edge of rocky outcrops. The section of the High Plains south of the Arkansas River piedmont area is quite different from the plains to the north. During the Cenozoic era (70 million YBP), volcanic activity near the vicinity of the Spanish Peaks covered much of this area with lava. The lava also intruded into weakened surface fissures. Volcanic rocks are more resistant to erosion than the surrounding materials are, thus radiat-

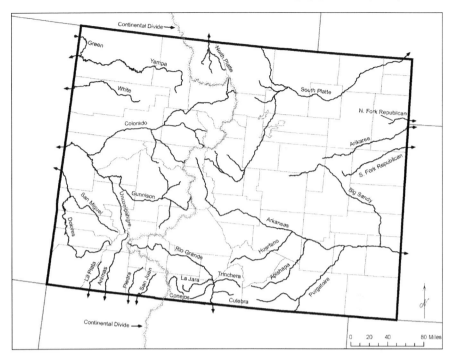

Figure 4. Major river systems and location of the Continental Divide in Colorado (Griffiths and Rubright 1983).

ing vertical dikes are a prominent feature in the area. Mesas protected by lava caps rise several hundred feet above the plains. Deep canyons are formed along the southern tributaries (Purgatorie, Huerfano, and Apishapa) of the Arkansas River (Chronic and Chronic 1972; Griffiths and Rubright 1983).

Piedmont

Two principal river systems, the South Platte and Arkansas, have eroded broad valleys several hundred feet below the level of the High Plains. These areas are known as the Piedmont (Figures 3 and 4). The South Platte has its origin in the mountains surrounding South Park. Several significant perennial mountain tributaries (Bear Creek, Clear Creek, Boulder Creek, St. Vrain, Big and Little Thompson, and Cache la Poudre) empty into the South Platte between Denver and Greeley, which greatly increases the river's volume and erosiveness. East of where the tributaries join, the South Platte Valley reaches over 50 miles (80 km) wide and nearly 900 ft (274 m) deep. Southeast of Greeley, prevailing northwesterly winds have built up a broad belt of curving, longitudinal sand dunes. The Arkansas Valley section of the Piedmont is much narrower and shallower than the South Platte Valley because of a lack of mountain tributaries enlarging the river. Also, sand dune development is diminished because of the narrowness of the valley along the Arkansas (Griffiths and Rubright 1983).

11

Southern Rocky Mountains

The Southern Rocky Mountains are the result of a global mountain-building episode that occurred between the end of the Cretaceous (70 million YBP) and the early Tertiary (10 million YBP). This major uplifting and folding, which formed the Rocky Mountain system, is called the Laramide Revolution or Orogeny. The mountains are not a single, connected barrier but rather a group of separate ranges (Figures 3 and 5). Generally, the ranges are oriented north-south, in many places they are parallel, and sometimes their ends overlap. Occasionally, the semi-isolated portions of larger ranges have individual names. Each distinct range has its own unique geologic history. The mountains were perhaps twice as tall as they are now, but the mountains as we see them are the result of the erosive power of water, wind, ice, and gravity (Chronic and Chronic 1972; Griffiths and Rubright 1983).

Intermountain Parks

Four large Intermountain Parks (North Park, Middle Park, South Park, and the San Luis Valley) were formed during the major mountain-building period (Figure 3). These basins are surrounded by mountain ranges and are the result of faulting and folding that created the ranges. North Park and Middle Park constitute one large basin divided by a ridge, the Rabbit Ear Range, running east to west. The North Platte River originates in North Park,

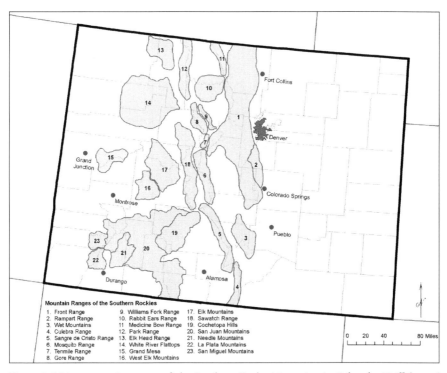

Mountain Ranges of the Southern Rockies

1. Front Range	9. Williams Fork Range	17. Elk Mountains
2. Rampart Range	10. Rabbit Ears Range	18. Sawatch Range
3. Wet Mountains	11. Medicine Bow Range	19. Cochetopa Hills
4. Culebra Range	12. Park Range	20. San Juan Mountains
5. Sangre de Cristo Range	13. Elk Head Range	21. Needle Mountains
6. Mosquito Range	14. White River Flattops	22. La Plata Mountains
7. Tenmile Range	15. Grand Mesa	23. San Miguel Mountains
8. Gore Range	16. West Elk Mountains	

Figure 5. Major mountain ranges of the Southern Rocky Mountains in Colorado (Griffiths and Rubright 1983).

and the headwaters of the Colorado River are in Middle Park (Figure 4). The South Platte River drainage begins in the mountains surrounding South Park. Two prominent features of South Park are the ridge (Red Hill) running north to south down the middle of the basin and the Elkhorn thrust fault along its eastern edge. The San Luis Valley is the largest and most southern of the Intermountain Parks. The Rio Grande, which begins in the San Juan Mountain Range, flows through the basin. Prevailing winds during the late Pleistocene (2 million YBP until 10,000 YBP) and Recent times have deposited sand and silt into large sand dunes (Great Sand Dunes National Park and Preserve) on the southwestern slopes of the Sangre de Cristo Range (Chronic and Chronic 1972; Griffiths and Rubright 1983).

Western Plateaus

Sedimentary beds cover most of the western flanks of the Southern Rockies. These sandstone, siltstone, and shale beds are ancient (mainly Jurassic Period [180–135 YBP] deposits) remains of what once covered most of the western portion of the North American continent. More resistant rocks cap the mesas and plateaus. The most prominent plateaus from north to south are the White River, Roan, Grand Mesa, Uncompahgre, and Mesa Verde. The more erodible substrates have been removed, forming wide valleys or basins (Green River, Uinta, and Paradox). The Uncompahgre, Gunnison, Colorado, White, Green, and Yampa rivers drain most of the area and have cut impressive canyons across the mountains and plateaus (Figures 3, 4) (Chronic and Chronic 1972; Griffith and Rubright 1983).

ECOREGIONS

Bailey's (1995) ecoregions are based on a hierarchical scheme using climate and vegetation to delineate ecosystems of regional extent. Three levels or categories to this hierarchy are the domains, divisions, and provinces. Domains and divisions are based primarily on the large ecological climate zones defined by Köppen (1931) and refined by Trewartha (1968). Each division is divided into provinces based on vegetative macrofeatures. Mountainous regions, denoted by an M, exhibit altitudinal zonation of the climatic regime from the adjacent lowlands and are distinguished according to the zonation. Figure 6 delineates the ecoregion hierarchical scheme for Colorado.

Divisions

Three ecoregion divisions are found in Colorado: Tropical/Subtropical Steppe (310), Temperate Steppe (330), and Temperate Desert (340) (Figure 6). The Temperate Steppe Division covers nearly two-thirds of the state (Figure 7). The Tropical/Subtropical Steppe is slightly warmer than the Temperate Steppe, while the Temperate Desert is drier than both.

Tropical/Subtropical Steppe (310). This division is characterized by a hot, semiarid climate in which potential evaporation exceeds precipitation and all months have temperatures above 32° F (0° C). Vegetation is typically grasslands composed of short grasses, with locally developed shrub land, woodland, or both. Soils are usually Mollisols and Aridisols.

Temperate Steppe (330). This division is characterized by a semiarid continental climate with maximum summer rainfall, but evaporation still exceeds precipitation. A Temperate Steppe differs from a Tropical/Subtropical Steppe in having a cooler climate, with at least

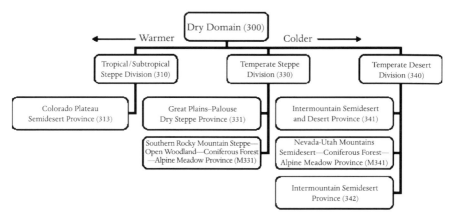

Figure 6. Schematic illustrating the hierarchical classification (domains, divisions, and provinces) of the ecoregions found in Colorado (Bailey 1995).

one month of average temperatures below 32° F (0° C). Winters are cold and dry, while summers are hot. Soils are predominantly Aridisols and Mollisols.

Temperate Desert (340). This division has low rainfall and strong temperature contrasts between winter and summer. In contrast to the Temperate Steppe, most precipitation falls as snow in the Temperate Desert, creating a short humid season and pronounced drought throughout the remainder of the year. Aridisols occur in the lowlands and Mollisols in the uplands.

Provinces

Colorado Plateau Semidesert (313). This province, the only member of the Tropical/Subtropical Steppe Division to occur in Colorado, covers a very small area near the Four Corners region of the state. The Colorado Plateau consists of tablelands and canyonlands, with moderate relief in our area. There are a number of streams with extremely fluctuating flows. The climate is characterized by cold winters, hot summer days, and cool summer nights. Annual temperatures are 40°–50° F (4°–13° C) and decrease with increasing elevation. Annual precipitation is about 10–20 in (260–510 mm). Summer rains are thunderstorms, and predictable rains occur in the winter. Vegetation zonation is conspicuous but not uniform, with arid grasslands with shrubs at lower elevations, woodlands, and coniferous forests at the higher elevations. Succulents are prominent at lower elevations.

Great Plains–Palouse Dry Steppe (331). This province is a member of the Temperate Steppe Division and covers most of the eastern half of the state (Figure 7). It includes the High Plains, Piedmont, and San Luis Valley landforms. The plains are typically flat, with valleys and canyons created by rivers and streams that dissect the area. The plains lie in the rain shadow of the Rocky Mountains, and the climate is a semiarid continental regime. There are typically six to seven arid months per year. Winters are cold and summers hot. Average annual temperatures vary from 45° to 60° F (7° to 16° C), while precipitation

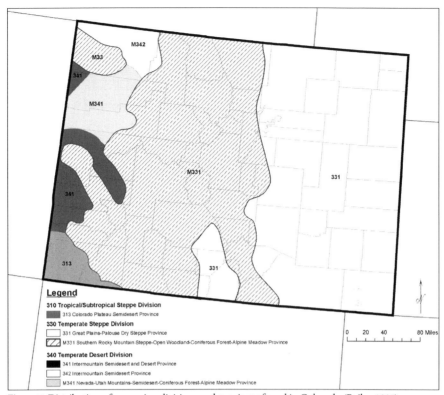

Figure 7. Distribution of ecoregion divisions and provinces found in Colorado (Bailey 1995).

ranges from 10 to 25 in (260–640 mm). Rainfall is usually greatest in the summer. The vegetation is shortgrass steppe, with low grasses sparsely spaced and a significant amount of bare ground.

Southern Rocky Mountain Steppe–Open Woodland–Coniferous Forest–Alpine Meadow (M331). This province is a member of the Temperate Steppe Division and covers most of the central portion of the state. It includes the Southern Rocky Mountains and North, Middle, and South Parks of the Intermountain Park landforms (Figures 3, 7). Obviously, the terrain is rugged and rocky, with elevations exceeding 14,000 ft (4,300 m). The climate is temperate semiarid, with average annual temperatures varying from 35° to 45° F (2°–7° C). Climate is influenced by prevailing westerly winds and the predominantly north-south orientation of most of the mountain ranges (Figure 5). The Eastern Slope of the Southern Rocky Mountains is typically more arid than the Western Slope. Winter precipitation, as snow, varies with elevation and exposure. Mountain bases may only receive 10 in (240 mm) of precipitation annually, while mountaintops may exceed 40 in (1,020 mm) in precipitation. A pronounced vegetative zonation is caused by a combination of elevation, latitude, prevailing winds, and slope exposure. The various zones are typically higher in the south than in the north, and they extend downward on east-facing and north-facing slopes into

15

ravines and valleys. The lowest zone is usually an arid grassland steppe with numerous shrubs that integrates upslope, with an open woodland composed primarily of oak (*Quercus*), juniper (*Juniperus*), and piñon (*Pinus*) and an abundance of shrubs. The coniferous forest types from lowest to highest are generally the ponderosa pine (*Pinus ponderosa*), Douglas fir (*Pseudotsuga menziesii*), lodgepole pine (*P. contorta*), and spruce-fir (*Picea-Abies*). The familiar aspen *(Populus tremuloides)* occurs in riparian and disturbed sites in all the coniferous forest types. The extremely diverse alpine meadow is the uppermost zone and is characterized by a lack of trees.

Intermountain Semidesert / Desert (341). This province is a member of the Temperate Desert Division and occupies a very small area of western Colorado along the border with Utah. It is considered an extension of the Great Basin and northern Colorado Plateau. Summers are hot, and winters are moderately cold. Average annual temperatures range from 40° to 50° F (4° to 13° C). Annual precipitation is only 5–20 in (130–490 mm), and most of that moisture comes in the form of snow, with little summer precipitation except in the mountains. Big sagebrush (*Artemisia tridentata*) and a variety of other shrubs dominate at lower elevations. On soils with high saline or alkaline content, the vegetation is dominated by greasewood (*Sarcobatus vermiculatus*) and saltgrass (*Distichlis spicata*).

Nevada-Utah Mountains Semidesert–Coniferous Forest–Alpine Meadow (M341). This province is a member of the Temperate Desert Division and represents a small area of mountains that arise in the northern portion of the Colorado Plateau (Figure 7). The majority of this province is west of Colorado, as the name implies. This area is a high-elevation variation of the Intermountain Semidesert / Desert Province, with a very pronounced drought season and a short humid period. Most precipitation occurs as snow in winter. Average annual temperatures range from 38° to 50° F (3° to 10° C). Average precipitation varies from 5–8 in (130–200 mm) in the lower areas to 25–35 in (640–890 mm) at higher elevations. Big sagebrush dominates at lower elevations, and piñon and juniper are the dominant woodland vegetative type. The montane zone is generally ponderosa pine and Douglas fir grading into spruce-fir at the highest elevations. Few mountains rise high enough to support an alpine meadow in this province.

Intermountain Semidesert (342). This province is a member of the Temperate Desert Division and is a small portion of the Wyoming Basin that extends into northwestern Colorado (Figure 7). Winters are typically cold and summers hot. Average annual temperatures range from 40° to 52° F (4° to 11° C). Average annual precipitation is 5–14 in (130–360 mm) and is evenly distributed throughout the year. Various types of sagebrush with interspersed short and mid-grasses constitute the characteristic vegetation of the Wyoming Basin portion of this province. Greasewood and saltgrass dominate alkaline areas.

The GRASS PLANT

The individual flowering grass tiller consists of extremely simple structures. A phenomenal amount of diversity, however, occurs within these structures, making up the characteristics that differentiate over 700 genera and 10,000 species of grasses.

ROOTS

At the time of germination, the grass seed, or caryopsis, extends along two axes. The radicle, protected by the coleorhiza, produces the primary root system, which begins to stabilize the seedling and absorbs water and nutrients from the soil. At the other end of the embryo the plumule, protected by the coleoptile, elongates upward to become the primary shoot system (Figure 8). This aboveground portion is responsible for photosynthesis, supplying the plant with energy to grow and reproduce.

The primary root system functions for only a few short weeks or months and soon withers and dies. There is no taproot in grasses, as is typically found in herbaceous forbs or woody plants. For the remainder of the life of the grass, belowground function is performed by a series of adventitious roots that compose the secondary root system. These adventitious roots are fibrous and arise from the lowermost nodes of the grass culm (stem). Occasionally, some grasses (*Zea*) develop stout "prop" roots from the lower aboveground culm nodes. These roots are necessary for mechanical support of the large aboveground portion of the plant.

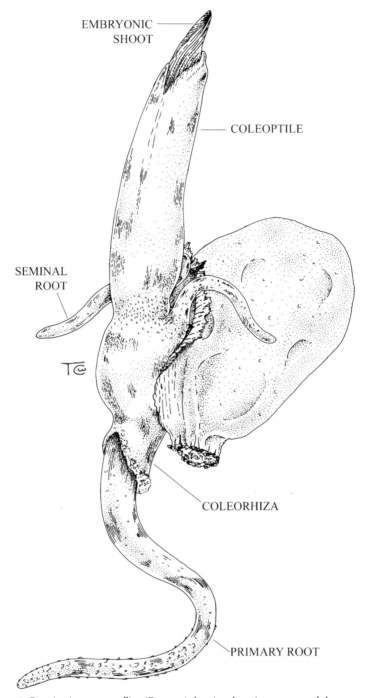

EMBRYONIC SHOOT

COLEOPTILE

SEMINAL ROOT

COLEORHIZA

PRIMARY ROOT

Figure 8. Germinating corn seedling (*Zea mays*) showing the primary root and shoot system.

CULMS (STEMS)

Grass culms are made up of swollen nodes, while the internodes are typically smooth, round, or elliptical in cross-section and elongated. Nodes are always solid, while the internodes may be solid, semisolid, or hollow. Grass leaves, branches, and adventitious roots arise from meristematic tissue near the nodes. It is convenient to think of the aboveground portion of a grass plant as composed of a series of "building blocks," which agrostologists call phytomeres. A phytomere, or basic unit of the grass shoot, is composed of a node, internode, meristematic tissue, and leaf (Figure 9). Usually, the internodes are very short during the major portion of the life cycle, particularly in grasses with a cespitose growth form. This low growth form also maintains the apical meristem (growing point) of the plant near the soil surface, protecting it from grazing animals and wildfires. In this stage the grass plant appears to be nothing more than a clump of leaves. However, when flowering occurs the internodes elongate, pushing the reproductive portion of the plant upward. Meristematic tissue (vegetative bud), located just above the basal node of the phytomere in the axil of the leaf, can develop into several structures. In most cases it produces a tiller or lateral branch; less frequently a horizontal culm develops. If the horizontal culm is underground it is called a rhizome, but if the horizontal culm is located aboveground it is referred to as a stolon (Figure 9). Some grasses may have both stolons and rhizomes (*Cynodon dactylon*).

LEAVES

Grass leaves are two-ranked and alternate. Thus, at each node a leaf is oriented 180° from the previous one. The grass leaf is differentiated into a basal portion (sheath) that encloses the culm and an upper portion (blade or lamina). The sheath typically has free margins, but in some genera the sheath margins are connate, forming a tubular-like structure (*Bromus, Glyceria,* and *Melica*). The blade is characteristically flat and elongated. In arid environments the blade may be terete, very narrow, or have variously rolled margins, while in forested areas the blades may become broad and ovate. Generally, at the inside (adaxial) junction of the leaf sheath and the leaf blade there is a ligule. The ligule may be a ring of hairs, membranous, or a ciliated membrane (Figures 9, 10). Rarely is it absent (*Echinochloa*). The ligule is the most often used and reliable vegetative character for grass identification. Occasionally, auricles, or finger-like appendages, are present at the top of the sheath or at the base of the blade.

INFLORESCENCE

The inflorescence, or flowering portion of the plant, is an excellent diagnostic characteristic in grass identification. The stalk upon which the inflorescence sits is called the peduncle, while the central axis of the grass inflorescence is the rachis (Figure 9). There are three basic types of grass inflorescences: the spike, the raceme, and the panicle.

The spike is the simplest of the inflorescence types (Figure 11). It consists of a central axis (rachis) and sessile flowering bodies. The next level of complexity occurs in the raceme, which has a rachis and flowering bodies borne upon a pedicel. A spicate raceme is a combination of the first two in which both sessile and pediceled flowers occur on the same rachis. The most common inflorescence type in grasses is the panicle. A panicle can be defined as an inflorescence that is at least branched and rebranched, or as all inflorescences in which the flowering bodies are not sessile or individually pediceled on the main axis

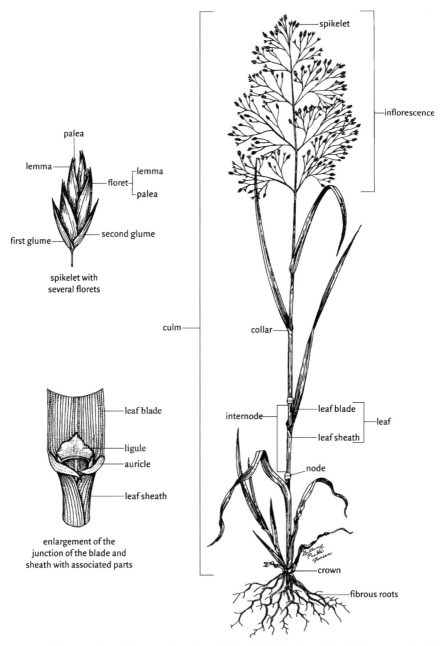

Figure 9. Characteristics of the grass plant (from Stubbendieck, Hatch, and Landholt 2003, used with permission of University of Nebraska Press).

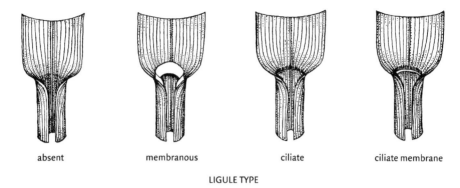

absent   membranous   ciliate   ciliate membrane

LIGULE TYPE

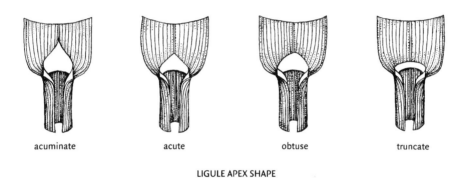

acuminate   acute   obtuse   truncate

LIGULE APEX SHAPE

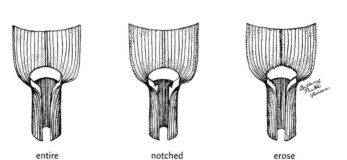

entire   notched   erose

LIGULE MARGIN

Figure 10. Variation in ligule type, apex shape, and margins (from Stubbendieck, Hatch, and Landholt 2003, used with permission of University of Nebraska Press).

(rachis). Figure 12 illustrates the enormous variation that occurs within this inflorescence type. Considerable variation in inflorescence form also is seen within the tribe Andropogoneae (Figure 13). Inflorescences range from multiple flowering culms composed of a single raceme (*Schizachyrium*) to complex panicles (*Sorghastrum*).

Caution should be used when deciding upon inflorescence type, especially with extremely contracted panicles. Always examine the lowermost flowering bodies to determine if they are indeed single and sessile on the rachis. Many misidentifications occur when genera with a reduced or contracted panicle (*Alopecurus, Phleum, Cenchrus, Pennisetum,* and others) are keyed as if they had a spike inflorescence.

SPIKELETS

The flowering bodies in grasses are called spikelets and are the basic unit of the grass inflorescence (Figure 14). Typically, the grass spikelet consists of a short axis (rachilla) upon which two series of floral bracts (modified leaves) subtend the grass flower. The first (lower) and second (upper) glumes compose the initial series of bracts. Both glumes are usually present, but occasionally the first may be reduced in size (*Digitaria, Paspalum,* and others) or completely absent (*Eriochloa*). The lower and upper glumes may be partially fused (*Alopecurus* and some species of *Polypogon*). The glumes become setaceous in *Critesion* and some *Elymus*. Very rarely, both glumes are lacking (*Leersia*).

The second series of bracts consists of the lemma and palea (Figure 14). The lemma is probably the most reliable and frequently used character in grass identification. It is always present and has a high degree of stability within a genus. Lemma texture, shape, nervation, awn development, and surface features are used extensively in identification.

The palea is usually two-nerved and two-keeled. It is less variable than the lemma and not as useful in differentiating between taxa. Together the two glumes, lemma, and palea enclose and protect the grass flower.

The grass flower is terminal on a short axis above the subtending lemma and palea. The lowermost organs of the grass flower are the lodicules, of which there are usually two. The lodicules are considered reduced perianth structures, and they function to help open the grass flower at anthesis. Above the lodicules is the anthecium (male floral parts), composed of the stamens. There are generally three stamens, consisting of a long, slender filament and a pollen-producing anther. Occasionally, stamen number is reduced to 1 or 2; conversely, they can reach as many as 120 or more in some bamboos. The gynecium (female portion of the flower) is a one-locular ovary with a single ovule and usually two styles and stigmas. The grass fruit, a caryopsis, is a dry, indehiscent, one-seeded structure in which the ovary wall and pericarp are fused. In some genera (*Sporobolus, Eleusine*) the pericarp does not fuse with the ovary wall, producing a fruit called an achene.

Spikelet sexuality is variable in the Poaceae. For the most part, grass spikelets are hermaphroditic or perfect (both male and female parts are present and function appropriately); however, occasionally they are unisexual or imperfect (lacking either the male or female reproductive parts). If the staminate and pistillate spikelets are present on the same plant, as in *Zea*, the plant is monoecious. If the staminate and pistillate spikelets are on separate plants, the plant is dioecious. *Buchloë dactyloides, Scleropogon brevifolius,* and *Distichlis spicata* are examples of dioecious grasses. Complete sterility of some spikelets in an inflorescence also occurs. In some species of *Critesion*, setaceous glumes are all that remain of lateral spikelets. Within the tribe Andropogoneae, many members have paired spikelets, one sessile

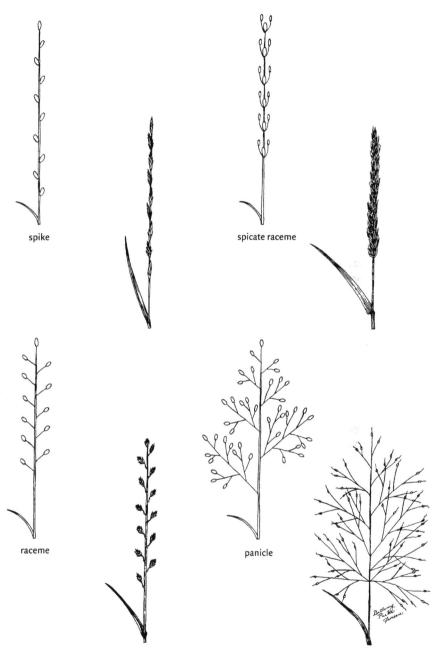

Figure 11. The four common grass inflorescence types (from Stubbendieck, Hatch, and Landholt 2003, used with permission of University of Nebraska Press).

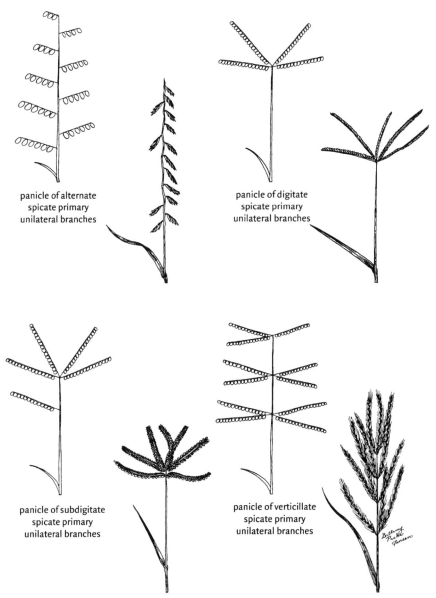

panicle of alternate
spicate primary
unilateral branches

panicle of digitate
spicate primary
unilateral branches

panicle of subdigitate
spicate primary
unilateral branches

panicle of verticillate
spicate primary
unilateral branches

Figure 12. Variation in the panicle-type grass inflorescence (from Stubbendieck, Hatch, and Landholt 2003, used with permission of University of Nebraska Press).

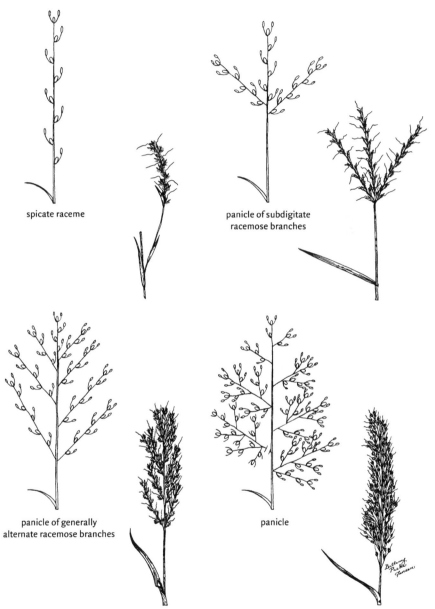

spicate raceme

panicle of subdigitate
racemose branches

panicle of generally
alternate racemose branches

panicle

Figure 13. Variation in inflorescences within the Andropogoneae (from Stubbendieck, Hatch, and Landholt 2003, used with permission of University of Nebraska Press).

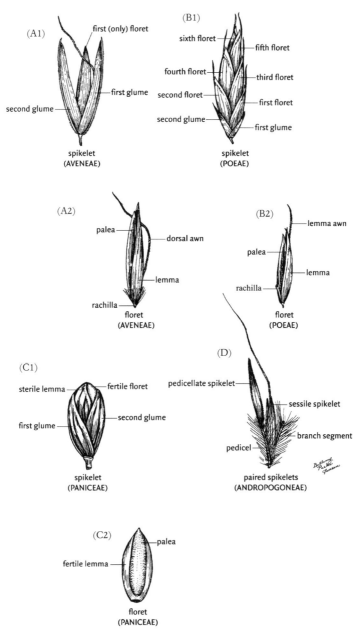

Figure 14. Variation in the grass spikelet. (A1) spikelet with one fertile floret, (A2) single (only) fertile floret, (B1) spikelet with more (in this case six) than one floret per spikelet, (B2) one of the six fertile florets, (C1) typical "panicoid" spikelet with a reduced (sterile) floret below and one fertile floret above, (C2) fertile floret from typical "panicoid" spikelet, (D) paired spikelets typical of the Andropogoneae (from Stubbendieck, Hatch, and Landholt 2003, used with permission of University of Nebraska Press). (Note: the Aveneae is not recognized in this work.)

and one pedicellate. Often the pedicellate spikelet is staminate, sterile, or in some cases completely lacking, leaving only a naked pedicel.

Disarticulation, or how the spikelets separate from the plant at maturity, has long been a diagnostic characteristic used in grass taxonomy. There are two general types of disarticulation: above the glumes and below the glumes. Disarticulation above the glumes means the glumes remain in the inflorescence and the lemma, palea, and flower separate from the plant. Conversely, disarticulation below the glumes means the glumes fall with the lemmas, paleas, and flowers, leaving nothing in the inflorescence except a naked pedicel or rachis. Rarely, the entire inflorescence will break away from the plant at maturity (*Schedonnardus* and some species of *Eragrostis* and *Aristida*). Also, in genera such as *Cenchrus* and *Pennisetum,* several spikelets are clustered together, subtended by an involucre of bristles or spines, and disarticulation is below the involucre. Members of the Triticeae tribe have spike or spicate raceme inflorescences, and disarticulation in some species occurs at the rachis nodes. Additionally, several members of the Cynodonteae tribe have spikelets that fall in clusters on short inflorescence branches (*Buchloë, Bouteloua,* and *Chondrosum*). Theoretically, these extreme forms of disarticulation are considered below the glumes because no glumes remain on the plant.

FLORETS

The basic unit of the grass spikelet is the floret (Figure 14). A floret consists of a lemma, palea, and flower. Thus the simplest grass spikelet that has all the parts would consist of a pair of glumes, lemma, palea, and perfect flower. This would be called a spikelet with one perfect floret. Because grass flowers may be perfect, staminate, pistillate, or sterile, these terms are also applied to spikelets and florets. Often the term "fertile flower," "fertile spikelet," or "fertile floret" will be used. This refers to a flower that has a functioning pistil and is capable of producing a fruit, regardless of the presence or absence of stamens. The next level of complexity after a spikelet with one perfect floret would be a spikelet with two perfect florets. This would consist of a pair of glumes; one floret consisting of a lemma, palea, and flower; plus a second floret with a lemma, palea, and flower. Complexity can continue until one sees spikelets with dozens of perfect florets. As in spikelets, reduction in the reproductive capacity of florets also occurs. Reduced florets would be either staminate or sterile. A staminate floret would have a lemma, palea, and functioning stamen, while a sterile floret would lack any functioning reproductive organs. Often, reduced florets consist of "empty" lemmas and paleas, lacking flowering parts within the protective envelope created by the lemma and palea. Occasionally, reduced florets lack the lemma and palea and are recognized by a naked rachilla extending past the uppermost fertile floret. In the Paniceae and Andropogoneae tribes, the spikelet consists of two florets. The upper floret is fertile, while the lower is reduced and either staminate or sterile. In most other cases the reduced florets are typically above the fertile florets.

Number of florets per spikelet, presence or absence of reduced florets, and location of reduced florets (above or below the fertile ones) are important diagnostic characteristics. Unfortunately, these are some of the most frequently misinterpreted and overlooked features in grass identification.

GRASSES *of* COLORADO

The classification for the grasses of Colorado follows the subfamilies and tribes presented by the Grass Phylogeny Working Group (2001). The current trend in grass classification has been to "split," or increase, the number of subfamilies. Within Colorado, seven of the fifteen grass subfamilies are present (Table 3). As throughout North America, the grass flora of Colorado is dominated by three large subfamilies. The Pooideae is the largest and exceeds the total number of tribes, genera, and species of the two next-largest subfamilies combined (Chloridoideae and Panicoideae). The Pooideae are cool-season grasses and are most common in the state's mountainous regions. They do, however, extend into the more arid regions of the east and west slopes in mesic riparian habitats or occur as cool-season plants that complete their life cycle prior to the onset of hotter and drier conditions. The second-largest subfamily, the Chloridoideae, is composed of warm-season grasses generally found in the arid habitats of the southeastern and southwestern areas of the state. Members of the Panicoideae are more characteristic of mesic to arid habitats and are most common in the eastern grasslands. Four of the seven subfamilies are represented by a single tribe and genus. *Aristida,* with eight species, is the only genus in the Aristidoideae, while *Danthonia,* with five species, is the only genus in the Danthonioideae. The only member of the Arundinoideae is *Phragmites australis*, and the only species in the Ehrhartoideae is *Leersia oryzoides*.

Opposite to the trend of "splitting" subfamilies, tribes have been "lumped," or decreased in number, over the last several decades. More recent modifications to tribal delineations are:

1. The Aveneae is now included in the Poeae, while the bromegrasses (*Anisantha, Bromopsis, Bromus, Ceratochloa*) have been placed in the Bromeae.
2. The Eragrosteae and Chlorideae have been merged and are now included in the Cynodonteae.

Table 3. Number of Grass Tribes, Genera, Species, Subspecies, and Varieties by Subfamily Found in Colorado.

	Subfamily	Tribes	Genera	Species	Subspecies	Varieties
I.	Aristidoideae	1	1	8	0	6
II.	Arundinoideae	1	1	1	0	0
III.	Chloridoideae	2	27	81	4	12
IV.	Danthonioideae	1	1	5	0	0
V.	Ehrhartoideae	1	1	1	0	0
VI.	Panicoideae	2	15	41	11	5
VII.	Pooideae	6	67	198	47	14
	Totals	14	113	335	62	37

Treated in this work are 113 genera, including a number of "new" genera. My generic concept is rather narrow and follows the trend of some agrostologists to elevate many of the subgenera found in Hitchcock's (1935, 1951) grass manual to generic level. I support, in general, the Colorado grass genera put forth by Weber and Wittmann (2001a, 2001b). Allred (2005) would label me a "liberal" agrostologist in my generic concepts. The most recent modifications are:

1. *Agropyron* to *Agropyron, Elymus, Leymus, Pascopyrum, Pseudoroegneria, Thinopyrum*
2. *Agrostis* to *Agrostis, Podagrostis*
3. *Bouteloua* to *Bouteloua, Chondrosum*
4. *Bromus* to *Anisantha, Bromopsis, Bromus, Ceratochloa*
5. *Festuca* to *Argillochloa, Festuca, Schedonorus*
6. *Hordeum* to *Critesion, Hordeum*
7. *Melica* to *Bromelica, Melica*
8. *Phalaris* to *Phalaris, Phalaroides*
9. *Stipa* to *Achnatherum, Hesperostipa, Jarava, Nassella*
10. *Trisetum* to *Graphephorum, Trisetum*

There are 335 species represented in this treatment. The genera with the largest number of species are *Poa* (24), *Muhlenbergia* (23), *Festuca* (15), *Eragrostis* (14), *Sporobolus* (11), and *Achnatherum* (11). These six genera represent 25 percent of the grasses found in the state.

It is up to the user's discretion to determine what nomenclature and classification scheme he or she wishes to follow. I am frequently asked "Who is right?" and "Which one should I follow?" There are no right or wrong answers as to what constitutes a specific taxon, whether a subfamily, tribe, genus, species, subspecies, or variety. There are some traditional criteria among taxonomists. However, rarely do they totally agree. An expert in a particular group or field is generally accepted for the most part, although not always. Personality, traditions, and resistance to change all play a part in what nomenclature is "popular" and what classification scheme is accepted. Many who learned a particular classification scheme see little or no reason to change to a new system. Also, new information is constantly published, concepts and schemes modified (some for the good, some revert back to the original later, and some are completely ignored). Finally, it is the user's responsibility to be able to support the decision as to which scheme and nomenclature are used. Regardless of which authority is followed, one should be prepared to justify one's choice.

CHECKLIST

Following are the classification and a checklist of the grasses of Colorado. Bold Roman numerals designate subfamilies (which end in -oideae); bold Arabic numerals designate tribes, followed by genus, species, subspecies, and variety. The checklist is alphabetized by subfamily, tribe within subfamily, genera within tribe, species within a genus, and subspecies or varieties within species. Subfamilies, tribes, genera, and species are numbered consecutively, but subspecies and varieties are not.

I. Aristidoideae
  1. Aristideae
    1. *Aristida*
      1. *A. adscensionis*
      2. *A. arizonica*
      3. *A. basiramea*
      4. *A. dichotoma*
        var. *curtissii*
      5. *A. divaricata*
      6. *A. havardii*
      7. *A. oligantha*
      8. *A. purpurea*
        var. *fendleriana*
        var. *longiseta*
        var. *purpurea*
        var. *nealleyi*
        var. *wrightii*

II. Arundinoideae
  2. Arundineae
    2. *Phragmites*
      9. *P. australis*

III. Chloridoideae
  3. Cyndonteae
    3. *Blepharoneuron*
      10. *B. tricholepis*
    4. *Bouteloua*
      11. *B. curtipendula*
        var. *caespitosa*
        var. *curtipendula*
    5. *Buchloë*
      12. *B. dactyloides*
    6. *Calamovilfa*
      13. *C. gigantea*

      14. *C. longifolia*
    7. *Chloris*
      15. *C. verticillata*
      16. *C. virgata*
    8. *Chondrosum*
      17. *C. barbatum*
        var. *barbatum*
      18. *C. eriopodum*
      19. *C. gracile*
      20. *C. hirsutum*
        subsp. *hirsutum*
      21. *C. simplex*
    9. *Crypsis*
      22. *C. alopecuroides*
    10. *Cynodon*
      23. *C. dactylon*
    11. *Dactyloctenium*
      24. *D. aegyptium*
    12. *Dasyochloa*
      25. *D. pulchella*
    13. *Distichlis*
      26. *D. spicata*
    14. *Eleusine*
      27. *E. indica*
    15. *Eragrostis*
      28. *E. barrelieri*
      29. *E. cilianensis*
      30. *E. curtipedicellata*
      31. *E. curvula*
      32. *E. hypnoides*
      33. *E. lutescens*
      34. *E. mexicana*
        subsp. *virescens*

35. *E. minor*
36. *E. pectinacea*
  var. *miserrima*
  var. *pectinacea*
37. *E. pilosa*
  var. *perplexa*
  var. *pilosa*
38. *E. secundiflora*
  subsp. *oxylepis*
39. *E. spectabilis*
40. *E. tef*
41. *E. trichodes*
16. *Erioneuron*
42. *E. pilosum*
  var. *pilosum*
17. *Leptochloa*
43. *L. dubia*
44. *L. fusca*
  subsp. *fascicularis*
18. *Lycurus*
45. *L. phleoides*
46. *L. setosus*
19. *Muhlenbergia*
47. *M. andina*
48. *M. arenacea*
49. *M. arenicola*
50. *M. asperifolia*
51. *M. brevis*
52. *M. cuspidata*
53. *M. depauperata*
54. *M. filiculmis*
55. *M. filiformis*
56. *M. glomerata*
57. *M. mexicana*
  var. *filiformis*
  var. *mexicana*
58. *M. minutissima*
59. *M. montana*
60. *M. pauciflora*
61. *M. porteri*
62. *M. pungens*
63. *M. racemosa*
64. *M. ramulosa*
65. *M. richardsonis*
66. *M. schreberi*
67. *M. thurberi*

68. *M. torreyi*
69. *M. wrightii*
20. *Munroa*
70. *M. squarrosa*
21. *Pleuraphis*
71. *P. jamesii*
22. *Redfieldia*
72. *R. flexuosa*
23. *Schedonnardus*
73. *S. paniculatus*
24. *Scleropogon*
74. *S. brevifolius*
25. *Spartina*
75. *S. gracilis*
76. *S. pectinata*
26. *Sporobolus*
77. *S. airoides*
78. *S. compositus*
  var. *compositus*
79. *S. contractus*
80. *S. cryptandrus*
81. *S. flexuosus*
82. *S. giganteus*
83. *S. heterolepis*
84. *S. nealleyi*
85. *S. neglectus*
86. *S. pyramidatus*
87. *S. texanus*
27. *Tridens*
88. *T. muticus*
  var. *elongatus*
28. *Triplasis*
89. *T. purpurea*
4. Pappophoreae
29. *Enneapogon*
90. *E. desvauxii*

IV. Danthonioideae
5. Danthonieae
30. *Danthonia*
91. *D. californica*
92. *D. intermedia*
93. *D. parryi*
94. *D. spicata*
95. *D. unispicata*

V. Ehrhartoideae

146. *B. lanatipes*
147. *B. mucroglumis*
148. *B. porteri*
149. *B. pubescens*
150. *B. pumpelliana*
     subsp. *pumpelliana*
151. *B. richardsonii*
50. *Bromus*
152. *B. briziformis*
153. *B. commutatus*
154. *B. hordeaceus*
     subsp. *hordeaceus*
155. *B. japonicus*
156. *B. racemosus*
157. *B. secalinus*
158. *B. squarrosus*
51. *Ceratochloa*
159. *C. carinata*

11. Meliceae
52. *Bromelica*
160. *B. bulbosa*
161. *B. spectabilis*
162. *B. subulata*
53. *Glyceria*
163. *G. borealis*
164. *G. elata*
165. *G. grandis*
     var. *grandis*
166. *G. striata*
     var. *stricta*
54. *Melica*
167. *M. porteri*
     var. *porteri*
55. *Schizachne*
168. *S. purpurascens*

12. Poeae (Aveneae)
56. *Agrostis*
169. *A. exarata*
170. *A. gigantea*
171. *A. idahoensis*
172. *A. mertensii*
173. *A. scabra*
174. *A. stolonifera*
175. *A. variabilis*
57. *Alopecurus*

176. *A. aequalis*
     var. *aequalis*
177. *A. arundinaceus*
178. *A. carolinianus*
179. *A. geniculatus*
     var. *geniculatus*
180. *A. magellanicus*
181. *A. pratensis*
58. *Anthoxanthum*
182. *A. odoratum*
59. *Apera*
183. *A. interrupta*
60. *Argillochloa*
184. *A. dasyclada*
61. *Arrhenatherum*
185. *A. elatius*
     subsp. *bulbosum*
     subsp. *elatius*
62. *Avena*
186. *A. fatua*
187. *A. sativa*
188. *A. sterilis*
63. *Avenula*
189. *A. hookeri*
64. *Beckmannia*
190. *B. syzigachne*
65. *Briza*
191. *B. maxima*
192. *B. media*
66. *Calamagrostis*
193. *C. canadensis*
194. *C. montanensis*
195. *C. purpurascens*
     var. *purpurascens*
196. *C. rubescens*
197. *C. scopulorum*
198. *C. stricta*
     subsp. *inexpansa*
     subsp. *stricta*
67. *Catabrosa*
199. *C. aquatica*
     var. *aquatica*
68. *Cinna*
200. *C. latifolia*
69. *Dactylis*
201. *D. glomerata*

70. *Deschampsia*
    202. *D. brevifolia*
    203. *D. cespitosa*
       subsp. *cespitosa*
    204. *D. elongata*
    205. *D. flexuosa*
71. *Festuca*
    206. *F. arizonica*
    207. *F. baffinensis*
    208. *F. brachyphylla*
       subsp. *coloradensis*
    209. *F. calligera*
    210. *F. campestris*
    211. *F. earlei*
    212. *F. hallii*
    213. *F. idahoensis*
    214. *F. minutiflora*
    215. *F. rubra*
       subsp. *arenaria*
       subsp. *rubra*
    216. *F. saximontana*
       var. *saximontana*
    217. *F. sororia*
    218. *F. subulata*
    219. *F. thurberi*
    220. *F. trachyphylla*
72. *Graphephorum*
    221. *G. wolfii*
73. *Helictotrichon*
    222. *H. mortonianum*
74. *Hierochloë*
    223. *H. hirta*
       subsp. *arctica*
75. *Holcus*
    224. *H. lanatus*
76. *Koeleria*
    225. *K. macrantha*
77. *Leucopoa*
    226. *L. kingii*
78. *Lolium*
    227. *L. multiflorum*
    228. *L. perenne*
79. *Phalaris*
    229. *P. canariensis*
    230. *P. caroliniana*
    231. *P. minor*

80. *Phalaroides*
    232. *P. arundinacea*
81. *Phippsia*
    233. *P. algida*
82. *Phleum*
    234. *P. alpinum*
    235. *P. pratense*
83. *Poa*
    236. *P. abbreviata*
       subsp. *pattersonii*
    237. *P. alpina*
       subsp. *alpina*
    238. *P. annua*
    239. *P. arctica*
       subsp. *aperta*
       subsp. *arctica*
       subsp. *grayana*
    240. *P. arida*
    241. *P. bigelovii*
    242. *P. bulbosa*
    243. *P. compressa*
    244. *P. cusickii*
       subsp. *epilis*
       subsp. *pallida*
    245. *P. fendleriana*
       subsp. *fendleriana*
       subsp. *longiligula*
    246. *P. glauca*
       subsp. *rupicola*
    247. *P. interior*
    248. *P. laxa*
       subsp. *banffiana*
    249. *P. leptocoma*
    250. *P. lettermanii*
    251. *P. occidentalis*
    252. *P. palustris*
    253. *P. pratensis*
       subsp. *agassizensis*
       subsp. *pratensis*
    254. *P. reflexa*
    255. *P. secunda*
       subsp. *juncifolia*
       subsp. *secunda*
    256. *P. stenantha*
       var. *stenantha*
    257. *P. tracyi*

312. *E. curvatus*
313. *E. elymoides*
   subsp. *brevifolius*
   subsp. *elymoides*
314. *E. glaucus*
   subsp. *glaucus*
315. *E. lanceolatus*
   subsp. *lanceolatus*
316. *E. multisetus*
317. *E. repens*
318. *E. scribneri*
319. *E. trachycaulus*
   subsp. *trachycaulus*
320. *E. violaceus*
   subsp. *banffiana*
321. *E. virginicus*
   var. *virginicus*
105. *Eremopyrum*
322. *E. triticeum*
106. *Hordeum*
323. *H. vulgare*
   subsp. *vulgare*

107. *Leymus*
324. *L. ambiguus*
325. *L. cinereus*
326. *L. racemosus*
327. *L. salina*
   subsp. *salina*
328. *L. triticoides*
108. *Pascopyrum*
329. *P. smithii*
109. *Psathyrostachys*
330. *P. juncea*
110. *Pseudoroegneria*
331. *P. spicata*
111. *Secale*
332. *S. cereale*
112. *Thinopyrum*
333. *T. intermedium*
   subsp. *barbulatum*
   subsp. *intermedium*
334. *T. ponticum*
113. *Triticum*
335. *T. aestivum*

KEYS AND DESCRIPTIONS

All keys presented here are artificial, and unrelated taxa are often grouped together. The keys are solely for identification and are not intended to express relationships (although in many cases related genera are grouped together in the key). Tribal and generic relationships are most easily viewed in the checklist. An exception is in the keys to the Andropogoneae and Paniceae (Keys 2 and 3), which group related genera based on the unique and consistent spikelet and inflorescence morphology of these tribes. Dichotomous keys are numbered and indented. The first series of keys leads to groups of genera with similar morphology. Keys within the groups then lead to a specific genus. Illustrations of inflorescences, spikelets, florets, and occasionally plants are included with the keys to genera. It is hoped that the illustrations will reduce the amount of page turning required during keying and increase the speed of initial identification. The illustrations are just an aid, and many genera exhibit considerable variation in spikelet morphology, so proceed with caution when comparing your specimen with a single example of a spikelet, floret, or both representing an entire genus. The page number on which the genus description and species key (if the genus has more than one species) appear is provided.

A brief generic description giving distinguishing morphological characteristics is provided. Also included are basic chromosome number (x) and photosynthetic pathway (C_3, C_4). Genera are numbered consecutively throughout the book.

A numbered and indented species key is provided for genera with more than one species. In monotypic genera, the species description immediately follows the generic description. Like genera, species are numbered consecutively throughout the book. The accepted species is italicized. A line drawing for each species is included. A Colorado county map (see inside front and rear covers for county names) showing documented distribution for the species is located immediately under the species name. The shaded areas indicate the county where a specimen for that particular species has been collected. Just because the county where a species is found is not shaded *does not* mean the species cannot occur there. It only indicates that no specimen for that species from that county is in any of the herbaria or databases surveyed. *Do not base species identification solely on distribution.* If you notice an obvious county distribution omission, properly collect the species and deposit a specimen in a recognized herbarium. Thus future editions will have more complete distributional information. A quick review of the distribution maps illustrates the lack of sufficient collections from the eastern counties, and further detailed documentation of the grass flora from that area is needed.

The descriptions contain a significant amount of species information. Included are a partial synonymy, vernacular name (sometimes referred to as common name), life span (annual, perennial, biennial), origin (native, introduced), season (cool [C_3], warm [C_4]), floral characteristics (inflorescence, spikelets, glumes, lemmas, paleas, awns), vegetative characteristics (growth habit, culms, sheaths, ligules, auricles, blades), habitat, and comments. A key to subspecies or varieties, if recognized within the species and if more than one occur within the state, is provided at the end of the species description.

KEY to the GENERA

1. Spikelets unisexual, staminate and pistillate conspicuously different . . . . KEY 1
1. Spikelets perfect, or, if unisexual, then staminate and pistillate not conspicuously different
  2. Spikelets with a single fertile floret, with or without reduced florets
    3. Spikelets in pairs of 1 sessile and 1 pediceled (2 pediceled at branch tips); sessile spikelet fertile; pedicellate spikelet staminate, rudimentary, or absent with pedicel still obvious; lower glume large and firm, tightly clasping, or enclosing the entire spikelet . . . . KEY 2
    3. Spikelets not as above
      4. Reduced floret or florets present below fertile floret
        5. Reduced floret 1; lemma of reduced floret similar to upper glume in shape, size, and texture; disarticulation below the glumes . . . . KEY 3
        5. Reduced floret 1 or 2; lemma of reduced florets not similar to upper glume in shape, size, or texture; disarticulation above the glumes . . . . KEY 4
      4. Reduced florets absent or above fertile floret
        6. Inflorescence a panicle, primary branches spreading or contracted but never spicate . . . . KEY 5
        6. Inflorescence a spike, spicate raceme, or with 2 to several spicate primary branches
          7. Inflorescence of 1 to several unilateral spicate primary branches . . . . KEY 6
          7. Inflorescence a terminal, bilateral spike or spicate raceme . . . . KEY 7
  2. Spikelets with 2 or more fertile florets
    8. Inflorescence a spike or spicate raceme, or with 2 to several spicate primary branches
      9. Inflorescence with 2 (rarely only 1) to several unilateral primary branches . . . .
      . . . . KEY 6
      9. Inflorescence a terminal, bilateral spike or spicate raceme . . . . KEY 7
    8. Inflorescence an open or contracted panicle or raceme with spikelets on well-developed pedicels . . . . KEY 8

Key 1 (Spikelets unisexual and conspicuously different) (Figure 15)

1. Plants monoecious; leaf blades 3 or more cm broad; cultivated corn . . . .
. . . . 37. *Zea* (page 238)
1. Plants dioecious; leaf blades much less than 3 cm broad, not cultivated
  2. Staminate and pistillate spikelets markedly different in structure; low, stoloniferous, mat-forming perennials
    3. Pistillate and staminate spikelets awnless, pistillate spikelets in burr-like clusters hidden in the leafy portion of the plant, staminate spikelets sessile on 1–4 short spicate branches of a well-exserted inflorescence . . . . 5. *Buchloë* (page 88)
    3. Pistillate spikelets long awned (3-awned from the nerves of each lemma), staminate spikelets awnless; both pistillate and staminate spikelets in contracted, usually spike-like racemes . . . . 24. *Scleropogon* (page 180)
  2. Staminate and pistillate spikelets similar in structure; erect plants over 15 cm tall
    4. Florets 8–15 per spikelet; lemma 5- to 11-nerved; plants of alkaline flats . . . .
. . . . 13. *Distichlis* (page 108)
    4. Florets 3–7 per spikelet; lemma 3- to 5-nerved; plants of ponderosa pine woods . . . .
. . . . 77. *Leucopoa* (page 430)

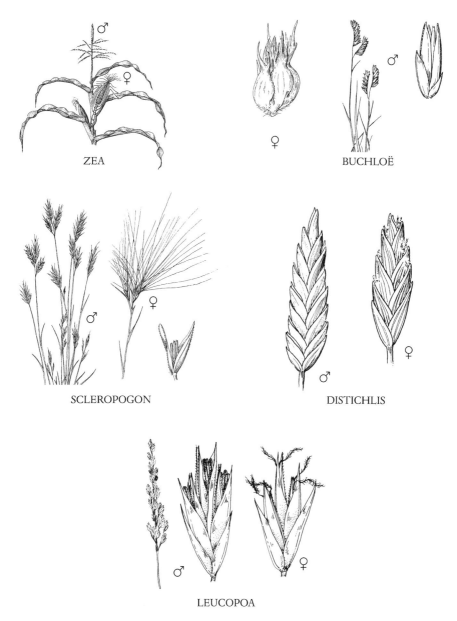

ZEA

BUCHLOË ♀ ♂

SCLEROPOGON ♂ ♀

DISTICHLIS ♂ ♀

LEUCOPOA ♂ ♀

Figure 15. Illustrations of genera found in Key 1.

Key 2 (Spikelets in pairs of 1 sessile and 1 pediceled; lower glume large and enclosing spikelet. Tribe Andropogoneae, in part) (Figure 16)

1. Inflorescence of one to several spicate primary branches
   2. Inflorescence composed of single spicate branches on long, slender peduncles . . . .
   . . . . 34. *Schizachyrium* (page 229)
   2. Inflorescence of 2 or more spicate branches per peduncle
      3. Upper pedicels and rachis internodes with a central groove or membranous area . . .
      . . . . 33. *Bothriochloa* (page 222)
      3. Upper pedicels and rachis internodes without a central groove . . . .
      . . . . 32. *Andropogon* (page 218)
1. Inflorescence an open or contracted panicle
   4. Panicle contracted; pedicels and rachis internodes covered with a gold, silky pubescence; pedicellate spikelet absent . . . . 35. *Sorghastrum* (page 231)
   4. Panicle open; pedicels and rachis internodes not covered with a gold, silky pubescence; pedicellate spikelet present . . . . 36. *Sorghum* (page 233)

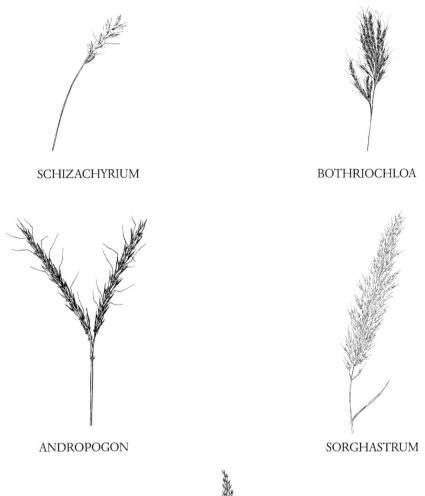

SCHIZACHYRIUM

BOTHRIOCHLOA

ANDROPOGON

SORGHASTRUM

SORGHUM

Figure 16. Illustrations of genera found in Key 2.

Key 3 (Two florets per spikelet, lower one reduced, upper one fertile; upper glume and lemma of lower floret similar in size, color, texture, and shape; disarticulation below the glumes. Tribe Paniceae) (Figure 17)

1. Spikelets in fascicles of bristles or flattened spines, which disarticulate with the spikelets
    2. Bristles glabrous, smooth, retrorsely scabrous, or strigose; usually at least some bristles fused for more than half their length . . . . 38. *Cenchrus* (page 240)
    2. Bristles plumose or antrorsely scabrous, free or fused for less than half their length . . . .
    . . . . 45. *Pennisetum* (page 280)
1. Spikelets not in spiny involucres; bristles persistent when present
    3. Spikelets subtended by 1 to several bristles . . . . 46. *Setaria* (page 282)
    3. Spikelets not subtended by bristles
        4. Lemma of lower floret awned or with a mucronate tip . . . .
        . . . . 41. *Echinochloa* (page 256)
        4. Lemma of lower floret not awned or mucronate tipped
            5. Lower glume absent or minute on some or all of the spikelets
                6. Spikelet with cuplike structure at base; lemma of upper floret awned or with a mucronate tip . . . . 42. *Eriochloa* (page 263)
                6. Spikelet without cuplike structure at base; lemmas of upper floret not awned
                    7. Spikelets plano-convex, lemma chartaceous to rigid when mature, blunt; margins of upper lemma not hyaline, often enrolled . . . .
                    . . . . 44. *Paspalum* (page 272)
                    7. Spikelets dorsally compressed, acuminate to acute; lemma cartilaginous and flexible when mature, margins of upper lemma hyaline, typically not enrolled
                    . . . . 40. *Digitaria* (page 251)
            5. Lower glume present, perhaps reduced but obvious
                8. Spikelets usually glabrous; basal and culm leaves similar, not forming winter rosette; panicle usually over 10 cm long, with numerous spikelets . . . .
                . . . . 43. *Panicum* (page 264)
                8. Spikelets usually pubescent; basal and culm leaves different, usually forming a winter rosette; panicle usually not over 10 cm long, with few spikelets . . . .
                . . . . 39. *Dichanthelium* (page 243)

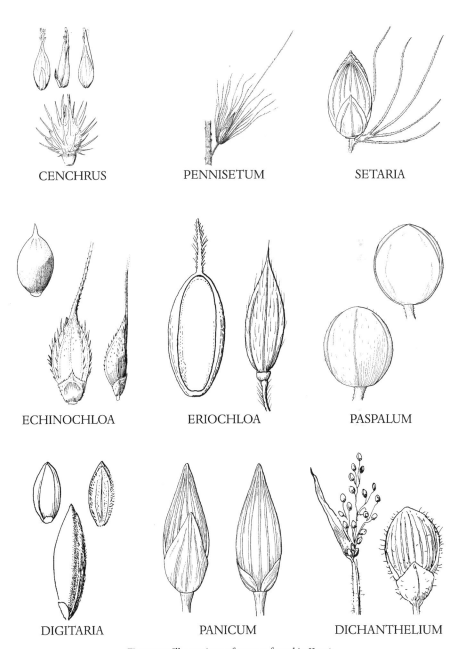

CENCHRUS PENNISETUM SETARIA

ECHINOCHLOA ERIOCHLOA PASPALUM

DIGITARIA PANICUM DICHANTHELIUM

Figure 17. Illustrations of genera found in Key 3.

Key 4 (One or 2 reduced florets below fertile floret; lemmas of reduced florets and glume of different shape, size, color, and texture; disarticulation above the glumes) (Figure 18)

1. Spikelets with 3 well-developed spikelets, the lower 2 staminate
    2. Glumes and florets about equal in length . . . . 74. *Hierochloë* (page 425)
    2. Glumes unequal, the second much larger than the first and the florets . . . .
        . . . . 58. *Anthoxanthum* (page 359)
1. Spikelets with 1–2 well-developed spikelets
    3. Well-developed floret 1; lemma awnless
        4. Rhizomes present; perennial . . . . 80. *Phalaroides* (page 438)
        4. Rhizomes absent; annual . . . . 79. *Phalaris* (page 433)
    3. Well-developed florets 2; upper or lower lemma awned
        5. Lower lemma awnless; upper lemma awned from between the lobes or teeth of a cleft apex . . . . 75. *Holcus* (page 427)
        5. Lower lemma with a stout, geniculate awn; upper lemma awnless . . . .
            . . . . 61. *Arrhenatherum* (page 364)

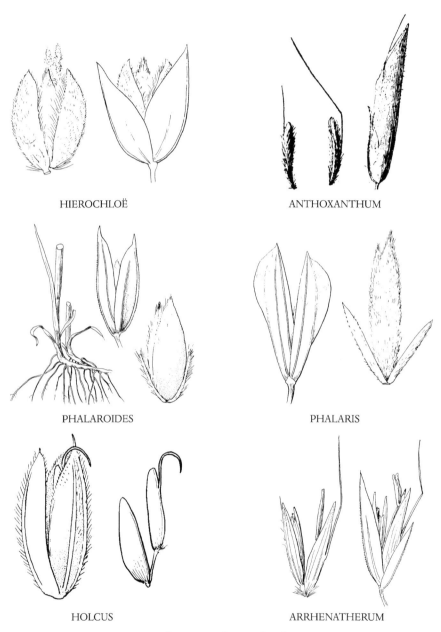

HIEROCHLOË

ANTHOXANTHUM

PHALAROIDES

PHALARIS

HOLCUS

ARRHENATHERUM

Figure 18. Illustrations of genera found in Key 4.

Key 5 (One fertile floret per spikelet; reduced absent or above; inflorescence a panicle) (Figure 19a, b, c)

1. Glumes absent or rudimentary . . . . 31. *Leersia* (page 215)
1. Glumes, at least the second well developed
   2. Spikelets suborbicular, laterally compressed, subsessile, and crowded on short branches of a narrow, elongate panicle . . . . 64. *Beckmannia* (page 374)
   2. Spikelets not as above
      3. Glumes and lemmas awnless (Check members of Key 4, which sometimes key here if the reduced florets are misinterpreted.)
         4. Lemma with tuft of hair at base; spikelets 5 mm or more in length . . . . 6. *Calamovilfa* (page 90)
         4. Lemma without tuft of hair at base when spikelets 5 mm or more in length
            5. Glumes as long as lemma
               6. Palea at least half as long as lemma . . . . 84. *Podagrostis* (page 482)
               6. Palea less than half as long as lemma . . . . 56. *Agrostis* (page 338)
            5. Glumes, at least the first, shorter than lemma
               7. Lemma 1-nerved
                  8. Low mat-forming annual; panicle dense . . . . 9. *Crypsis* (page 104)
                  8. Perennial or, if annual, not mat-forming; panicle not dense . . . . 26. *Sporobolus* (page 186)
               7. Lemma 3-nerved
                  9. Nerves of lemma densely pubescent . . . . 3. *Blepharoneuron* (page 84)
                  9. Nerves of lemma glabrous or scabrous
                     10. Lower glume minute or wanting; lemma about 1.5 mm long, broadly acute at apex; low tufted alpine perennial . . . . . 81. *Phippsia* (page 439)
                     10. Lower glume usually well developed, if minute, either lemma greater than 1.5 mm long or plant annual . . . . 19. *Muhlenbergia* (page 139)

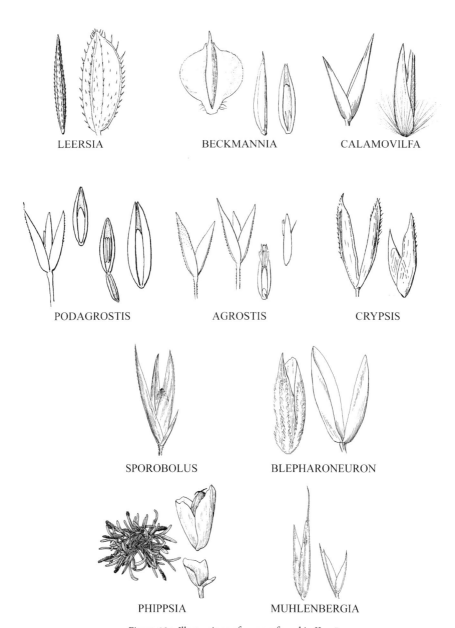

LEERSIA

BECKMANNIA

CALAMOVILFA

PODAGROSTIS

AGROSTIS

CRYPSIS

SPOROBOLUS

BLEPHARONEURON

PHIPPSIA

MUHLENBERGIA

Figure 19a. Illustrations of genera found in Key 5.

3. Glumes or lemmas awned
   11. Lower glume usually 2- or 3-awned, upper glume usually 1-awned . . . .
   . . . . 18. *Lycurus* (page 136)
   11. Lower glume not as above
      12. Disarticulation below the glumes . . . . KEY A
      12. Disarticulation above the glumes . . . . KEY AA

Key A (Disarticulation below the glumes)

1. Glumes awned . . . . 85. *Polypogon* (page 485)
1. Glumes awnless
  2. Lemma awned from the middle or below . . . . 57. *Alopecurus* (page 351)
  2. Lemma awned from or near the tip
    3. Awn 2 mm or less long; strong perennial . . . . 68. *Cinna* (page 389)
    3. Awn 6 mm or more long; weak annual . . . . 18. *Lycurus* (page 136)

Key AA (Disarticulation above the glumes)

1. Lemma not indurate or permanently enclosing palea
  2. Glumes equal, broad, abruptly short-awned . . . . 82. *Phleum* (page 441)
  2. Glumes not equal or, if nearly so, then not abruptly awned
    3. Lemma awned from the back, base, or cleft apex
      4. Lemma firm, awned from near the apex; awn straight or flexuous, 3 to 4 times as
      long as lemma . . . . 59. *Apera* (page 362)
      4. Lemma thin, awned from the back or base; awn usually geniculate when long
        5. Floret with tuft of hair at base; rachilla prolonged above insertion of palea . . . .
        . . . . 66. *Calamagrostis* (page 378)
        5. Floret usually without tuft of hair at base; rachilla not prolonged above insertion of palea
          6. Palea poorly developed, absent, or less than half the length of lemma . . . . . .
          . . . . 56. *Agrostis* (page 338)
          6. Palea well developed, half or more the length of lemma . . . .
          . . . . 84. *Podagrostis* (page 482)
    3. Lemma awned from an entire apex . . . . 19. *Muhlenbergia* (page 139)
1. Lemma indurate and enclosing the palea
  7. Awn column bearing 3 awns . . . . 1. *Aristida* (page 69)
  7. Awn single

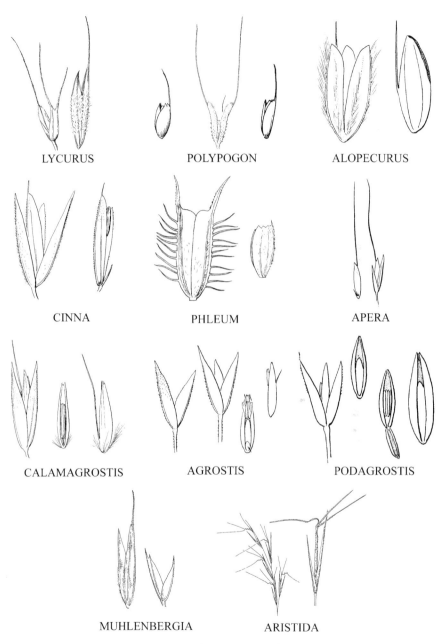

LYCURUS   POLYPOGON   ALOPECURUS

CINNA   PHLEUM   APERA

CALAMAGROSTIS   AGROSTIS   PODAGROSTIS

MUHLENBERGIA   ARISTIDA

Figure 19b. Key 5 continued.

8. Lemma with maximum thickness of 0.5 mm, mostly pubescent with hairs shorter than 0.5 mm on lower portion; distal quarter narrowed, bearing strongly divergent hairs 1.5–2.5 mm long below the apices . . . . 96. *Jarava* (page 538)
8. Lemma greater than 0.5 mm thick, glabrous or hairy, never with strongly divergent hairs solely on distal quarter, sometimes with divergent apical hairs
    9. Palea glabrous, a quarter to half the length of lemma; lemmas grayish brown to black, often tuberculate for most of their length . . . . 97. *Nassella* (page 538)
    9. Palea usually pubescent, from a third to slightly longer than lemma; lemma usually tan to chestnut brown, never black and tuberculate
        10. Callus sharp, 1.5–6 mm long; awns 4–30 cm long . . . .
        . . . . 95. *Hesperostipa* (page 531)
        10. Callus blunt or sharp, 0.1–2 mm long; awns 1–8 cm long
            11. Cauline leaf blades reduced, the uppermost less than 1 cm long; callus hairs dense, pilose, in a ring at top of the callus . . . . 98. *Oryzopsis* (page 543)
            11. Cauline leaf blades not reduced, more than 1 cm long; callus absent or not as above
                12. Blades no more than 0.5 mm wide when flat; awn plumose . . . .
                . . . . 100. *Ptilagrostis* (page 548)
                12. Blades more than 1 mm wide when flat; awn not plumose
                    13. Florets dorsally compressed, callus 0.1–0.6 mm long, rounded . . . .
                    . . . . 99. *Piptatherum* (page 545)
                    13. Florets terete to laterally compressed; callus 0.2–2 mm long, pointed but ranging from blunt to sharp . . . . 94. *Achnatherum* (page 514)

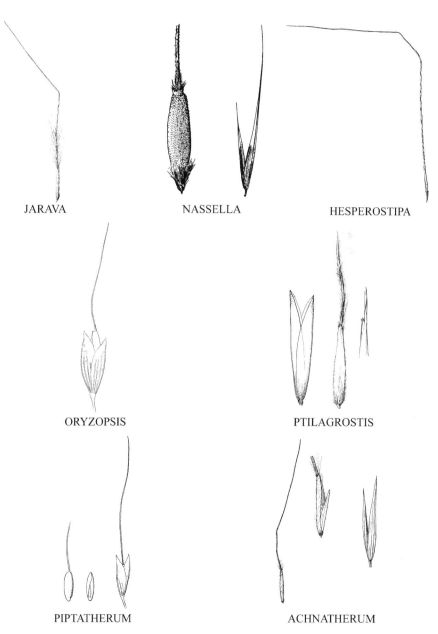

JARAVA

NASSELLA

HESPEROSTIPA

ORYZOPSIS

PTILAGROSTIS

PIPTATHERUM

ACHNATHERUM

Figure 19c. Key 5 continued.

Key 6 (Spikelet with 1 or more fertile florets; inflorescence a spike, spicate raceme, or with 2 to several spicate primary branches) (Figure 20)

1. Spikelets with 2 or more fertile florets
    2. Inflorescence branches paired, verticillate, or clustered at the culm apex
        3. Lemma awnless . . . . 14. *Eleusine* (page 111)
        3. Lemma awned
            4. Upper glume short-awned or mucronate; rachis projecting stiffly beyond terminal spikelet . . . . 11. *Dactyloctenium* (page 106)
            4. Upper glume not awned or mucronate; rachis not extended beyond terminal spikelet . . . . 7. *Chloris* (page 93)
    2. Inflorescence branches distributed along culm axis, seldom more than 1 at each rachis node
        5. Lemma 3-nerved
            6. Lemma acute and awnless; spikelets widely spaced and not overlapping, on stiffly spreading branches . . . . 15. *Eragrostis* (page 112)
            6. Lemma usually awned or mucronate; spikelets closely placed and overlapping . . . . 17. *Leptochloa* (page 132)
        5. Lemma 5- or more nerved
            7. Plant perennial, with tall culms and slender, flexuous inflorescence branches . . . . 53. *Glyceria* (page 329)
            7. Plant annual, with short, tufted culms, stiff inflorescence branches . . . . 88. *Sclerochloa* (page 498)
1. Spikelets with 1 floret
    8. Spikelets on main inflorescence axis (rachis) as well as on branches . . . . 23. *Schedonnardus* (page 179)
    8. Spikelets all on branches
        9. Inflorescence branches 2 or more, digitate, clustered, or in 2 or 3 verticels at culm apex
            10. Rudimentary floret absent or a minute scale; spikelets awnless . . . . 10. *Cynodon* (page 104)
            10. Rudimentary floret or florets present above fertile one; spikelets usually awned . . . . 7. *Chloris* (page 93)
        9. Inflorescence branches 1 to several, not digitate or clustered
            11. Spikelets without reduced florets . . . . 25. *Spartina* (page 181)
            11. Spikelets with 1 or more reduced florets above fertile one
                12. Spicate branches 10–50 per culm, falling entire at maturity, each branch bearing 1–5 spikelets . . . . 4. *Bouteloua* (page 85)
                12. Spicate branches no more than 8 per culm, florets falling from persistent glumes, each branch bearing 10–80 spikelets . . . . 8. *Chondrosum* (page 96)

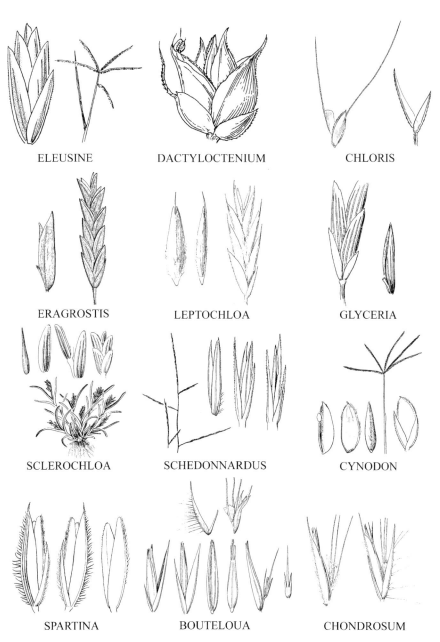

ELEUSINE

DACTYLOCTENIUM

CHLORIS

ERAGROSTIS

LEPTOCHLOA

GLYCERIA

SCLEROCHLOA

SCHEDONNARDUS

CYNODON

SPARTINA

BOUTELOUA

CHONDROSUM

Figure 20. Illustrations of genera found in Key 6.

Key 7 (Inflorescence a terminal, bilateral spike or spicate raceme) (Figure 21a, b)

1. Spikelets in capitate clusters, these subsessile in leafy portion of plant; lemmas 3-nerved; low, tufted, or sod-forming
    2. Disarticulation below glumes, spikelets falling in thick, indurate burr-like clusters; spikelets pistillate . . . . 5. *Buchloë* (page 88)
    2. Disarticulation above the glumes, spikelet not in clusters; spikelets perfect
        3. Glumes much longer than lemmas; lemmas deeply bifid at apex . . . .
        . . . . 12. *Dasyochloa* (page 108)
        3. Glumes shorter than lemma; lemmas acuminate at apex, not bifid . . . .
        . . . . 20. *Munroa* (page 174)

1. Spikelets not in capitate clusters, exserted above leafy portion of the plant
    4. Spikelets 3 at each node, at least central spikelet with a single fertile floret
        5. Spikelets seated in a cuplike structure, subtended by long, white hairs; rhizomatous and/or stoloniferous . . . . 21. *Pleuraphis* (page 175)
        5. Spikelets not as above; cespitose annual or perennial
            6. Central (sessile) spikelet fertile, lateral spikelets pediceled and sterile . . . .
            . . . . 103. *Critesion* (page 555)
            6. Central as well as lateral spikelets all fertile . . . . 106. *Hordeum* (page 588)
            (Check *Elymus elymoides*, which occasionally has the second floret of the central spikelet reduced, and the lateral spikelets have only a single, sterile floret.)
    4. Spikelets not 3 at each node, spikelets with 2 or more fertile florets
        7. Spikelets oriented edgewise to rachis, lower glume absent except on terminal spikelet
        . . . . 78. *Lolium* (page 431)
        7. Spikelets not oriented edgewise to rachis; lower glume present on all spikelets
            8. Annual or biennial; introduced, cultivated cereal or weed . . . . KEY B
            8. Perennial; native or introduced . . . . KEY BB

Key B (Annual or biennial)

1. Inflorescence a spicate raceme . . . . 47. *Brachypodium* (page 290)
1. Inflorescence a terminal, bilateral spike
    2. Glumes ovate with 3 or more nerves at midlength
        3. Spikelet sunken or embedded in rachis; spike narrow, less than 5 mm wide; rachis disarticulating at maturity . . . . 101. *Aegilops* (page 551)
        3. Spikelets not sunken into rachis; spike thick, more than 5 mm wide; rachis not disarticulating at maturity . . . . 113. *Triticum* (page 608)
    2. Glumes subulate to lanceolate with only 1 nerve at midlength
        4. Spike less than 2.5 cm long; lemmas 5–7.5 mm long . . . . 105. *Eremopyrum* (page 585)

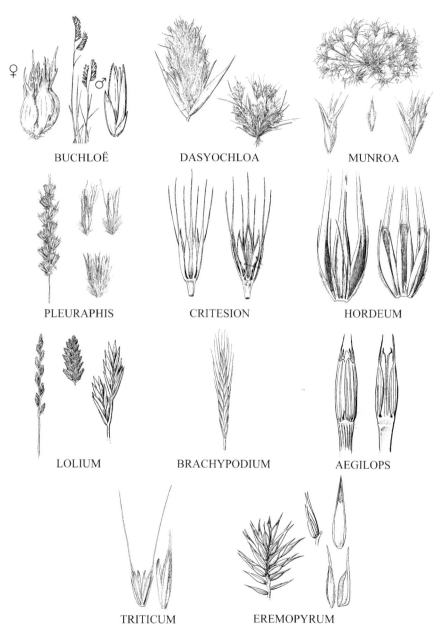

BUCHLOË DASYOCHLOA MUNROA

PLEURAPHIS CRITESION HORDEUM

LOLIUM BRACHYPODIUM AEGILOPS

TRITICUM EREMOPYRUM

Figure 21a. Illustrations of genera found in Key 7.

4. Spike more than 4 cm long; lemma more than 8 mm long . . . .
. . . . 111. *Secale* (page 603)

Key BB (Perennial)

1. Spikelets solitary at all (or almost all) rachis nodes
   2. Rachis internodes less than 5 mm long; spikelets closely imbricate, usually more than 3 times the length of the internodes, diverging . . . . 102. *Agropyron* (page 552)
   2. Rachis internodes more than 5 mm long; spikelets 1–3 times the length of the internodes, appressed or ascending
      3. Glumes subulate to narrowly lanceolate, stiff, 0- to 1-nerved at midlength
         4. Glumes lanceolate, tapering to an acuminate tip, slightly curved to one side apically . . . . 108. *Pascopyrum* (page 596)
         4. Glumes subulate from near base, straight . . . . 107. *Leymus* (page 590)
      3. Glumes usually ovate, rectangular, or lanceolate
         5. Glumes flexible, acute to awned or subulate throughout
            6. Spikelets distant, scarcely reaching base of spikelet above . . . .
            . . . . 110. *Pseudoroegneria* (page 601)
            6. Spikelets closely spaced, usually reaching at least midlength of spikelet above
            . . . . 104. *Elymus* (page 562)
         5. Glumes stiff, truncate, or obtuse . . . . 112. *Thinopyrum* (page 603)

1. Spikelets 2–7 at all (or most) rachis nodes
   7. Glumes 4–18 mm long, subulate to narrowly lanceolate, 0- to 1-nerved at midlength
      8. Ligules 0.2–0.3 mm long . . . . 109. *Psathyrostachys* (page 599)
      8. Ligules 0.3–8 mm long . . . . 107. *Leymus* (page 590)
   7. Glumes flat and with 3 nerves at midlength or, if subulate and 1-nerved, less than 4 mm long or more than 18 mm long . . . . 104. *Elymus* (page 562)

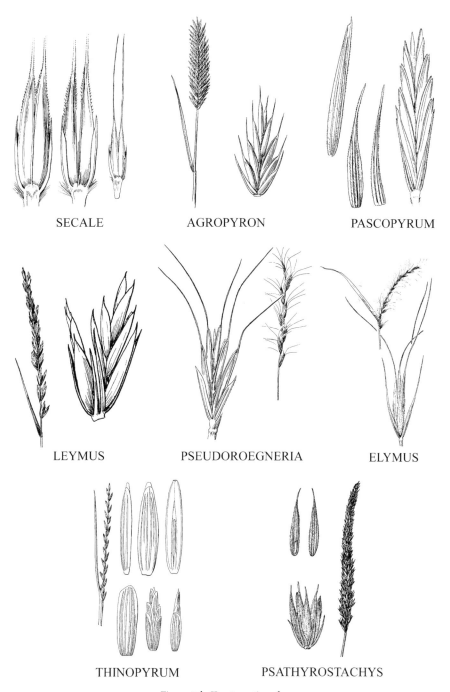

SECALE   AGROPYRON   PASCOPYRUM

LEYMUS   PSEUDOROEGNERIA   ELYMUS

THINOPYRUM   PSATHYROSTACHYS

Figure 21b. Key 7 continued.

59

Key 8 (Inflorescence an open or contracted panicle or raceme with spikelets on long pedicels) (Figure 22a, b, c, d, e)

1. Plant 2–6 m tall . . . . 2. *Phragmites* (page 83)
1. Plant less than 2 m tall
  2. Lemma 3-nerved, these generally conspicuous
    3. Nerves of lemma pubescent or base of lemma long-hairy
      4. Plant with rhizomes; lemma glabrous on nerves, with tuft of hair at base . . . .
      . . . . 22. *Redfieldia* (page 176)
      4. Plant without rhizomes; lemma pubescent or puberulent on nerves, without tuft of hair at base
        5. Palea densely long-ciliate on upper half; annual . . . . 28. *Triplasis* (page 205)
        5. Palea not densely long-ciliate on upper half; perennial
          6. Panicle narrow, 7–20 cm long, lemma not awned . . . . 27. *Tridens* (page 202)
          6. Panicle 1.5–4 cm long, lemma awned . . . . 16. *Erioneuron* (page 131)
    3. Nerves of lemma not pubescent, base of lemma not long-hairy
      7. Lemma 3-awned . . . . 24. *Scleropogon* (page 180)
      7. Lemma awnless or with a single awn
        8. Lemma awned from a bifid apex
          9. Setae present on lemma apex, awn conspicuous . . . . 91. *Trisetum* (page 504)
          9. Setae absent on lemma apex, awn inconspicuous . . . .
          . . . . 72. *Graphephorum* (page 422)
        8. Lemma awnless
          10. Glumes flat, the first and often second nerveless, irregularly toothed at broad apex; spikelets usually with 2 florets; on stream banks and marsh habitats . . .
          . . . . 67. *Catabrosa* (page 386)
          10. Glumes rounded or keeled, 1- to 3-nerved, tapering to an entire, acute apex; spikelets usually 3- to many-flowered, typically mesic to arid environments . . .
          . . . . 15. *Eragrostis* (page 112)
  2. Lemmas 5- to 13-nerved, occasionally obscure
    11. One or all lemmas in a spikelet awned . . . . KEY C
    11. All lemmas awnless . . . . KEY CC

Key C (Lemmas awned)

1. Lemma with 9 subequal, plumose awns . . . . 29. *Enneapogon* (page 207)
1. Lemma with a single awn
  2. Lemma 14–40 mm long; glumes 7- to 11-nerved . . . . 62. *Avena* (page 366)
  2. Lemma 1.3–16 mm long; glumes 1- to 9-nerved

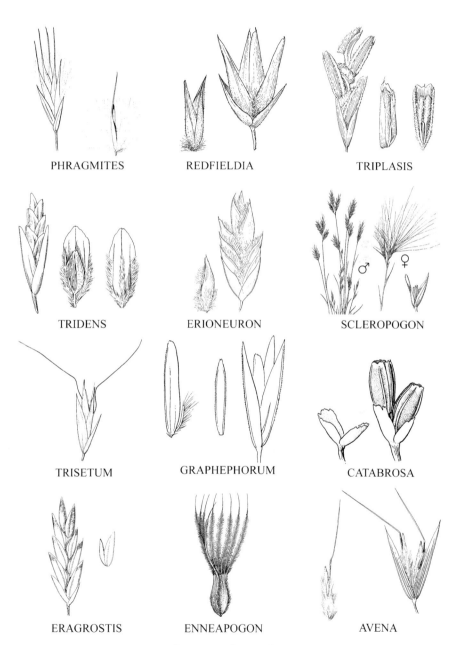

PHRAGMITES

REDFIELDIA

TRIPLASIS

TRIDENS

ERIONEURON

SCLEROPOGON

TRISETUM

GRAPHEPHORUM

CATABROSA

ERAGROSTIS

ENNEAPOGON

AVENA

Figure 22a. Illustrations of genera found in Key 8.

3. Lemma awned from the back or base

    4. Lemma 1.3–7 mm long

        5. Callus hairs about half as long as lemma; rachilla not prolonged (or up to 0.5 mm) beyond base of lower floret . . . . 92. *Vahlodea* (page 511)

        5. Callus glabrous or hairs much shorter than half as long as lemma; rachilla prolonged more than 0.5 mm beyond distal floret . . . . 70. *Deschampsia* (page 392)

    4. Lemma 7–16 mm long

        6. Adaxial leaf surface ribbed; rachillas pilose on all sides; ligules 0.5–1.5 mm long . . . . 73. *Helictotrichon* (page 422)

        6. Adaxial leaf surface not ribbed; rachillas glabrous on side adjacent to paleas, hairy elsewhere; ligules 0.5–7 mm long . . . . 63. *Avenula* (page 371)

3. Lemma awned from an entire or bifid apex

    7. Lower glume longer than lowermost floret; awn flattened below . . . . 30. *Danthonia* (page 208)

    7. Lower glume shorter than lowermost floret; awn not flattened below

        8. Lower glume with 3–5 nerves, indistinct . . . . KEY D

        8. Lower glume with 1–3 nerves, distinct . . . . KEY DD

Key D (Lower glume 3- or 5-nerved, indistinct)

1. Lemma long-pilose on callus; margins of leaf sheath free to base . . . . 55. *Schizachne* (page 336)

1. Lemma not long-pilose on callus; margins of leaf sheath connate, at least below

    2. Plants annual . . . . 50. *Bromus* (page 310)

    2. Plants perennial . . . . 51. *Ceratochloa* (page 321)

Key DD (Lower glume 1- or 3-nerved, distinct)

1. Palea colorless

    2. Upper glume obovate, broadest below the middle; disarticulation below the glumes . . . . 89. *Sphenopholis* (page 498)

    2. Upper glume broadest below the middle; disarticulation above the glumes . . . . 76. *Koeleria* (page 427)

1. Palea colored green or brown, at least on nerves

    3. Spikelets generally 1.5 cm or more long (rarely 1.2 cm long)

        4. Plants annual . . . . 48. *Anisantha* (page 292)

        4. Plants perennial . . . . 49. *Bromopsis* (page 297)

    3. Spikelets less than 1.2 cm long

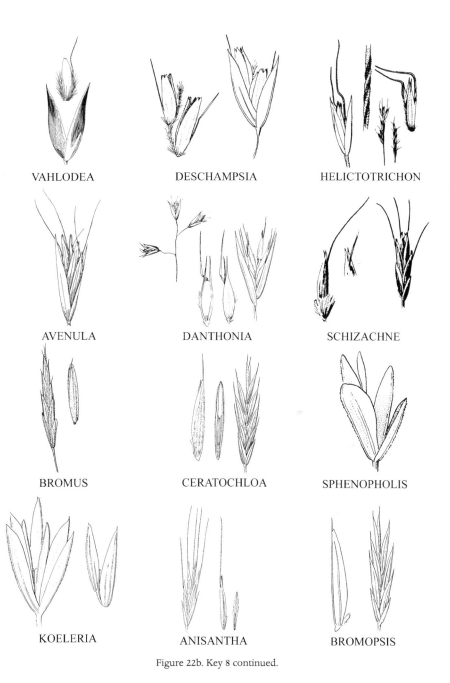

VAHLODEA   DESCHAMPSIA   HELICTOTRICHON

AVENULA   DANTHONIA   SCHIZACHNE

BROMUS   CERATOCHLOA   SPHENOPHOLIS

KOELERIA   ANISANTHA   BROMOPSIS

Figure 22b. Key 8 continued.

5. Lemma awned from a distinctly bifid apex, awn straight or geniculate; upper glume equaling or exceeding lowermost floret
    6. Setae present on lemma apex, awn conspicuous . . . . 91. *Trisetum* (page 504)
    6. Setae absent on lemma apex, awn inconspicuous . . . .
    . . . . 72. *Graphephorum* (page 422)
5. Lemma awned from an entire or minutely notched apex, awn straight; upper glume usually shorter than lowermost floret
    7. Spikelets laterally compressed, more or less asymmetrical, subsessile in dense clusters at tips of stiff, erect, or spreading branches; glumes and lemmas acute or irregularly short-awned . . . . 69. *Dactylis* (page 389)
    7. Spikelets not laterally compressed or asymmetrical, not in dense clusters at tip of branches
        8. Plants annual . . . . 93. *Vulpia* (page 513)
        8. Plants perennial
            9. Auricles present . . . . 87. *Schedonorus* (page 493)
            9. Auricles absent
                10. Secondary panicle branches strongly divaricate at maturity . . . .
                . . . . 60. *Argillochloa* (page 364)
                10. Secondary panicle branches not strongly divaricate at maturity . . . .
                . . . . 71. *Festuca* (page 397)

Key CC (Lemmas awnless)

1. Nerves of lemma strongly and uniformly developed and evenly spaced
    2. Lower glume 3- to 5-nerved
        3. Culms bulbous at base; disarticulation above the glumes . . . .
        . . . . 52. *Bromelica* (page 324)
        3. Culms not bulbous at base; disarticulation below the glumes . . . .
        . . . . 54. *Melica* (page 336)
    2. Lower glume 1-nerved, rarely 3-nerved
        4. Leaf sheath margins free
            5. Plant rhizomatous; not growing in saline or alkaline habitats . . . .
            . . . . 90. *Torreyochloa* (page 501)
            5. Plant not rhizomatous, sometimes rooting at buried lower nodes; growing in saline or alkaline habitats . . . . 86. *Puccinellia* (page 490)
        4. Leaf sheath margins connate, at least below . . . . 53. *Glyceria* (page 329)

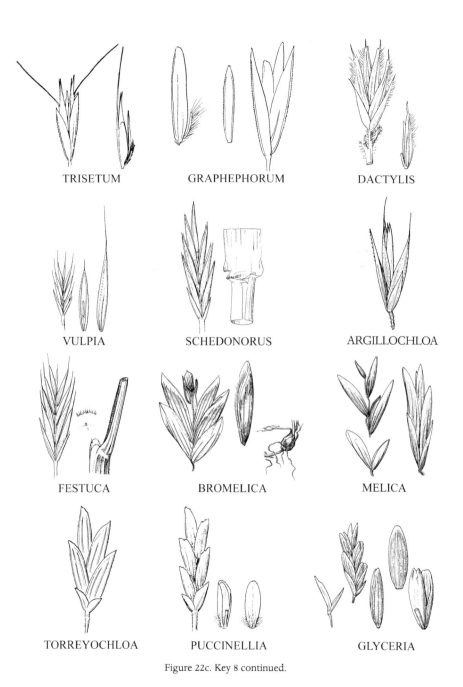

TRISETUM   GRAPHEPHORUM   DACTYLIS

VULPIA   SCHEDONORUS   ARGILLOCHLOA

FESTUCA   BROMELICA   MELICA

TORREYOCHLOA   PUCCINELLIA   GLYCERIA

Figure 22c. Key 8 continued.

1. Nerves of lemma not strongly and uniformly developed or, if so, then not equally spaced
    6. Glumes and lemmas spreading at right angles to rachilla, inflated and papery; spikelets on slender pedicels . . . . 65. *Briza* (page 374)
    6. Glumes and lemmas not as above
        7. Palea colorless; lateral nerves of lemma indistinct
            8. Upper glume obovate, usually abruptly narrowing to an obtuse or broadly acute apex; disarticulation below glumes . . . . 89. *Sphenopholis* (page 498)
            8. Upper glume not broadened above the middle, acute at apex; disarticulation above glumes . . . . 76. *Koeleria* (page 427)
        7. Palea green or brown, at least on nerves
            9. Lemma 7- to 13-nerved . . . . KEY E
            9. Lemma 5-nerved . . . . KEY EE

Key E (Lemmas 7- to 13-nerved)

1. Spikelets unisexual, the staminate and pistillate in separate inflorescences and usually on separate plants; glumes and lemmas thick, firm, indistinctly nerved . . . .
    . . . . 13. *Distichlis* (page 108)
1. Spikelets perfect
    2. Palea adhering to caryopsis
        3. Plants annual
            4. Lemma long and narrow, tapering to a long awn . . . . 48. *Anisantha* (page 292)
            4. Lemma broad and rounded at apex, abruptly awned . . . . . . 50. *Bromus* (page 310)
        3. Plants perennial
            5. Spikelets flattened; glumes and lemmas strongly keeled . . . .
                . . . . 51. *Ceratochloa* (page 321)
            5. Spikelets more or less terete; lemmas rounded on the back . . . .
                . . . . 49. *Bromopsis* (page 297)
    2. Palea not adhering to caryopsis . . . . 54. *Melica* (page 336)

Key EE (Lemmas 5-nerved)

1. Lemmas narrowly acute or attenuate at apex, not scarious on margins
    2. Spikelets unisexual, plants dioecious . . . . 77. *Leucopoa* (page 430)
    2. Spikelets perfect
        3. Secondary panicle branches strongly divaricate at maturity . . . .
            . . . . 60. *Argillochloa* (page 364)
        3. Secondary panicle branches not strongly divaricate at maturity . . . .
            . . . . 71. *Festuca* (page 397)

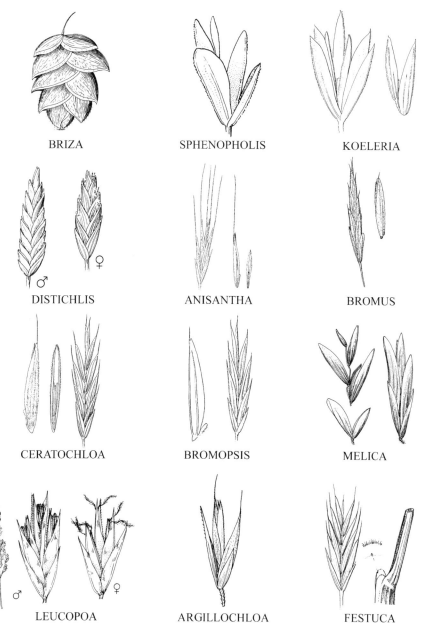

BRIZA   SPHENOPHOLIS   KOELERIA

DISTICHLIS   ANISANTHA   BROMUS

CERATOCHLOA   BROMOPSIS   MELICA

LEUCOPOA   ARGILLOCHLOA   FESTUCA

Figure 22d. Key 8 continued.

POA

TORREYOCHLOA

PUCCINELLIA

Figure 22e. Key 8 continued.

1. Lemmas obtuse or broadly acute at apex, usually scarious on margins above
    4. Lemmas moderately to strongly keeled, often with tuft of cobwebby hairs at base . . .
    . . . . 83. *Poa* (page 444)
    4. Lemmas rounded on back, glabrous or puberulent at base
        5. Nerves prominent; not growing in saline or alkaline soils; rhizomatous . . . .
        . . . . 90. *Torreyochloa* (page 501)
        5. Nerves obscure; growing in saline or alkaline soils; not rhizomatous but occasionally
        rooting at buried lower nodes . . . . 86. *Puccinellia* (page 490)

I. ARISTIDOIDEAE

Annual or perennial; cespitose. Sheaths open; auricles absent; ligules membranous and ciliate or a ring of hairs. Inflorescence terminal, panicle, or rarely a raceme; disarticulation above the glumes. Spikelets bisexual, with 1 floret; rachilla extension absent. Glumes 2, longer than floret, acute to acuminate; lemma 1–3-nerved, coriaceous, margins overlapping at maturity and concealing palea, 3-awned (or sometimes interpreted as a single awn that is 3-parted); lateral ones sometimes reduced or absent; paleas less than half as long as lemma.

1. ARISTIDEAE

See subfamily descriptions.

1. *Aristida* L.

Annuals or perennials, cespitose, lacking rhizomes or stolons. Inflorescence an open or contracted panicle of usually large, 1-flowered spikelets. Disarticulation above the glumes. Glumes thin, lanceolate, with a strong central nerve and occasionally 2 lateral nerves. Lemma indurate, terete, 3-nerved, with a hard, sharp-pointed callus at base, tapering gradually or less frequently abruptly to an awn column bearing usually 3 stiff awns, lateral awns partially or totally reduced in a few species. Caryopses long and slender, permanently enclosed in the firm lemma. Basic chromosome number, $x = 11$. Photosynthetic pathway, C_4.

Represented in Colorado by 8 species and 5 varieties.

1. Plants annual
  2. Lower glume 3- to 7-nerved . . . . 7. *A. oligantha*
  2. Lower glume 1- to 2(3)-nerved
    3. Central awn coiled
      4. Lateral awns 5–10 mm long, generally curved and twisted basally (not coiled), distally strongly divergent . . . . 3. *A. basiramea*
      4. Lateral awns 1–4 mm long, straight, erect . . . . 4. *A. dichotoma*
    3. Central awn not coiled . . . . 1. *A. adscensionis*
1. Plants perennial
  5. Inflorescence divaricate to ascending with axillary swellings
    6. Apex of lemma with 4 or more twists . . . . 5. *A. divaricata*
    6. Apex of lemma without twists or no more than 2 twists . . . . 6. *A. havardii*
  5. Inflorescence closed or, if open, without axillary swellings
    7. Glumes equal or nearly so . . . . 2. *A. arizonica*
    7. Glumes unequal, the first shorter than the second . . . . 8. *A. purpurea*

1. *Aristida adscensionis* L. (Fig. 23).

Synonyms: *Aristida adscensionis* L. var. *abortiva* Beetle, *Aristida adscensionis* L. var. *modesta* Hack., *Aristida fasciculata* Torr. **Vernacular Names:** Sixweeks threeawn, common needlegrass. **Life Span:** Annual. **Origin:** Native. **Season:** C_4, warm season. **FLORAL CHARACTERISTICS: Inflorescence:** narrow panicle, 5–15 cm long, 0.5–3 cm wide, loose to dense; branches 1–4 cm long, ascending to erect. **Spikelets:** 1-flowered, 8–10 mm long, disarticulation

Figure 23. *Aristida adscensionis*.

above the glumes. **Glumes:** unequal, 1-nerved, generally glabrous, keel occasionally scabrous, apex acute to acuminate; lower 4–8 mm long; upper 6–11 mm long, generally not more than 1.5 times as long as lower, occasionally nerve extending beyond glume as short mucro. **Lemmas:** slightly keeled, 6–9 mm long, obscurely 3-nerved, scabrous on keel, awn junction not evident. **Awns:** lemma topped by awn column bearing 3 awns, awns straight to slightly curved at base, not disarticulating when mature; central awn 7–15 mm long, not coiled; lateral awns generally shorter. **VEGETATIVE CHARACTERISTICS: Growth Habit:** cespitose, distinctly tufted. **Culms:** ascending to geniculate and branched below, 10–50 cm tall, glabrous. **Sheaths:** striate, shorter than internodes, occasionally lateral auricles present at summit. **Ligules:** densely ciliate membrane, 0.4–1 mm long. **Blades:** flat to involute, 2–14 cm long, 1–2.5 mm wide, upper surface scaberulous, lower surface glabrous. **HABITAT:** Found on sandy, dry, open ground and waste areas in central and southeastern counties. **COMMENTS:** This species is being used in revegetation efforts in eastern Colorado.

2. *Aristida arizonica* Vasey (Fig. 24).

Synonyms: None. **Vernacular Name:** Arizona threeawn. **Life Span:** Perennial. **Origin:** Native. **Season:** C_4, warm season. **FLORAL CHARACTERISTICS: Inflorescence:** narrow, spikelike panicles with appressed branches, 10–25 cm long, 1–3 cm wide, erect to slightly nodding, nodes glabrous to pubescent; branches 2–6 cm long. **Spikelets:** 1-flowered, 10–15 cm long, often reddish to purple, disarticulation above the glumes. **Glumes:** subequal to equal, 10–15 mm long, narrowly lanceolate, apex acuminate to awned, 1- to 2-nerved. **Lemmas:** narrowly lanceolate, glabrous, rarely sparsely pilose, 12–18 mm long, awn junction not evident. **Awns:** glumes when awned with awns to 3 mm; lemma topped by 3–6 mm twisted awn column bearing 3 awns, awns straight to slightly curved at base, ascending distally, not disarticulating when mature; central awn 20–35 mm long; lateral awns slightly shorter. **VEGETATIVE CHARACTERISTICS: Growth Habit:** generally cespitose but occasionally rhizomatous. **Culms:** erect, simple, 30–80 cm tall, glabrous. **Sheaths:** glabrous, occasionally scaberulous to hairy at summit (collar), generally longer than internodes. **Ligules:** ciliate membrane, 0.2–0.4 mm long. **Blades:** generally flat, curling as mature, 10–25 cm long, 1–3 mm wide, upper surface scaberulous, lower glabrous. **HABITAT:** Typically found in pine-juniper woodlands (Barkworth et al. 2003). **COMMENTS:** Uncommon in Colorado. It has curly leaves and longer awns than *A. purpurea* var. *nealleyi*, with which it can be confused (Barkworth et al. 2003).

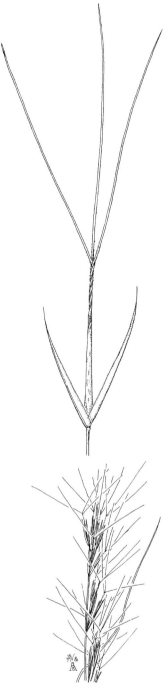

Figure 24. *Aristida arizonica.*

3. *Aristida basiramea* Engelm. *ex* Vasey (Fig. 25).

Synonyms: None. **Vernacular Names:** Forked three-awn, forked triple-awned grass, triple-awned grass. **Life Span:** Annual. **Origin:** Native. **Season:** C_4, warm season. **FLORAL CHARACTERISTICS: Inflorescence:** narrow raceme or panicle, 4–10 cm long, 1–2 cm wide, stiff to flexuous; branches and pedicels appressed to ascending. **Spikelets:** 1-flowered, 12–14 mm long, overlapping only slightly, disarticulation above the glumes. **Glumes:** unequal, 1-nerved, acute to acuminate, awned, purplish to brownish in color; lower 8–10 mm long; upper 12–14 mm long. **Lemmas:** mottled gray, 8–9 mm long. **Awns:** glumes with awns 1–2 mm long; lemma topped by 3 awns, awns erect to divergent; central awn 10–15 mm long, with 2–3 spiral coils at base; lateral awns 5–10 mm long, generally curved and twisted basally (not coiled), distally strongly divergent. **VEGETATIVE CHARACTERISTICS: Growth Habit:** cespitose. **Culms:** ascending to erect, 25–45 cm tall, branching at base, slender, glabrous to scaberulous. **Sheaths:** glabrous to slightly pilose, shorter than internodes. **Ligules:** ciliate membrane, to 0.3 mm long. **Blades:** flat to folded to involute as aging, pale green, 3–8 cm long, 1–1.5 mm wide, upper surface with occasional pilose hairs, margins with thickened ridges. **HABITAT:** On sandstone on hogbacks in the Front Range and Montezuma County. **COMMENTS:** Infrequent. Weber and Wittmann (2001a) and Wingate (1994) include *A. dichotoma* here; however, *A. dichotoma* has shorter lateral awns. The relationship between these taxa warrants further study (Barkworth et al. 2003).

4. *Aristida dichotoma* Michx. (Fig. 26).

Synonyms: *Aristida basiramea* Engelm. *ex* Vasey var. *curtissii* (Gray *ex* S. Wats. & Coult.) Shinners, *Aristida curtissii* (Gray *ex* S. Wats. & Coult.) Nash. **Vernacular Names:** Churchmouse threeawn, Shinners' threeawn, poverty grass. **Life Span:** Annual. **Origin:** Native. **Season:** C_4, warm season. **FLORAL CHARACTERISTICS: Inflorescence:** narrow raceme or panicle, 2–11 cm long, to 1 cm wide; branches and pedicels appressed. **Spikelets:** 1-flowered, 12–14 mm long, overlapping slightly, often found paired (1 subsessile, 1 pedicellate), disarticulation above the glumes. **Glumes:** subequal to unequal, 1-nerved, acute to acuminate, awned, pale gray to purplish to brownish in color; lower 3–8 mm long; upper 4–13 mm long. **Lemmas:** generally mottled gray to purplish, 3–11 mm long, glabrous to strigose with scabrous keel, awn junction obscure. **Awns:** lemma topped by 3 awns, central awn 3–8 mm long, coiled at base, deflexed; lateral awns 1–4 mm long, erect, straight. **VEGETATIVE CHARACTERISTICS: Growth Habit:** cespitose. **Culms:** erect to geniculate at base, 15–60 cm tall, wiry, usually branching at nodes, glabrous to scabrous. **Sheaths:** glabrous to scabrous, generally shorter than internodes. **Ligules:** ciliate membrane, generally less than 0.5 mm long. **Blades:** flat to folded to involute distally, 3–10 cm long, 1–2 mm wide, upper surface scabrous and

Figure 25. *Aristida basiramea*.

Figure 26. *Aristida dichotoma.*

occasionally thinly pilose near collar, lower surface scabrous. **HABITAT:** On sandstone on hogbacks in the Front Range. **COMMENTS:** Infrequent. Weber and Wittmann (2001a) and Wingate (1994) include this taxon in *A. basiramea*. However, *A. dichotoma* has shorter lateral awns. The relationship between these taxa warrants further study (Barkworth et al. 2003). The variety of this species found in Colorado is var. *curtissii* Gray *ex* S. Wats. & Coult., which differs from the type with glumes unequal, lemma smooth to scabridulous, and 6–11 mm long.

5. *Aristida divaricata* Humb. & Bonpl. *ex* Willd. (Fig. 27).

Synonyms: None. **Vernacular Name:** Poverty three-awn. **Life Span:** Perennial. **Origin:** Native. **Season:** C$_4$, warm season. **FLORAL CHARACTERISTICS: Inflorescence:** open, diffuse panicle, 10–30 cm long, 6–25 cm wide; primary branches stiffly divaricate to ascending, with axillary swellings, generally second-ary branches, and appressed pedicels. **Spikelets:** over-lapping, 12–13 mm long, 1-flowered, disarticulation above the glumes. **Glumes:** equal to subequal, 1- to obscurely 3-nerved, 8–12 mm long, keel scabrous, apex acuminate to short-awned. **Lemmas:** slender, 8–13 mm long, apex becoming narrowed and twisting 4 or more times in the distal 2–3 mm before juncture with awns, awn junction obscure. **Awns:** glumes when awned, with awns to 4 mm long; lemma topped by 3 awns, 10–20 mm long, persisting after maturity; central awn curved to straight basally, becoming ascending to slightly divergent distally; lateral awns slightly shorter or more than central awn, slender, ascending to diver-gent. **VEGETATIVE CHARACTERISTICS: Growth Habit:** cespitose, leaves generally basal. **Culms:** prostrate to erect, 25–70 cm tall, glabrous to retrorsely scaberulous. **Sheaths:** gla-brous to scabrous, collar densely pilose, longer than internodes. **Ligules:** ciliate membrane, 0.5–1 mm long. **Blades:** flat to involute, glabrous to scaberulous, 5–20 cm long, 1–2 mm wide. **HABITAT:** Found in grasslands and piñon-juniper woodlands of eastern Colorado. **COMMENTS:** Scattered to locally common. This species intergrades with *A. havardii*.

6. *Aristida havardii* Vasey (Fig. 28).

Synonym: *Aristida barbata* Fourn. **Vernacular Name:** Havard's threeawn. **Life Span:** Perennial. **Origin:** Native. **Season:** C$_4$, warm season. **FLORAL CHAR-ACTERISTICS: Inflorescence:** open panicle, 8–18 cm long, 4–12 cm wide; central axis curved to sinuous, slender; branchlets and pedicels flexuous with axil-lary pulvini. **Spikelets:** generally divergent, 1-flow-ered, disarticulation above the glumes. **Glumes:** equal, 1-nerved, 8–12 mm long, apex acuminate or awned. **Lemmas:** glabrous to scabrous, 8–13 mm long, apex 2–3 mm straight or twisted 1–2 times, awn junction obscure. **Awns:** glumes when awned to 4 mm long; lemma topped by 3 awns, persisting after maturity, ranges from straight to slightly curved basally and ascend-

Figure 27. *Aristida divaricata*.

Figure 28. *Aristida havardii.*

ing to divergent distally; central awn 10–22 mm long; lateral awns more slender and slightly shorter. **VEGETATIVE CHARACTERISTICS: Growth Habit:** cespitose. **Culms:** ascending to erect, 15–40 cm tall, wiry, hollow. **Sheaths:** generally glabrous, except occasionally villous near ligule. **Ligules:** ciliate membrane, 0.5–1 mm long. **Blades:** flat to involute, 5–20 cm long, 1–2 mm wide, glabrous to scabrous above, glabrous to scaberulous below. **HABITAT:** Found on dry hills, plains, and piñon-juniper woodlands at lower elevations. **COMMENTS:** This species is often included in *Aristida purpurea*. It was previously known as *A. barbata* in older texts. This species has been known to intergrade with *A. divaricata* (Barkworth et al. 2003).

7. *Aristida oligantha* Michx. (Fig. 29).

Synonyms: *Aristida ramosissima* Engelm. *ex* Gray var. *chaseana* Henr. **Vernacular Names:** Prairie three-awn, oldfield threeawn. **Life Span:** Annual. **Origin:** Native. **Season:** C_4, warm season. **FLORAL CHARACTERISTICS: Inflorescence:** spicate to racemose panicle; 7–20 cm long, 2–4 cm wide, few-flowered. **Spikelets:** large, generally divergent, 1-flowered, disarticulation above the glumes, pedicels short with axillary pulvini. **Glumes:** subequal to unequal, to 3 cm long, narrowly lanceolate, brownish green with purple tinge; lower 12–22 mm long, 3- to 7-nerved; upper 11–20 mm long, 1-nerved. **Lemmas:** 12–22 mm long, generally light-colored with mottling, shorter than glumes; callus 0.5–2 mm long, pilose. **Awns:** lower glume with midnerve extended into awn, 12–22 mm long surrounded by 1 seta on each side; upper has a central awn 4–10 mm long with lateral aristate teeth; lemma with short awn column and 3 awns generally 2.5–6 cm long, divergent to bent; lateral awns subequal to slightly shorter than central one. **VEGETATIVE CHARACTERISTICS: Growth Habit:** solitary or in small clumps. **Culms:** erect to geniculate, 2–6 dm tall, scabrous, branched freely. **Sheaths:** rounded, glabrous to scabrous-pubescent, shorter than the internodes, lowermost appressed-pilose basally. **Ligules:** a fringe of fine hairs, less than 0.5 mm long; collar often pubescent. **Blades:** flat to loosely involute, lax, 4–12 cm long, 0.5–1.5 mm wide, usually slightly pilose basally. **HABITAT:** Tends to grow in dry and sandy soils where disturbance is present. **COMMENTS:** Rare in Colorado, but one of the most widely distributed and common species of *Aristida* in the central and eastern United States.

8. *Aristida purpurea* Nutt. (Fig. 30).

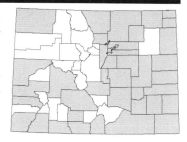

Synonyms: Var. *fendleriana*: *Aristida fendleriana* Steud.; var. *longiseta*: *Aristida longiseta* Steud., *Aristida longiseta* Steud. var. *rariflora* Hitchc., *Aristida longiseta* Steud. var. *robusta* Merr.; var. *nealleyi*: *Aristida nealleyi* Vasey; var. *purpurea*: *Aristida purpurea* Nutt. var. *robusta* (Merr.) Piper; *Aristida purpurea* Nutt. var. *laxiflora* Merr., *Aristida brownii* Warnock; var. *wrightii*: *Aristida wrightii* Nash. **Vernacular Names:** Purple threeawn,

Figure 29. *Aristida oligantha*.

Figure 30. *Aristida purpurea.*

Fendler threeawn. **Life Span:** Perennial. **Origin:** Native. **Season:** C$_4$, warm season. **FLORAL CHARACTERISTICS: Inflorescence:** panicle, occasionally reduced to raceme with 2 or more spikelets per node, 3–30 cm long, 2–12 cm wide; branches with wide variation: appressed to widely diverging, flexuous to stiff, pulvini absent. **Spikelets:** divergent to appressed, 1-flowered, disarticulation above the glumes. **Glumes:** unequal to occasionally subequal, glabrous to scabridulous, brownish to purplish, 1-nerved, rarely 2-nerved, unawned to short-awned, lower 4–15 mm long, upper 8–26 mm long. **Lemmas:** firm, 6–16 mm long, glabrous to scabridulous to tuberculate, purplish to whitish, awn junction obscure. **Awns:** glumes when awned to 1 mm long; lemma topped by 3 ascending to divaricate awns, 13–140 mm long, persisting after maturity; central awn thicker than lateral ones; lateral awns generally subequal to central awn, sometimes as much as one-third shorter than central awn. **VEGETATIVE CHARACTERISTICS: Growth Habit:** densely cespitose. **Culms:** ascending to erect, 10–100 cm tall, glabrous. **Sheaths:** glabrous to slightly scabrous, shorter to longer than internodes, summit glabrous to sparsely pilose. **Ligules:** ciliate membrane, to 0.5 mm long. **Blades:** flat to involute, 4–25 cm long, 1–1.5 mm wide, generally glabrous but occasionally scaberulous below. **HABITAT:** Found at low to mid-elevations on plains, dry grasslands, and valleys up to foothills over Colorado. **COMMENTS:** Extremely common species with numerous varieties, 5 of which occur in Colorado [var. *fendleriana* (Steud.) Vasey, var. *longiseta* (Steud.) Vasey, var. *nealleyi* (Vasey) Allred, var. *purpurea*, and var. *wrightii* (Nash) Allred]. The varieties can be distinguished by these characteristics:

1. Awns 4–10 cm long
   2. Awns 4–5 cm long, upper glume mostly less than 16 mm . . . . .var. *purpurea*
   2. Awns 4–10 cm long, upper glume 14–25 mm long . . . . .var. *longiseta*
1. Awns 1–3.5 cm long
   3. Lemma apices 0.1–0.3 mm wide; awns 0.1–0.3 mm wide at base
      4. All or most panicle branches straight; pedicels straight, appressed to ascending . . .
      . . . . var. *nealleyi*
      4. All or most panicle branches and pedicels drooping, flexuous, or wavy . . . .
      . . . . var. *purpurea*
   3. Lemma apices 0.2–0.3 mm wide; awns stout, 0.2–0.3 mm wide at base
      5. Panicle branches drooping or flexuous . . . . var. *purpurea*
      5. Panicle branches stiffly erect
         6. Panicles generally 3–15 cm long . . . . var. *fendleriana*
         6. Panicles generally 14–30 cm long, leaf blades 10–25 cm long . . . . . . var. *wrightii*

II. ARUNDINOIDEAE

In ours, plants perennial; rhizomatous. Culms woody, "reedlike," internodes hollow. Sheaths usually open, auricles absent, ligule membranous and usually ciliate. Inflorescence terminal, paniculate. Spikelets laterally compressed, more than 1 floret; distal florets often reduced, disarticulation above the glumes. Glumes 2, usually longer than the florets; lemmas 5- to 7-nerved, rarely awned; paleas shorter than lemmas.

2. ARUNDINEAE

See subfamily description.

2. *Phragmites* Adans.

Tall, rhizomatous, and stoloniferous perennial reed grasses with broad leaves and large, plumose panicles. Ligule a ring of short hairs. Spikelets several-flowered, disarticulating above the glumes and between the florets. Lower floret staminate or sterile, terminal floret also reduced. Rachilla villous with long hairs. Glumes and lemmas thin, acute or acuminate, glabrous, the lower 1-nerved, the upper and lemmas usually 3-nerved. Basic chromosome number, $x = 6$. Photosynthetic pathway, C_3.

Represented in Colorado by a single species.

9. *Phragmites australis* (Cav.) Trin. *ex* Steud. (Fig. 31).

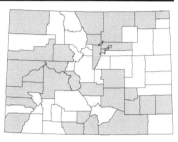

Synonyms: *Phragmites australis* (Cav.) Trin. *ex* Steud. var. *berlandieri* (Fourn.) C. F. Reed, *Phragmites communis* Trin., *Phragmites communis* Trin. subsp. *berlandieri* (Fourn.) Á. & D. Löve, *Phragmites communis* Trin. var. *berlandieri* (Fourn.) Fern., *Phragmites phragmites* (L.) Karst. **Vernacular Name:** Common reed. **Life Span:** Perennial. **Origin:** Native. **Season:** C_3, cool season. **FLORAL CHARACTERISTICS: Inflorescence:** dense panicle, 15–35 cm long, generally 2–8 cm wide, often purplish, becoming straw-colored with age; rachilla hairy, hairs 6–10 mm long. **Spikelets:** narrowly lanceolate, 10–15 mm long, 3- to 10-flowered. **Glumes:** unequal, thin, lanceolate to elliptic; lower 3–7 mm long; upper 5–10 mm long. **Lemmas:** glabrous, thin, linear-lanceolate, 8–15 mm long, margin slightly inrolled, size of lemmas decreases upward. **Awns:** lemma tip long-acuminate and appearing like an awn. **VEGETATIVE CHARACTERISTICS: Growth Habit:** rhizomatous with stout, creeping rhizomes. **Culms:** erect, 2–6 m tall, 0.5–1.5 cm thick, glabrous. **Sheaths:** open, glabrous, margins hyaline, occasionally minutely ciliate. **Ligules:** ciliate membrane, to 1 mm long. **Blades:** flat, glabrous, long-acuminate, 15–40 cm long, 2–4 cm wide, constricted basally, disarticulate from sheath at maturity. **HABITAT:** Found in moist or wet areas along irrigation ditches and rivers at low elevations over Colorado. **COMMENTS:** The culms have been used to make arrow shafts (Weber and Wittmann 2001a, 2001b). Occasionally, some populations are sterile, and all florets are abortive (Welsh et al. 1993).

III. CHLORIDOIDEAE

Plants annual or perennial. Sheaths usually open; ligules membranous, often ciliate, cilia often longer than membranous base, or a ring of hairs; auricles absent. Inflorescences paniculate, racemose, or spicate; disarticulation usually beneath the florets, sometimes at base of the panicle. Spikelets usually bisexual, occasionally monoecious or dioecious, usually lateral-compressed, 1 to many florets, distal florets often reduced. Glumes usually 2, shorter or longer than the floret(s); lemmas 1- to 3- or 7- to 13-nerved, awned or awnless.

3. CYNODONTEAE

Plants annual or perennial. Sheaths open; ligules membranous, often ciliate, cilia usually longer than membranous base, or a ring of hairs; auricles absent. Inflorescence varied,

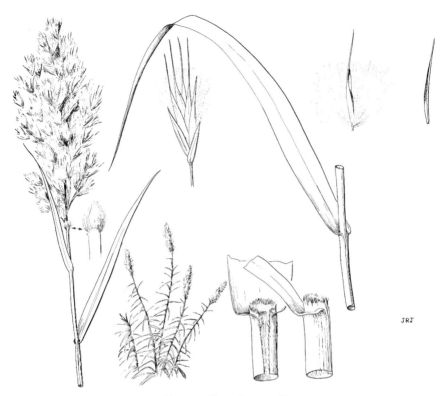

Figure 31. *Phragmites australis.*

simple panicle, panicle of 1 to many spicate branches, spicate racemes, or spike. Disarticulation beneath fertile florets or glumes, sometimes at base of the panicle or panicle branches. Spikelets with 1 to many florets, if present, reduced floret above. Glumes 2, shorter than lowest lemma, to exceeding all florets; lemmas 1- to 3-nerved or 7- to 13-nerved, rarely 5-nerved, awned or awnless.

3. *Blepharoneuron* Nash

Perennials, cespitose, with culms 20–70 cm tall. Leaves mostly in basal tuft. Sheaths open; ligule membranous, short, rounded, appearing as a continuation of the sheath margins; auricles absent. Inflorescence a loosely contracted panicle. Spikelets with 1 floret, grayish green, borne on slender pedicels, pedicels glandular. Disarticulation above the glumes. Glumes 2, broad, subequal, 1-nerved. Lemma firmer than the glumes and slightly longer, rounded on the back, 3-nerved, nerves usually densely ciliate-pubescent, apex acute, occasionally mucronate. Palea slightly exceeding lemma, pubescent on the 2 nerves. Basic chromosome number, $x = 8$. Photosynthetic pathway, C_4.

Represented in Colorado by a single species.

10. *Blepharoneuron tricholepis* (Torr.) Nash (Fig. 32).

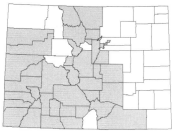

Synonym: *Vilfa tricholepis* Torr. **Vernacular Names:** Pine dropseed, hairy dropseed. **Life Span:** Perennial. **Origin:** Native. **Season:** C$_4$, warm season. **FLORAL CHARACTERISTICS: Inflorescence:** loosely contracted panicle, branches ascending, 3–25 cm long, 1–10 cm wide. **Spikelets:** 1-flowered, slightly laterally compressed, disarticulation above the glumes; pale to olivaceous. **Glumes:** broad, subequal, faintly 1-nerved, rounded dorsally, broadly acute or obtuse; lower 1.5–3 mm long, ovate, obtuse; upper 1.5–3.5 mm long or rarely longer, broader and more blunt, occasionally rounded apically. **Lemmas:** firmer than glumes, 2–4 mm long, acute to obtuse at apex, 3-nerved, silky hairs on keel and marginal nerves, scarious toward the tip. **Paleas:** densely hairy between the 2 nerves. **Awns:** none. **VEGETATIVE CHARACTERISTICS: Growth Habit:** densely tufted. **Culms:** slender, erect, 2.5–6.5 dm tall or rarely to 8.5 dm tall. **Sheaths:** open, crowded near base, broad, glabrous on back, ciliate on scarious margins. **Ligules:** 0.5–2 mm long, short, rounded, appearing as a continuation of the sheath margins. **Blades:** narrow, 1–15 cm long, 0.5–2.5 mm wide, flat to involute, scabrous. **HABITAT:** Dry, open slopes, rocky meadows, and gravelly open pine to spruce forests; montane. **COMMENTS:** An important forage species and one of the highest-quality grasses in timbered areas (Stubbendieck, Hatch, and Landholt 2003).

4. *Bouteloua* Lag.

Annuals and perennials of diverse habit, some cespitose, others with rhizomes or stolons. Leaves mostly basal, with flat or folded blades. Ligule usually a ring of hairs. Inflorescence of several to numerous short, spicate branches, closely or distantly spaced along the main axis and bearing relatively few (1–15) large, non-pectinate, sessile spikelets in two rows along the margins of an angular or flattened rachis. Disarticulation at base of the branch rachis. Spikelets with 1 fertile floret and 1–2 rudimentary florets above. Glumes lanceolate, 1-nerved, unequal to nearly equal, awnless or short-awned. Lemmas membranous, 3-nerved, midnerve often extending into an awn, lateral nerves occasionally short-awned. Paleas membranous, the 2 nerves occasionally awn-tipped. Basic chromosome number, $x = 10$. Photosynthetic pathway, C$_4$.

Represented in Colorado by a single species and 2 varieties. All other species of *Bouteloua* are now placed in the genus *Chondrosum*.

11. *Bouteloua curtipendula* (Michx.) Torr. (Fig. 33).

Synonyms: Var. *curtipendula*: *Atheropogon curtipendulus* (Michx.) Fourn., *Chloris curtipendula* Michx. **Vernacular Name:** Sideoats grama. **Life Span:** Perennial. **Origin:** Native. **Season:** C$_4$, warm season. **FLORAL CHARACTERISTICS: Inflorescence:** panicle with reflexed, often secund spicate branches, 10–30 cm long, of 30–80 short, deciduous, pendulous spikes, 0.8–3 cm long. **Spikelets:** 1–15 spikelets per branch, disarticulating at base of branches, subterete, closely appressed to the short rachis, with or without a single rudimentary floret above

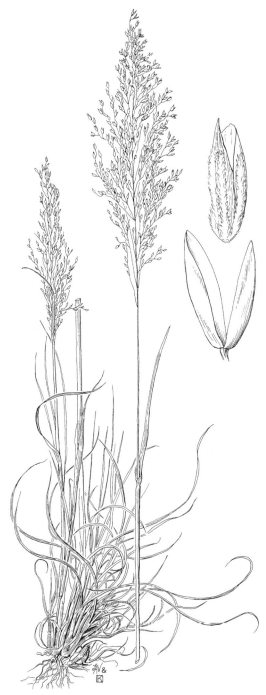

Figure 32. *Blepharoneuron tricholepis.*

Figure 33. *Bouteloua curtipendula*.

the fertile one. Anthers orange or red, less frequently yellow or purple. **Glumes:** 1-nerved, unequal, glabrous to scabrous; lower 2.5–6 mm long, filiform and scarious; upper 4.5–8 mm long, lanceolate, glabrous to scabrous, sharply acute to awn-tipped, purpletinged. **Lemmas:** fertile lemma 3–6.5 mm long, 3-nerved, rudimentary lemma variable. **Awns:** fertile lemma with 3 awn tips; rudimentary lemma, when present, 3-awned, unequal, 3–6 mm long. **VEGETA-**

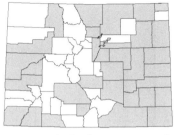

TIVE CHARACTERISTICS: Growth Habit: short rhizomatous, with slender to stout, scaly rhizomes (var. *curtipendula*) to densely tufted with no rhizomes (var. *caespitosa* Gould & Kapadia), leaves mostly basal. **Culms:** decumbent to erect, slender, 0.2–1 m tall, smooth. **Sheaths:** open, glabrous to sparsely long-hairy above; collar pilose on margin, sometimes persistent at base. **Ligules:** fringe of hairs, 0.2–1 mm long, truncate. **Blades:** 2–30 cm long, 2–7 mm wide, flat to loosely involute, scabrous adaxially, glabrous abaxially, commonly with papillose-based hairs along margin. **HABITAT:** Wingate (1994) reports this on both the Eastern and Western slopes at low elevations. This species is found on the plains and outwash mesas of the Front Range and in sagebrush and piñon-juniper on the Western Slope (Weber and Wittmann 2001b). Stubbendieck and colleagues (2003) state that it is most abundant on fine-textured soils and is better adapted to calcareous to moderately alkaline soils than to those that are neutral or acidic. **COMMENTS:** Common species. Forage value is good for both livestock and wildlife through the summer, into the fall, and somewhat into the winter (Stubbendieck, Hatch, and Landholt 2003). Often used in restoration projects and as an ornamental. Two varieties occur in Colorado, the typical variety (*curtipendula*) and var. *caespitosa* Gould & Kapadia.

1. Plants with short rhizomes . . . . 11. var. *curtipendula*
1. Plants cespitose, without rhizomes . . . . 11. var. *caespitosa*

5. *Buchloë* Engelm.

Perennials, low mat-forming, stoloniferous. Sheaths open; ligule a fringe of short hairs; auricles absent. Staminate and pistillate spikelets in separate inflorescences, usually on different plants (dioecious) but not infrequently on the same plant (monoecious). Staminate spikelets with 2 florets, sessile, and closely crowded on 1–4 short, spicate inflorescence branches, these well exserted above the leafy portion of the plant. Pistillate spikelets with 1 fertile floret, in deciduous, capitate, burr-like clusters of 2–4, these present in the leafy portion of the plant and usually partially included in expanded leaf sheaths. Rachis and lower two-thirds of upper glume of pistillate spikelets thickened, indurate. Lemma of pistillate spikelet membranous, 3-nerved, usually glabrous and awnless. Basic chromosome number, $x = 10$. Photosynthetic pathway, C_4.

Represented in Colorado by a single species.

12. *Buchloë dactyloides* (Nutt.) Englem. (Fig. 34).

Synonyms: *Bulbilis dactyloides* (Nutt.) Raf. *ex* Kuntze, *Sesleria dactyloides* Nutt., *Bouteloua dactyloides* (Nutt.) J. T. Columbus. **Vernacular Name:** Buffalograss. **Life Span:** Perennial.

Origin: Native. **Season:** C_4, warm season. **FLORAL CHARACTERISTICS: Inflorescence:** staminate with panicle of 1–4 racemosely arranged spicate branches, 0.5–1.5 cm long, usually exceeding the leaves, spikelets 6–12 in 2 rows on each branch, disarticulating at the branch and falling as 1 unit; pistillate with panicle of burr-like clusters partially hidden within bracteate leaf sheaths or near the mid-culm, clusters up to 1 cm long, spikelets 3–5 per cluster, usually falling as 1 unit.

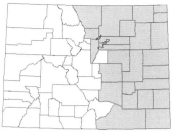

Spikelets: staminate 2-flowered, 4–6 mm long, 1.3–1.8 mm wide; pistillate 1 floret enclosed by glumes, to 7 mm long and 2.5 mm wide, yellowish. **Glumes:** staminate unequal, glabrous, 1- to 2-nerved, acute; lower 1–3 mm long; upper 2–4.5 mm long; pistillate unequal; lower reduced and irregular; upper indurate, white, enclosing lemma and bearing 3 awn-tipped lobes. **Lemmas:** staminate 3-nerved, glabrous, longer than glumes; pistillate 3-nerved, glabrous, unawned to shortly 3-awned. **Awns:** pistillate upper glume awn-tipped, lemma occasionally awn-tipped. **VEGETATIVE CHARACTERISTICS: Growth Habit:** dioecious, infrequently monoecious, rarely synoecious; stoloniferous, forming extensive mats by way of wiry stolons. **Culms:** erect, 1–30 cm tall, those of the pistillate plants much shorter, nodes glabrous. **Sheaths:** open, rounded, sparsely pilose near collar. **Ligules:** 0.5–1 mm long, ciliate membrane. **Blades:** flat, usually curling when dry, glabrous to sparsely pilose, 1–15 cm long, 1–2.5 mm wide. **HABITAT:** This species is a dominant plant of the shortgrass prairie, found in dry areas on the eastern plains and lower foothills. **COMMENTS:** Forage value is good for all livestock and fair for wildlife. This plant dries well and provides forage in winter when not covered by snow (Stubbendieck, Hatch, and Landholt 2003). In addition, it is drought-resistant and can aid in the control of erosion of heavy soils. Frequently used as an alternative to bluegrass or fescue lawns where water conservation is an issue. A monotypic genus that some taxonomists suggest merging with *Bouteloua* (Columbus 1999).

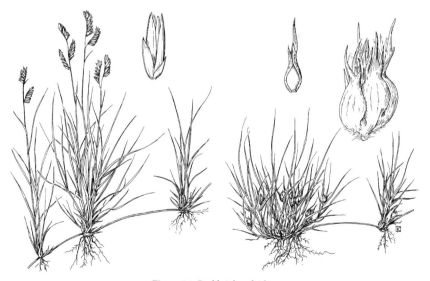

Figure 34. *Buchloë dactyloides.*

6. *Calamovilfa* Hack.

Tall, stout perennials, with short or widely spreading rhizomes and large, open or contracted panicles of small spikelets. Spikelets with 1 floret; rachilla disarticulating above the glumes, not prolonged beyond the floret. Glumes firm, unequal, 1-nerved, acute, the second nearly as long as the lemma, the first shorter. Lemma firm, 1-nerved, rounded on the back, glabrous or pubescent, bearded on the callus. Palea usually greatly reduced. Basic chromosome number, $x = 10$. Photosynthetic pathway, C_4.

Represented in Colorado by 2 species.

1. Lemma and/or palea pubescent . . . . 13. *C. gigantea*
1. Lemma and palea glabrous . . . . 14. *C. longifolia*

13. *Calamovilfa gigantea* (Nutt.) Scribn. & Merr. (Fig. 35).

Synonym: *Calamagrostis gigantea* Nutt. **Vernacular Name:** Giant sandreed. **Life Span:** Perennial. **Origin:** Native. **Season:** C_4, warm season. **FLORAL CHARACTERISTICS: Inflorescence:** open panicle 20–80 cm long, 20–60 cm wide, branches stiffly spreading and the lowermost sometimes reflexed. **Spikelets:** yellowish to purple-tinged, 7–11 mm long, 1-flowered; disarticulation above the glumes. **Glumes:** straight, 1-nerved, keeled, glabrous; lower 4.5–10.5 mm long; 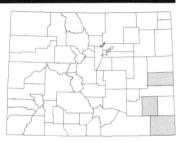 upper 6.4–10.1 mm long. **Lemmas:** straight, 1-nerved, 6–10 mm long, villous to sparsely villous and rarely glabrous. **Paleas:** subequal to lemma, villous between nerves and often villous to tip. **Awns:** none. **VEGETATIVE CHARACTERISTICS: Growth Habit:** rhizomatous, rhizomes covered with leaflike scales. **Culms:** erect, to 2.5 m tall, robust, usually solitary. **Sheaths:** open, glabrous except for some pubescence at the throat. **Ligules:** ciliate, fringed membrane, 0.7–2 mm long. **Blades:** to 90 cm long, 5–12 mm broad at base, tapering at tip, flat to involute, glabrous. **HABITAT:** Found on sand dunes and wet sand flats (Weber and Wittmann 2001a). Wingate (1994) records locations in Baca and Cheyenne counties, while Weber and Wittmann (2001a) have locations in the lower Arkansas and Cimarron river drainages in eastern Colorado. **COMMENTS:** This species is too coarse to have much forage value, but it is a good sand binder, which may be helpful in controlling soil erosion. This genus is unusual in the Chloridoideae with its 1-nerved lemma. Additionally, a fruit in which the seed coat is free from the pericarp makes this an achene rather than a true caryopsis.

14. *Calamovilfa longifolia* (Hook.) Scribn. (Fig. 36).

Synonyms: *Calamagrostis longifolia* Hook., *Vilfa rigida* Buckley, *Ammophila longifolia* Benth. *ex* Vasey, *Alternotus longifolia* Lunell. **Vernacular Name:** Prairie sandreed. **Life Span:** Perennial. **Origin:** Native. **Season:** C_4, warm season. **FLORAL CHARACTERISTICS: Inflorescence:** panicle narrow, 15–78 cm long, 2–26 cm wide, branches erect to strongly divergent, lowermost sometimes reflexed. **Spikelets:** 1-flowered, disarticulation above the glumes, 5–8 mm long. **Glumes:** straight; lower 3.5–6.5 mm long; upper 5–8.2 mm long. **Lemmas:** glabrous,

Figure 35. *Calamovilfa gigantea*.

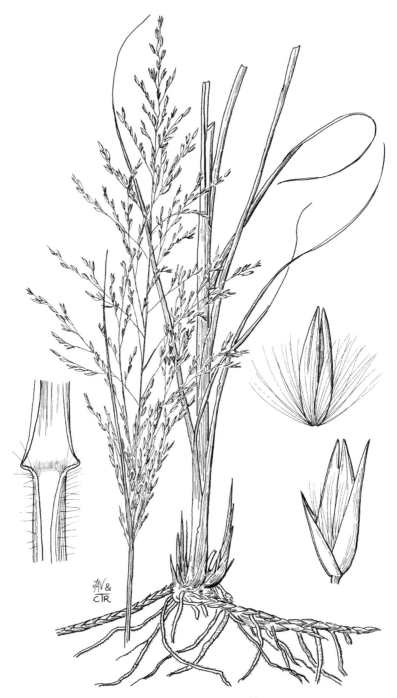

Figure 36. *Calamovilfa longifolia*.

4.5–7.1 mm long, straight, callus with long hairs. **Paleas:** 4.4–6.9 mm long, glabrous. **Awns:** none. **VEGETATIVE CHARACTERISTICS: Growth Habit:** rhizomatous, rhizomes covered with leaflike scales. **Culms:** erect, to 2.4 m tall. **Sheaths:** open, glabrous to densely pubescent. **Ligules:** fringe of hairs, 0.7–2.5 mm long. **Blades:** to 64 cm long and 12 mm wide. **HABITAT:** Found mostly in sandy areas on the eastern plains and piedmont valleys. **COMMENTS:** Common. This genus is unusual in the Chloridoideae with its 1-nerved lemma. Additionally, a fruit in which the seed coat is free from the pericarp makes this an achene rather than a true caryopsis.

7. *Chloris* Sw.

Annuals and perennials with flowering culms from erect or decumbent and stoloniferous bases. Lower portion of culms often flattened, and laterally compressed and sharply keeled leaf sheaths. Inflorescence with few to several persistent, spicate branches, these mostly digitate or clustered at the culm apex but loosely distributed along the main axis in a few species. Spikelets with a single perfect floret and 1–3 staminate or rudimentary florets above. Disarticulation above the glumes. Glumes thin, lanceolate, 1-nerved (except in cleistogamous, subterranean spikelets), strongly unequal to nearly equal. Lemmas broadly to narrowly ovate or oblong, awnless or more frequently awned from a minutely bifid apex, 1- to 5-nerved but usually with 3 strong nerves, glabrous or ciliate-pubescent or puberulent on the nerves. Palea well developed, strongly 2-nerved. Basic chromosome number, $x = 10$. Photosynthetic pathway, C_4.

Represented in Colorado by 2 species.

1. Plants perennial; lemma sparsely pubescent with minute hairs; panicle branches in several whorls . . . . 15. *C. verticillata*
1. Plants annual; lemma densely pubescent with long hairs; panicle branches digitate . . . . 16. *C. virgata*

15. *Chloris verticillata* Nutt. (Fig. 37).

Synonyms: None. **Vernacular Name:** Tumble windmill grass. **Life Span:** Perennial. **Origin:** Native. **Season:** C_4, warm season. **FLORAL CHARACTERISTICS: Inflorescence:** panicle of 10–20 separate spicate branches, 5–16 cm long, in 2–5 separate digitate whorls and 1 vertical terminal branch; branches spikelet-bearing to base. **Spikelets:** spikelets in 1 row, sessile to short-pedicellate, 2-flowered with 1 fertile and 1 sterile floret, 2–4 mm long excluding awns; disarticulation above the glumes. **Glumes:** unequal, 1-nerved; lower 2–3 mm long; upper 2.8–3.5 mm long. **Lemmas:** 3-nerved, fertile lemma strongly laterally compressed, lanceolate to elliptic, firm, 2–3.5 mm long, 1.5–1.9 mm wide, keels sparsely appressed pubescent

Figure 37. *Chloris verticillata.*

to glabrous, hairs minute; sterile lemma 1–2.3 mm long, oblong and sometimes inflated, glabrous, truncate apically with awn. **Awns:** glumes awn-tipped; fertile lemma awn 4–9 mm long, attached just below apex of lemma; sterile lemma with awn 3–7 mm long. **VEGETA-TIVE CHARACTERISTICS: Growth Habit:** cespitose. **Culms:** erect to decumbent, occasionally rooting at lower nodes, 14–40 cm tall. **Sheaths:** open, generally glabrous, with hairs up to 3 mm long near ligule. **Ligules:** ciliate with membranous base, 0.7–1.3 mm long. **Blades:** glabrous to scabrous, to 15 cm long and 2–3 mm wide, with hairs at base, flat to folded. **HABITAT:** At low elevations, found in lawns and urban areas, gardens, and roadsides in both eastern and less frequently western Colorado (Weber and Wittmann 2001a, 2001b). **COMMENTS:** The entire inflorescence will separate from the plant and become a "tumbleweed," considered an adaptation for seed dispersal.

16. *Chloris virgata* Sw. (Fig. 38).

Synonyms: None. **Vernacular Names:** Feather fingergrass, blackseed grass, feather windmill grass. **Life Span:** Annual. **Origin:** Native. **Season:** C₄, warm season. **FLORAL CHARACTERISTICS: Inflorescence:** panicle of 4–20 digitate branches in a single whorl, branches ascending, 5–10 cm long. **Spikelets:** spikelets in 2 rows, strongly imbricate, mostly subsessile, 2- to 3-flowered with 1 fertile and 1–2 sterile reduced florets, 2–4.2 mm long excluding awns; disarticula-

tion above the glumes. **Glumes:** unequal, narrowly lanceolate, 1-nerved; lower 1.5–2.5 mm long, sometimes aristate; upper 2.5–4.3 mm long, tapering into a short awn. **Lemmas:** fertile

Figure 38. *Chloris virgata*.

lemma 2.5–4.2 mm long, compressed, obovate with prominent gibbous base, keel densely appressed pilose, margins with long-pilose hairs, especially near apex margin, 3-nerved; sterile lemma(s) 1.4–3 mm, narrowly club-shaped, glabrous, truncate apically. **Awns:** upper glume short-awned; fertile lemma awns 2.5–15 mm long; sterile lemma awn 3–9.5 mm. **VEGETATIVE CHARACTERISTICS: Growth Habit:** tufted, occasionally stoloniferous. **Culms:** 10–100 cm tall or more, ascending, occasionally geniculate at lower nodes and rooting. **Sheaths:** open, keeled, generally glabrous, occasionally long-hairy near ligule. **Ligules:** membranous, to 1 mm long, occasionally lacking, truncate, lacerate to ciliate. **Blades:** flat or folding, glabrous to scaberulous, occasionally pilose near the collar, to 30 cm long and 15 mm wide. **HABITAT:** Disturbed areas, especially roadsides at low elevations, lawns and gardens, and other urban areas in both eastern and western Colorado (Weber and Wittmann 2001a, 2001b). **COMMENTS:** This species is widespread in the United States and grows in many habitats where the summers are hot.

8. *Chondrosum* Desv.

Annuals and perennials of diverse habit, some cespitose, others with rhizomes or stolons. Ligule usually a ring of hairs. Inflorescence of 1 to numerous short, spicate branches, these closely or distantly spaced along the main axis and bearing numerous (20–80) small, closely placed, pectinate, sessile spikelets in 2 rows along the margins of an angular or flattened rachis. Disarticulation above the glumes; branches persistent. Spikelets with 1 fertile floret and 2–3 reduced or rudimentary florets above. Glumes lanceolate, 1-nerved, unequal to nearly equal, awnless or short-awned. Lemmas membranous, 3-nerved, midnerve often extending into an awn, lateral nerves occasionally short-awned. Paleas membranous, the 2 nerves occasionally awn-tipped. Basic chromosome number, $x = 10$. Photosynthetic pathway, C_4.

Represented in Colorado by 5 species. A segregate of *Bouteloua*.

1. Culm internodes densely pubescent . . . . 18. *C. eriopodum*
1. Culm internode glabrous
  2. Plant annual
    3. Inflorescence reduced to a single branch . . . . 21. *C. simplex*
    3. Inflorescence 2 or more branches . . . . 17. *C. barbatum*
  2. Plant perennial
    4. Rachis prolonged beyond spikelets as a naked point . . . . 20. *C. hirsutum*
    4. Rachis not prolonged beyond spikelets as a naked point . . . . 19. *C. gracile*

17. *Chondrosum barbatum* (Lag.) Clayton (Fig. 39).

Synonyms: *Bouteloua arenosa* Vasey, *Bouteloua barbata* Lag. *Chondrosum exile* Fourn., *Chondrosum microstachyum* Fourn., *Chondrosum polystachyum* Benth., *Chondrosum subscorpioides* C. Muell. **Vernacular Name:** Sixweeks grama. **Life Span:** Annual. **Origin:** Native. **Season:** C_4, warm season. **FLORAL CHARACTERISTICS: Inflorescence:** panicle of 1–9 solitary, spikelike branches, rarely more, exceeding the leaves, 0.5–3 cm long. **Spikelets:** pectinate, 2–5 mm long, 6–50 per branch; with 1 perfect floret and 2 rudimentary florets, disarticulating above the glumes. **Glumes:** unequal, glabrous to scaberulous, 1-nerved; lower 1–2 mm long, hya-

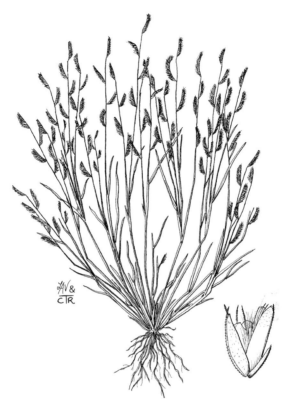

Figure 39. *Chondrosum barbatum.*

line; upper 1.5–2.5 mm long, glabrous, scabrous, or strigose, hairs not papillose-based, purplish with scarious to hyaline margins above. **Lemmas:** fertile lemma small, 1.5–2.5 mm long, densely pilose, at least on the margins; sterile lemmas 0.5–1 mm long, globose, empty. **Awns:** glume awns up to 0.7 mm long; fertile lemma 3-awned, central awn 0.5–3 mm long; sterile lemmas 3-awned, 0.5–4 mm long; palea 2-awned, 1–2 mm long. **VEGETATIVE CHARACTERISTICS: Growth Habit:** tufted, at times stoloniferous; leaves mostly basal. **Culms:** numerous, slender stems, 0.2–3 dm, rarely to 6 dm tall, geniculate to sprawling to occasionally erect, internodes glabrous. **Sheaths:** open, glabrous, often ciliate with long hairs near collar. **Ligules:** a fringe of hairs sometimes membranous at base to approximately 1 mm long. **Blades:** 5–10 cm long, 1–4 mm wide, flat to involute, often curly, glabrous to sparsely hairy with some papillose-based hairs basally. **HABITAT:** Dry to moist open, mostly disturbed sites. According to Weber and Wittmann (2001a, 2001b), this species is found in the lower Arkansas Valley on the Eastern Slope and in the desert steppe community on the Western Slope. **COMMENTS:** Infrequent. Colorado specimens belong to the typical variety (var. *barbatum*).

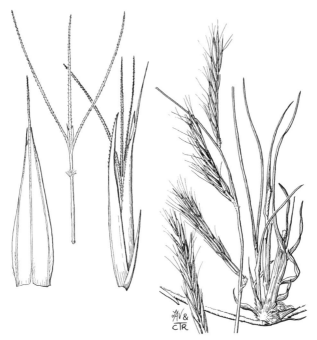

Figure 40. *Chondrosum eriopodum.*

18. *Chondrosum eriopodum* Torr. (Fig. 40).

Synonym: *Bouteloua eriopoda* (Torr.) Torr. **Vernacular
Name:** Black grama. **Life Span:** Perennial. **Origin:**
Native. **Season:** C_4, warm season. **FLORAL CHAR-
ACTERISTICS: Inflorescence:** panicle, 4–15 cm long,
of 2–8 loosely ascending spicate branches, these 2–5
cm long, each bearing 8–20 uncrowded, appressed
spikelets in 2 rows, appearing pectinate, rachis
densely bearded at base. **Spikelets:** 6–10 mm long,
2-flowered, 1 perfect floret with rudimentary floret

above; disarticulating above the glumes. **Glumes:** unequal, 1-nerved; lower 2–4.5 mm long,
narrow, glabrous to sometimes scaberulous on the nerves; upper 6–9 mm long, acute to
acuminate, firm, glabrous to minutely scaberulous, stramineous to sometimes purplish.
Lemmas: fertile lemma 4–7 mm long, linear, 3-nerved, purplish, usually bearded at base;
rudimentary lemma 0.5–1.2 mm long, sometimes reduced to 3 awns. **Awns:** fertile lemma
3-awned, central awn 0.5–4 mm long, lateral awns 0.5–1.5 mm long; rudimentary lemma
1- to 3-awned, 3.5–9 mm long. **VEGETATIVE CHARACTERISTICS: Growth Habit:** stoloniferous, leaves mostly basal. **Culms:** decumbent to ascending from a hard, knotty base, wiry,
2–6 dm tall, lower internodes densely woolly pubescent. **Sheaths:** glabrous, sometimes
ciliate at the summit, striate. **Ligules:** ring of hairs less than 1 mm long, sometimes membranous at base. **Blades:** flat at base to involute and twisted at the tips, 2–7 cm long, 0.5–2

mm wide, papillose-based hairs on lower margins. **HABITAT:** Weber and Wittmann (2001a) and Wingate (1994) place this species on rocky slopes on Mesa de Maya on the Eastern Slope. Shaw and colleagues (1989a, 1989b) report it from canyon rims, slopes, and bottoms along the Purgatoire River. **COMMENTS:** Forage value is excellent throughout the year for all livestock (Stubbendieck, Hatch, and Landholt 2003). Only Colorado grass species with woolly internodes.

19. *Chondrosum gracile* Kunth (Fig. 41).

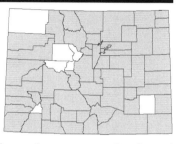

Synonyms: *Bouteloua gracilis* (Kunth) Lag. *ex* Griffiths var. *stricta* (Vasey) Hitchc., *Bouteloua oligostachya* (Nutt.) Torr. *ex* Gray, *Chondrosum oligostachyum* (Nutt.) Torr. **Vernacular Names:** Blue grama, graceful grama, eyelash grass. **Life Span:** Perennial. **Origin:** Native. **Season:** C_4, warm season. **FLORAL CHARACTERISTICS: Inflorescence:** panicle of 1–3 and rarely up to 6 spicate branches, the branches 1.5–5 cm long, pectinate, ascending to spreading, rachis usually pubescent at base and glabrous to scabrous above, not extending beyond the spikelet; spikelets 40–90, crowded in 2 rows on each branch to the end. **Spikelets:** 2- to 3-flowered, 1 fertile floret with 1–2 sterile florets above; disarticulation above the glumes. **Glumes:** unequal, 1-nerved; lower 1.5–3.5 mm long, linear to linear-lanceolate, hyaline, glabrous or with hairs on the nerve; upper subequal to fertile lemma, 3.5–6 mm long, pale to purple-tinged, with papillose-based hairs on the nerve. **Lemmas:** fertile lemma 3.5–6 mm long, 3-nerved, narrow-lanceolate, slightly keeled, pubescent below and callus densely hairy; rudimentary lemma(s); first 3-lobed surpassed by rachilla hairs, 1–3 mm long; second, when present, obovate, 1–2 mm long, awnless. **Awns:** fertile lemma 3-awned, terminal awn 0.5–1.5 mm long, lateral awns 1–2 mm long; lower rudimentary lemma 3-awned, 2.5–5 mm long. **VEGETATIVE CHARACTERISTICS: Growth Habit:** densely tufted, often with short rhizomes and mat-forming, leaves mostly basal. **Culms:** erect to decumbent to geniculate below, sometimes rooting at lower nodes, 2–7 dm tall, glabrous or nodes minutely hairy, internodes glabrous. **Sheaths:** glabrous or sparsely long-hairy, round, long-hairy at collar. **Ligules:** 0.1–0.5 mm long, ciliate, sometimes with a membranous base, truncate. **Blades:** flat or loosely involute at maturity, 2–25 cm long, 0.5–4 mm wide, often curly, scaberulous above, glabrous below. **HABITAT:** Dominant grass of the shortgrass steppe in eastern Colorado. Also in grasslands at lower elevations on the Western Slope. Can be found in the foothills, to mid-elevation forests, and mountain park grasslands (Weber and Wittmann 2001a, 2001b). **COMMENTS:** One of the most common grasses in the state. Forage quality is good for all livestock and retains its value when dry, thereby able to support fall and winter grazing (Stubbendieck, Hatch, and Landholt 2003).

20. *Chondrosum hirsutum* (Lag.) Sw. (Fig. 42).

Synonyms: *Bouteloua glandulosa* (Cerv.) Swallen, *Bouteloua hirsuta* Lag. var. *glandulosa* (Cerv.) Gould. **Vernacular Name:** Hairy grama. **Life Span:** Perennial. **Origin:** Native. **Season:** C_4, warm season. **FLORAL CHARACTERISTICS: Inflorescence:** panicle 1–7 cm long, rarely to 15

Figure 41. *Chondrosum gracile.*

cm, consisting of 1–4 spicate branches, these 1–5 cm long, ascending or spreading, straight to occasionally curved, bearing 18–50 spikelets in 2 rows, crowded, rachis extending beyond terminal floret 5–10 mm; disarticulation above the glumes. **Spikelets:** 2- to 3-flowered, 1 fertile floret and 1–2 rudimentary florets above, green to dark purple. **Glumes:** 1-nerved, small, unequal; lower 1.4–3.5 mm long, very narrow; upper 3–6 mm long, broadly lanceolate, subequal to lemma, purplish with tuberculate-hirsute spreading hairs that darken and become

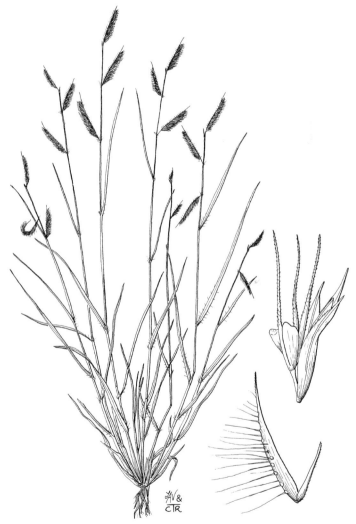

Figure 42. *Chondrosum hirsutum*.

visible to the naked eye at maturity. **Lemmas:** fertile lemma 3-nerved, pubescent, 2–4.5 mm long, deeply 3-cleft, callus bearded; rudimentary lemma 0.5–2 mm, callus naked; upper rudiment, when present, obovate to deltoid, glabrous. **Awns:** upper glume acuminate to awn-tipped; fertile lemma 1- to 3-awned, 0.2–2.5 mm long, lateral awns absent or as long as terminal awn; rudimentary lemma(s) 3-awned, 2–4 mm long, rarely up to 6 mm long; upper rudimentary lemma, when present, may be awnless. **VEGETATIVE CHARACTERISTICS: Growth Habit:** densely tufted, leaves basal and cauline, occasionally stoloniferous. **Culms:** erect to sprawling to geniculate at lower nodes, 15–75 cm tall, internodes glabrous. **Sheaths:** glabrous, scabrous, or pilose mainly near the ligules. **Ligules:** fringe of hairs 0.2–0.4 mm long, base occasionally membranous, truncate. **Blades:** flat to involute, 2–18 cm long, rarely longer, 1–3 mm wide, both surfaces smooth to scabrous, margins with papillose-based hairs mostly near base. **HABITAT:** Found on the plains of eastern Colorado and the mesas and grasslands of western portions of the state. **COMMENTS:** An often overlooked species frequently confused with *C. gracile*. The Colorado specimens belong to the typical subspecies (subsp. *hirsutum*).

21. *Chondrosum simplex* (Lag.) Kunth (Fig. 43).

Synonyms: *Bouteloua simplex* Lag., *Chondrosum procumbens* Desv. *ex* P. Beauv., *Chondrosum prostratum* (Lag.) Sweet. **Vernacular Name:** Matted grama. **Life Span:** Annual. **Origin:** Native. **Season:** C$_4$, warm season. **FLORAL CHARACTERISTICS: Inflorescence:** panicle with a single spikelike branch, 1–2.5 cm long, rarely with 2–4 subdigitate branches; branch curving, bearing 20–80 spikelets, with terminal spikelet sterile. **Spikelets:** 2- to 3-flowered, 1 fertile floret and 1–2 rudimentary florets above, pectinate, 4–6 mm long including the short awns; disarticulation above the glumes. **Glumes:** unequal, keeled with 1 nerve, scaberulous on the nerve; lower 2–3 mm long; upper commonly purple-tinged, 3.5–5.5 mm long. **Lemmas:** fertile lemma 2.5–3.5 mm long, 3-nerved, appressed pilose adjacent to midnerve at least near base, with a bearded callus; rudimentary lemma 1–1.3 mm long, with a hairy callus; second rudimentary lemma, when present, obovate to flabellate scale. **Awns:** fertile lemma 3-awned, awns flattened, central awn 1–2 mm long, lateral awns shorter than central awn; rudimentary lemma reduced to an awn column with 3 awns 3–6 mm long. **VEGETATIVE CHARACTERISTICS: Growth Habit:** densely tufted. **Culms:** numerous, 1–30 cm long, spreading to decumbent with some branching, geniculate at base, internodes glabrous. **Sheaths:** glabrous to pilose near the collar, deeply striate. **Ligules:** up to 0.2 mm long, hairs with a membranous base. **Blades:** flat, becoming involute at base, 2–8 cm long, 0.5–1.5 mm wide, glabrous to scaberulous and frequently pilose at base. **HABITAT:** Weber and Wittmann (2001a, 2001b) say this plant has become established along roadsides on the Western Slope and on the eastern plains from Denver southeastward and in the San Luis Valley. **COMMENTS:** Common. Weber and Wittmann (2001a, 2001b) have this species as *C. prostratum* (Lag.) Sweet.

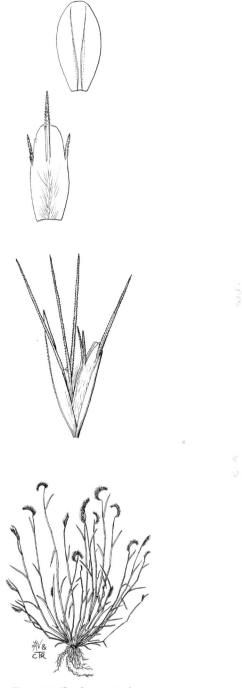

Figure 43. *Chondrosum simplex*.

9. *Crypsis* Aiton

Annuals, low and tufted, with many weakly spreading, many-noded culms. Sheaths open; ligule a fringe of hairs; auricles absent. Inflorescences of short, dense panicles, these terminal on the main shoots and also in the axis of the upper leaves. Panicles remaining partially enclosed by expanded leaf sheaths. Spikelets with 1 fertile floret. Disarticulation below or above the glumes. Glumes 2, subequal, narrow, acute, 1-nerved. Lemma thin, broad, 1-nerved, acute, awnless. Palea broad and about as long as the lemma, splitting between the nerves at maturity. Basic chromosome numbers, $x = 8$ and 9. Photosynthetic pathway, C_4.

Represented in Colorado by a single species.

22. *Crypsis alopecuroides* (Piller & Mitterp.) Schrad. (Fig. 44).

Synonym: *Phleum alopecuroides* (Piller & Mitterp.) *Heleochloa alopecuroides* (Piller & Mitterp.) Host *ex* Roem. **Vernacular Name:** Foxtail pricklegrass. **Life Span:** Annual. **Origin:** Introduced. **Season:** C_4, warm season. **FLORAL CHARACTERISTICS: Inflorescence:** spikelike, cylindrical panicle, 1.5–6.5 cm long, 4–6 mm wide, lower portion often enclosed in upper sheath at anthesis, often purplish. **Spikelets:** 1-flowered, compressed; 1.8–2.8 mm long, often black-tinged; disarticulation above or below the glumes. **Glumes:** subequal or unequal, 1-nerved, strongly keeled, glabrous to minutely ciliate on keel; lower 1.8–2.3 mm long, narrowly lanceolate; upper 1.4–2.4 mm long, lanceolate. **Lemmas:** lanceolate, 1-nerved, keeled, glabrous to sometimes scabrous on keel, 1.7–2.8 mm long. **Awns:** none. **VEGETATIVE CHARACTERISTICS: Growth Habit:** tufted, herbage purple to nearly black. **Culms:** prostrate to erect, branching at base, 5–75 cm long, nodes reddish brown. **Sheaths:** open, glabrous, margins membranous. **Ligules:** ring of hairs, 0.2–1 mm. **Blades:** flat to involute, 1–12 cm long, 1–2.5 mm wide, glabrous to scaberulous. **HABITAT:** Moist to wet areas, reported from Stanley Lake area near Broomfield and in the Denver and Boulder areas (Weber and Wittmann 2001a). **COMMENTS:** Uncommon.

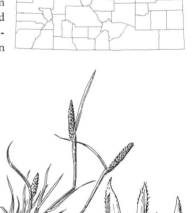

Figure 44. *Crypsis alopecuroides.*

10. *Cynodon* Rich.

Low, mostly mat-forming stoloniferous and rhizomatous perennials. Culms much-branched, mostly with short internodes. Leaf blades flat, short, narrow, soft, and succulent. Ligule a fringe of hairs. Inflorescence with 2 to several slender, spicate branches, these digitately arranged at culm apex. Spikelets sessile in 2 rows on a narrow, somewhat triangular branch rachis. Spikelets with 1 fertile floret, the rachilla prolonged beyond the palea as a bristle and occasionally bearing a rudimentary lemma. Glumes slightly unequal, lanceolate, awnless, 1-nerved, the upper nearly as long as the lemma. Lemma firm, laterally compressed, awnless, 3-nerved, usually puberulent on the midnerve. Palea narrow, 2-nerved, as long as the lemma. Basic chromosome number, $x = 9$. Photosynthetic pathway, C_4.

Represented in Colorado by a single species.

23. *Cynodon dactylon* (L.) Pers. (Fig. 45).

Synonyms: *Capriola dactylon* (L.) Kuntze, *Cynodon aristiglumis* Caro & Sánchez, *Cynodon incompletus Nees, Panicum dactylon* L. **Vernacular Names:** Bermuda grass, Bahama grass. **Life Span:** Perennial. **Origin:** Introduced. **Season:** C_4, warm season. **FLORAL CHARACTERISTICS: Inflorescence:** panicle of 2–7 digitate spicate branches; branches 2–6 cm long, ascending to spreading, numerous spikelets sessile in 2 rows. **Spikelets:** laterally compressed, 1-flowered, 2–3.2 mm long; rachilla prolonged behind palea and sometimes a vestigial lemma

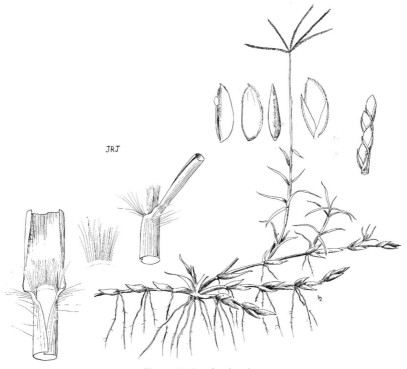

JRJ

Figure 45. *Cynodon dactylon.*

present; disarticulation above the glumes. **Glumes:** subequal, 1-nerved, lanceolate; lower 1.5–2 mm long, curved; upper 1.4–2.3 mm long, straight, tapered into an awn tip. **Lemmas:** laterally compressed, 3-nerved, 1.9–3.1 mm long, pubescent or not on nerves. **Awns:** none. **VEGETATIVE CHARACTERISTICS: Growth Habit:** stoloniferous and rhizomatous with wide-ranging, creeping rhizomes, much-branched, forming mats, stolons flat. **Culms:** much-branched, 5–40 cm long, creeping and only flowering culms erect, rooting at lower nodes, leafy. **Sheaths:** rounded, glabrous but for hair tufts on either side of collar and ligule. **Ligules:** truncate ciliate membrane 0.2–0.5 mm long. **Blades:** flat at maturity, folded in bud, 3–12 cm long, 1–4 mm wide, glabrous to sometimes puberulent above, scaberulous on margins. **HABITAT:** Used as a lawn grass at low elevations where winters are milder, sometimes escaping. **COMMENTS:** This species is a native of Africa. Much more common than collections would indicate, particularly in the southern counties, but specimens of turf grasses are traditionally poorly represented in most herbarium collections. Used as a pasture grass in warmer areas.

11. *Dactyloctenium* Willd.

Annuals with thick, weak culms that are often decumbent and rooting at the lower nodes. Sheaths open; ligule ciliate with a membranous base; auricles absent. Inflorescence of usually 2 to several digitately arranged, unilateral spicate branches. Spikelets closely placed and pectinate in 2 rows on one side of a short, stout rachis projecting as a point beyond insertion of uppermost spikelet. Spikelets with 2 to several florets, laterally compressed. Disarticulation often between the glumes, the lower remaining on the rachis. Glumes 2, keeled, subequal, 1-nerved, the lower awnless while the upper is mucronate or with a short, stout awn. Lemmas firm, broad, 3-nerved, abruptly narrowing to a beaked, usually short-awned tip. Palea well developed, about as long as the lemma. Basic chromosome number, $x = 10$. Photosynthetic pathway, C_4.

Represented in Colorado by a single species.

24. *Dactyloctenium aegyptium* (L.) Willd. (Fig. 46).

Synonym: *Cynosurus aegyptius* L. **Vernacular Names:** Egyptian grass, crowfoot grass, Durban crowfoot. **Life Span:** Annual, short-lived perennial. **Origin:** Introduced. **Season:** C_4, warm season. **FLORAL CHARACTERISTICS: Inflorescence:** panicle of 1–8 digitately arranged spicate branches; branches 1.5–6 cm long; branch axis extends beyond distal spikelet 1–6 mm. **Spikelets:** unequal, 1-nerved, 3–7 fertile florets with additional sterile florets above, 3–4.5
mm long, to 3 mm wide; disarticulation is between the glumes, the lower often persisting. **Glumes:** unequal, 1.5–2 mm long, 1-nerved, keeled; lower ovate, acute; upper oblong elliptic, obtuse. **Lemmas:** glabrous, 3-nerved, 2.5–3.5 mm long, ovate. **Awns:** upper glume awns 1–2.5 mm long; lemma awns 0.5–1 mm long, extension of midnerve, curved. **VEGETATIVE CHARACTERISTICS: Growth Habit:** tufted to shortly stoloniferous. **Culms:** ascending, geniculate at base and usually rooting at nodes, 10–35 cm tall and rarely to 100 cm. **Sheaths:**

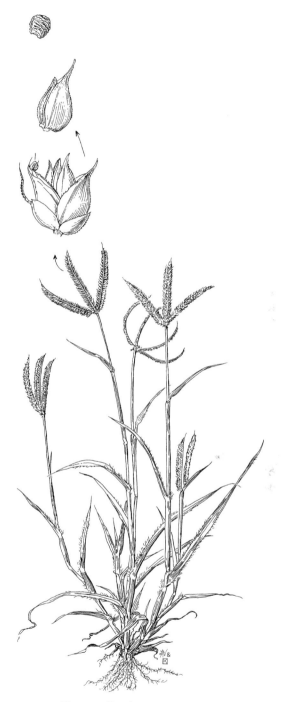

Figure 46. *Dactyloctenium aegyptium*.

open, keeled, papillose-based hairs near collar. **Ligules:** ciliate with membranous base, 0.5–1.5 mm tall. **Blades:** flat, sparsely papillose-pubescent, margins scabrous, 5–22 cm long, 2–8 (12) mm wide. **HABITAT:** Weed of disturbed sites (Barkworth et al. 2003). **COMMENTS:** This taxon has been used as a sand stabilizer along the coast in Australia (Barkworth et al. 2003).

12. *Dasyochloa* Willd. *ex* Rydb.

Perennial, stoloniferous. Culms composed of elongated internodes topped by a fascicle of leaves, the fascicle finally touching the ground and rooting. Sheaths open; ligules a short, ciliated ring of hairs; auricles absent. Inflorescences a small and compact panicle, capitate, usually not exceeding blades of the fascicle, white-woolly. Spikelets with 5–15 florets. Glumes 2, subequal, 1-nerved, often tinged with purple, acuminate, awn-tipped. Lemmas deeply cleft, long-pilose on lower half, often purple-tinged, awn from bifid apex and barely exceeding the lobes. Palea about half the lemma, long-pilose below, nerves extending into awns. Basic chromosome number, $x = 8$. Photosynthetic pathway, C_4.

Represented in Colorado by a single species in the genus. This species has been included in *Erioneuron* and *Tridens*.

25. *Dasyochloa pulchella* (Kunth) Willd. *ex* Rydb. (Fig. 47).

Synonyms: *Erioneuron pulchellum* (Kunth) Tateoka, *Tridens pulchellus* (Kunth) Hitchc., *Triodia pulchella* Kunth. **Vernacular Names:** Low woollygrass, fluffgrass. **Life Span:** Perennial. **Origin:** Native. **Season:** C_4, warm season. **FLORAL CHARACTERISTICS: Inflorescence:** compact panicle, 1.5–4 cm long, 1–1.5 cm wide, usually not exceeding blades of fascicle, light green to purple-tinged, white pubescent. **Spikelets:** 4–10 florets, 5–10 mm long, 3–6 mm wide; paleas 2–3.5 mm long, keels ciliate distally, long-pilose proximally; disarticulates above the glumes. **Glumes:** subequal, 1-nerved, glabrous, purple-tinged; lower 6–8.5 mm long; upper 6.5–9 mm long. **Lemmas:** 3-nerved, 2-lobed, 2.5–5 mm long, long-pilose on lower half, often purple-tinged. **Awns:** lemma awns 1.5–4 mm long, barely extending beyond lobes of lemma. **VEGETATIVE CHARACTERISTICS: Growth Habit:** densely tufted, stoloniferous, sometimes mat-forming. **Culms:** scabrous or puberulent erect initially, inflorescence eventually falling over and rooting at base, 2–15 cm long. **Sheaths:** open, striate, margins scarious, pilose at base, bearing a tuft of hairs at apex. **Ligules:** 3–5 mm long, ciliate fringe of hairs. **Blades:** involute, 1–6 cm long, scabrid above, scaberulous below. **HABITAT:** Piñon-juniper and desert steppe in western Colorado (Weber and Wittmann 2001b) and other dry, open areas. **COMMENTS:** This taxon has been included in *Erioneuron* [*E. pulchellum* (Kunth) Tateoka] and earlier in *Tridens* [*T. pulchellus* (Kunth) Hitchc.].

13. *Distichlis* Raf.

Low, dioecious (rarely monoecious) perennials with stout rhizomes and firm, glabrous culms and leaves. Culms with many nodes and internodes, usually completely covered to the

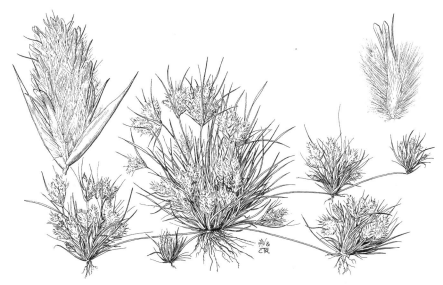

Figure 47. *Dasyochloa pulchella.*

inflorescence by the distichous and closely overlapping leaf sheaths. Ligule a short, fringed membrane. Spikelets large, several-flowered, laterally flattened—in short, contracted panicles or racemes. Disarticulation above the glumes and between the florets. Glumes unequal, acute, 3- to 5-nerved. Lemmas broad, indistinctly 9- to 11-nerved, awnless, laterally compressed, those of the pistillate spikelets coriaceous, thicker than those of the staminate spikelets. Palea large, strongly 2-keeled. Basic chromosome number, $x = 10$. Photosynthetic pathway, C_4.

Represented in Colorado by a single species.

26. *Distichlis spicata* (L.) Greene (Fig. 48).

Synonyms: *Distichlis spicata* (L.) Greene subsp. *stricta* (Torr.) Thorne, *Distichlis spicata* (L.) Greene var. *borealis* (J. Presl) Beetle, *Distichlis spicata* (L.) Greene var. *divaricata* Beetle, *Distichlis spicata* (L.) Greene var. *nana* Beetle, *Distichlis spicata* (L.) Greene var. *stolonifera* Beetle, *Distichlis spicata* (L.) Greene var. *stricta* (Torr.) Scribn., *Distichlis stricta* (Torr.) Rydb., *Distichlis stricta* (Torr.) Rydb. var. *dentata* (Rydb.) C. L. Hitchc., *Uniola spicata* L. **Vernacular Names:** Salt-

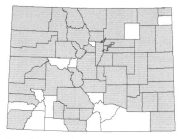

grass, inland saltgrass. **Life Span:** Perennial. **Origin:** Native. **Season:** C_4, warm season. **FLORAL CHARACTERISTICS: Inflorescence:** contracted panicle, pedicels short, 1–7 cm long. **Spikelets:** dioecious, rarely bisexual, 5–20 florets, 5–20 mm long; pistillate spikelets green, disarticulation above the glumes and below each floret; staminate spikelets straw-colored,

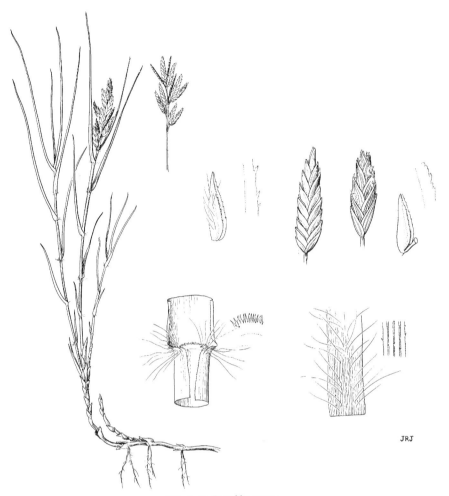

Figure 48. *Distichlis spicata.*

spikelet does not disarticulate. **Glumes:** unequal, acute, glabrous; lower 2–3 mm long, 3- to 9-nerved; upper 3–4 mm long, 5- to 11-nerved. **Lemmas:** acute, 3–6 mm long with yellow, coarse margin. **Awns:** none. **VEGETATIVE CHARACTERISTICS: Growth Habit:** strongly rhizomatous; rhizomes scaly, creeping, extensive. **Culms:** prostrate to decumbent to erect with numerous short internodes, 10–60 cm tall, glabrous. **Sheaths:** glabrous, open; lower sheaths reduced; upper sheaths strongly overlapping. **Ligules:** short ciliate membrane, truncate, 0.2–0.6 mm long, often with ciliate hairs. **Blades:** flat to involute, distichous, glabrous, stiffly spreading, equal to exceeding the panicle, 1–12 cm long, rarely up to 20 cm, 1–3.5 mm wide. **HABITAT:** Found in alkaline swales and borrow pits and lower valleys at low elevations. **COMMENTS:** This species and *D. stricta* (Torr.) Rydb. have been merged. A common and frequently dominant species of alkaline soils.

14. *Eleusine* Gaertn.

Low, spreading annuals with thick, weak culms and soft, flat or folded, succulent leaf blades. Ligule a short, lacerate membrane. Spikelet sessile in 2 rows on 2 (occasionally 1) to several branches digitately arranged at culm apex. One or 2 branches frequently developed below apical cluster. Spikelets 3- to several-flowered, disarticulating above the glumes and between the florets. Glumes firm, acute, unequal; the lower short, 1-nerved; the upper 3- to 7-nerved. Lemmas acute, awnless or mucronate, broadly keeled, 3-nerved, lateral nerves very close to midnerve. Palea shorter than lemma. Grain plump, with minutely transversely rugose seed loosely enclosed in a thin pericarp. Basic chromosome number, x = 9. Photosynthetic pathway, C_4.

Represented in Colorado by a single species.

27. *Eleusine indica* (L.) Gaertn. (Fig. 49).

Synonym: *Cynosurus indicus* L. **Vernacular Names:** Indian goosegrass, fowlfoot grass, yard grass, crowfoot grass. **Life Span:** Annual. **Origin:** Introduced. **Season:** C_4, warm season. **FLORAL CHARACTERISTICS: Inflorescence:** panicle of 4–10 and rarely up to 17 digitally arranged branches, sometimes with 1–2 branches attached below; branches linear, 3–16 cm long, 3–5.5 mm wide. **Spikelets:** 4–8 mm long, 2–3 mm wide, 5- to 8-flowered; disarticulation above  the glumes. **Glumes:** unequal; lower 1-nerved, 1.1–2.3 mm long; upper 3- to 5- and rarely

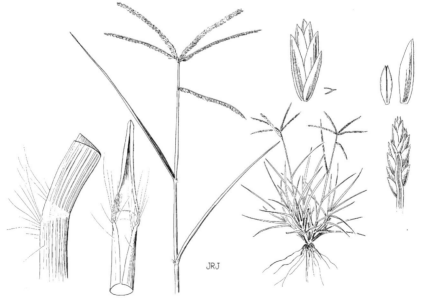

Figure 49. *Eleusine indica*.

to 7-nerved, 2–2.9 mm long. **Lemmas:** 3-nerved, rarely 5- to 7-nerved, glabrous, keeled, 2.4–4 mm long, mucronate-tipped. **Awns:** none. **VEGETATIVE CHARACTERISTICS: Growth Habit:** cespitose. **Culms:** erect to ascending, much-branched at base, 30–90 cm tall, slightly compressed. **Sheaths:** open, margins and throat often with papillose-based pilose hairs 2–6 mm long, conspicuously keeled. **Ligules:** membranous, truncate, erose, 0.2–1.2 mm long. **Blades:** flat to loosely involute, 15–40 cm long, 3–7 mm wide, have prominent white midveins; margins, lower surface, or both, often with papillose-based hairs at base. **HABITAT:** Found as a ruderal weed in both eastern and western Colorado at lower elevations. **COMMENTS:** Occurs infrequently as a lawn weed.

15. *Eragrostis* Wolf

Cespitose annuals and perennials, a few with rhizomes. Inflorescence an open (infrequently contracted) panicle. Spikelets 3- to many-flowered, awnless. Glumes 2, unequal, 1- to 3-nerved, shorter than lemmas. Lemmas 3-nerved, acute or acuminate at apex, keeled or rounded on the back. Paleas strongly 2-nerved and usually 2-keeled, often ciliolate on the keels, as long or nearly as long as lemmas. Glumes, lemmas, and mature caryopsis usually early deciduous, paleas persistent on rachilla. Basic chromosome number, $x = 10$. Photosynthetic pathway, C_4.

Represented in Colorado by 14 species, 2 subspecies, and 4 varieties.

1. Plant annual
    2. Plant stoloniferous, mat-forming . . . . 32. *E. hypnoides*
    2. Plant neither stoloniferous nor mat-forming
        3. Plant with glandular depressions on panicle branches, keel of glumes and lemma, margins of leaf blades and sheath keel, and below the inflorescence
            4. Glands located below the inflorescence, on vegetative portions of the plant, along the blade midnerves, and on sheaths . . . . 37. *E. pilosa*
            4. Glands located on panicle branches, keel of glumes and lemma, or margins of leaf blades and sheath keel, not limited to vegetative portions of the plant
                5. Panicle contracted, 0.5–2 cm wide; primary panicle branches generally appressed, on occasion diverging from the rachis by up to 30°. . . . . 33. *E. lutescens*
                5. Panicle open to somewhat contracted, 2–18 cm wide; primary panicle branches generally diverging from the rachis 20°–110°
                    6. Lemmas 2 mm or more long, glands prominent on keel of lemma; disarticulation below florets; rachillas persistent . . . . 29. *E. cilianensis*
                    6. Lemmas less than 1.9 mm long, glands prominent on panicle branches and leaf blades, rarely on keel of glumes and lemma; disarticulation below lemmas; with both paleas and rachillas generally persistent
                        7. Glandular regions in panicle usually as a dull greenish gray to stramineous spot, or crateriform pit (not a ring), below nodes; pedicels usually with a ring of glands distally . . . . 35. *E. minor*
                        7. Glandular regions in panicle usually as a shiny or yellowish band or ring below nodes; pedicels lacking glandular bands . . . . 28. *E. barrelieri*
        3. Plant without glandular depressions
            8. Spikelets to 1.4 mm wide, linear, slender

9. Caryopsis shallowly to deeply grooved on ventral surface; ligule a ring of hairs; blades 2–7 mm wide . . . . 34. *E. mexicana*

9. Caryopsis lacking a ventral groove; ligule a ciliate membrane; blades 1–2.5 mm wide . . . . 37. *E. pilosa*

8. Spikelets 1.5 mm or more wide, linear to ovate

10. Spikelets grayish green to purplish, 4–16 florets; lower glume 1–2 mm long; upper glume 1.5–2.8 mm long; lemma 1.6–3.3 mm long; caryopsis 0.7–1.3 mm long . . . . 40. *E. tef*

10. Spikelets yellowish brown to dark reddish purple, 6–22 florets; lower glume 0.5–1.5 mm long; upper glume 1–1.7 mm long; lemma 1–2.2 mm long; caryopsis 0.5–1.1 mm long . . . . 36. *E. pectinacea*

1. Plant perennial

11. Spikelets with at least some pedicels 1 mm or less

12. Panicle large, becoming a tumbleweed, viscid . . . . 30. *E. curtipedicellata*

12. Panicle narrow, not a tumbleweed or viscid . . . . 38. *E. secundiflora*

11. Spikelets with pedicels more than 1 mm long

13. Panicle branches villous at conjunction with central axis

14. Rhizomatous, with short, knotty rhizomes; ligules 0.1–0.2 mm long; spikelets generally reddish purple . . . . 39. *E. spectabilis*

14. Cespitose, lacking rhizomes; ligules 0.6–1.3 mm long; spikelets generally grayish green . . . . 31. *E. curvula*

13. Panicle branches naked in axils, occasionally a few short hairs present at conjunction with central axis . . . . 41. *E. trichodes*

28. *Eragrostis barrelieri* Daveau (Fig. 50).

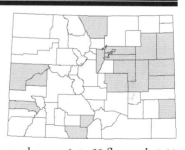

Synonyms: None. **Vernacular Name:** Mediterranean lovegrass. **Life Span:** Annual. **Origin:** Introduced. **Season:** C$_4$, warm season. **FLORAL CHARACTERISTICS: Inflorescence:** open to contracted panicle, 4–20 cm long, 2–10 cm wide; panicle branches generally diverging from rachis 20°–100°, with yellowish to shiny glandular rings or spots below nodes; pulvini present, glabrous; pedicels 1–4 mm long, stiff, glandular bands absent. **Spikelets:** compressed, narrowly ovate to lanceolate-linear, reddish purple to greenish, rarely gray, 6- to 20-flowered, 4–11 mm long, 1–2.2 mm wide; glumes, lemmas, and caryopsis early deciduous, while paleas persistent on rachis. **Glumes:** subequal, 1-nerved, acute, scabrous on keel; lower 0.9–1.4 mm long; upper 1.4–1.8 mm long. **Lemmas:** broadly ovate, 3-nerved, acute to obtuse apex, 1.4–2 mm long, scabrous on upper keel. **Awns:** none. **VEGETATIVE CHARACTERISTICS: Growth Habit:** tufted, branching near base. **Culms:** erect to decumbent, 5–60 cm tall, ring or partial ring of yellow glandular tissue below inflorescence and nodes. **Sheaths:** glabrous, long-hairy at apex. **Ligules:** ciliate, 0.2–0.5 mm long. **Blades:** flat to involute, 1.5–10 cm long, 2–5 mm wide, upper surface glabrous, lower surface glabrous to occasionally scabrous, margins lacking crater-like glands. **HABITAT:** Scattered on the eastern plains and western plateaus. **COMMENTS:** A European introduction that has become a naturalized weedy species. Frequent around rail yards according to Barkworth and colleagues (2003).

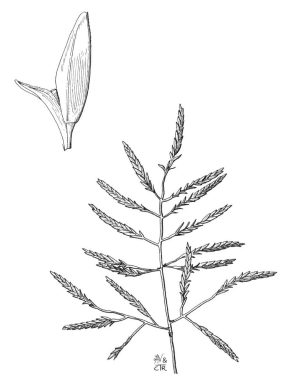

Figure 50. *Eragrostis barrelieri*.

29. *Eragrostis cilianensis* (All.) Vignolo *ex* Janch. (Fig. 51).

Synonyms: *Eragrostis major* Host, *Eragrostis mega-stachya* (Koel.) Link, *Poa cilianensis* All. **Vernacular Names:** Stinkgrass, Grey lovegrass. **Life Span:** Annual. **Origin:** Introduced. **Season:** C$_4$, warm season. **FLORAL CHARACTERISTICS: Inflorescence:** contracted to open panicle, congested, 5–16 cm long, 2–8.5 cm wide; primary panicle branches generally diverging from rachis 20°–80°; pulvini glabrous to hairy. **Spikelets:** ovate-oblong, flattened, grayish to

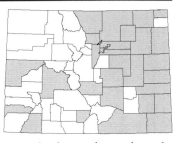

greenish in color, 10–40 florets, 6–20 mm long, 2–4 mm wide; disarticulation above the glumes with florets falling as a unit. **Glumes:** keeled, broadly ovate to lanceolate, usually glandular; lower 1-(3)-nerved, 1.2–2 mm long; upper 3-nerved, 1.2–2.6 mm long. **Lemmas:** with crater-like glands on keel, 2–2.8 mm long, strongly 3-nerved, scaberulous near margins, closely imbricate. **Awns:** none. **VEGETATIVE CHARACTERISTICS: Growth Habit:** tufted. **Culms:** erect to decumbent, 10–45 cm tall, occasionally glands present below nodes; malodorous when fresh. **Sheaths:** glabrous, open, pilose at apex. **Ligules:** ciliate, 0.4–0.8

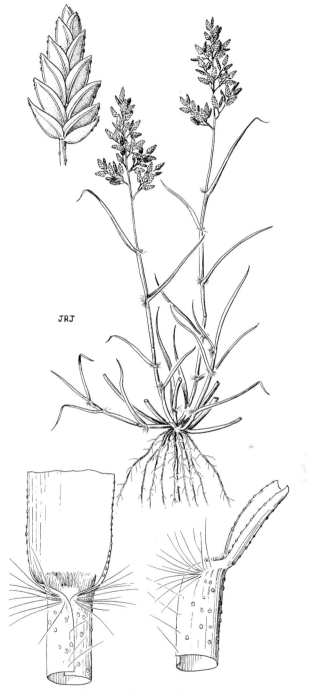

Figure 51. *Eragrostis cilianensis.*

mm long. **Blades:** flat to involute, 5–20 cm long, 2–6 mm wide, upper surface scabrous, lower surface glabrous and occasionally glandular. **HABITAT:** Found in disturbed areas at lower elevations over the state. **COMMENTS:** A common weed introduced from Europe and now naturalized. Fresh plants have a foul odor.

30. *Eragrostis curtipedicellata* Buckley (Fig. 52).

Synonyms: None. **Vernacular Name:** Gummy lovegrass. **Life Span:** Perennial. **Origin:** Native. **Season:** C_4, warm season. **FLORAL CHARACTERISTICS: Inflorescence:** open panicle, broadly ovate, 18–35 cm long, 10–30 cm wide; branches diverging 10°–90° from rachises, stiff, viscid; pedicels 0.2–1 mm long. **Spikelets:** linear-lanceolate, 4- to 12-flowered, 3–6 mm long, 1–1.5 mm wide; reddish purple to stramineous; disarticulation above the glumes. **Glumes:**

lanceolate, subequal, scabrous on midnerve; lower 0.9–1.8 mm long; upper 1.2–2 mm long, 1- to 3-nerved. **Lemmas:** membranous, lanceolate, 3-nerved, 1.5–2.2 mm long. **Awns:** none. **VEGETATIVE CHARACTERISTICS: Growth Habit:** cespitose, with short, knotty rhizomes. **Culms:** erect to spreading, viscid, 20–65 cm tall. **Sheaths:** viscid, hairy at apex, on collar, and along margins. **Ligules:** membranous, 0.1–0.5 mm long. **Blades:** flat to involute, sometimes viscid, 5–18 cm long, 2–5 mm wide. **HABITAT:** Found in sandy ground in the southeastern corner of Colorado in Baca County. **COMMENTS:** The vernacular name, gummy lovegrass, comes from the "stickiness" of the inflorescence. After anthesis the panicle often breaks off at base of the inflorescence and becomes a "tumbleweed."

Figure 52. *Eragrostis curtipedicellata.*

31. *Eragrostis curvula* (Schrad.) Nees (Fig. 53).

Synonyms: *Poa curvula* Schrad. *Eragrostis chloromelas* Steud., *Eragrostis curvula* (Schrad.) Nees var. *conferta* Stapf, *Eragrostis curvula* (Schrad.) Nees var. *curvula* (Schrad.) Nees, *Eragrostis robusta* Stent. **Vernacular Name:** Weeping lovegrass. **Life Span:** Perennial. **Origin:** Introduced. **Season:** C$_4$, warm season. **FLORAL CHARACTERISTICS: Inflorescence:** open panicle, oblong to ovate, 16–40 cm long, 8–24 cm wide, branches solitary to paired, villous in axils; generally grayish green to occasionally yellowish in appearance; pedicels 0.5–5 mm long. **Spikelets:** compressed, linear-lanceolate, 5- to 12-flowered, 4–10 mm long, 1.2–2 mm wide; disarticulation irregular. **Glumes:** unequal, lanceolate, hyaline; lower 1-nerved, acute to acuminate, 1.5–2 mm long; upper 1- to 3-nerved, 2–3 mm long. **Lemmas:** lanceolate, 1.8–3 mm long, membranous, 3-nerved, acute apex. **Awns:** none. **VEGETATIVE CHARACTERISTICS: Growth Habit:** cespitose, lacking rhizomes. **Culms:** erect to geniculate at base, glabrous, 60–150 cm. **Sheaths:** keeled, lowermost densely appressed hairy at base, upper glabrous or hairy, pilose at throat. **Ligules:** ciliate membrane, 0.6–1.3 mm long, longer hairs behind. **Blades:** involute, 12–50 cm long, 1–3 mm wide, upper surface scattered hairs basally, lower surface glabrous, occasionally scabrous. **HABITAT:** Found on the eastern plains. **COMMENTS:** A common ornamental or pasture grass that easily escapes. It was introduced from southern Africa.

32. *Eragrostis hypnoides* (Lam.) Britton (Fig. 54).

Synonym: *Poa hypnoides* Lam. **Vernacular Names:** Teal lovegrass, creeping lovegrass. **Life Span:** Annual. **Origin:** Native. **Season:** C$_4$, warm season. **FLORAL CHARACTERISTICS: Inflorescence:** contracted to open panicle, 1–3.5 cm long, 0.7–2.5 cm wide, terminal and axillary; branches ascending to spreading. **Spikelets:** compressed, linear-oblong, sometimes curving, 7- to 35-flowered; glumes, lemmas, and caryopsis early deciduous, while paleas persistent on rachis, 4–13 mm long, 1–1.5 mm wide. **Glumes:** unequal, hyaline, linear-lanceolate to lanceolate; lower 0.4–0.7 mm long, 1-nerved; upper 0.8–1.2 mm long, 3-nerved. **Lemmas:** ovate, 3-nerved, 1.5–2 mm long, apex acuminate, nerves greenish. **Awns:** none. **VEGETATIVE CHARACTERISTICS: Growth Habit:** tufted to stoloniferous and mat-forming. **Culms:** decumbent to prostrate, rarely erect, upright portion 5–12 cm tall, rooting at lower nodes. **Sheaths:** hairy on margins, summit, and collar, hairs to 0.6 mm. **Ligules:** ring of hairs 0.3–0.6 mm. **Blades:** flat to involute, lower surface glabrous, upper surface appressed pubescent, 0.5–2.5 cm long, 1–2 mm wide. **HABITAT:** Found on sandy areas adjacent to streams, ponds, or lakes and as an occasional weed in lawns in eastern Colorado. **COMMENTS:** An unusual grass in that it is both stoloniferous and an annual.

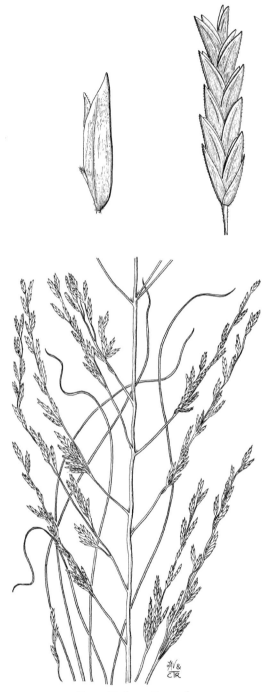

Figure 53. *Eragrostis curvula*.

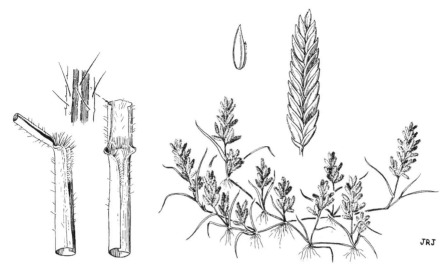

Figure 54. *Eragrostis hypnoides.*

33. *Eragrostis lutescens* Scribn. (Fig. 55).

Synonyms: None. **Vernacular Name:** Sixweeks love-grass. **Life Span:** Annual. **Origin:** Native. **Season:** C_4, warm season. **FLORAL CHARACTERISTICS: Inflorescence:** numerous contracted dense panicles 4–10 cm long, 0.5–2 cm wide, glandular pits on rachises and branches. **Spikelets:** narrowly ovate, 3.6–7.5 mm long, 1.2–2 mm wide, 6- to 11-flowered; glumes, lemmas, and caryopsis early deciduous, while paleas persistent on rachis. **Glumes:** subequal, hyaline,
ovate to lanceolate, 1-nerved; lower 0.9–1.4 mm long; upper 1.2–1.8 mm long. **Lemmas:** ovate, stramineous, conspicuously 3-nerved, scabrous primarily on nerves, 1.5–2.2 mm long. **Awns:** none. **VEGETATIVE CHARACTERISTICS: Growth Habit:** tufted. **Culms:** erect to ascending to rarely decumbent, 6–25 cm tall, glabrous with yellowish glandular pits below nodes. **Sheaths:** elliptical glandular pits present, sparsely hairy at collar. **Ligules:** dense row of whitish hairs, 0.2–0.5 mm long. **Blades:** flat to involute, 2–12 cm long, 1–3 mm wide, lower surface scabrous, glandular pits present at base. **HABITAT:** Reported near Boulder in a piedmont valley on a drying pond (Weber and Wittmann 2001a). **COMMENTS:** Poorly collected, but should be expected along streams, lakes, and alkaline flats.

34. *Eragrostis mexicana* (Hornem.) Link (Fig. 56).

Synonyms: *Poa Mexicana* Hornem., *Eragrostis orcuttiana* Vasey, *Eragrostis virescens* J. Presl. **Vernacular Name:** Mexican lovegrass. **Life Span:** Annual. **Origin:** Native. **Season:** C_4,

119

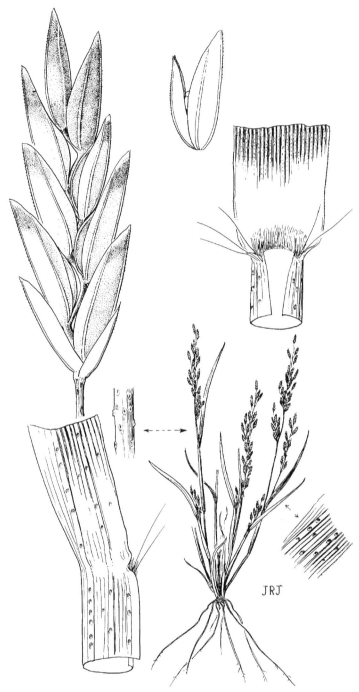

Figure 55. *Eragrostis lutescens.*

Figure 56. *Eragrostis mexicana.*

warm season. **FLORAL CHARACTERISTICS: Inflorescence:** open panicle, ovoid to narrowly oblong, 10–
40 cm long, 4–18 cm wide. **Spikelets:** ovate to linear-
lanceolate, 5- to 11- flowered, 5–10 mm long, 0.7–2.4
mm wide, gray-green to purplish in color; glumes,
lemmas, and caryopsis early deciduous, while paleas
persistent on rachis; caryopsis with a shallow to deep
groove on adaxial surface. **Glumes:** subequal, ovate
to lanceolate, 1-nerved; 0.7–2 mm long. **Lemmas:**

ovate to lanceolate, 3-nerved, 1.2–2.4 mm long, membranous, glabrous. **Awns:** none. **VEG-
ETATIVE CHARACTERISTICS: Growth Habit:** cespitose. **Culms:** erect or geniculate, generally glabrous, occasionally a ring of glandular tissue found below nodes, 10–130 cm tall.
Sheaths: occasionally with glandular pits, pilose at summit and collar, hairs papillose-based,
to 4 mm long. **Ligules:** ring of hairs, 0.2–0.5 mm long. **Blades:** flat, 5–25 cm long, 2–7 mm
wide, upper surface scabrous, occasionally pubescent near base; lower surface glabrous,
margins scabrous. **HABITAT:** Disturbed open sites, along roadsides, and at Bonny Reservoir
and in the Colorado River Valley (Weber and Wittmann 2001a, 2001b). **COMMENTS:** Colorado has subsp. *virescens* (J. Presl) S. D. Koch & Sánchez Vega with linear to linear-lanceolate
spikelets 0.7–1.4 mm wide. The lower glume is 0.7–1.7 mm long, and the culms and sheaths
lack glandular pits.

35. *Eragrostis minor* Host (Fig. 57).

Synonyms: *Eragrostis eragrostis* (L.) P. Beauv., *Eragrostis
poaeoides* P. Beauv. *ex* Roem. & Schult., *Eragrostis sua-
veolens* Becker *ex* Claus. **Vernacular Name:** Little
lovegrass. **Life Span:** Annual. **Origin:** Introduced.
Season: C$_4$, warm season. **FLORAL CHARACTERIS-
TICS: Inflorescence:** open panicle, ovoid to narrowly
oblong, 4–20 cm long, 2–8 cm wide; rachises typically with glands or glandular pits below nodes (not
a ring); primary panicle branches diverging 20°–100°

from rachis; dull grayish green to stramineous in appearance; pedicels usually with a ring of
glands distally. **Spikelets:** compressed, 4–7 mm long, 1.1–2 mm wide, 5- to 12-flowered and
rarely 20-flowered; glumes, lemmas, and caryopsis early deciduous, while paleas persistent
on rachis. **Glumes:** subequal, 1-nerved, broadly ovate, membranous, occasionally 1–2 glands
on midnerve; lower 0.9–1.4 mm long, upper 1.2–1.6 mm long. **Lemmas:** broadly lanceolate
to ovate, membranous, 3-nerved, 1.4–1.8 mm long, occasionally 1–2 glands on midnerve,
apex acute to obtuse. **Awns:** none. **VEGETATIVE CHARACTERISTICS: Growth Habit:** tufted.
Culms: erect to decumbent, occasionally with a ring of glandular tissue below nodes, 10–45
cm tall. **Sheaths:** long-hairy at summit, occasionally glandular on midveins. **Ligules:** ring
of hairs, 0.2–0.5 mm long. **Blades:** flat to folded, 1.5–10 cm long, 1–4 mm wide, smooth
to scabrous to sparsely white hairy, usually with glands along margins. **HABITAT:** Found in
disturbed, open areas at low elevations over most of Colorado. **COMMENTS:** A European
introduction that is now naturalized and occurring as a weedy species.

Figure 57. *Eragrostis minor.*

36. *Eragrostis pectinacea* (Michx.) Nees (Fig. 58).

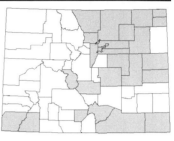

Synonyms: Var. *miserrima: Eragrostis arida* Hitchc., *Eragrostis tephrosanthos* Schult.; var. *pectinacea: Eragrostis caroliniana* auct. non (Spreng.) Scribn., *Eragrostis diffusa* Buckley, *Eragrostis purshii* Hort. *ex* Schrad., *Poa pectinacea* Michx. **Vernacular Name:** Tufted lovegrass. **Life Span:** Annual. **Origin:** Native. **Season:** C₄, warm season. **FLORAL CHARACTERISTICS: Inflorescence:** open panicle, pyramidal to ovoid, 5–30 cm long, 3–15 cm wide; branches solitary or paired at lowest 2 nodes, primary branches spreading, mostly bare at base, branchlets and pedicels remaining appressed to primary branches with appressed spikelets. **Spikelets:** compressed, 4.5–11 mm long, 1.2–2.5 mm broad, 6- to 22-flowered; glumes, lemmas, and caryopsis early deciduous, while paleas persistent on rachis, appear yellowish brown to dark reddish purple; caryopsis 0.5–1.1 mm long. **Glumes:** subequal, hyaline, 1-nerved, acute to acuminate; lower 0.5–1.5 mm long; upper 1–1.7 mm long. **Lemmas:** ovate-lanceolate, 3-nerved, 1–2.2 mm long, scabrid on nerves. **Awns:** none. **VEGETATIVE CHARACTERISTICS: Growth Habit:** tufted. **Culms:** erect to decumbent or geniculate below, 10–80 cm tall, no glands below nodes. **Sheaths:** glabrous with hairs at summit. **Ligules:** ring of hairs, 0.2–0.5 mm. **Blades:** flat to involute, 2–20 cm long, 0.5–4.5 mm wide, upper surface scabrous, lower surface glabrous, smooth. **HABITAT:** Found in weedy, open areas at lower elevations in the late summer. **COMMENTS:** Common. Weber and Wittmann (2001a, 2001b) include this species in *Eragrostis pilosa* (L.) Beau. There are two varieties in Colorado, var. *miserrima* (Fourn.) J. Reeder and var. *pectinacea*, which can be separated by these characteristics:

1. Spikelet pedicels appressed to branches, generally diverging 20° or less from branches . . .
. . . . var. *pectinacea*
1. Spikelet pedicels widely divergent, generally diverging 20°–60° or more from branches . . .
. . . . var. *miserrima*

37. *Eragrostis pilosa* (L.) P. Beauv. (Fig. 59).

Synonyms: Var. *perplexa: Eragrostis perplexa* L. H. Harvey; var. *pilosa: Eragrostis multicaulis* Steud., *Poa pilosa* L. **Vernacular Name:** Indian lovegrass. **Life Span:** Annual. **Origin:** Introduced. **Season:** C₄, warm season. **FLORAL CHARACTERISTICS: Inflorescence:** open panicle becoming diffuse at maturity, 4–20 cm long, 2–15 cm wide, ellipsoid to ovoid. **Spikelets:** linear-oblong to narrowly ovate, 3.5–6 mm long, to 1 mm wide (rarely to 1.3 mm); glumes, lemmas, and caryopsis early deciduous, while paleas tardily dehiscent on rachis; 3- to 17-flowered; caryopsis lacking a ventral groove. **Glumes:** unequal, hyaline, 1-nerved, narrowly ovate to lanceolate; lower 0.3–0.6 mm long; upper 0.7–1.2 mm long. **Lemmas:** ovate-lanceolate, 3-nerved, lateral nerves obscure, 1.2–1.8 mm long, apex acute, hyaline to membranous. **Awns:** none. **VEGETATIVE CHARACTERISTICS: Growth Habit:** tufted. **Culms:** erect to geniculate, 8–50

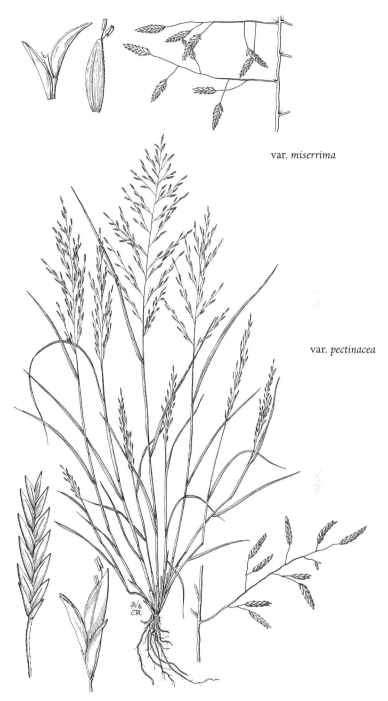

var. *miserrima*

var. *pectinacea*

Figure 58. *Eragrostis pectinacea.*

Figure 59. *Eragrostis pilosa*.

cm tall, glabrous, with few to no glandular depressions. **Sheaths:** generally glabrous, no glandular depressions present or glandular depressions on the sheaths and blades, usually pilose at collar. **Ligules:** ciliate membrane, 0.1–0.3 mm long. **Blades:** flat, 2–15 cm long, 1–2.5 mm wide, glabrous on lower surface, upper surface scabrous. **HABITAT:** Found in weedy, open areas in the late summer (Weber and Wittmann 2001a, 2001b). **COMMENTS:** Common weed. Two varieties are found in Colorado [the typical one (var. *pilosa*) and var. *perplexa* (L. H. Harv.) S. D. Koch]. They differ in this character:

1. Sheaths without glandular depressions . . . . var. *pilosa*
1. Sheaths with glandular depressions . . . . var. *perplexa*

38. *Eragrostis secundiflora* J. Presl (Fig. 60).

Synonym: *Eragrostis oxylepis* (Torr.) Torr. **Vernacular Name:** Red lovegrass. **Life Span:** Perennial. **Origin:** Native. **Season:** C$_4$, warm season. **FLORAL CHAR-ACTERISTICS: Inflorescence:** contracted interrupted panicle, 5–30 cm long, 1–15 cm wide, branches stiffly ascending and densely flowered, spikelets in dense aggregates on branchlets on pedicels 1 mm or less long, not viscid. **Spikelets:** strongly compressed, 6–20 mm long, 2.4–5 mm wide, 8–40-flowered; usually reddish brown; disarticulation above the glumes with florets falling intact. **Glumes:** subequal, membranous, ovate-lanceolate to lanceolate, 1-nerved, scabrous on keel; lower 1.7–3 mm long; upper 2.2–4 mm long, apex acuminate. **Lemmas:** membranous, ovate, strongly 3-nerved, 2–6 mm long, scabrous on keel. **Awns:** none. **VEGETATIVE CHARACTERISTICS: Growth Habit:** cespitose. **Culms:** erect, 30–75 cm tall, glabrous below. **Sheaths:** glabrous, but pilose at summit when young. **Ligules:** 0.2–0.3 mm long. **Blades:** flat to involute near tip, 5–25 cm long, 1–5 mm wide, tapering to a fine point, upper surface scabrous, lower surface glabrous. **HABITAT:** Found scattered in eastern Colorado. This species prefers sandy soils, dunes, usually dry soils (Barkworth et al. 2003). **COMMENTS:** The subsp. *oxylepis* (Torr.) S. D. Koch is the only subspecies reported from Colorado.

39. *Eragrostis spectabilis* (Pursh) Steud. (Fig. 61).

Synonyms: *Eragrostis spectabilis* (Pursh) Steud. var. *sparsihirsuta* Farw., *Poa spectabilis* Pursh. **Vernacular Names:** Purple lovegrass, tumble grass. **Life Span:** Perennial. **Origin:** Native. **Season:** C$_4$, warm season. **FLORAL CHARACTERISTICS: Inflorescence:** open panicle, 25–45 cm long, 15–35 cm wide, broadly ovate to oblong, purplish to reddish purple in color; branches 12–20 cm long, villous at junction with rachis; pedicels 1.5–17 mm long, divergent to appressed. **Spikelets:** linear-lanceolate, 3–7.5 mm long, 1–2 mm wide, 6- to 12-flowered, generally reddish purple in appearance, disarticulation above the glumes. **Glumes:** subequal to equal, 1.3–2.3 mm long, 1-nerved, apex acute. **Lemmas:** lanceolate to ovate,

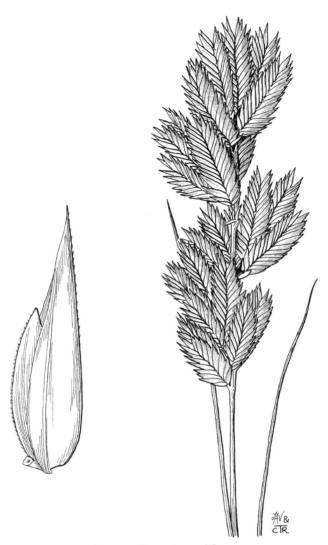

Figure 60. *Eragrostis secundiflora.*

slightly scabrous near apex, 1.3–2.5 mm long, 3-nerved. **Awns:** none. **VEGETATIVE CHAR-ACTERISTICS: Growth Habit:** cespitose, sometimes with short, knotty delicate rhizomes. **Culms:** glabrous, erect, 30–70 cm tall, conspicuously hairy at collar. **Sheaths:** glabrous to sparsely hairy, hairy on margins and at summit, hairs to 7 mm long. **Ligules:** short, 0.1–0.2 mm long. **Blades:** flat to involute, 10–32 cm long, 3–8 mm wide, upper and lower surfaces generally pilose, although sometimes glabrous; often with a row of hairs behind the ligule. **HABITAT:** Found in the southeastern and southwestern corners of Colorado in sandy disturbed areas or sometimes in the shortgrass steppe. **COMMENTS:** Occasionally grown as an ornamental.

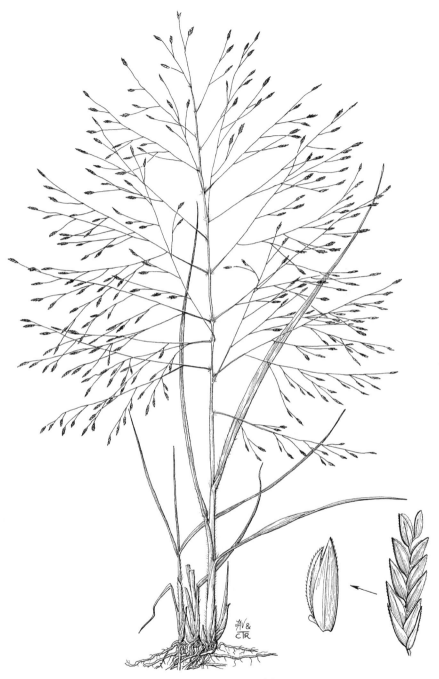

Figure 61. *Eragrostis spectabilis.*

40. *Eragrostis tef* (Zucc.) Trotter (Fig. 62).

Synonyms: *Eragrostis abyssinica* (Jacq.) Link, *Poa tef* Zucc. **Vernacular Name:** Tef. **Life Span:** Annual. **Origin:** Introduced. **Season:** C$_4$, warm season. **FLORAL CHARACTERISTICS: Inflorescence:** open to contracted panicle, 10–45 cm long, 2.5–22 cm wide, ovate; primary branches flexible, appressed to diverging from central rachis; pulvini present, glabrous to hairy, when hairy, hairs to 5 mm long; pedicels appressed to divergent, 2.5–17 mm long. **Spikelets:** ovate to linear-lan-

ceolate, 4–11 mm long, 1.3–2.5 mm wide, 4- to 16-flowered, grayish green to purplish to occasionally stramineous; disarticulation beginning at distal tip of spikelet, eventually below the glumes with paleas persistent on rachis; caryopsis 0.7–1.3 mm long. **Glumes:** unequal, hyaline to membranous, lanceolate; lower 1–2 mm long; upper 1.5–2.8 mm long. **Lemmas:** lanceolate with acute apex, 1.6–3.3 mm long, membranous. **Awns:** none. **VEGETATIVE CHARACTERISTICS: Growth Habit:** loosely tufted. **Culms:** erect, 25–60 cm tall, glabrous. **Sheaths:** generally glabrous, although with hairs to 5 mm long near collar. **Ligules:** ciliate membrane, 0.2–0.4 mm long. **Blades:** flat to involute, 10–30 cm long, 2–5.5 mm wide, lower surface scabridulous, upper surface glabrous. **HABITAT:** Cultivated, occasionally escaped and found along roadsides and in old fields. **COMMENTS:** Grain used for food, beverage, and fodder in Ethiopia; sometimes grown for similar purposes in the United States (Barkworth et al. 2003).

Figure 62. *Eragrostis tef.*

Figure 63. *Eragrostis trichodes.*

41. *Eragrostis trichodes* (Nutt.) Wood (Fig. 63).

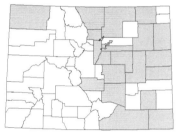

Synonyms: *Eragrostis pilifera* Scheele, *Eragrostis trichodes* (Nutt.) Wood var. *pilifera* (Scheele) Fern., *Poa trichodes* Nutt. **Vernacular Name:** Sand lovegrass. **Life Span:** Perennial. **Origin:** Native. **Season:** C$_4$, warm season. **FLORAL CHARACTERISTICS: Inflorescence:** open, diffuse panicle, oblong to ovoid, 35–80 cm long, 6–30 cm wide; color generally purple to red; branches capillary, in groups of 3–4, naked in axils, although occasionally with a few short hairs at junction with central axis; pedicels 2–22 mm long, capillary. **Spikelets:** compressed, purplish, 4- to 18-flowered, 3–15 mm long, 1.5–3.5 mm wide; lemmas and caryopsis early deciduous, while paleas persistent on rachis. **Glumes:** subequal, narrowly ovate to linear-lanceolate, 1-nerved, 1.8–4 mm long. **Lemmas:** broadly ovate to lanceolate, 3-nerved, apex acute, 2.2–3.5 mm long. **Awns:** none. **VEGETATIVE CHARACTERISTICS: Growth Habit:** cespitose. **Culms:** glabrous, erect, 30–150 cm tall, non-glandular below nodes. **Sheaths:** round, sometimes hairy along margins, throat pilose. **Ligules:** ciliate membrane, 0.3–0.5 mm long. **Blades:** flat to involute, 15–50 cm long, 2–8 mm wide, lower surface glabrous, upper surface scabridulous, pilose at base near ligule. **HABITAT:** Found on the plains, usually in sandy areas and along roadsides. Grows in association with *Redfieldia flexuosa*. On the Western Slope there is one record from Moffat County (Weber and Wittmann 2001b). **COMMENTS:** Occasionally grown as an ornamental.

16. *Erioneuron* Nash

Low, tufted perennials with narrow, often involute, cartilaginous-margined leaf blades. Sheath open; ligule a ring of hairs; auricles absent. Inflorescence a short, usually capitate raceme or panicle of several-flowered spikelets, these disarticulating above the glumes and between the florets. Glumes large, membranous, subequal, 1-nerved. Lemmas broad, rounded on the back, 3-nerved, conspicuously long-hairy along the nerves, at least below; midnerve short-awned, each lateral nerve often prolonged as a short mucro. Paleas slightly shorter than lemmas, ciliate on the keels, long-hairy on lower part between the nerves. Basic chromosome number, $x = 8$. Photosynthetic pathway, C_4.

Represented in Colorado by a single species.

42. *Erioneuron pilosum* (Buckley) Nash (Fig. 64).

Synonyms: *Tridens pilosus* (Buckley) A. S. Hitchc., *Uralepis pilosa* Buckley. **Vernacular Names:** Hairy woollygrass, hairy tridens. **Life Span:** Perennial. **Origin:** Native. **Season:** C_4, warm season. **FLORAL CHARACTERISTICS: Inflorescence:** panicle to raceme in depauperate specimens, 1–4.5 cm long, 3–9 spikelets per branch. **Spikelets:** 6–12 mm long, 6- to 12-flowered, pale to purple-tinged; disarticulation above the glumes. **Glumes:** subequal, 4–7 mm long, ovate to lanceolate, 1-nerved, sharply acute to awn-tipped, shorter than lowest florets. **Lemmas:** lanceolate to ovate, 3–7 mm long, densely pubescent on nerves at base, pubescence extends up the margins, 3-nerved, green to purplish green when immature, stramineous when mature, apex acute to bidentate. **Paleas:** shorter than lemma with ciliate keel. **Awns:** lemma awn 0.5–2.5 mm long, as an extension of midnerve. **VEGETATIVE CHARACTERISTICS: Growth Habit:** densely tufted, leaves basal. **Culms:** erect, glabrous to hispidulous, 10–30 cm tall. **Sheaths:** generally glabrous with hairs at summit. **Ligules:** ring of hairs with or without a membranaceous base, 2–3.5 mm long. **Blades:** flat to folded, both surfaces glabrous to sparsely pilose, margins thickened and whitish, 1–6 cm long, 1–2 mm wide. **HABITAT:** Found at low elevations in both eastern and western Colorado, concentrated more on limestone, gypsum hills, and rimrock. **COMMENTS:** Moved from *Tridens* because of numerous differences in the caryopsis, stigma color, shape and pubescence of the lemma and palea, leaf anatomy, plant habit, and panicle type. Colorado specimens belong to the typical variety, var. *pilosum*.

17. *Leptochloa* P. Beauv.

Annuals and perennials, cespitose. Sheaths open; ligule a ciliate membrane; auricles absent. Inflorescence a panicle, with few to numerous unbranched primary branches distributed along upper portion of the culm or clustered near the tip. Spikelets with 2 to several florets, overlapping and closely spaced on branches. Disarticulation above the glumes and between the florets. Glumes 2, thin, 1-nerved or the second occasionally 3-nerved, acute, awnless or with a mucronate tip. Lemmas 3-nerved, frequently puberulent on the nerves, apex acute to obtuse or notched, awnless, awned, or with a mucro. Palea well developed and frequently puberulent on the nerves. Caryopsis apex 2-horned. Basic chromosome number, $x = 10$. Photosynthetic pathway, C_4.

Represented in Colorado by 2 species and 1 subspecies.

1. Lemma apices obtuse, truncate, and often emarginate; lemmas membranous; plants with secondary panicles concealed in lower leaf sheaths . . . . 43. *L. dubia*
1. Lemma apices acute; lemmas chartaceous; plants without secondary panicles . . . . 44. *L. fusca*

Figure 64. *Erioneuron pilosum*.

Figure 65. *Leptochloa dubia.*

43. *Leptochloa dubia* (Kunth) Nees (Fig. 65).

Synonyms: *Chloris dubia* Kunth, *Diplachne dubia* (Kunth) Scribn. **Vernacular Name:** Green sprangletop. **Life Span:** Perennial. **Origin:** Native. **Season:** C$_4$, warm season. **FLORAL CHARACTERISTICS: Inflorescence:** 5–25 cm long, panicle with 1-sided racemose primary branches, 4–12 cm long; secondary panicles with cleistogamous spikelets frequently found in lowest leaf sheaths. **Spikelets:** light brown to dark olive green, 4–12 mm long, 2- to 10-flowered, subsessile, spreading at maturity; disarticulation above the glumes. **Glumes:** unequal, 1-nerved, narrowly triangular to lanceolate, acute, translucent with scabrous green nerves; lower 2.3–4.8 mm long; upper 3.3–6 mm long. **Lemmas:** membranous, ovate to obovate, 3-nerved, 3–5 mm long, truncate to emarginate, sericeous hairs over nerves. **Awns:** none. **VEGETATIVE CHARACTERISTICS: Growth Habit:** cespitose. **Culms:** erect, 3–11 dm tall, unbranched above base, round to compressed, nodes dark. **Sheaths:** generally glabrous with a pilose collar, keeled. **Ligules:** truncate, erose, ciliate membrane, 0.3–2 mm long. **Blades:** flat to involute when dry, 8–35 cm long, 2–10 mm wide, glabrous to pilose. **HABITAT:** Found on rocky hills, canyons, and prairies; most frequent on well-drained rocky or sandy soils (Stubbendieck, Hatch, and Butterfield 1992). Only a single collection in Las Animas County (Weber and Wittmann 2001a). **COMMENTS:** This species constitutes good livestock forage and is frequently used in various seed mixtures.

44. *Leptochloa fusca* (L.) Kunth (Fig. 66).

Synonyms: *Diplachne acuminata* Nash, *Diplachne fascicularis* (Lam.) P. Beauv., *Diplachne maritima* Bickn., *Leptochloa acuminata* (Nash) Mohlenbrock, *Leptochloa fascicularis* (Lam.) Gray, *Leptochloa fascicularis* (Lam.) Gray var. *acuminata* (Nash) Gleason, *Leptochloa fascicularis* (Lam.) Gray var. *maritima* (Bickn.) Gleason, *Leptochloa fusca* (L.) Kunth var. *fascicularis* (Lam.) Dorn. **Vernacular Name:** Malabar sprangletop. **Life Span:** Annual. **Origin:**

Figure 66. *Leptochloa fusca*.

Native. **Season:** C$_4$, warm season. **FLORAL CHAR-ACTERISTICS: Inflorescence:** panicle 7–70 cm long, 4–22 cm wide, usually partially enclosed in upper sheath to exserted from sheath, secondary panicles absent, with 3–35 racemose branches; branches 3–12 cm long, commonly spreading. **Spikelets:** 5–12 mm long, 5- to 9-flowered; pedicels less than 1 mm long, appressed, overlapping to remotely spaced, linear, subterete; disarticulation above the glumes. **Glumes:** 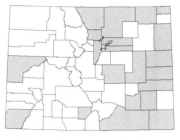 unequal, 1-nerved; lower 2–3 mm long, narrowly lanceolate to linear; upper 2.5–5 mm long, lanceolate to elliptic. **Lemmas:** lanceolate to elliptic, 3-nerved, 3–5 mm long, chartaceous, often with a dark spot near base, apex acute, bifid. **Awns:** lower glume sometimes minutely awn-tipped; upper glume with awn 0.5–1.5 mm long; lemma with awn to 3.5 mm from bifid apex. **VEGETATIVE CHARACTERISTICS: Growth Habit:** cespitose. **Culms:** compressed, erect to ascending and occasionally prostrate, often branching above base, 5–110 cm tall, internodes hollow. **Sheaths:** glabrous to scabrous, strongly keeled, somewhat inflated below and tapering above. **Ligules:** membranous, 2–8 mm long, acute to obtuse, entire to lacerate at maturity, often decurrent. **Blades:** flat to involute, 3–50 cm long, 2–7 mm wide, glabrous to scabrous, flag leaves sometimes exceeding panicles. **HABI-TAT:** Found at low elevations in eastern and western Colorado on muddy shores of ponds and oxbows in lower valleys (Weber and Wittmann 2001a, 2001b). **COMMENTS:** Subsp. *fascicularis* (Lam.) N. Snow occurs in Colorado. Weber and Wittmann (2001a, 2001b) call this taxon *Diplachne fascicularis* (Lam.) P. Beauv. It differs from the typical variety in that the uppermost leaf blades are longer than the panicles, and the panicles are generally partially enclosed in the uppermost leaf sheaths.

18. *Lycurus* Kunth

Tufted perennial with dense, spikelike panicles of 1-flowered spikelets. Sheaths laterally compressed and sharply keeled. Blades narrow, flat, or folded, usually with a whitish midnerve and margins. Inflorescence mostly 3–8 cm long and 5–8 mm thick. Spikelets short-pediceled, deciduous in pairs together with the pedicels, the lowermost spikelet of the pair often sterile or staminate. Glumes shorter than lemma, the lower 2- or 3-nerved and with 2–3 short awns, the upper similar but 1-nerved and 1-awned. Palea similar in texture to lemma and about as long, enclosed by lemma only at base. Basic chromosome number, $x = 10$. Photosynthetic pathway, C$_4$.

Represented in Colorado by 2 species.

1. Distal portion of leaf terminating with a fragile, awnlike tip 4–7 mm long; ligules generally in 1 piece, 3–10 mm long, elongate, acute to acuminate; culms erect . . . . 46. *L. setosus*
1. Distal portion of leaf acute and midnerve extending beyond end of the leaf as a short mucro or bristle 1–3 mm long; ligules 1.5–3 mm long with narrow triangular lobes to each side, also 1.5–3 mm long; culms ascending to erect, often geniculate at base . . . .
. . . . 45. *L. phleoides*

45. *Lycurus phleoides* Kunth (Fig. 67).

Synonyms: None. **Vernacular Name:** Common wolfstail or wolfstail. **Life Span:** Perennial. **Origin:** Native.
Season: C_4, warm season. **FLORAL CHARACTER-
ISTICS: Inflorescence:** spikelike panicle, cylindric,
terminal and axillary, 4–8 cm long, 7–8 mm wide,
densely flowered. **Spikelets:** subterete, 1-flowered,
3–4 mm long, borne on short, mostly fused paired
pedicels; disarticulation at base of this fused pedicel.
Glumes: subequal, 1–1.8 mm long, apex scabrous;
lower 2-nerved; upper 1-nerved. **Lemmas:** narrowly lanceolate, 3-nerved, 3–4 mm long.
Paleas: subequal to lemma, 2-nerved, acute or occasionally 2 nerves extending into short
mucros. **Awns:** lower glume 2-awned, these are unequal, scabrous, ranging from 1 to 1.5
mm for the shorter one and 1.5 to 3 mm for the longer one; upper glume with 1 flexuous awn 2.5–4.5 mm long and, occasionally, a second shorter, more delicate awn; lemma
awn scabrous, 1.5–2 mm long. **VEGETATIVE CHARACTERISTICS: Growth Habit:** densely
cespitose. **Culms:** ascending to erect, commonly geniculate, 2–5 dm tall, usually branched.
Sheaths: open, glabrous to scabrous, compressed keeled. **Ligules:** generally acute to acuminate, 1.5–3 mm, with narrow subequal to equal triangular lobes extending from sides of
sheaths. **Blades:** folded to flat, stiff, 4–8 cm long, 1–1.5 mm wide, acute to mucronate with
midnerve sometimes extending to 3 mm as a bristle beyond the blade. **HABITAT:** Grows
on rocky hills and open slopes. **COMMENTS:** Weber and Wittmann (2001a, 2001b) place
all *Lycurus* specimens into *L. setosus* (Nutt.) C. Reeder, stating that *L. phleoides* Kunth is
chiefly a Mexican species. These two species are only reliably separated by their vegetative
characters.

46. *Lycurus setosus* (Nutt.) C. Reeder (Fig. 68).

Synonym: *Pleopogon setosus* Nutt. **Vernacular Name:**
Bristly wolfstail or wolfstail. **Life Span:** Perennial.
Origin: Native. **Season:** C_4, warm season. **FLORAL
CHARACTERISTICS: Inflorescence:** spikelike panicle,
cylindric, terminal and axillary, 4–8 cm long, 7–8 mm
wide, densely flowered. **Spikelets:** subterete, 1-flow-
ered, 3–4 mm long; borne on short, mostly fused
paired pedicels; disarticulation is at base of fused
pedicel. **Glumes:** subequal, 1–1.8 mm long, apex sca-
brous; lower 2-nerved; upper 1-nerved. **Lemmas:** narrowly lanceolate, 3-nerved, 3–4 mm
long. **Paleas:** subequal to lemma, 2-nerved, acute or occasionally 2 nerves exserted into
short mucros. **Awns:** lower glume 2-awned, these unequal, scabrous, ranging from 1 to 1.5
mm for the shorter one and 1.5 to 3 mm for the longer one; upper glume with a single, flexuous awn 2.5–4.5 mm long; lemma awn scabrous, 1.5–2 mm long. **VEGETATIVE CHARAC-
TERISTICS: Growth Habit:** densely cespitose. **Culms:** erect, 3–5 dm tall, sparingly branched.
Sheaths: open, glabrous to scabrous, compressed keeled. **Ligules:** acuminate, hyaline, 3–10
mm long, elongate, sometimes shortly cleft on sides. **Blades:** folded to flat, 4–9 cm long,

Figure 67. *Lycurus phleoides.*

Figure 68. *Lycurus setosus.*

1–2 mm wide, upper leaves narrowing to a fragile awnlike tip 4–7 or more mm in length. **HABITAT:** Found at low elevations on mesas, foothills, hogbacks, and canyon sides (Weber and Wittmann 2001a, 2001b; Wingate 1994). **COMMENTS:** Weber and Wittmann (2001a, 2001b) only recognize this species in Colorado. The two *Lycurus* species are reliably separated only by their vegetative characters.

19. *Muhlenbergia* Schreb.

Plants of diverse habits, from delicate, tufted annuals to large, coarse, cespitose perennials. Several species with creeping rhizomes. Sheaths open; ligules usually a prominent membrane; auricles absent. Inflorescence an open or contracted panicle, spikelike in a few species. Spikelets typically 1-flowered, a second or third floret occasionally produced. Disarticulation above the glumes. Glumes usually 1-nerved or nerveless, occasionally 3-nerved, mostly shorter than lemma, obtuse, acute, acuminate, or short-awned. Lemma as firm as, or firmer than, the glumes, 3-nerved (nerves indistinct in some species), with a short, usually bearded callus at base and a single flexuous awn at the apex, less frequently mucronate or awnless. Palea well developed, shorter than or about equaling the lemma. Caryopsis usually not falling free from lemma and palea. Basic chromosome number, $x = 10$. Photosynthetic pathway, C_4.

Represented in Colorado by 23 species and 2 varieties.

1. Plant annual
   2. Lemma awned
      3. Glumes equal to or slightly longer than floret . . . . 53. *M. depauperata*
      3. Glumes shorter than floret, sometimes minute . . . . 51. *M. brevis*

2. Lemma awnless
    4. Glumes minutely pilose . . . . 58. *M. minutissima*
    4. Glumes glabrous
        5. Panicle open; branches closely appressed at maturity; lemmas mottled with patches of greenish black, greenish white, or pale yellow to orangish or reddish yellow . . . . 64. *M. ramulosa*
        5. Panicle contracted; branches spreading at maturity; lemmas dark green . . . . 55. *M. filiformis*
1. Plant perennial
    6. Rhizomes well developed, scaly, creeping
        7. Panicle open
            8. Lemma awned, awn 1–1.5 mm long . . . . 62. *M. pungens*
            8. Lemma unawned to mucronate, mucro to 0.3 mm long
                9. Ligule 1–2 mm long, ligule with lateral lobes . . . . 48. *M. arenacea*
                9. Ligule less than 1 mm long, ligule without lateral lobes . . . . . 50. *M. asperifolia*
        7. Panicle contracted
            10. Lemma awned, usually 4 mm or more long, awn flexuous
                11. Blade less than 1.8 mm wide . . . . 60. *M. pauciflora*
                11. Blade 2 mm or more wide . . . . 47. *M. andina*
            10. Lemma awnless to awn-tipped, generally to 3 mm long
                12. Glumes awn-tipped to awned, awns to 5 mm long, much exceeding the awnless lemma
                    13. Internodes smooth and polished, strongly keeled; culms much branched above base . . . . 63. *M. racemosa*
                    13. Internodes dull, puberulent, usually terete; culms not branched above base . . . . 56. *M. glomerata*
                12. Glumes acuminate or, if awn-tipped, then not exceeding lemma
                  14. Lemma glabrous to occasionally scabrous or with a few appressed hairs above . . . . 65. *M. richardsonis*
                14. Lemma hairy
                    15. Lemma and palea pilose on at least lower half . . . . 67. *M. thurberi*
                    15. Lemma with hairs less than 0.7 mm long, occasionally only on callus . . . . 57. *M. mexicana*
    6. Rhizomes wanting or very short and knotty
        16. Culms geniculate and rooting at lower nodes
            17. Glumes minute, lower sometimes absent . . . . 66. *M. schreberi*
            17. Glumes evident
                18. Glumes acuminate to acute . . . . 60. *M. pauciflora*
                18. Glumes obtuse to broadly rounded . . . . 55. *M. filiformis*
        16. Culms erect or spreading, not rooting at nodes
            19. Upper glume 3-toothed
                20. Lemma awn less than 6 mm long . . . . 54. *M. filiculmis*
                20. Lemma awn more than 6 mm long . . . . 59. *M. montana*
            19. Upper glume not 3-toothed
              21. Panicle spikelike
                22. Culms hispid, ligule 1–3 mm long; glumes acute . . . . 69. *M. wrightii*

22. Culms minutely pubescent, ligule about 0.5 mm long; glumes acute to acuminate . . . . 52. *M. cuspidata*

21. Panicle open or, if narrow, not spikelike

    23. Plant widely spreading, wiry, knotty at base . . . . 61. *M. porteri*

    23. Plant erect

        24. Blades 1–3 cm long . . . . 68. *M. torreyi*

        24. Blades 4–10 cm long . . . . 49. *M. arenicola*

47. *Muhlenbergia andina* (Nutt.) Hitchc. (Fig. 69).

Synonym: *Calamagrostis andina* Nutt., *Muhlenbergia comata* (Thurb.) Thurb. *ex* Benth. **Vernacular Name:** Foxtail muhly. **Life Span:** Perennial. **Origin:** Native. **Season:** C$_4$, warm season. **FLORAL CHARACTERISTICS: Inflorescence:** contracted panicle, dense, spikelike, 2–15 cm long, 0.5–2.8 cm wide, appears interrupted to lobed. **Spikelets:** whitish to gray-green, often purplish, 2–4 mm long, 1-flowered; disarticulation above the glumes. **Glumes:** subequal to equal, 2–4 mm long, 1-nerved, equal to or longer than florets, often gibbous at base, keel ciliate-scabrous, apex acuminate to awn-tipped. **Lemmas:** lanceolate, 2–3.5 mm long, calluses and lemma bases with dense, long silky hairs 2–3.5 mm long, 3-nerved. **Paleas:** lanceolate, 2–3.5 mm long, silky hairs at base between nerves. **Awns:** lemma awn flexuous, capillary, 1–10 mm long. **VEGETATIVE CHARACTERISTICS: Growth Habit:** rhizomatous with well-developed scaly rhizomes. **Culms:** erect, sparsely branched, 25–80 cm tall, internodes hollow, puberulent to scaberulous below nodes, otherwise glabrous. **Sheaths:** keeled, glabrous to puberulent. **Ligules:** membranous, 0.5–1.5 mm long, truncate, erose-ciliate. **Blades:** flat at maturity, 4–16 cm long, 2–4 mm wide, upper surface pubescent, lower surface scabrous. **HABITAT:** Found in the foothills and montane areas of Colorado around damp areas and water seeps at the base of rock outcrops and cliffs (Weber and Wittmann 2001a, 2001b). **COMMENTS:** This species has been collected up to elevations of 10,000 ft (3,000 m) (Barkworth et al. 2003).

48. *Muhlenbergia arenacea* (Buckley) Hitchc. (Fig. 70).

Synonym: *Sporobolus arenaceus* Buckley. **Vernacular Name:** Ear muhly. **Life Span:** Perennial. **Origin:** Native. **Season:** C$_4$, warm season. **FLORAL CHARACTERISTICS: Inflorescence:** open panicle, broadly ovoid, diffuse; branches capillary, straight to slightly flexuous. **Spikelets:** usually purplish, 1- to occasionally 2-flowered, 1.5–2.6 mm long, pedicels 1–11 mm long; disarticulation above the glumes. **Glumes:** equal, 1-nerved, glabrous to scabrous, 0.9–2 mm long; usually acute to acuminate, but sometimes apex erose with mucronate tip to 0.2 mm. **Lemmas:** lanceolate to oblong-elliptic, 3-nerved, sparsely appressed pubescent on lower half

Figure 69. *Muhlenbergia andina.*

Figure 70. *Muhlenbergia arenacea*.

of margins and midnerve, 1.5–2.5 mm long, lemma mucro to 0.3 mm. **Paleas:** lanceolate, glabrous, 1.5–2.6 mm long, apex acute to obtuse. **Awns:** none. **VEGETATIVE CHARACTERISTICS: Growth Habit:** rhizomatous. **Culms:** decumbent, ascending or erect, 1–3 dm tall, often scabrous. **Sheaths:** glabrous to scabrous, open, about half the length of internodes. **Ligules:** membranous, hyaline, 0.5–2 mm long, central portion long, but margins usually splitting and forming 2 auricle-like, erect, lateral lobes up to 2 mm long. **Blades:** flat at base to involute at the sharp tip, 0.7–4 cm long, 0.5–1.7 mm wide, upper surface strigose, lower surface glabrous to scabrous, midnerve and margins white-thickened. **HABITAT:** Found on the southeastern plains on sandy flats, alluvial fans, washes, depressions, and alkaline mesas (Weber and Wittmann 2001a). **COMMENTS:** Infrequent.

49. *Muhlenbergia arenicola* Buckley (Fig. 71).

Synonyms: None. **Vernacular Name:** Sand muhly. **Life Span:** Perennial. **Origin:** Native. **Season:** C$_4$, warm season. **FLORAL CHARACTERISTICS: Inflorescence:** diffuse panicle, 12–30 cm long, 5–20 cm wide; branches naked at base. **Spikelets:** 2.5–4.2 mm long on pedicels as long or longer; disarticulation above the glumes. **Glumes:** subequal to equal, 1-nerved, 1.4–2.5 mm long, apex entire, acute to acuminate, scabrous. **Lemmas:** narrowly elliptic, 3-nerved, pur-

Figure 71. *Muhlenbergia arenicola.*

plish, distally scabrous, appressed pubescent on lower half to three-quarters of margins and nerves, apex acuminate. **Awns:** glumes can be awned, awn up to 1 mm long; lemma tapering to a delicate awn 0.5–4.2 mm long. **VEGETATIVE CHARACTERISTICS: Growth Habit:** cespitose. **Culms:** erect to ascending from a decumbent base, 2–6 dm tall, glabrous to hispid below nodes. **Sheaths:** open, glabrous to scabrous, margins hyaline. **Ligules:** hyaline, acute, 2–9 mm long, acute to lacerate, often with lateral lobes. **Blades:** flat to folded to involute, 4–10 cm long, 1–2.2 mm wide, scabrous, sometimes glaucous. **HABITAT:** Found in the southeastern portion of the state in sandy areas. **COMMENTS:** Superficially similar to *M. torreyi, M. arizonica,* and *M. pungens.*

50. *Muhlenbergia asperifolia* (Nees & Meyen *ex* Trin.) Parodi (Fig. 72).

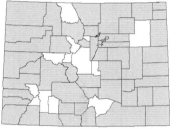

Synonyms: *Sporobolus asperifolius* (Nees & Meyen *ex* Trin.) Nees, *Vilfa asperifolia* Nees & Meyen *ex* Trin.. **Vernacular Name:** Scratchgrass. **Life Span:** Perennial. **Origin:** Native. **Season:** C$_4$, warm season. **FLORAL CHARACTERISTICS: Inflorescence:** large, open, diffuse panicle, 6–21 cm long, 4–16 cm wide, ovoid, breaking away at maturity; branches capillary; pedicels 3–14 mm long, longer than spikelets. **Spikelets:** often deep purple, 1- to sometimes 2- to 3-flowered, 1.2–2.1 mm long; disarticulation above the glumes. **Glumes:** equal to subequal, 0.6–1.7 mm long, purplish, puberulent-scabrous on keel, margins sometimes scarious, 1-nerved, apex acute. **Lemmas:** thin, lanceolate to oblong-elliptic, 3-nerved, 1.2–2 mm long, apex acute to mucronate, mucro to 0.3 mm. **Paleas:** as long as or slightly longer than lemma. **Awns:** none. **VEGETATIVE CHARACTERISTICS: Growth Habit:** rhizomatous to occasionally stoloniferous. **Culms:** decumbent to ascending, compressed at base, 1–6 dm tall, spreading, branching at base, pale to glaucous. **Sheaths:** overlapping, glabrous, margins hyaline. **Ligules:** truncate, firm, erose-ciliate with a membranous, somewhat decurrent base, 0.2–1 mm long. **Blades:** flat to folded, smooth to scabrous, chiefly cauline, crowded, 2–7 cm long, 1–2.8 mm wide, apex acute. **HABITAT:** Found in alkaline areas, along roadsides in ditches and swales in valleys at low elevations throughout the state. **COMMENTS:** Common.

51. *Muhlenbergia brevis* C. O. Goodd. (Fig. 73).

Synonyms: None. **Vernacular Name:** Short muhly. **Life Span:** Annual. **Origin:** Native. **Season:** C$_4$, warm season. **FLORAL CHARACTERISTICS: Inflorescence:** contracted panicle, 3–11.5 cm long, 0.8–1.8 cm wide, often included in uppermost leaf sheath, densely flowered; branches erect. **Spikelets:** usually in subsessile-pedicellate pairs; 2.5–6 mm long; disarticulation beneath spikelet pairs. **Glumes:** unequal, scabrous, shorter than lemma; lower 2–3.5 mm long, 2-nerved, minutely to deeply bifid and having 2 aristate teeth or awns; upper entire, 1-nerved, 2.4–4 mm long, acuminate to setaceous. **Lemmas:** narrowly lanceolate, scabrous on nerves and

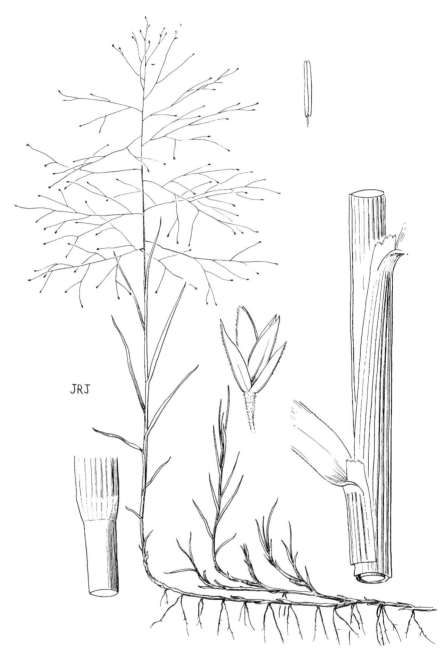

JRJ

Figure 72. *Muhlenbergia asperifolia*.

Figure 73. *Muhlenbergia brevis.*

appressed pubescent between nerves at base, 3.5–6 mm long, 3-nerved, apex acuminate, often bifid. **Paleas:** narrowly lanceolate, 4–6 mm long, apex acuminate. **Awns:** lower glume sometimes has 2 awns, to 1.8 mm; upper glume with awn to 2 mm; lemma awns 10–20 mm long, stiff. **VEGETATIVE CHARACTERISTICS: Growth Habit:** tufted, much branched below. **Culms:** erect, slender, 3–20 cm tall, scabrous to hispid below nodes. **Sheaths:** compressed-keeled, slightly inflated, glabrous to scabrous, usually longer than internodes. **Ligules:** membranous, 1–3 mm long, truncate to obtuse, lacerate with lateral lobes longer than central portion. **Blades:** flat to involute, 1–4.5 cm long, 0.8–2 mm wide, with white-thickened margins and midrib. **HABITAT:** Found in 1 location in rocky hills in Conejos County (Harrington 1954), while Barkworth and colleagues (2003) report it from Rio Grande and Saguache counties as well. **COMMENTS:** Rare. This species and *M. depauperata* are very similar and closely related to the genus *Lycurus.*

52. *Muhlenbergia cuspidata* (Torr. *ex* Hook.) Rydb. (Fig. 74).

Synonym: *Muhlenbergia brevifolia* (Nutt.) M. E. Jones. **Vernacular Names:** Plains muhly, prairie muhly, prairie satin grass. **Life Span:** Perennial. **Origin:** Native. **Season:** C$_4$, warm season. **FLORAL CHARACTERISTICS: Inflorescence:** contracted, spikelike panicle, 4– 14 cm long, 0.1–0.8 cm wide, loosely flowered. **Spikelets:** dark green to grayish, 1- to 2-flowered, 2.5–3.6 mm long; disarticulation above the glumes. **Glumes:** subequal, 1.2–3 mm long, 1-nerved, scabrous on keel, gradually acute to acuminate, mucronate with mucro to 0.3 mm long. **Lemmas:** lanceolate, 2.5–3.6 mm long, scaberulent to minutely pubescent on back, apex acuminate to occasionally mucronate with mucro to 0.6 mm long. **Paleas:** lanceolate 2.4–3.6 mm long, glabrous, apex acuminate. **Awns:** none. **VEGETATIVE CHARACTERISTICS: Growth Habit:** cespitose with knotty bases, not rooting at nodes. **Culms:** erect, 2–6 dm tall, 1–2 mm thick, wiry, glabrous to minutely pubescent. **Sheaths:** smooth to scabrous, keeled, laterally compressed, shorter than internodes. **Ligules:** membranous, truncate, to 0.5 mm long, rarely to 0.8 mm long. **Blades:** flat to folded to involute-setaceous when dry, 2–22 cm long, 0.5–2.7 mm wide, upper surface strigose, lower surface smooth to scabrous. **HABITAT:** Grows in dry gravelly prairies, rocky slopes, rocky limestone outcrops, and sandy drainages (Barkworth et al. 2003). **COMMENTS:** This species is often confused with *M. richardsonis*, but *M. cuspidata* lacks rhizomes and has a shorter ligule (Barkworth et al. 2003).

53. *Muhlenbergia depauperata* Scribn. (Fig. 75).

Synonyms: None. **Vernacular Name:** Sixweeks muhly. **Life Span:** Annual. **Origin:** Native. **Season:** C$_4$, warm season. **FLORAL CHARACTERISTICS: Inflorescence:** spikelike panicle, 2.5–8.5 cm long, 0.5–0.7 cm wide; branches with spikelets to base. **Spikelets:** appressed, 1-flowered, 2.5–5.1 mm long, usually in subsessile-pedicellate pairs; disarticulation beneath each spikelet pair. **Glumes:** narrowly lanceolate, subequal, scabrous, equal to or slightly longer than floret, 2.3–5.1 mm long; lower 2-nerved, subulate apex minutely to deeply bifid; upper lanceolate, 1-nerved, entire, acuminate. **Lemmas:** narrowly lanceolate, 2.5–4.5 mm long, 3-nerved, minutely pubescent with appressed hairs between nerves, apex acute to awned. **Paleas:** about as long as lemma, commonly minutely hairy between nerves. **Awns:** lower glume with awns to 1.3 mm on bifid apex; lemma awns 6–15 mm, stiff. **VEGETATIVE CHARACTERISTICS: Growth Habit:** tufted, much branched from lower nodes. **Culms:** ascending to erect, 3–15 cm tall, scabrous to pubescent internodes. **Sheaths:** glabrous to puberulent, sometimes longer than internodes, smooth to scabrous, slightly inflated, keeled, margins scabrous. **Ligules:** membranous, 1.4–2.5 mm long, acute, lateral lobes present. **Blades:** flat to involute on drying, 1–3 cm long, 0.6–1.5 mm wide, scabrous to strigose, midnerve and margins white-thickened. **HABITAT:** Found in Colorado National Monument in small, seasonally

Figure 74. *Muhlenbergia cuspidata*.

Figure 75. *Muhlenbergia depauperata*.

wet depressions in sandstone rimrock (Weber and Wittmann 2001b) and Mesa Verde National Park in a bedrock drainage on Cliff House sandstone in a canyon. **COMMENTS:** This species and *M. brevis* are very similar and closely related to the genus *Lycurus*.

54. *Muhlenbergia filiculmis* Vasey (Fig. 76).

Synonyms: None. **Vernacular Name:** Slimstem muhly. **Life Span:** Perennial. **Origin:** Native. **Season:** C_4, warm season. **FLORAL CHARACTERISTICS: Inflorescence:** dense, contracted panicle 1.5–7 cm long, 0.4–2 cm wide; branches erect. **Spikelets:** 1-flowered, 2.2–3.5 mm long, pedicels 0.1–2 mm long; disarticulation above the glumes. **Glumes:** subequal, 0.8–2.5 mm long; lower 1-nerved, hyaline to scarious, apex mucronate or erose; upper 3-nerved, apex truncate to acute, 3-toothed, teeth occasionally with awns. **Lemmas:** lanceolate, 3-nerved, appressed pubescent on lower portion, scabrous above, apex acute to acuminate. **Paleas:** lanceolate, acute to acuminate, pubescent below, glabrous above. **Awns:** lower glume awns to 1.6 mm long; upper glume awns, when present, to 0.6 mm long; lemma awns flexuous or straight, 1–5 mm long. **VEGETATIVE CHARACTERISTICS: Growth Habit:** cespitose, leaves in dense basal cluster, old clumps with persistent sheaths. **Culms:** erect, hollow, glabrous, 5–30 cm tall, rounded near base. **Sheaths:** glabrous to minutely roughened, strongly nerved, longer than lower internodes. **Ligules:** acute, hyaline, entire, decurrent, 1.5–4.2 mm long. **Blades:** flat to folded to involute, tapering, midnerve and margins white-thickened, 0.7–4 cm long, 0.5–1.7 mm wide, sharp-tipped, lower surface scabrous, upper surface sparsely hirtellous. **HABITAT:** Found in open, gravelly, grassy, dry slopes and foothills. **COMMENTS:** Common. The differences between *M. montana* and this species are sometimes difficult to distinguish. This species has shorter spikelets and shorter lemma awns. Additionally, the leaf blades are sharply tipped and more involute than *M. montana* (Barkworth et al. 2003).

55. *Muhlenbergia filiformis* (Thurb. *ex* S. Watson) Rydb. (Fig. 77).

Synonyms: *Muhlenbergia filiformis* (Thurb. *ex* S. Wats.) Rydb. var. *fortis* E. H. Kelso, *Muhlenbergia idahoensis* St. John, *Muhlenbergia simplex* (Scribn.) Rydb. **Vernacular Name:** Pull-up muhly. **Life Span:** Annual. **Origin:** Native. **Season:** C_4, warm season. **FLORAL CHARACTERISTICS: Inflorescence:** spikelike panicle, 1.6–6 cm long, 0.2–0.5 cm wide, long-exserted, few-flowered, interrupted near base. **Spikelets:** pedicels 1–3 mm long, stout, 1.5–3.2 mm long; 1-flowered, sometimes purple-tinged; disarticulation above the glumes. **Glumes:** subequal, glabrous, 1-nerved, greenish gray, rounded to fan-shaped, apex rounded to subacute; lower 0.6–1.4 mm long; upper 0.7–1.7 mm long. **Lemmas:** lanceolate, minutely pubescent, apex scabrous, acute to acuminate, unawned, but occasionally mucronate with short mucro less than 1 mm. **Paleas:** lanceolate, 1.6–2.6 mm long, scabrous distally. **Awns:** none. **VEGETATIVE CHAR-**

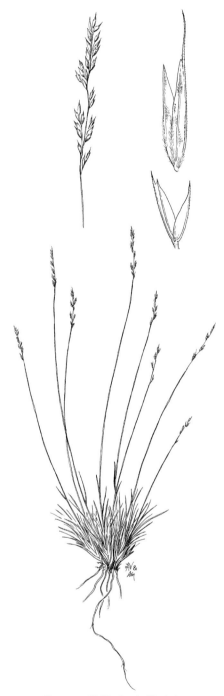

Figure 76. *Muhlenbergia filiculmis*.

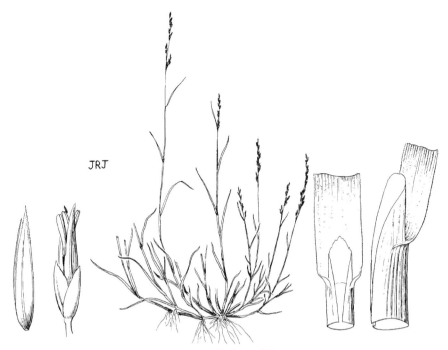

JRJ

Figure 77. *Muhlenbergia filiformis.*

ACTERISTICS: Growth Habit: tufted, sometimes rooting at lower nodes. **Culms:** erect to geniculate, commonly rooting at lower nodes, branching at lower nodes, 5–20 cm tall, solid, slightly flattened. **Sheaths:** glabrous to scabrous, shorter or longer than internodes. **Ligules:** hyaline to membranous, 1–3.5 mm long, rounded to acute, decurrent, entire. **Blades:** flat to involute near apex, 1–4 cm long, 0.6–1.6 mm wide, upper surface scabrous to puberulent, lower surface smooth to scabrous, midnerve thickened. **HABITAT:** Found in wet areas, swampy woodlands, meadows, and boggy streamsides in subalpine over Colorado (Weber and Wittmann 2001a, 2001b). **COMMENTS:** This species resembles *M. richardsonis* but differs in having glabrous internodes and subacute glume apices (Barkworth et al. 2003).

56. *Muhlenbergia glomerata* (Willd.) Trin. (Fig. 78).

Synonyms: *Muhlenbergia glomerata* (Willd.) Trin. var. *cinnoides* (Link) F. J. Herm., *Muhlenbergia racemosa* (Michx.) B.S.P. var. *cinnoides* (Link) Boivin. **Vernacular Name:** Spiked muhly. **Life Span:** Perennial. **Origin:** Native. **Season:** C_4, warm season. **FLORAL CHARACTERISTICS: Inflorescence:** narrow spikelike panicle, dense with lobes, 1.5–12 cm long, 0.3–1.8 mm wide; spikelets densely clustered on panicle branches 0.2–2.5 cm long; pedicels 0–1.2 mm long. **Spikelets:**

Figure 78. *Muhlenbergia glomerata.*

green to purplish, 1-flowered, 3–8 mm long; disarticulation above the glumes. **Glumes:** subequal, 3–8 mm long, longer than lemma, narrow, tapering, 1-nerved. **Lemmas:** lanceolate with acuminate apex, 1.9–3.1 mm long, pubescent on callus, midnerve, and margin; hairs to 1.2 mm long. 3-nerved. **Paleas:** subequal to lemma, 1.9–3.1 mm long, apical tip acuminate, sparsely pilose between nerves. **Awns:** lemma when awned, with short awn tip to 1 mm. **VEGETATIVE CHARACTERISTICS: Growth Habit:** rhizomatous with long, creeping scaly rhizomes. **Culms:** slender, erect, unbranched above, 3–12 dm tall, internodes generally dull and puberulent to some degree, hollow, slightly keeled. **Sheaths:** scabrous to glabrous, slightly keeled. **Ligules:** erose-ciliate, membranous, truncate, 0.2–0.6 mm long. **Blades:** flat when mature, rolled in bud, 2–15 cm long, 2–6 mm wide, lax, generally scabrous, occasionally smooth. **HABITAT:** Found in wet meadows, moist soils, bogs, fens, along streams and ponds and gravelly slopes. **COMMENTS:** Sometimes confused with *M. racemosa*, but *M. glomerata* has broader leaf blades and dull, puberulent, terete culms.

57. *Muhlenbergia mexicana* (L.) Trin. (Fig. 79).

Synonyms: *Agrostis mexicana* L., *Muhlenbergia ambigua* Torr., *Muhlenbergia foliosa* (Roem. & Schult.) Trin., *Muhlenbergia foliosa* (Roem. & Schult.) Trin. subsp. *ambigua* (Torr.) Scribn., *Muhlenbergia foliosa* (Roem. & Schult.) Trin. subsp. *setiglumis* (S. Wats.) Scribn., *Muhlenbergia mexicana* (L.) Trin. var. *filiformis* (Willd.) Scribn. **Vernacular Name:** Mexican muhly. **Life**

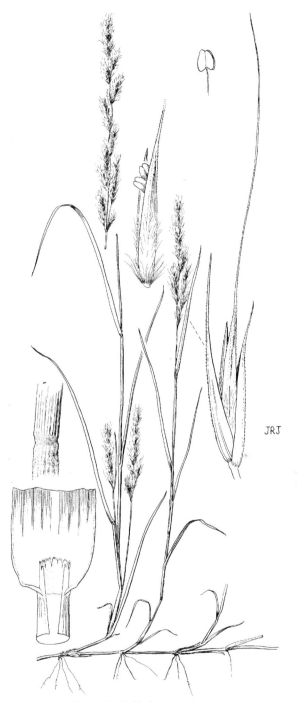

JRJ

Figure 79. *Muhlenbergia mexicana*.

Span: Perennial. **Origin:** Native. **Season:** C_4, warm
season. **FLORAL CHARACTERISTICS: Inflorescence:**
narrow panicle, terminal and axillary, 2–21 cm long,
0.3–3 cm wide, dense; terminal panicle short exserted
or partially included in upper sheath; axillary panicles
from upper 2–3 nodes, exserted on long peduncles.
Spikelets: pale green to purple-tinged, 1-flowered,
1.5–3.7 mm long; disarticulation above the glumes.
Glumes: subequal, 1.5–3.8 mm long, 1-nerved, equal

to or slightly shorter than lemma, narrowly lanceolate to elliptic tapering to acuminate apices, keel scabrous. **Lemmas:** lanceolate, barely exceeding glumes, 3-nerved, apex scabridulous, acuminate, pilose below and on calluses, hairs less than 0.7 mm long. **Paleas:** 1.5–3.8
mm, narrowly lanceolate, apex acuminate. **Awns:** glumes, if awned, to 2 mm; lemma, if
awned, to 3 mm long. **VEGETATIVE CHARACTERISTICS: Growth Habit:** rhizomatous from
creeping, scaly rhizomes; culms much branched above base. **Culms:** erect, 3–9 dm tall, hollow. **Sheaths:** smooth to scabrous, keeled. **Ligules:** membranous, truncate to acute, 0.4–1
mm long, lacerate-ciliolate. **Blades:** flat, lax, scabrous, 2–20 cm long, 2–6 mm wide. **HABI-
TAT:** Found in moist thickets, low woods, and low open ground (Harrington 1954). **COM-
MENTS:** Two varieties are found in Colorado, the typical (var. *mexicana*) and var. *filiformis*
(Torr.) Scribn. They can be separated by these characters:

1. Lemma with awn 3–10 mm long . . . . var. *filiformis*
1. Lemma awnless or awn less than 3 mm . . . . var. *mexicana*

58. *Muhlenbergia minutissima* (Steud.) Swallen (Fig. 80).

Synonyms: *Sporobolus confusus* (Fourn.) Vasey, *Agrostis
minutissima* Steud., *Sporobolus minutissima* Hitchc.
Vernacular Name: Annual muhly. **Life Span:** Annual.
Origin: Native. **Season:** C_4, warm season. **FLORAL
CHARACTERISTICS: Inflorescence:** open, diffuse
panicle, 5–16 cm long, 1.5–6.5 cm wide; pedicels capillary and flexuous, 2–7 mm long. **Spikelets:** greenish
to straw-colored, often purplish, 1-flowered, 0.8–1.5
mm long; disarticulation above the glumes. **Glumes:**

minutely pilose, at least near apex, usually finely ciliate, 1-nerved, subequal; lower 0.5–0.8
mm long, acute to obtuse; upper 0.6–0.9 mm long, obtuse. **Lemmas:** lanceolate, 0.8–1.5 mm
long, obscurely 3-nerved, glabrous to midveins and margins appressed pubescent, apex acute
to obtuse. **Paleas:** shortly pubescent to glabrous, 0.8–1.4 mm long. **Awns:** none. **VEGETA-
TIVE CHARACTERISTICS: Growth Habit:** weakly tufted, slender. **Culms:** erect to spreading,
geniculate-branched near base, internodes glabrous to scabrous, puberulent below nodes,
5–40 cm tall, solid. **Sheaths:** smooth to scabrous, strongly striate, shorter or longer than
internodes. **Ligules:** truncate, hyaline, quickly becoming lacerate and splitting, giving the
appearance of lateral lobes, 1–2.6 mm long. **Blades:** flat to involute near apex with age,
0.5–4 cm long, 0.8–2 mm wide, minutely pubescent above and scabrous below. **HABITAT:**
Found on sandy flats, roadsides, borrow pits, and areas without much competition from
other plants. **COMMENTS:** Common. A handsome, delicate, and diminutive annual.

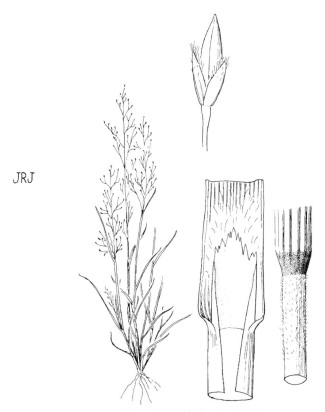

JRJ

Figure 80. *Muhlenbergia minutissima.*

59. *Muhlenbergia montana* (Nutt.) Hitchc. (Fig. 81).

Synonyms: *Muhlenbergia trifida* Hack., *Calycodon montanum* Nutt. **Vernacular Name:** Mountain muhly. **Life Span:** Perennial. **Origin:** Native. **Season:** C_4, warm season. **FLORAL CHARACTERISTICS: Inflorescence:** contracted panicle, 4–25 cm long, 2–6 cm wide, loosely flowered; branches ascending, floriferous to base. **Spikelets:** 3–7 mm long, 1-flowered; disarticulation above the glumes. **Glumes:** subequal, thin, smooth to scabrous distally; lower 1.5–3 mm

long, 1-nerved, ovate; upper 2–3.3 mm long, 3-nerved, apex 3-toothed, sometimes awned. **Lemmas:** lanceolate, 3-nerved, 3–4.5 mm long, apex acute to acuminate, sparsely to densely appressed pubescent on lower portion of midnerve and margins. **Paleas:** lanceolate, 3–4.5 mm long, apex acute to acuminate. **Awns:** upper glume sometimes awned, to 1.6 mm long; lemma awned, 6–25 mm long, flexuous. **VEGETATIVE CHARACTERISTICS: Growth Habit:** cespitose, leaves generally in dense basal clusters. **Culms:** erect, 1–8 dm, stout, glabrous.

157

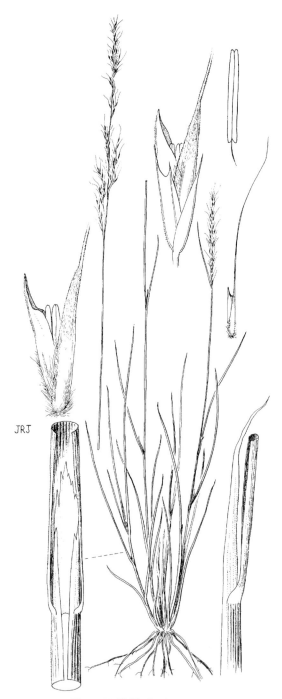

JRJ

Figure 81. *Muhlenbergia montana.*

Sheaths: smooth to scabrous, becoming papery, flattened, and sometimes spirally coiled with age, rounded, longer than internodes. **Ligules:** acute to acuminate, membranous, entire, 4–14 mm long. **Blades:** flat to involute with aging, 6–25 cm long, 1–2.5 mm wide, scabrous. **HABITAT:** Found in dry areas from foothills to subalpine, usually in or adjacent to forested areas, in eastern Colorado and on dry montane hillsides in western Colorado. **COMMENTS:** The differences between *M. filiculmis* and this species are sometimes difficult to distinguish. This species has longer spikelets and longer lemma awns. Additionally, the leaf blades are not sharply tipped, are flatter, and have wide, loose leaf sheaths (Weber and Wittmann 2001a; Barkworth et al. 2003). A good forage species of montane areas.

60. *Muhlenbergia pauciflora* Buckley (Fig. 82).

Synonyms: None. **Vernacular Name:** New Mexico muhly. **Life Span:** Perennial. **Origin:** Native. **Season:** C_4, warm season. **FLORAL CHARACTERISTICS: Inflorescence:** narrow, densely flowered panicle, interrupted below, 2–15 cm long, 0.5–2.8 cm wide; branches 0.5–4 cm long, appressed to slightly spreading, bearing spikelets to base. **Spikelets:** 3.5–5.5 mm long, 1- to occasionally 2-flowered; disarticulation above the glumes. **Glumes:** subequal to equal, 1.5–3.5 mm long, 1-nerved, nerves scabrous, occasionally scabrous across the back, apex acuminate to acute, often erose. **Lemmas:** lanceolate, 3–5.5 mm long, 3-nerved, callus with few short hairs, apex acuminate, scabrous. **Paleas:** lanceolate, glabrous, apex acuminate, 3–5.5 mm long. **Awns:** glumes unawned or awned, awn to 2.2 mm long; lemma awn flexuous, 7–25 mm long. **VEGETATIVE CHARACTERISTICS: Growth Habit:** tufted to rhizomatous, rhizomes knotty, very short. **Culms:** erect, geniculate and rooting at lower nodes, 3–7 dm tall, solid. **Sheaths:** smooth to scabrous, striate, generally shorter than internodes, becoming flat and spreading at maturity. **Ligules:** membranous, 0.5–2.5 mm, rarely to 5 mm, often split with lateral lobes 1.5–3 mm longer than central portion. **Blades:** flat to involute, 5–12 cm long, 0.5–1.5 mm wide, glabrous to scabrous, often seen with dark brown necrotic spots. **HABITAT:** Usually located in open or closed forests on steep locations with soils and rocks of granitic or calcareous origin (Barkworth et al. 2003). **COMMENTS:** Generally a southwestern species, found only in La Plata County (Weber and Wittmann 2001b).

61. *Muhlenbergia porteri* Scribn. *ex* Beal (Fig. 83).

Synonyms: None. **Vernacular Name:** Bush muhly. **Life Span:** Perennial. **Origin:** Native. **Season:** C_4, warm season. **FLORAL CHARACTERISTICS: Inflorescence:** open diffuse panicle, 4–14 cm long, 6–15 cm wide, few-flowered; branches brittle, pedicels 2–13 mm long. **Spikelets:** purplish, 3–4.5 mm long, 1-flowered; disarticulation above the glumes. **Glumes:** subequal, 2–3 mm long, 1-nerved, acute to acuminate and rarely mucronate. **Lemmas:** lanceolate, 3–4.5

Figure 82. *Muhlenbergia pauciflora*.

Figure 83. *Muhlenbergia porteri*.

mm long, 3-nerved, sparsely pubescent on lower portion. **Paleas:** lanceolate, 3–4.5 mm long, apex acuminate. **Awns:** glumes rarely mucronate, mucro to 0.4 mm long; lemma awned, 5–13 mm long, straight. **VEGETATIVE CHARACTERISTICS: Growth Habit:** loosely cespitose from a knotty base. **Culms:** ascending to widely spreading to erect, geniculate, 2–10 dm tall, wiry, freely branched from nodes, nodes often swollen, glabrous above, scabrous below. **Sheaths:** round, shorter than internodes, loosening from culm and flattening as mature, glabrous. **Ligules:** membranous, 1–2.5 mm long, lacerate to erose, truncate, splitting to form lateral lobes longer than central portion, decurrent. **Blades:** flat to involute at maturity, 2–8 cm long, 0.5–2 mm wide, scabrous, early deciduous from sheath. **HABITAT:** Usually occurring on dry hills, plains, and mesas. **COMMENTS:** A very palatable species to livestock but rarely abundant.

62. *Muhlenbergia pungens* Thurb. *ex* A. Gray (Fig. 84).

Synonyms: None. **Vernacular Name:** Sandhill muhly. **Life Span:** Perennial. **Origin:** Native. **Season:** C$_4$, warm season. **FLORAL CHARACTERISTICS: Inflorescence:** panicles, 8–16 cm long, 4–14 cm wide, slow to open until entirely out of subtending leaf sheath; primary branches capillary, widespread, and few-flowered; pedicels 10–25 mm long. **Spikelets:** red-purple to brownish, 1- or occasionally 2-flowered, 2.6–4.5 mm long; disarticulation above the glumes. **Glumes:** equal, lanceolate to ovate, 1.2–3 mm long, scaberulous above or throughout, 1-nerved, acute to acuminate to erose to 3-toothed. **Lemmas:** narrow-lanceolate, 2.6–4.5 mm long, scabrous above and on margins, purplish, apex acuminate. **Paleas:** 2.6–4.5 mm long, glabrous, lanceolate, equal to or longer than lemma. **Awns:** glumes awn-tipped to awned, awn to 1 mm; lemma awned, awns 1–1.5 mm long, straight; palea 2-awned, awns to 1 mm. **VEGETATIVE CHARACTERISTICS: Growth Habit:** rhizomatous with coarse scaly rhizomes. **Culms:** erect to decumbent-ascending, much branched at base, 1–7 cm tall, lanate below nodes. **Sheaths:** longer than internodes, margins much overlapping, woolly-hairy at base, glabrous above. **Ligules:** densely ciliate, 0.2–1 mm long, obtuse with membranous, ciliate lateral lobes. **Blades:** flat to tightly involute, stiff, pungent, 2–8 cm long, 1–2.2 mm wide, glabrous to minutely puberulent and occasionally scabrous, occasionally deciduous at blade-sheath junction. **HABITAT:** Found in sandy areas and blowouts over Colorado at low elevations. **COMMENTS:** *M. pungens* often grows in large rounded clumps, the center eventually dying out, leaving a circle of plants. Similar in general appearance to *M. arenicola, M. arizonica,* and *M. torreyi.*

63. *Muhlenbergia racemosa* (Michx.) Britton, Sterns, & Poggenb. (Fig. 85).

Synonym: *Agrostis racemosa* Michx. **Vernacular Names:** Marsh muhly, upland wild-Timothy. **Life Span:** Perennial. **Origin:** Native. **Season:** C$_4$, warm season. **FLORAL CHARACTERISTICS: Inflorescence:** narrow spikelike panicles, 0.8–16 cm long, 0.3–1.8 cm wide, lobed or interrupted, with dense clusters of spikelets; pedicels, when present, to 1 mm long. **Spikelets:** light green, often sessile, 1-flowered; disarticulation above the glumes. **Glumes:**

Figure 84. *Muhlenbergia pungens.*

subequal, 3–8 mm long including the awn, usually exceeding lemma, 1-nerved, narrow. **Lemmas:** lanceolate, 2.2–3.8 mm long, 3-nerved, pilose on lower half, callus short-bearded. **Paleas:** lanceolate, 2.2–3.8 mm long, ascending-pilose on keels, apex acuminate. **Awns:** glumes awn-tipped to awned, awns to 5 mm long. **VEGETATIVE CHARACTERISTICS: Growth Habit:** rhizomatous, with long, creeping, scaly rhizomes. **Culms:** erect, 3–11 dm tall, stiff, much

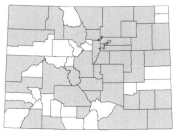

branched above the middle, hollow, slightly to strongly keeled (appearing elliptic in cross-section below nodes), internodes appearing smooth and polished for most of their length. **Sheaths:** scabrous to glabrous, slightly keeled. **Ligules:** membranous, erose-ciliate, 0.6–1.5 mm long, truncate. **Blades:** flat, 2–17 cm long, 2–5 mm wide, scabrous to scabridulous, occasionally smooth. **HABITAT:** Found over Colorado in the sagebrush and aspen zone on the Western Slope, on seasonally wet meadows, along streambanks and on rocky slopes generally at cliff bases, in gulches and similar sites on the Front Range and plains. **COMMENTS:** Widespread. Differs from the closely related *M. glomerata* in having smooth and shiny, often keeled internodes and narrower leaf blades.

Figure 85. *Muhlenbergia racemosa.*

64. *Muhlenbergia ramulosa* (Kunth) Swallen (Fig. 86).

Synonyms: *Muhlenbergia wolfii* (Vasey) Rydb., *Sporobolus ramulosus* (Kunth) Kunth. **Vernacular Names:** Green muhly, red muhly. **Life Span:** Annual. **Origin:** Native. **Season:** C$_4$, warm season. **FLORAL CHARACTERISTICS: Inflorescence:** open panicle, 2–9 cm long, 0.6–2.7 cm wide, sparsely flowered; branches ascending to spreading, 1–3.2 cm long; pedicels 1–3 mm long, stout, 0.5–1.5 mm thick. **Spikelets:** appressed to divaricate, 0.8–1.3 mm long, 1-flowered. **Glumes:** equal, 0.4–0.7 mm long, obtuse to subacute, generally glabrous, 1-nerved. **Lemmas:** obtuse, mottled with patches of greenish black, greenish white, or yellowish coloring; 3-nerved, lateral nerves sometimes obscure, 0.8–1.3 mm long, generally glabrous to occasionally short appressed pubescent along margins and midveins, apex acute. **Awns:** none. **VEGETATIVE CHARACTERISTICS: Growth Habit:** tufted. **Culms:** erect, 5–25 cm tall, not rooting at nodes, branching below. **Sheaths:** glabrous to scabrous, generally shorter than internodes. **Ligules:** hyaline, 0.2–0.5 mm long, truncate, lateral lobes lacking. **Blades:** flat to involute, 0.5–3 cm long, 0.8–1.2 mm wide, puberulent above, glabrous below. **HABITAT:** Found on open ground and forest clearings on moist soils. **COMMENTS:** Rare, but that may be a result of it being poorly collected because of its small size (Weber and Wittmann 2001a). Weber and Wittmann (2001a, 2001b) consider this a separate species from *M. wolfii* (Vasey) Rydb.

65. *Muhlenbergia richardsonis* (Trin.) Rydb. (Fig. 87).

Synonym: *Muhlenbergia squarrosa* (Trin.) Rydb. **Vernacular Names:** Mat muhly, soft-leaf muhly. **Life Span:** Perennial. **Origin:** Native. **Season:** C$_4$, warm season. **FLORAL CHARACTERISTICS: Inflorescence:** narrow to spikelike panicle, 1–15 cm long, 0.1–1.7 cm wide, exserted from upper sheath; branches short; pedicels 0.2–2 mm long. **Spikelets:** green to nearly black, 1- to occasionally 2-flowered, 1.7–3.1 mm long; disarticulation above the glumes. **Glumes:** subequal, green, ovate, apex acute, 0.6–2 mm long, sometimes mucronate with mucro less than 0.2 mm long; lower 1-nerved; upper 3-nerved, lateral nerves often obscure. **Lemmas:** narrow, lanceolate, rounded on back, glabrous, 1.7–2.6 mm long, 3-nerved, acute to acuminate, occasionally mucronate, mucro to 0.5 mm long. **Paleas:** glabrous, 1.2–2.4 mm long, lanceolate, acute. **Awns:** none. **VEGETATIVE CHARACTERISTICS: Growth Habit:** rhizomatous, mat-forming, rhizomes slender, hard, scaly, creeping. **Culms:** slender, decumbent to geniculate to erect, becoming much branched at base, 5–30 cm tall, solid, slightly flattened, and minutely nodulose, although not always noticeable. **Sheaths:** glabrous, striate, shorter or longer than internodes. **Ligules:** membranous, erose, 0.8–3 mm long, strongly decurrent, acute to truncate. **Blades:** flat to involute, 0.4–6.5 cm long, 0.5–4.2 mm wide, glabrous to puberulent above and scabrous below. **HABITAT:** Found on rocky ledges and gravelly meadows, from piñon-juniper to montane and the Intermountain Parks. **COMMENTS:** On

Figure 86. *Muhlenbergia ramulosa.*

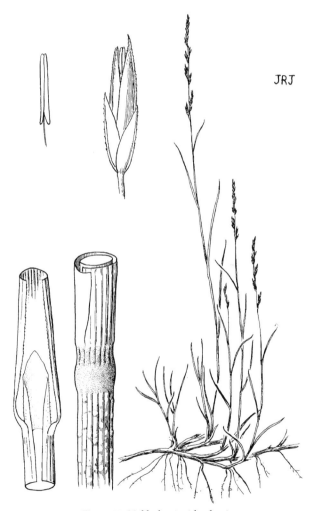

JRJ

Figure 87. *Muhlenbergia richardsonis.*

occasion this plant will have bulbils from modified florets in the inflorescence, and these plants have been called *M. squarrosa* (Trin.) Rydb. This species is frequently confused with *M. cuspidata* and *M. filiformis* (Barkworth et al. 2003).

66. *Muhlenbergia schreberi* J. F. Gmel. (Fig. 88).

Synonyms: *Muhlenbergia palustris* Scribn., *Muhlenbergia schreberi* J. F. Gmel. var. *palustris* (Scribn.) Scribn. **Vernacular Name:** Nimblewill. **Life Span:** Perennial. **Origin:** Native. **Season:** C$_4$, warm season. **FLORAL CHARACTERISTICS: Inflorescence:** contracted panicle, often interrupted below, 3–15 cm long, 1–1.6 cm wide; branches appressed to slightly spreading, spikelet-bearing to base; pedicels 0.1–4 mm long. **Spikelets:** 1-flowered, 1.8–2.8 mm long; disarticulation above the glumes. **Glumes:** unequal, thin, membranous; lower rudimentary or lacking; if present, nerveless and rounded; upper nerveless, 0.1–0.3 mm long. **Lemmas:** oblong-elliptic to narrowly lanceolate, 3-nerved, 1.8–2.8 mm long, mostly scabrous to puberulent, at least near base. **Paleas:** oblong-elliptic, 1.8–2.8 mm long, apex acute to acuminate. **Awns:** lemma awned, 1.5–5 mm long, straight. **VEGETATIVE CHARACTERISTICS: Growth Habit:** generally cespitose but occasionally stoloniferous. **Culms:** decumbent to geniculate, often rooting at lower nodes, 10–45 cm tall, simple to branched above base, glabrous to puberulent below nodes. **Sheaths:** glabrous or pubescent at summit, shorter than internodes. **Ligules:** truncate, 0.2–0.5 mm long, erose-ciliate or occasionally absent or only short hairs. **Blades:** flat, glabrous to sparsely hairy adjacent to ligule, 3–10 cm long, 1–4.5 mm wide. **HABITAT:** Found in Boulder County as a lawn weed. **COMMENTS:** Considered a noxious weed in California.

67. *Muhlenbergia thurberi* (Scribn.) Rydb. (Fig. 89).

Synonym: *Sporobolus thurberi* Scribn. **Vernacular Name:** Thurber's muhly. **Life Span:** Perennial. **Origin:** Native. **Season:** C$_4$, warm season. **FLORAL CHARACTERISTICS: Inflorescence:** contracted panicle, 0.7–5.5 cm long, 0.2–0.7 cm wide, long exserted, occasionally interrupted; branches appressed; pedicels short, 0.1–4 mm long. **Spikelets:** greenish white, occasionally purple-tinged, 1- to rarely 2-flowered, 2.6–4 mm long; disarticulation above the glumes. **Glumes:** subequal, lanceolate, apex acute, 1.6–3 mm long, 1-nerved, occasionally scabrous toward apex, often pilose on margins below. **Lemmas:** lanceolate, 2.6–4 mm long, pilose on margins and lower half, 3-nerved. **Paleas:** lanceolate, 2.6–4 mm long, pilose between nerves. **Awns:** lemma unawned or awned, awns to 1 mm. **VEGETATIVE CHARACTERISTICS: Growth Habit:** rhizomatous, often clumped, with long, creeping rhizomes. **Culms:** erect to decumbent at base, solid, 12–36 cm tall, essentially glabrous, strigose at times below nodes. **Sheaths:** somewhat keeled, shorter than internodes, hirtellous near margins. **Ligules:**

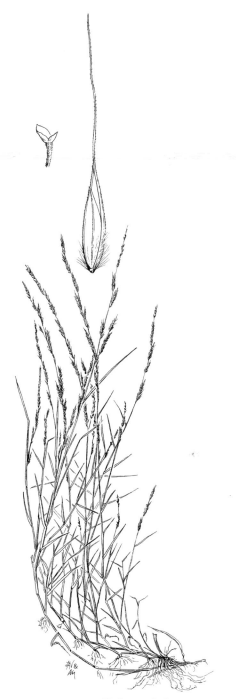

Figure 88. *Muhlenbergia schreberi.*

Figure 89. *Muhlenbergia thurberi.*

membranous, ciliate, 0.9–1.2 mm long, truncate to obtuse, erose, decurrent. **Blades:** flat at base to tightly involute at tip, 0.2–3.7 cm long, 0.2–1 mm wide, pubescent on both surfaces to minutely hirsute above and scabrous below, tip pungent. **HABITAT:** This species likes dry to moist sites on sandstone rimrock and in piñon-juniper on the Western Slope (Weber and Wittmann 2001b; Wingate 1994). **COMMENTS:** This species is very similar to *M. curtifolia*, which occurs in adjacent Utah and might be expected in Colorado.

68. *Muhlenbergia torreyi* (Kunth) Hitchc. *ex* Bush (Fig. 90).

Synonyms: *Muhlenbergia gracillima* Torr., *Agrostis torreyi* Kunth. **Vernacular Name:** Ring muhly. **Life Span:** Perennial. **Origin:** Native. **Season:** C$_4$, warm season. **FLORAL CHARACTERISTICS: Inflorescence:** diffuse open panicle, 7–21 cm long, 3–15 cm wide; branches and pedicels appressed, spreading at maturity; pedicels 1–8 mm long. **Spikelets:** 1-flowered, 2–3.5 mm long; disarticulation above the glumes. **Glumes:** equal to subequal, 1.3–2.5 mm long, 1-nerved, scabrous, apex acute to acuminate to awn-tipped or irregularly toothed. **Lemmas:** narrowly elliptic to lanceolate, 2–3.2 mm long, 3-nerved, appressed-pubescent below, apex acuminate to minutely bifid. **Paleas:** narrowly elliptic, 2–3.2 mm long, equaling to exceeding lemma, area between nerves sparsely pubescent, apex acuminate. **Awns:** glumes unawned to awned, awn to 1.1 mm; lemma awns 0.5–4 mm long. **VEGETATIVE CHARACTERISTICS: Growth Habit:** cespitose, forming a ring as tillers expand outward and center dies; leaves generally in basal cluster. **Culms:** erect with decumbent base, 1–4 dm tall, branching below, nodes concealed by sheaths. **Sheaths:** shorter than internodes, rounded, smooth to scaberulent, margins hyaline. **Ligules:** hyaline, 2–7 mm long, acute, entire initially becoming lacerate and splitting with age, often appearing as lateral lobes. **Blades:** folded to tightly involute, arcuate, 1–3 cm long, 0.3–0.9 mm wide, apex sharp-pointed; leaves forming a mat or cushion. **HABITAT:** Found on sandy soils on the eastern plains, Arkansas Valley, and San Luis Valley. **COMMENTS:** Abundant. Similar in general appearance to *M. arenicola*, *M. arizonica,* and *M. pungens.*

69. *Muhlenbergia wrightii* Vasey *ex* J. M. Coult. (Fig. 91).

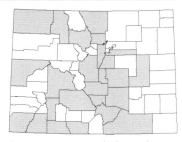

Synonyms: none. **Vernacular Name:** Spike muhly. **Life Span:** Perennial. **Origin:** Native. **Season:** C$_4$, warm season. **FLORAL CHARACTERISTICS: Inflorescence:** spikelike panicle, 5–16 cm long, 0.2–1.2 cm wide, densely flowered, often interrupted; branches appressed; pedicels 0.1–1.4 mm long. **Spikelets:** greenish to dark gray, 2–3 mm long, 1- to occasionally 2-flowered; disarticulation above the glumes. **Glumes:** equal, 0.5–1.6 mm long, ovate to lanceolate, 1-nerved, scabrous, apex obtuse to acute to awn-tipped. **Lemmas:** narrowly lanceolate, 3-nerved, lateral nerves sometimes obscure, 2–3 mm long, appressed-pubescent below, apex

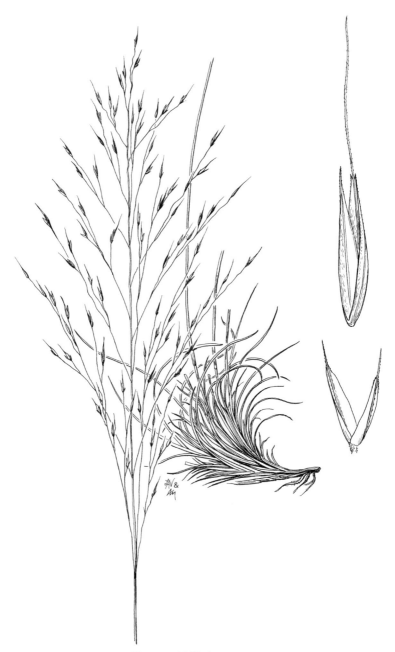

Figure 90. *Muhlenbergia torreyi*.

Figure 91. *Muhlenbergia wrightii*.

acuminate to acute. **Paleas:** lanceolate, pubescent between nerves, 1.9–3 mm long, apex acute to acuminate. **Awns:** glumes awn-tipped, awn 0.5–1 mm long; lemma mucronate, mucro 0.3–1 mm. **VEGETATIVE CHARACTERISTICS: Growth Habit:** cespitose with knotty base. **Culms:** erect to rarely decumbent, hispid, wiry, 1–6 dm tall, not rooting at nodes, branching below. **Sheaths:** glabrous to scabrous, generally shorter than internodes, compressed-keeled. **Ligules:** membranous, 1–3 mm long, often truncate on lower leaves and rounded to acute on upper leaves. **Blades:** flat to folded, 1.4–12 cm long, 1–3 mm wide, scabrous above, glabrous below. **HABITAT:** Found at lower elevations on mesas and rocky slopes in the lower foothills, especially in the Front Range, and on sandstone rimrock ledges in southwestern Colorado (Weber and Wittmann 2001a, 2001b). **COMMENTS:** This species is variable and grows at a wide range of elevations and on a number of different substrates.

20. *Munroa* Torr.

Low, tufted annuals with spreading, much-branched culms. Leaves short, blades stiff, pungent, and flat. Inflorescences of small, subsessile clusters of spikelets, these almost hidden in fascicles of leaves at branch tips. Spikelets few-flowered, lower ones of a cluster 3- or 4-flowered, upper ones 2- or 3-flowered. Typically lower florets pistillate or perfect, distal florets sterile and rudimentary. Disarticulation above the glumes and sometimes between florets. Glumes of lower spikelets subequal, narrow, acute, 1-nerved, slightly shorter than lowermost lemma. Glumes of upper spikelets unequal, the lower reduced or even absent. Lemmas prominently 3-nerved, those of lower spikelets coriaceous, usually with tufts of hairs on margins near the middle, gradually narrowing above, abruptly mucronate or short-awned from a slender, spreading tip. Palea narrow, membranous. Basic chromosome number, $x = 8$. Photosynthetic pathway, C_4. In the original publication of the genus, the name was misspelled *Monroa*, and some suggest that should be the correct spelling for the genus.

Represented in Colorado by a single species.

70. *Munroa squarrosa* (Nutt.) Torr. (Fig. 92).

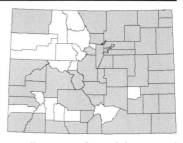

Synonyms: *Monroa squarrosa* (Nutt.) Torr. var. *floccuosa* Vasey *ex* Beal, *Munroa squarrosa* (Nutt.) Torr. var. *floccuosa* Vasey *ex* Beal. **Vernacular Name:** False buffalograss. **Life Span:** Annual. **Origin:** Native. **Season:** C_4, warm season. **FLORAL CHARACTERISTICS: Inflorescence:** capitate panicle much reduced, with spikelike branches bearing 2–4 subsessile to pedicellate spikelets; branches often hidden in subtending leaves. **Spikelets:** irregular, lowest one usually sessile, while upper ones pedicellate in a fascicle; lower florets pistillate or perfect, while terminal ones sterile and rudimentary, 6–10 mm long, 3- to 5-flowered; disarticulation above the glumes or below subtending leaves. **Glumes:** irregular, those of lower 1–2 spikelets subequal, 1-nerved, 2.5–4.2 mm long, acute; upper spikelet glumes unequal and reduced with lower sometimes absent. **Lemmas:** lanceolate, scabrous, 3.5–5 mm long, 3-nerved, tuft of

Figure 92. *Munroa squarrosa.*

hairs present along margins at midlength, apex emarginate or 2-lobed. **Paleas:** glabrous, subequal to lemma, conspicuously 2-nerved. **Awns:** lemma midnerve excurrent, forming a stout, scabrous awn, 0.5–2 mm long. **VEGETATIVE CHARACTERISTICS: Growth Habit:** stoloniferous, mat-forming. **Culms:** decumbent, spreading, highly branched, 3–15 cm, internodes scabrous to puberulent. **Sheaths:** open, short, broad, with a tuft of stiff hairs at apex, 1–2 mm long, lowermost sheaths greatly reduced. **Ligules:** ring of hairs, 0.5 and rarely to 1 mm long. **Blades:** linear, generally involute but occasionally flat to folded, arise in pairs from each node, 1–5 cm long, 1–2.5 mm wide, margins white-thickened, apex pointed. **HABITAT:** Found in dry areas, plains, and along sandy roadsides and rimrock depressions over Colorado. **COMMENTS:** Common. Prophyll highly developed, and nerves extending into short awns.

21. *Pleuraphis* Torr.

Perennials, mostly rhizomatous or occasionally stoloniferous. Sheaths open; ligule a short membrane; auricles absent. Inflorescence a two-sided spike or spikelike panicle, spikelets in clusters of 3 at each node of a zigzag rachis, clusters deciduous as a whole. Disarticulation at base of branches, leaving the zigzag rachis. Spikelets of cluster dissimilar. The 2 lateral spikelets of each branch short pedicellate, with 1–5 staminate or sterile florets. Glumes nearly as long as florets, deeply cleft into 2 or more lobes, with 1 or more dorsal awns. The central spikelet sessile, with 1 fertile floret. Glumes firm, flat, indurate, and usually asymmetrical, bearing an awn on one side from about the middle. Lemmas thin, 3-nerved, awned or awnless. Palea about as large as lemma and similar in texture. Basic chromosome number, $x = 9$. Photosynthetic pathway, C_4.

Represented in Colorado by a single species. A segregate of *Hilaria*.

71. *Pleuraphis jamesii* Torr. (Fig. 93).

Synonym: *Hilaria jamesii* (Torr.) Benth. **Vernacular Name:** James's galleta. **Life Span:** Perennial. **Origin:** Native. **Season:** C$_4$, warm season. **FLORAL CHARACTERISTICS: Inflorescence:** 2-sided spike, 2–8 cm long, dense. **Spikelets:** 3 spikelets per node (lateral 2 staminate and central 1 fertile), 6–9 mm long subtended by hairs; staminate spikelets 2-flowered; fertile spikelets 1- to 2-flowered, lower fertile; upper floret, when present, staminate; disarticulation at

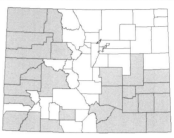

base of each spikelet cluster. **Glumes:** central spikelet glumes subequal, 4–5.5 mm long, irregularly cleft, 3- to 9-nerved; lateral spikelet glumes subequal, 5–7 mm long, narrowly lanceolate to elliptic tapered to an acute or obtuse tip. **Lemmas:** lanceolate, blunt-tipped, scabrous at apex; central lemma narrowly lanceolate, 5.8–7.6 mm long, bifid; lateral lemmas 4.4–7 mm long, apex ciliate and inrolled. **Awns:** lateral spikelets; lower glume with awn coming off displaced midnerve midway down glume, 3–5.5 mm long; upper glume no awn to awn-tipped; central spikelet glumes with each of 5 nerves excurrent into irregular awns 1.5–5 mm long; fertile lemma with dorsal awn from below bifid apex 1–2.5 mm long. **VEGETATIVE CHARACTERISTICS: Growth Habit:** strongly rhizomatous to stoloniferous, leaves mostly basal. **Culms:** erect to decumbent with bases much branched, 2–7 dm tall, nodes villous. **Sheaths:** open, glabrous, round, striate, villous at collar along margins. **Ligules:** membranous, often laciniate, 1–5 mm long, truncate. **Blades:** flat to involute and curled when dry, sparsely villous behind ligule. **HABITAT:** Found at low elevations over Colorado on river benches and sagebrush stands, dry and rocky slopes. **COMMENTS:** *P. jamesii* is most frequent on heavy clay soils but is occasionally found on sandy to gravelly soils as well.

22. *Redfieldia* Vasey

Perennials, long creeping or vertical rhizomes. Sheaths open, longer than internodes; ligules a fringe of hairs; auricles absent. Inflorescence terminal, an open panicle, one-third to one-half as long as culms. Spikelets (1) 2- to 6-flowered, rachilla pronounced between florets, reduced floret (when present) above fertile florets, on long flexuous pedicels, laterally compressed, grayish in color. Disarticulation above the glumes and between florets. Glumes 2, unequal, lanceolate, 1-nerved, upper shorter than lowermost lemma. Lemmas 3-nerved, nerves forming 3 minute teeth, hairy, chartaceous, callus with a tuft of hairs. Palea glabrous, slightly shorter than lemma. Basic chromosome number, $x = 10$. Photosynthetic pathway, C$_4$.

Represented in Colorado by the only species in the genus.

72. *Redfieldia flexuosa* (Thurb. *ex* A. Gray) Vasey (Fig. 94).

Synonym: *Graphephorum flexuosa* Thurb. *ex* A. Gray. **Vernacular Name:** Blowout grass. **Life Span:** Perennial. **Origin:** Native. **Season:** C$_4$, warm season. **FLORAL CHARACTERISTICS: Inflorescence:** open to diffuse panicle, 20–50 cm long, 8–25 cm wide, with flexuous,

Figure 93. *Pleuraphis jamesii.*

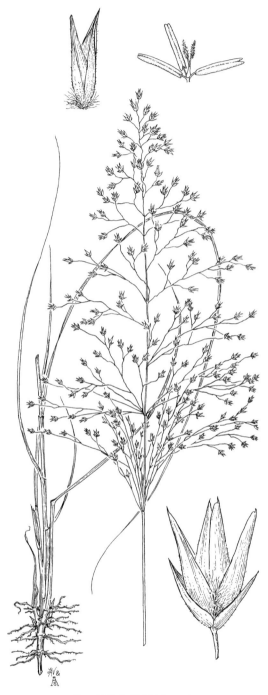

Figure 94. *Redfieldia flexuosa*.

capillary branches; branches and branchlets wide-spreading, hairlike to slender, dichotomous; pedicels flexuous, hairlike, generally 1–5 cm long. **Spikelets:** ovate to lanceolate, 2- to 6-flowered, 5–8 mm long, 3–5 mm wide, V-shaped, widely spreading at maturity; disarticulation above the glumes. **Glumes:** sub-equal, glabrous, lanceolate, apex acuminate; lower 1-nerved, 3–4 mm long; upper 3.5–4.5 mm long, 1- to 3-nerved. **Lemmas:** lanceolate to falcate, 4.5–6 mm long, 3-nerved, glabrous to puberulent, tuft of hairs at base of floret. **Paleas:** glabrous, stout, 2-nerved, folded in center. **Awns:** none. **VEGETATIVE CHARACTERISTICS: Growth Habit:** rhizomatous, with extensive horizontal as well as vertical taproot-like rhizomes. **Culms:** erect to ascending, 5–13 dm tall, simple. **Sheaths:** open, overlapping, glabrous, nearly round, ribbed, shorter than internodes, shredding with age. **Ligules:** ring of hairs with membranous base, to 1.5 mm long, truncate to rounded. **Blades:** loosely involute, 15–45 cm long, 2–8 mm wide, glabrous to scabrous. **HABITAT:** Found in sand blowout areas, sandhills, and dunes in eastern Colorado and San Luis Valley. **COMMENTS:** Common. *R. flexuosa* is used as a sand binder for erosion control (Barkworth et al. 2003).

23. *Schedonnardus* Steud.

Low, tufted perennial with slender, wiry, erect, or decumbent culms. Leaves mostly in a basal clump, blades short, flat, usually spirally twisted and wavy-margined, about 1 mm broad. Ligule membranous, 2–3 mm long, decurrent. Panicles half to three-fourths the entire length of culm, with a stiff, curved, wiry central axis and slender, spreading, undivided primary branches. At maturity the panicle breaks off at base and becomes a "tumbleweed." Spikelets slender, 1-flowered, sessile, widely spaced, and appressed on branches and at the tip of main axis. Disarticulation above the glumes. Glumes narrow, lanceolate or acuminate, 1-nerved, the upper about as long as the lemma, the lower shorter. Lemma narrow, rigid, 3-nerved, awnless or with a minute awn tip. Palea similar to lemma and about as long. Basic chromosome number, $x = 10$. Photosynthetic pathway, C_4.

Represented in Colorado by the only species in the genus.

73. *Schedonnardus paniculatus* (Nutt.) Trel. (Fig. 95).

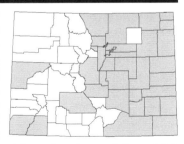

Synonym: *Lepturus paniculatus* Nutt. **Vernacular Name:** Tumblegrass. **Life Span:** Perennial. **Origin:** Native. **Season:** C_4, warm season. **FLORAL CHARACTERISTICS: Inflorescence:** panicle 5–50 cm long, with 3–13 widely spaced, spicate primary unilateral branches; branches becoming curved as they mature, 2–20 cm long, spreading. **Spikelets:** sessile, 3–5.5 mm long, 1-flowered, slender, embedded and appressed to branches, laterally compressed; disarticulation at base of panicle and above the glumes. **Glumes:** unequal, lanceolate, 1-nerved; lower 1.5–3 mm long; upper 1.5–4 mm long. **Lemmas:** glabrous to scabrous, 3–5 mm long, 3-nerved.

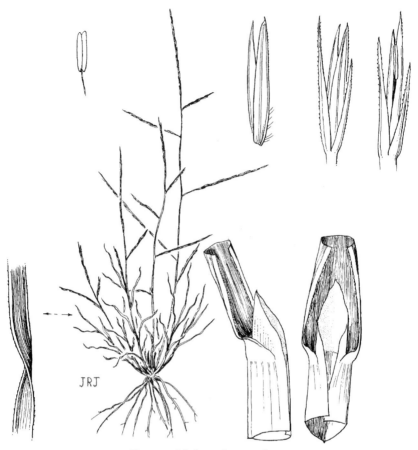

Figure 95. *Schedonnardus paniculatus.*

Awns: upper glume awn-tipped or mucronate, minute; lemma awn-tipped or mucronate, minute. **VEGETATIVE CHARACTERISTICS: Growth Habit:** cespitose, leaves mostly basal. **Culms:** ascending to erect, 8–55 cm tall, decumbent to geniculate at base, branching basally, internodes minutely retrorsely pubescent. **Sheaths:** open, glabrous, laterally compressed, margins hyaline. **Ligules:** membranous, 1–3.5 mm long, acuminate to obtuse, erose. **Blades:** glabrous, flat to folded, becoming twisted on drying, 1–12 cm long, 0.7–2 mm wide, margins white-thickened, scabrous. **HABITAT:** Found on dry sites from low elevations to montane areas, most common on plains and outwash mesas. **COMMENTS:** This is a monotypic genus. *S. paniculatus* is often found on disturbed areas. When the plant reaches maturity the panicle is deciduous and acts like a "tumbleweed," dispersing seed as it is blown around.

24. *Scleropogon* Phil.

Low mat-forming perennials, stoloniferous. Leaves tufted with flat, firm blades. Sheaths open; ligules a fringe of hairs; auricles absent. Plants staminate or pistillate (dioecious) or less frequently both sexes present in different spikelets but on the same plant (monoecious). Spikelets large, in spicate racemes or reduced panicles, pistillate ones bristly with long awns. Staminate spikelets mostly with 5–10 (–20) persistent florets. Glumes 2, 1-nerved or rarely 3-nerved, thin, pale, awnless, lower and upper separated by a short internode. Lemmas of staminate spikelets similar to the glumes. Palea obtuse, shorter than lemma. Pistillate spikelets with 3–5 florets and 1 to several rudiments (reduced to awns) above. Rachilla disarticulating above the glumes, florets falling together. Glumes of pistillate spikelets 2, unequal, usually 3-nerved. Lemmas of pistillate spikelet firm, rounded on back, 3-nerved, nerves extending to slender, spreading awns. Palea narrow, the 2 nerves extending into short awns. Basic chromosome number, $x = 10$. Photosynthetic pathway, C_4.

Represented in Colorado by the only species in the genus.

74. *Scleropogon brevifolius* Phil. (Fig. 96).

Synonym: *Scleropogon longisetus* Beetle. **Vernacular Name:** Burrograss. **Life Span:** Perennial. **Origin:** Native. **Season:** C_4, warm season. **FLORAL CHARACTERISTICS: Inflorescence:** spikelike racemes or contracted panicles; staminate panicle 1–6 cm long, loosely spikelike; pistillate panicle 10–15 cm long including awns. **Spikelets:** staminate 5–10 florets, 20–30 mm long; pistillate 2–3 cm long excluding awns, 1- to 4-flowered with 1 to many sterile rudimentary florets reduced to awns. **Glumes:** staminate glumes subequal, 3–10 mm long, thin, membranous, 1- to 3-nerved, separated by a short internode; pistillate glumes unequal, strongly 3-nerved, thin, lanceolate, lower 1–2 cm long; upper 1.6–3 cm long, subtended by a glumelike bract. **Lemmas:** staminate lemmas similar to glumes, 3-nerved; pistillate lemmas narrow, 3-nerved, nerves extending beyond lemma to form awns. **Awns:** staminate lemmas unawned or awned to 3 mm long; pistillate lemmas 3-awned, awns 5–10 cm, spreading to reflexed at maturity. **VEGETATIVE CHARACTERISTICS: Growth Habit:** tightly tufted and stoloniferous with wiry, creeping stolons forming mats, leaves mostly basal. **Culms:** erect, 1–2 dm tall. **Sheaths:** smooth, short, strongly veined, upper glabrous, lower hispid to villous, margins sometimes scarious. **Ligules:** ciliate membrane, to 1 mm long. **Blades:** flat to folded, 2–8 cm long, 1–2 mm wide, firm, margins sometimes scabrous. **HABITAT:** Occasional to locally abundant in the grasslands of lower Colorado. **COMMENTS:** Rarely, some plants of *S. brevifolius* are monoecious and have a bisexual panicle, staminate spikelets below pistillate spikelets in the inflorescence (gynaecandrous).

25. *Spartina* Schreb.

Perennials, frequently rhizomatous. Leaves tough and firm; blades long, flat or involute. Ligule a ring of long or short hairs. Inflorescence of few to numerous, racemosely arranged, short, usually appressed branches, bearing closely placed sessile spikelets. Disarticulation below the glumes. Spikelets 1-flowered, laterally flattened. Glumes unequal, keeled, usually 1-nerved or the upper with 3 closely placed nerves, acute or the second short-awned. Lemma firm, keeled, strongly 1- or 3-nerved and often with additional indistinct

Figure 96. *Scleropogon brevifolius.*

lateral nerves, tapering to a narrow but usually rounded awnless tip. Palea as long as or longer than lemma, with broad, membranous margins on either side of the closely placed nerves. Basic chromosome number, $x = 10$. Photosynthetic pathway, C_4.

Represented in Colorado by 2 species.

1. Plant over 1 m tall; lower glume nearly as long as floret . . . . 76. *S. pectinata*
1. Plant less than 1 m tall; lower glume noticeably shorter than floret . . . . 75. *S. gracilis*

75. *Spartina gracilis* Trin. (Fig. 97).

Synonyms: none. **Vernacular Name:** Alkali cordgrass. **Life Span:** Perennial. **Origin:** Native. **Season:** C_4, warm season. **FLORAL CHARACTERISTICS: Inflorescence:** panicle, 8–25 cm long, with 3–12 racemosely arranged appressed branches; branches alternate 1.5–8 cm long, appressed to ascending, 10–30 spikelets per branch. **Spikelets:** sessile, 6–11 mm long, ovate to lanceolate, 1-flowered, strongly compressed, appressed; disarticulation below the

glumes. **Glumes:** unequal, margins glabrous to hispid; lower 3–7 mm long, shorter than floret, glabrous to sparsely pubescent, linear, 1-nerved, keels glabrous to strigose; upper 6–10 mm long, keels strigose, narrowly lanceolate, 3- to 5-nerved, lateral nerves adjacent to midnerve, nerves scabrous to pectinate, mucronate. **Lemmas:** glabrous to sparsely hirsute, 6.2–7.5 mm long, lanceolate, 1- to 3-nerved, lateral nerves usually obscure, apex blunt. **Paleas:** subequal to lemma, glabrous, thin and papery. **Awns:** upper glume awnless to mucronate, mucro to 1 mm long. **VEGETATIVE CHARACTERISTICS: Growth Habit:** strongly rhizomatous, with elongated pale rhizomes 3–5 mm thick and overlapping scales. **Culms:** erect, solitary, 4–10 dm tall, glabrous, 2–3.5 mm thick. **Sheaths:** smooth to striate, glabrous, collars occasionally ciliate. **Ligules:** ciliate membrane, 0.5–1 mm long. **Blades:** flat to involute on drying, 6–30 cm long, 2.5–8 mm wide, upper surface scabrous, lower surface glabrous, margins scabrous. **HABITAT:** Found in moist to wet areas in sloughs and alkali flats on the plains. **COMMENTS:** This species is more common in the western part of the state.

76. *Spartina pectinata* Link (Fig. 98).

Synonyms: *Spartina michauxiana* Hitchc., *Spartina pectinata* Bosc *ex* Link var. *suttiei* (Farw.) Fern. **Vernacular Names:** Prairie cordgrass, slough grass. **Life Span:** Perennial. **Origin:** Native. **Season:** C_4, warm season. **FLORAL CHARACTERISTICS: Inflorescence:** panicle, 10–50 cm long, with 5–50 racemosely arranged appressed branches; branches alternate 1.5–15 cm long, appressed to slightly spreading, 10–80 spikelets imbricate in 2 rows on abaxial side of each

branch. **Spikelets:** sessile, 10–25 mm long, ovate to lanceolate, 1-flowered, strongly compressed, appressed; disarticulation below the glumes. **Glumes:** unequal, margins glabrous

JRJ

Figure 97. *Spartina gracilis.*

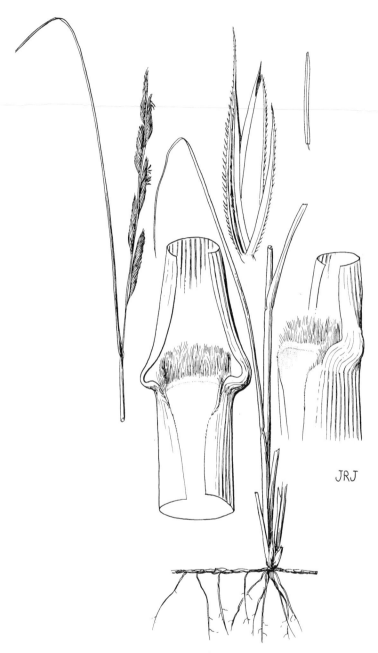

Figure 98. *Spartina pectinata*.

to sparsely hispid, keel pectinate; lower 5–10 mm long, usually as long as floret, glabrous to sparsely pubescent, linear, 1-nerved; upper 10–20 mm long, including the awn; exceeding floret, narrowly lanceolate, 3-nerved, lateral nerves adjacent to midnerve. **Lemmas:** glabrous to sparsely hirsute, 6–9 mm long, lanceolate, 1- to 3-nerved, lateral nerves usually obscure, keel scabrous, apex cleft, lobes 0.2–0.9 mm long. **Paleas:** subequal to lemma, glabrous, thin and papery. **Awns:** lower glume short-awned, 0.9–4 mm long; upper glume with stout awn, 3–10 mm long. **VEGETATIVE CHARACTERISTICS: Growth Habit:** strongly rhizomatous, with stout, elongated rhizomes 4–10 mm thick, light brown, with closely imbricate scales. **Culms:** erect, solitary or in small clumps, to 2.5 m tall, robust. **Sheaths:** smooth to slightly striate, glabrous, collars occasionally pilose. **Ligules:** ciliate membrane, 2–4 mm long, truncate. **Blades:** flat to involute on drying, 20–120 cm long, 6–15 mm wide, glabrous, margins scabrous to minutely scabrous. **HABITAT:** Found in moist to wet areas in irrigation ditches, sloughs, and around ponds. **COMMENTS:** This species is more common in the eastern part of the state.

26. *Sporobolus* R. Br.

Annuals and perennials, mostly cespitose but a few with creeping rhizomes. Sheaths open; ligule a fringe of hairs or a ciliate membrane; auricles absent. Inflorescence an open or less frequently contracted panicle. Spikelets with 1 fertile floret, small, awnless. Disarticulation above the glumes. Glumes 2, unequal, 1-nerved, usually shorter than lemma. Lemma 1-nerved, thin, awnless. Palea well developed, as long as to longer than lemma. Basic chromosome number, $x = 6$ and 9. Photosynthetic pathway, C_4.

Represented in Colorado by 11 species and 1 variety.

1. Plants annual or short-lived perennials
    2. Lower panicle nodes with 7–15 branches . . . . 86. *S. pyramidatus*
    2. Lower panicle nodes with 1–3 branches . . . . 85. *S. neglectus*
1. Plant long-lived perennial
    3. Spikelets 3–7 mm long
        4. Upper glume shorter than lemma; panicle tightly contracted, spikelike . . . .
        . . . . 78. *S. compositus*
        4. Upper glume as long as to longer than lemma; panicle open . . . . . . 83. *S. heterolepis*
    3. Spikelets less than 3 mm long or culms over 1 m tall
        5. Summit of sheaths naked or nearly so
            6. Pedicels 6–25 mm long, capillary . . . . 87. *S. texanus*
            6. Pedicels 1–2 mm long, stout . . . . 77. *S. airoides*
        5. Summit of sheaths with conspicuous tuft of hairs
            7. Culms 1–2 m tall . . . . 82. *S. giganteus*
            7. Culms generally less than 1 m tall
                8. Panicle closed, spikelike . . . . 79. *S. contractus*
                8. Panicle open, not spikelike
                    9. Culm bases hard and knotty . . . . 84. *S. nealleyi*
                    9. Culm bases not hard and knotty
                        10. Panicle branches flexuous; spikelets loosely arranged . . . . 81. *S. flexuosus*
                        10. Panicle branches appressed; spikelets crowded . . . . 80. *S. cryptandrus*

77. *Sporobolus airoides* (Torr.) Torr. (Fig. 99).

Synonym: *Agrostis airoides* Torr. **Vernacular Name:** Alkali sacaton. **Life Span:** Perennial. **Origin:** Native. **Season:** C$_4$, warm season. **FLORAL CHARACTERISTICS: Inflorescence:** open, diffuse panicle, subpyramidal in shape, 15–45 cm long, 15–25 cm wide, occasionally lower portion still included in uppermost leaf sheath; branches naked at base; pedicels 0.5–2 mm long, occasionally to 3 mm, stout. **Spikelets:** purplish to greenish, 1.3–2.8 mm long, 1-flowered, subterete; disarticulating both above and below the glumes. **Glumes:** unequal, lanceolate to ovate; lower 0.5–1.8 mm long, often no nerves present; upper 1.1–2.4 mm long, 1-nerved, apex acute. **Lemmas:** 1.2–2.5 mm long, glabrous, membranous, 1-nerved, apex acute. **Paleas:** 1.1–2.4 mm long, equal to lemma, ovate, membranous, glabrous. **Awns:** none. **VEGETATIVE CHARACTERISTICS: Growth Habit:** cespitose. **Culms:** decumbent to erect, 3–12 dm tall, glabrous, solid, shiny. **Sheaths:** rounded below, lower sheaths generally appear bleached out in appearance, apex glabrous to sparsely hairy. **Ligules:** ciliate membrane, 0.1–0.3 mm long. **Blades:** flat to involute with age, 10–45 cm long, 2–5 mm wide, scabrous above with prominent ridges, glabrous below, apex pointed. **HABITAT:** Found at low elevations on alkaline flats, forming large stands on the Western Slope and on the plains and San Luis Valley on the Eastern Slope. **COMMENTS:** It is an important forage species in alkaline flats and salty areas where other, higher-quality forage is lacking.

78. *Sporobolus compositus* (Poir.) Merr. (Fig. 100).

Synonyms: *Sporobolus asper* (P. Beauv.) Kunth, *Sporobolus asper* (P. Beauv.) Kunth var. *hookeri* (Trin.) Vasey. **Vernacular Name:** Composite dropseed, tall dropseed, rough dropseed. **Life Span:** Perennial. **Origin:** Native. **Season:** C$_4$, warm season. **FLORAL CHARACTERISTICS: Inflorescence:** contracted, spike-like panicles, both terminal and axillary, 5–30 cm long, 0.4–1.6 cm wide, partially included in uppermost sheath, upper culm nodes may give rise to axillary solitary panicles with cleistogamous spikelets that may be enclosed partially to totally by the subtending sheath; branches appressed and spikelet-bearing to base. **Spikelets:** stramineous to purple-tinged, glabrous, 4–6 mm long, 1-flowered, strongly compressed; disarticulating above glumes. **Glumes:** unequal to subequal, lanceolate, 1-nerved, scabrous on keel; lower 2–4 mm long; upper 2.5–5 mm long, shorter than lemma. **Lemmas:** lanceolate, strongly flattened, 3–6 mm long, 1- to occasionally 2- to 3-nerved, glabrous, apex boat-shaped. **Paleas:** ovate, 2.1–3 mm long, membranous. **Awns:** none. **VEGETATIVE CHARACTERISTICS: Growth Habit:** cespitose. **Culms:** stout, 3–13 dm tall, 2–5 mm thick, glabrous. **Sheaths:** open, margins overlapping, 2.6–6 mm wide, apex sparsely hairy. **Ligules:** ciliate membrane, 0.1–0.5 mm long, thick, truncate. **Blades:** flat to involute as mature, 5–70 cm long, 1.5–10 mm wide, upper surface glabrous to scabrous, lower surface glabrous, margins glabrous. **HABITAT:** Found in eastern Colorado in piedmont valleys and outwash mesas and

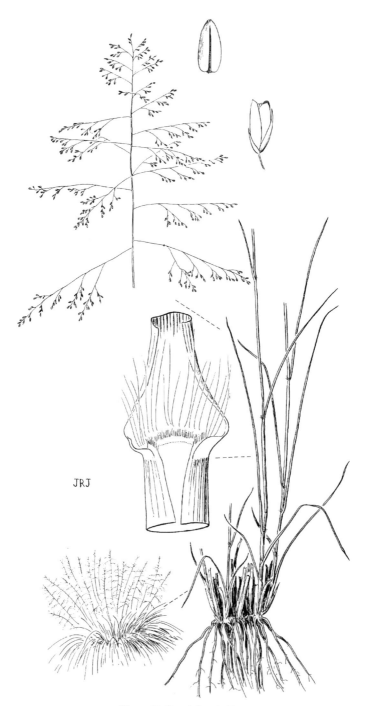

JRJ

Figure 99. *Sporobolus airoides.*

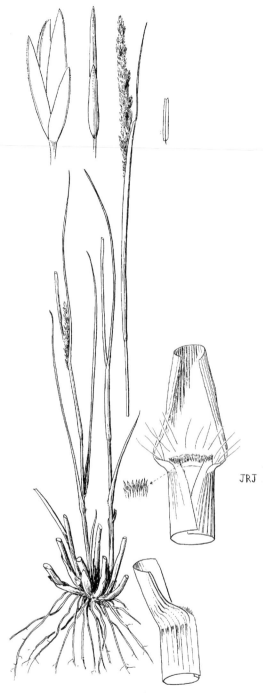

Figure 100. *Sporobolus compositus.*

often along roadsides and plains. **COMMENTS:** Weber and Wittmann (2001a) and Wingate (1994) call this plant *S. asper* (Michx.) Kunth. The plants in Colorado belong to the typical variety (var. *compositus*).

79. *Sporobolus contractus* Hitchc. (Fig. 101).

Synonym: *Sporobolus cryptandrus* (Torr.) Gray var. *strictus* Scribn. **Vernacular Name:** Spike dropseed. **Life Span:** Perennial. **Origin:** Native. **Season:** C_4, warm season. **FLORAL CHARACTERISTICS: Inflorescence:** contracted spikelike panicle, terminal, 15–45 cm long, 0.2–0.8 cm wide, dense, base usually included in uppermost sheath; branches at lower nodes usually 1–2 and rarely 3, appressed, with spikelets to base. **Spikelets:** grayish to whitish, 1.7–3.2 mm long, 1-flowered; disarticulation above the glumes. **Glumes:** unequal, 1-nerved, narrowly lanceolate, membranous, keels prominent; lower 0.7–1.7 mm long, apex acute to acuminate; upper 2–3.2 mm long. **Lemmas:** linear-lanceolate, 2–3.2 mm long, 1-nerved, glabrous, flattened, hyaline, keel scabrous, apex acute. **Paleas:** linear-lanceolate, glabrous, 1.8–3 mm long, membranous, apex acute. **Awns:** none. **VEGETATIVE CHARACTERISTICS: Growth Habit:** cespitose. **Culms:** erect, 4–10 dm tall, 2–4 mm thick near base. **Sheaths:** rounded below, margins hairy, apex with conspicuous tufts of hair. **Ligules:** line of hairs, 0.4–1 mm long. **Blades:** flat to involute, 4–35 cm long, 3–8 mm wide, glabrous, margins whitened, irregularly scabrous, tapering to fine point. **HABITAT:** Found in the San Luis Valley and northwestern Colorado. **COMMENTS:** Scattered in our area on sandy soils, but the species is very common in the more arid regions of the Southwest.

80. *Sporobolus cryptandrus* (Torr.) A. Gray (Fig. 102).

Synonyms: *Agrostis cryptandra* Torr., *Sporobolus cryptandrus* (Torr.) A. Gray subsp. *fuscicola* (Hook.) E. K. Jones & Fassett, *Sporobolus cryptandrus* (Torr.) A. Gray var. *fuscicola* (Hook.) Pohl, *Sporobolus cryptandrus* (Torr.) A. Gray var. *occidentalis* E. K. Jones & Fassett. **Vernacular Name:** Sand dropseed. **Life Span:** Perennial. **Origin:** Native. **Season:** C_4, warm season. **FLORAL CHARACTERISTICS: Inflorescence:** contracted to open panicle, 15–40 cm long, 2–12 cm wide; lower panicle nodes with 1–2 and rarely 3 branches; lower branches usually included in subtending leaf sheath; secondary branches appressed; pedicels 0.1–1.3 mm long, appressed. **Spikelets:** glabrous, lead gray to purplish, 1.5–2.5 mm long, 1-flowered, densely crowded on distal portion of panicle branches, imbricate; disarticulation above and sometimes below the glumes. **Glumes:** unequal, ovate to linear-lanceolate, membranous, apex acute; lower 0.6–1.1 mm long, nerveless to 1-nerved; upper 1.5–2.7 mm long, 1-nerved, thin. **Lemmas:** ovate to lanceolate, 1.4–2.5 mm long, glabrous, membranous, 1-nerved. **Paleas:** membranous, 1.2–2.4 mm long, lanceolate. **Awns:** none. **VEGETATIVE CHARACTERISTICS: Growth**

Figure 101. *Sporobolus contractus.*

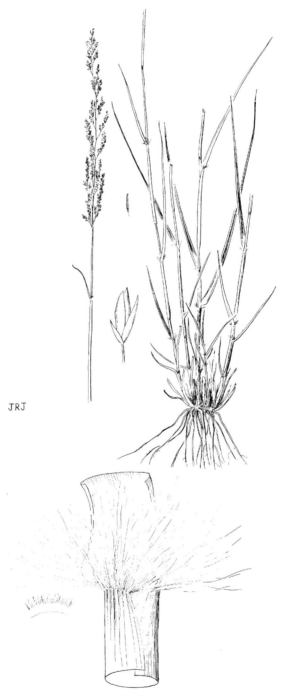

JRJ

Figure 102. *Sporobolus cryptandrus.*

Habit: cespitose. **Culms:** erect to decumbent, at times geniculate below, 3–10 cm tall, glabrous, solid. **Sheaths:** open, strongly overlapping, rounded below, glabrous to scabrous, margins ciliate distally, apex with conspicuous tufts of hairs to 4 mm long. **Ligules:** line of dense hairs, 0.5–1 mm long, rounded to truncate. **Blades:** flat to involute at drying, tending to be more involute nearer the slender tip, 5–26 cm long, 2–6 mm wide, margins scabrous; flag leaf blade nearly perpendicular to culm. **HABITAT:** Found on roadsides and in dry areas from low elevations to the lower foothills over Colorado. **COMMENTS:** Very common species now frequently used in seed mixtures for roadside stabilization and on restoration sites.

81. *Sporobolus flexuosus* (Thurb. *ex* Vasey) Rydb. (Fig. 103).

Synonym: *Sporobolus cryptandrus* (Torr.) A. Gray var. *flexuosus* Thurb. **Vernacular Name:** Mesa dropseed. **Life Span:** Perennial. **Origin:** Native. **Season:** C$_4$, warm season. **FLORAL CHARACTERISTICS: Inflorescence:** open panicle, 10–30 cm long, 4–12 cm wide, oblong to subovate, lower portion often included in uppermost leaf sheath; branches 1–2 at lower nodes, flexuous, tangled, pulvini pubescent. **Spikelets:** grayish, loosely arranged on branches, 1.8–2.5 mm long, 1-flowered; disarticulation above and sometimes below the glumes (glumes sometimes early deciduous). **Glumes:** unequal, 1-nerved, ovate, glabrous, membranous, keel scabrous; lower 0.9–1.5 mm long; upper 1.4–2.5 mm long. **Lemmas:** ovate to lanceolate, membranous, 1.4–2.5 mm long, glabrous, apex acute. **Paleas:** glabrous, 1.4–2.4 mm long, membranous. **Awns:** none. **VEGETATIVE CHARACTERISTICS: Growth Habit:** cespitose. **Culms:** decumbent to erect, 3–10 cm tall, slender, terete, solid. **Sheaths:** strongly overlapping, glabrous to scabrous, striate, distal margins and apex with conspicuous tufts of long white hairs, to 4 mm long. **Ligules:** dense ring of hairs with membranous base, 0.5–1 mm long. **Blades:** flat to involute, 5–24 cm long, 2–4 mm wide, glabrous below, occasionally scabrous above. **HABITAT:** Found on rimrock and in piñon-juniper communities in western Colorado and in Cheyenne County on the Eastern Slope (Weber and Wittmann 2001b). **COMMENTS:** *S. flexuosus* is an important grass in the stabilization of loose, sandy soils. It has the ability to survive in areas with low mean annual precipitation, and it reseeds easily (Barkworth et al. 2003).

82. *Sporobolus giganteus* Nash (Fig. 104).

Synonyms: none. **Vernacular Name:** Giant dropseed. **Life Span:** Perennial. **Origin:** Native. **Season:** C$_4$, warm season. **FLORAL CHARACTERISTICS: Inflorescence:** contracted spikelike panicle, occasionally opening slightly, 25–75 cm long, 1–4 cm wide, base usually included in subtending leaf sheath; branches 1–2 and occasionally 3 at lower nodes, appressed to slightly spreading, with spikelets to base; pulvini gla-

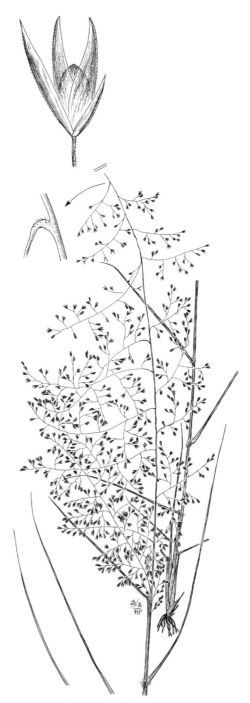

Figure 103. *Sporobolus flexuosus.*

Figure 104. *Sporobolus giganteus.*

brous; pedicels 0.5–2 mm long, appressed. **Spikelets:** pale to greenish gray, compressed, 2.5–3.5 mm long, 1-flowered; disarticulating above and eventually below the glumes. **Glumes:** unequal, 1-nerved, membranous, narrowly lanceolate, keel scabrous; lower 0.6–2 mm long; upper 2–3.5 mm long. **Lemmas:** linear-lanceolate, glabrous, 2.5–3.5 mm long, membranous, apex acute. **Paleas:** linear-lanceolate, glabrous, 2.4–3.4 mm long, membranous, apex acute. **Awns:** none. **VEGETATIVE CHARACTERISTICS: Growth Habit:** cespitose. **Culms:** erect, 10–20 dm tall, robust, glabrous, solid, 4–10 mm thick near base. **Sheaths:** strongly overlapping, striate, rounded below, margins with long, pilose hairs, apex with conspicuous tufts of hair to 2 (5) mm long that occasionally continues as a line across the collar behind ligule. **Ligules:** dense ring of hairs, 0.5–1.5 mm long. **Blades:** flat, becoming involute at tip, 10–50 cm long, 4–10 mm wide, glabrous, margins whitish, scabrous, blade sometimes early deciduous at sheath summit. **HABITAT:** Found on sandy banks of the San Juan River on the Western Slope and scattered on sand dunes and sandy areas on the Eastern Slope. **COMMENTS:** As the name implies, *S. giganteus* is one of the largest members of the genus, often with culms over 1 m tall and 1 cm thick.

83. *Sporobolus heterolepis* (A. Gray) A. Gray (Fig. 105).

Synonym: *Vilfa heterolepis* A. Gray. **Vernacular Names:** Prairie dropseed, northern dropseed. **Life Span:** Perennial. **Origin:** Native. **Season:** C$_4$, warm season. **FLORAL CHARACTERISTICS: Inflorescence:** open panicle, 5–22 cm long, 6–11 cm wide; lower nodes 1- to 2- and rarely 3-branched; branches usually naked below, few-flowered above, generally exserted from uppermost leaf sheath. **Spikelets:** gray, 3–6 mm long, membranous, 1-flowered; disarticulation above the glumes. **Glumes:** unequal, 1-nerved, lanceolate, keel glabrous to scabrous; lower 1.8–4.5 mm long; upper 2.4–6 mm long, generally longer than lemma, occasionally 3-nerved. **Lemmas:** ovate, glabrous, 3–4.3 mm long, apex acute. **Paleas:** ovate, 3.1–4.5 mm long, glabrous. **Awns:** none. **VEGETATIVE CHARACTERISTICS: Growth Habit:** cespitose, leaves mostly basal. **Culms:** erect, 3–8 dm tall, slender, glabrous. **Sheaths:** fibrous basally, glabrous to sparsely pilose below, contorted. **Ligules:** ring of hairs, 0.1–0.3 mm long. **Blades:** flat to folded to becoming involute near tip, 7–31 cm long, 1.2–2.5 mm wide, upper surface scabrous, lower surface glabrous, margin and midnerve scabrous. **HABITAT:** Found on the plains and tallgrass prairie remnants up to the foothills on the Eastern Slope. **COMMENTS:** Infrequent in the wild but often used as an ornamental.

84. *Sporobolus nealleyi* Vasey (Fig. 106).

Synonyms: none. **Vernacular Names:** Gyp dropseed, gypgrass. **Life Span:** Perennial. **Origin:** Native. **Season:** C$_4$, warm season. **FLORAL CHARACTERISTICS: Inflorescence:** panicle, open when mature, 3–10 cm long, 1–5 cm wide, delicate, few-flowered, base often included in uppermost leaf sheath; lower nodes 1- to 2-branched; lateral pedicels 4 mm or less long. **Spikelets:** purplish, 1.4–2.1 mm long, membranous, 1-flowered; disarticulation above the glumes. **Glumes:** unequal, ovate to linear-lanceolate; lower 0.5–1.1 mm long;

Figure 105. *Sporobolus heterolepis*.

Figure 106. *Sporobolus nealleyi.*

upper 1.3–2 mm long. **Lemmas:** ovate, glabrous, 1.4–2.1 mm long, apex acute. **Awns:** none. **VEGETATIVE CHARACTERISTICS: Growth Habit:** cespitose, with hard, knotty base. **Culms:** erect, slender, 10–50 cm tall, 0.7–1.2 mm thick near base. **Sheaths:** generally villous to tomentose along margins and back, with conspicuous tufts of hair at apex, hairs to 4 mm long. **Ligules:** ring of hairs, 0.2–0.4 mm long. **Blades:** involute, 1.5–6 cm long, 1–1.5 mm wide, stiff, spread-

ing at right angles to sheath, above scabrous, below glabrous, margins smooth. **HABITAT:** Found on gypsum flats on the Eastern Slope and in San Miguel County on the Western Slope. **COMMENTS:** This species is only found on sites derived from gypsum.

85. *Sporobolus neglectus* Nash (Fig. 107).

Synonym: *Sporobolus vaginiflorus* (Torr. *ex* Gray) Wood var. *neglectus* (Nash) Scribn. **Vernacular Names:** Puffsheath dropseed, small dropseed. **Life Span:** Annual. **Origin:** Native. **Season:** C$_4$, warm season. **FLORAL CHARACTERISTICS: Inflorescence:** contracted, cylindrical terminal and axillary panicles, 2–5 cm long, 0.2–0.5 cm wide, included in leaf sheaths; branches at lower nodes 1–3, appressed, bearing spikelets to base. **Spikelets:** pale to purple-tinged, 1.6–3 mm

long, plump, glabrous, 1-flowered; disarticulation above the glumes. **Glumes:** subequal, ovate to lanceolate, white-scarious with 1 greenish nerve; lower 1.5–2.4 mm long; upper 1.7–2.7 mm long. **Lemmas:** ovate, 1.6–2.9 mm long, chartaceous. **Paleas:** 1.6–3 mm long, ovate, chartaceous. **Awns:** none. **VEGETATIVE CHARACTERISTICS: Growth Habit:** tufted. **Culms:** decumbent to erect, delicate, slender, 1–5 dm tall, wiry, freely branching. **Sheaths:** generally glabrous, but apex with small hair tufts, inflated. **Ligules:** crown of short hairs, 0.1–0.3 mm long. **Blades:** flat to loosely involute, 1–12 cm long, 0.6–2 mm wide, leaf base occasionally with papillose-based hairs, upper surface scabrous, lower surface glabrous, margin smooth to scabrous. **HABITAT:** Usually found in moist depressions. **COMMENTS:** Scattered, but infrequent in our area.

86. *Sporobolus pyramidatus* (Lam.) Hitchc. (Fig. 108).

Synonyms: *Agrostis pyramidata* Lam., *Sporobolus argutus* (Nees) Kunth, *Sporobolus coromandelianus* (Retz.) Kunth, *Sporobolus patens* Swallen, *Sporobolus pulvinatus* Swallen. **Vernacular Names:** Whorled dropseed, Madagascar dropseed. **Life Span:** Annual to short-lived perennial. **Origin:** Native. **Season:** C$_4$, warm season. **FLORAL CHARACTERISTICS: Inflorescence:** panicle, contracted to open at maturity, 4–15 cm long, 0.3–6 cm wide, shape pyramidal; branch-

Figure 107. *Sporobolus neglectus.*

ing 7–12 at lower nodes, appearing distinctly whorled. **Spikelets:** gray to brownish, 1.2–1.8 mm long, glabrous, membranous, 1-flowered; disarticulation above the glumes. **Glumes:** unequal, ovate to obovate; lower 0.3–0.7 mm long, nerveless; upper 1.2–1.8 mm long, thin, 1-nerved. **Lemmas:** ovate to elliptic, 1.2–1.7 mm long, apex acute. **Paleas:** 1.1–1.6 mm long, elliptic to ovate. **Awns:** none. **VEGETATIVE CHARACTERISTICS: Growth Habit:** tufted to cespitose, leaves basal to low cauline. **Culms:** decumbent to erect, 7–35 cm tall, glabrous. **Sheaths:** generally glabrous, margins and apex hairy with hairs to 3 mm long. **Ligules:** fringe of white hairs, 0.3–1 mm long. **Blades:** flat, 2–12 cm long, 2–6 mm wide; lower surface glabrous, upper surface scabridulous to occasionally sparsely hispid. **HABITAT:** Found in disturbed soils, along roadsides and railroad tracks, and on alluvial slopes in many plant communities (Barkworth et al. 2003). **COMMENTS:** Some botanists consider this plant to be *S. coromandelianus* (Retz.) Kunth, a species from the Eastern Hemisphere.

Figure 108. *Sporobolus pyramidatus*.

87. *Sporobolus texanus* Vasey (Fig. 109).

Synonyms: none. **Vernacular Name:** Texas dropseed. **Life Span:** Perennial. **Origin:** Native. **Season:** C_4, warm season. **FLORAL CHARACTERISTICS: Inflorescence:** open, diffuse panicle, 10–35 cm long, 4.5–30 cm wide, subpyramidal in shape, partially included in uppermost leaf sheath; pedicels 6–25 mm long, capillary, spreading; inflorescence breaking off at maturity. **Spikelets:** purple-tinged, 2.3–3 mm long, membranous, glabrous, 1-flowered; disarticulation above the glumes. **Glumes:** unequal, lanceolate to linear-lanceolate; lower glume narrow, 0.5–1.7 mm long, apex acute, often nerveless; upper 1.7–3 mm long, 1-nerved, apex acute to acuminate. **Lemmas:** ovate to lanceolate, 1.8–3 mm long, apex acute. **Paleas:** ovate, 1.7–2.9 mm long. **Awns:** none. **VEGETATIVE CHARACTERISTICS: Growth Habit:** cespitose with fibrous roots. **Culms:** decumbent to spreading to erect, 2–7 dm tall, scabrous to glabrous below. **Sheaths:** glabrous to sparsely hairy, apex glabrous to occasionally with scattered, appressed, papillose-based hairs. **Ligules:** a ring of hairs 0.2–0.6 mm long. **Blades:** flat to involute on drying, 2.5–13 cm long, 1–4.2 mm wide, upper surface scabrous, lower surface glabrous, margins scabrous and sometimes with a few papillose-based hairs. **HABITAT:** Found in the lower Arkansas Valley and on alkali flats in eastern Colorado. **COMMENTS:** Infrequent.

27. *Tridens* Roem. & Schult.

Perennial, cespitose or infrequently rhizomatous. Sheaths open; ligules a ciliate fringe of hairs; auricles absent or occasionally with short, membranous lateral auricles. Inflorescence an open or closed panicle. Disarticulation above the glumes and between florets. Spikelets with 3–9 florets, florets strongly imbricate. Glumes 2, subequal, lower 1-nerved, upper 1- to 3-nerved. Lemma broad, thin, 3-nerved, short-hairy on nerves below, rounded at back, mostly bidentate at apex, midnerve and often lateral nerves extended as minute mucros. Palea slightly shorter than lemma. Basic chromosome number, $x = 10$. Photosynthetic pathway, C_4.

Represented in Colorado by a single species and a single variety.

88. *Tridens muticus* (Torr.) Nash (Fig. 110).

Synonyms: *Tridens elongatus* (Buckley) Nash, *Triodia elongata* (Buckley) Scribn. **Vernacular Name:** Slim tridens. **Life Span:** Perennial. **Origin:** Native. **Season:** C_4, warm season. **FLORAL CHARACTERISTICS: Inflorescence:** narrow panicle, occasionally appearing racemose, 7–20 cm long, 0.3–0.8 cm wide; branches erect. **Spikelets:** imbricate, terete, purple-tinged, 8–13 mm long, 5- to 11-flowered; disarticulation above the glumes. **Glumes:** glabrous, hyaline, thin; lower 3–8 mm long, ovate, 1- to 3-nerved; upper 5.5–10 mm long, 3- to 7 nerved,

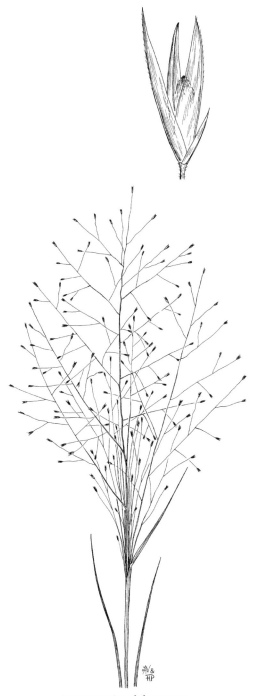

Figure 109. *Sporobolus texanus.*

Figure 110. *Tridens muticus.*

elliptic-ovate. **Lemmas:** hyaline, broadly obovate, 3.5–7 mm long, 3-nerved, nerves pilose below, apex broadly rounded to obtuse, sometimes cleft or mucronate. **Paleas:** shorter and narrower than lemma, with ciliate keels below, pilose along margins below. **Awns:** none. **VEGETATIVE CHARACTERISTICS: Growth Habit:** cespitose, occasionally with very short, knotty rhizomes. **Culms:** slender, erect, 4–8 dm tall, nodes frequently with soft pubescence. **Sheaths:** lower often strigose to pilose, while upper glabrous to scabrous, rounded. **Ligules:** ciliate fringe with a membranous base, 0.5–1 mm long, occasionally with lateral membranous lobes. **Blades:** flat to involute to loosely folded, 3–25 cm long, 3–4 mm wide, glabrous to scabrous to sparsely pilose. **HABITAT:** Grows on well-drained sandy and clayey soils (Barkworth et al. 2003). **COMMENTS:** The only variety reported for Colorado is var. *elongatus* (Buckley) Shinners.

28. *Triplasis* P. Beauv.

Annuals or perennials, cespitose or rarely rhizomatous. Culms with many nodes and short internodes, eventually breaking up at nodes. Sheaths open; ligules a fringe of hairs or a ciliated membrane; auricles absent. Inflorescences a panicle, open, exserted or partially included in upper sheath. Clusters of cleistogamous spikelets regularly borne in axils of upper leaf sheaths. Spikelets of terminal inflorescence with 2–4 florets, those of sheath axils usually with a single floret. Glumes 2, unequal, 1-nerved. Lemmas narrow, 3-nerved, nerves ciliate or densely pubescent, midnerve extended as a mucro or a short awn from a notched apex. Palea strongly 2-nerved and 2-keeled, silky-villous on keels. Basic chromosome number, $x = 10$. Photosynthetic pathway, C_4.

Represented in Colorado by a single species.

89. *Triplasis purpurea* (Walter) Chapm. (Fig. 111).

Synonym: *Triplasis intermedia* Nash. **Vernacular Name:** Purple sandgrass. **Life Span:** Annual. **Origin:** Native. **Season:** C_4, warm season. **FLORAL CHARACTERISTICS: Inflorescence:** panicle, 3–7 cm long, 1–6 cm wide; branches few, small axillary inflorescences often included in leaf sheaths; cleistogamous inflorescences occasionally present. **Spikelets:** laterally compressed, generally purple-tinged, 6.5–9 mm long, 3- to 4-flowered, 1-nerved, occasionally upper 2-nerved, 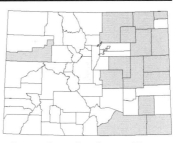 sterile florets above fertile ones, short-pediceled; disarticulation above the glumes. **Glumes:** equal, to 2 mm long, glabrous to scabrous, apex erose. **Lemmas:** 3–4 mm long, apex 2-lobed, lobes less than 1 mm long, rounded. **Paleas:** ciliate, to 2.5 mm long. **Awns:** lemma awns less than 2 mm, straight. **VEGETATIVE CHARACTERISTICS: Growth Habit:** tufted. **Culms:** ascending to erect to widely spreading, 1.5–10 dm tall, nodes pubescent, internodes glabrous. **Sheaths:** open, swollen at base, glabrous to pilose near base, occasionally villous near collar. **Ligules:** ciliate, to 1 mm long. **Blades:** flat to involute, 1–8 cm long, 1–5 mm wide, scabrid to hispid, papillose-ciliate near base. **HABITAT:** Found on dry, sandy soils on plains, generally on sand dunes and blowouts, in eastern Colorado (Weber and Wittmann 2001a; Wingate 1994) and in Garfield County. **COMMENTS:** *Triplasis* has a disarticulating

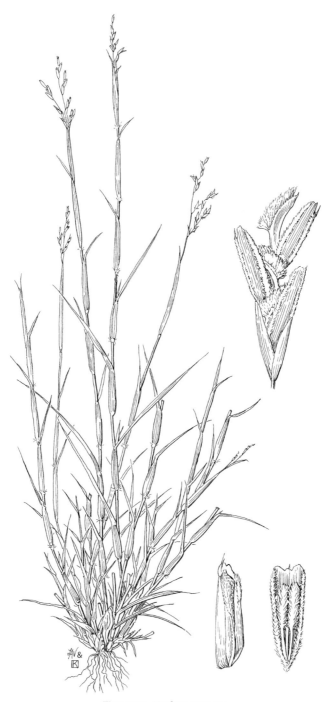

Figure 111. *Triplasis purpurea.*

culm, uncommon among grass genera. This allows the cleistogamous flowers, hidden in the sheaths, to be dispersed as the culm disarticulates.

4. PAPPOPHOREAE

Plants perennial, cespitose. Sheaths open, auricles absent, ligule a ring of hairs. Inflorescence terminal, spicate or an open panicle. Disarticulation above the glumes but not between florets, these falling united. Spikelets with 3–10 florets, sometimes only lowest florets bisexual. Lemmas with 5–13 nerves, nerves extending into awn, often with intermixed hyaline lobes.

29. *Enneapogon* Desv. *ex* P. Beauv.

Low, tufted perennials, with narrow, often spikelike panicles. Ligule a ring of hairs. Spikelets several-flowered, upper florets reduced. Disarticulation above the glumes and tardily between florets. Glumes subequal, lanceolate, with 5 to numerous nerves. Lemmas broad, much shorter than the glumes, firm, rounded on the back, strongly 9-nerved and with 9 equal, plumose awns. Palea slightly longer than body of lemma. Basic chromosome number, $x = 10$. Photosynthetic pathway, C_4.

Represented in Colorado by a single species.

90. *Enneapogon desvauxii* P. Beauv. (Fig. 112).

Synonym: *Pappophorum wrightii* S. Wats. **Vernacular Name:** Nineawn pappusgrass. **Life Span:** Perennial. **Origin:** Native. **Season:** C_4, warm season. **FLORAL CHARACTERISTICS: Inflorescence:** spikelike panicle, narrow, 2–10 cm long, grayish to grayish green. **Spikelets:** 5–7 mm long, 3-flowered, lowest floret fertile, while upper ones sterile; cleistogamous spikelets located in lower sheaths. **Glumes:** subequal to unequal, thin, strongly 3- to 7-nerved near middle, nerves irregular and pilose; lower 3.5–4.7 mm long; upper 5–6 mm long. **Lemmas:** pilose, 1.5–2.5 mm long, strongly 9-nerved, nerves exserted beyond lemma into 9 subequal plumose awns. **Paleas:** equal to longer than lemma, entire, 2-nerved on keels, keels pubescent. **Awns:** lemma awns 2–4.5 mm long, subequal, plumose. **VEGETATIVE CHARACTERISTICS: Growth Habit:** tufted. **Culms:** ascending to erect, branching and geniculate below from knotty base, 1–5 dm tall, finely pilose to villous, densely at nodes. **Sheaths:** generally shorter than internodes, somewhat pubescent, margins membranous that extend upward to form lateral auricles to ligule. **Ligules:** fringe of hairs 0.5–1 mm long. **Blades:** loosely involute, finely pilose to villous, 2–12 mm long, 1–2 mm wide. **HABITAT:** Found on dry soils at low elevations and in the sandy soils of slick-rock areas at Sewemup Mesa (Weber and Wittmann 2001b). Additionally, it is being used on some revegetation sites on the eastern plains (Wingate 1994). **COMMENTS:** Infrequent.

Figure 112. *Enneapogon desvauxii.*

IV. DANTHONIOIDEAE

Plants perennial. Sheaths open; auricles absent; ligule a ring of hairs or membranous and ciliate. Inflorescences terminal, simple panicle, sometimes racemose or spicate, occasionally a single spikelet; disarticulation usually above the glumes and between florets. Spikelets bisexual, with several florets, rachilla extension present. Glumes 2, equal, exceeding uppermost floret; lemmas 3- to 9-nerved, firm, rounded across back, midnerve extending into an awn, base of awn flattened and twisted; paleas well developed, shorter than lemma.

5. DANTHONIEAE

See subfamily description.

30. *Danthonia* DC.

Low to moderately tall cespitose perennials with few-flowered panicles. Inflorescence commonly reduced to 1 spikelet in *Danthonia unispicata*. Spikelets several-flowered; disarticulating above the glumes and between florets. Glumes about equal, 1- to 5-nerved, much longer than lemmas. Lemmas rounded and hairy on back, indistinctly several-nerved, with a well-developed callus and a 2-toothed apex, midnerve diverging as a stout, flat, twisted, geniculate awn at base of apical cleft. Palea broad, well developed. Basic chromosome number, $x = 6$. Photosynthetic pathway, C_3.

Represented in Colorado by 5 species.

1. Inflorescence of 1 (rarely 2–3) spikelets . . . . 95. *D. unispicata*
1. Inflorescence of 2 or more spikelets
  2. Inflorescence open, at least lower branches spreading or reflexed . . . . . 91. *D. californica*
  2. Inflorescence closed with erect branches
    3. Lemmas glabrous on back, pilose on margins . . . . .92. *D. intermedia*
    3. Lemmas pilose on back
      4. Glumes 15 mm or longer; lemmas 10 mm or longer . . . . 93. *D. parryi*
      4. Glumes 10–12 mm long; lemmas 4–5 mm long . . . . 94. *D. spicata*

91. *Danthonia californica* Bol. (Fig. 113).

Synonyms: *Danthonia americana* Scribn., *Danthonia californica* Boland. var. *americana* (Scribn.) Hitchc., *Danthonia californica* Boland. var. *palousensis* St. John, *Danthonia californica* Boland. var. *piperi* St. John. **Vernacular Name:** California oatgrass. **Life Span:** Perennial. **Origin:** Native. **Season:** C$_3$, cool season. **FLORAL CHARACTERISTICS: Inflorescence:** racemose panicle to raceme, 2–7 cm long; spikelets 3–6 per panicle; branches widely spreading, flexible, with pulvini at base. **Spikelets:** generally deep purple, 14–26 mm long, 3- to 8-flowered; disarticulation above the glumes; cleistogamous spikelets may be present in mid-culm sheaths. **Glumes:** subequal, generally exceeding florets, 14–18 mm long, glabrous to scaberulous, 1- to 7-nerved, keeled. **Lemmas:** glabrous to sparsely pilose, 5–10 mm long, 9- to 11-nerved with lateral nerves usually obscure, margins usually pubescent, apex bifid, teeth 4–6 mm long, aristate. **Paleas:** cleft, 2-nerved, keel ciliate. **Awns:** lemma awn arising from between teeth, geniculate, 8–12 mm long, lower portion flattened and twisted. **VEGETATIVE CHARACTERISTICS: Growth Habit:** tufted, sheaths persistent at base. **Culms:** erect to somewhat geniculate at base, robust, slender, 3–13 dm tall, glabrous, disarticulating at nodes at maturity. **Sheaths:** open, glabrous to pilose, apex slightly pilose on throat. **Ligules:** irregularly lacerate and ciliolate to a crown of hairs, 0.5–1 mm long. **Blades:** flat to rolled to loosely involute, 10–30 cm long, 2–5 mm wide, strongly striate, glabrous to pilose. **HABITAT:** Scattered in prairies, meadows, and open woods (Barkworth et al. 2003). **COMMENTS:** Widespread but infrequent, mostly on the Western Slope.

92. *Danthonia intermedia* Vasey (Fig. 114).

Synonyms: *Danthonia canadensis* Baum & Findlay, *Danthonia intermedia* Vasey var. *cusickii* Williams. **Vernacular Name:** Timber oatgrass. **Life Span:** Perennial. **Origin:** Native. **Season:** C$_3$, cool season. **FLORAL CHARACTERISTICS: Inflorescence:** narrow panicle to raceme with 5–10 spikelets, often secund, racemose; branches erect, appressed. **Spikelets:** generally deep purple, 11–15 mm long, 3- to 6-flowered; disarticulation above the glumes. **Glumes:** subequal,

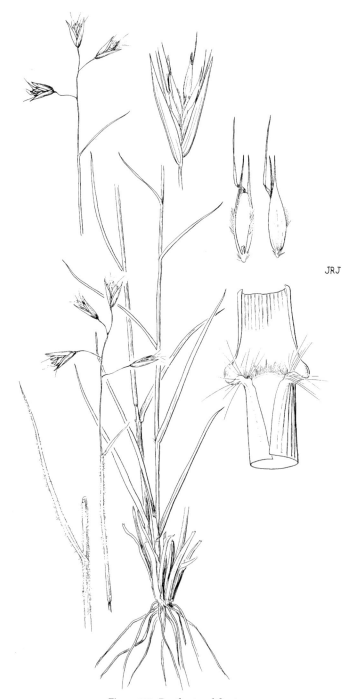

JRJ

Figure 113. *Danthonia californica.*

12–14 mm long, glabrous to sparsely scaberulous, irregularly 3- to 5-nerved, lateral nerves obscure. **Lemmas:** pilose on or near margins below, glabrous on back, 3–6 mm long, 7- to 9-nerved, apex bifid, teeth 1.5–2.5 mm long. **Paleas:** slightly shorter than lemma, broadly lanceolate, keels ciliolate near margins. **Awns:** lemma awn arising from between teeth, 6.5–8 mm long, lower portion flattened and twisted. **VEGETATIVE CHARACTERISTICS: Growth Habit:** densely tufted, leaves generally basal with some persistent sheaths. **Culms:** erect, 1–5 dm tall, glabrous. **Sheaths:** generally glabrous, pilose at apex near throat. **Ligules:** short, ciliolate crown, 0.1–0.5 mm long. **Blades:** flat to involute, 5–10 cm long, 1–3.5 mm wide, glabrous to slightly pilose. **HABITAT:** Typically found in alpine and subalpine grasslands and meadows, open woods, and rocky slopes (Barkworth et al. 2003). **COMMENTS:** Cleistogamous spikelets occasionally produced.

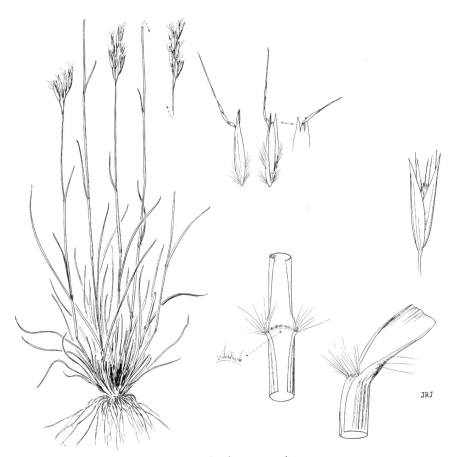

Figure 114. *Danthonia intermedia.*

93. *Danthonia parryi* Scribn. (Fig. 115).

Synonyms: None. **Vernacular Name:** Parry's oat-grass. **Life Span:** Perennial. **Origin:** Native. **Season:** C$_3$, cool season. **FLORAL CHARACTERISTICS: Inflorescence:** panicle, rarely a raceme, 3–7 cm long, with 4–11 spikelets; branches appressed to ascending, appressed pubescent. **Spikelets:** 16–24 mm long, 5- to 7-flowered; disarticulation above the glumes; cleistogamous spikelets may be present in mid-culm sheaths. **Glumes:** equal to subequal, 16–23 mm long, 5-nerved, but lateral nerves obscure. **Lemmas:** backs generally pilose, more so below; 10–15 mm long, 11-nerved, apex deeply bifid with acuminate teeth 2.5–8 mm long, aristate. **Paleas:** subequal to lemma, narrowed above, keel ciliolate. **Awns:** lemma awn arising from between teeth, geniculate, 12–15 mm long, lower portion flattened and twisted, scabrous. **VEGETATIVE CHARACTERISTICS: Growth Habit:** tufted, sheaths persistent. **Culms:** erect, glabrous, 3–8 dm tall, stout. **Sheaths:** glabrous to sparsely pubescent, round, apex pilose at throat. **Ligules:** ciliate membrane, 0.3–0.7 mm long. **Blades:** flat to involute, 15–25 cm long, 1–4 mm wide, glabrous to scabrous to rarely pilose, erect to slightly diverging. **HABITAT:** Found on dry, gravelly hillsides and forested areas in montane to subalpine communities and on buttes in the plains. **COMMENTS:** A frequent codominant in many mid-elevation montane grasslands; particularly abundant in open ponderosa pine communities.

94. *Danthonia spicata* (L.) P. Beauv. *ex* Roem. & Schult. (Fig. 116).

Synonyms: *Danthonia spicata* (L.) P. Beauv. *ex* Roem. & Schult. var. *longipila* Scribn. & Merr., *Danthonia spicata* (L.) P. Beauv. *ex* Roem. & Schult. var. *pinetorum* Piper, *Danthonia thermalis* Scribn. **Vernacular Names:** Poverty oatgrass, poverty danthonia, poverty grass. **Life Span:** Perennial. **Origin:** Native. **Season:** C$_3$, cool season. **FLORAL CHARACTERISTICS: Inflorescence:** contracted panicle with 5–10 spikelets. **Spikelets:** 7–15 mm long, 4- to 6-flowered; disarticulation above the glumes; cleistogamous spikelets may be present in mid-culm sheaths. **Glumes:** subequal, 10–12 mm long, apex acuminate, chartaceous, particularly on margins. **Lemmas:** sparsely villous to pilose on back, margins pilose below, 4–5 mm long, apex bifid with acute to aristate teeth 0.5–2 mm long. **Paleas:** flat, broad, apex obtuse, shorter than lemma. **Awns:** lemma awn arising from between teeth, geniculate, 5–8 mm long, lower portion flattened and twisted. **VEGETATIVE CHARACTERISTICS: Growth Habit:** tufted. **Culms:** slender, 1–7 dm tall, disarticulating at nodes at maturity. **Sheaths:** glabrous to pilose, apex with tuft of hairs at throat on lower sheaths. **Ligules:** ciliate membrane, to 0.5 mm long. **Blades:** flat to involute, 6–15 cm long, 0.8–3 mm wide, glabrous to sparsely pilose, generally involute and curling at maturity. **HABITAT:** Found on eastern plains, on outwashes, mesas, and open pine forests in the foothills. **COMMENTS:** Locally abundant.

Figure 115. *Danthonia parryi.*

Figure 116. *Danthonia spicata*.

95. *Danthonia unispicata* (Thurb.) Munro *ex* Vasey (Fig. 117).

Synonym: *Danthonia californica* Bol. var. *unispicata* Thurb. **Vernacular Names:** Onespike danthonia, onespike oatgrass. **Life Span:** Perennial. **Origin:** Native. **Season:** C$_3$, cool season. **FLORAL CHARAC-TERISTICS: Inflorescence:** raceme to panicle with 1–2 spikelets. **Spikelets:** purple-tinged, 12–26 mm long, 3- to 6-flowered; disarticulation above the glumes. **Glumes:** subequal, glabrous, narrow-lanceolate, 14–20 mm long; upper shorter, 5-nerved. **Lemmas:** glabrous to rarely with few spreading hairs over back, 5.5–11 mm long, margins generally pilose, apex bifid with acute to aristate teeth 1.5–7 mm long. **Paleas:** slightly shorter than lemma, keel ciliate. **Awns:** lemma awn arising from between teeth, geniculate, 5.5–13 mm long, lower portion flattened and twisted. **VEGETATIVE CHARACTERISTICS: Growth Habit:** tufted to matted. **Culms:** glabrous, 1.5–3 dm tall, disarticulating at nodes at maturity. **Sheaths:** generally densely pilose, hairs occasionally with papillose bases, at least on lower nodes, apex with dense tuft of spreading hairs on collar, light yellow in color. **Ligules:** irregularly lacerate to ciliate, to 0.5 mm long. **Blades:** flat to loosely involute, 3–8 cm long, 1–3 mm wide, strongly striate, pilose, rarely glabrous. **HABITAT:** Found on sandstone benches in Moffat and Routt counties (Weber and Wittmann 2001b; Wingate 1994). **COMMENTS:** Unusual among the grasses, the inflorescence on this species is often reduced to a single spikelet.

V. EHRHARTOIDEAE

Annuals or perennials (ours), often with rhizomes, usually of wet or moist areas. Sheath open; auricles sometimes present; ligule membranous. Inflorescence terminal, an open panicle. Disarticulation at base of spikelet. Spikelet strongly laterally compressed, a single floret, often subtended by 2 scales or bristles that represent reduced florets, often unisexual. Glumes absent or reduced to a small cuplike structure at top of pedicel; lemma strongly laterally compressed, 5- or more-nerved, awnless. Palea more than 2-nerved.

6. ORYZEAE

See subfamily description.

31. *Leersia* Sweet

Rhizomatous perennials with flat blades and open panicles. Ligule a short, firm membrane, often continued laterally as short sheath auricles. Spikelets 1-flowered, strongly compressed laterally, awnless, subsessile and crowded at branch tips. Disarticulation below spikelet. Glumes absent. Lemma firm or indurate, boat-shaped, 5-nerved, tightly enclosing the margins of a firm, narrow, usually 3-nerved palea. Stamen number varying from 1 to 6. Basic chromosome number, $x = 12$. Photosynthetic pathway, C$_3$.

Represented in Colorado by a single species.

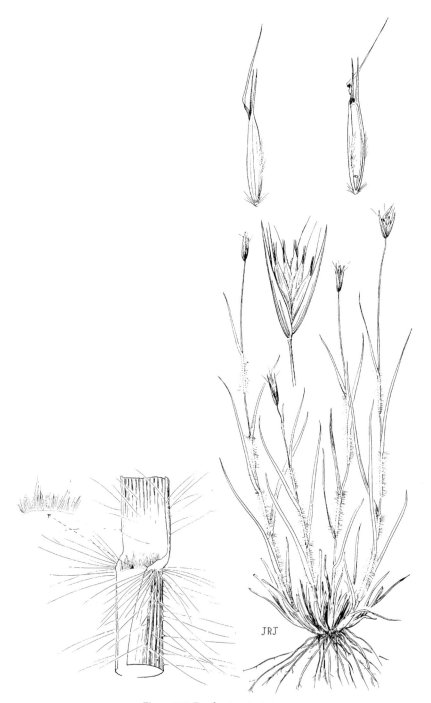

Figure 117. *Danthonia unispicata*.

96. *Leersia oryzoides* (L.) Sweet (Fig. 118).

Synonyms: *Homalocenchrus oryzoides* (L.) Pollich, *Phalaris oryzoides* L. **Vernacular Name:** Rice cutgrass. **Life Span:** Perennial. **Origin:** Native. **Season:** C$_3$, cool season. **FLORAL CHARACTERISTICS: Inflorescence:** open panicle, 10–20 cm long, nodding to erect, cleistogamous spikelets on axillary panicles often enclosed in sheaths. **Spikelets:** flat, 1-flowered, 1.5–2 mm long; disarticulation below spikelet. **Glumes:** lacking. **Lemmas:** strongly compressed,

keel and marginal nerves hispid-ciliate, 4–5 mm long, internerve surface hispid. **Paleas:** equal to or longer than lemma, lanceolate, keel stiffly ciliate. **Awns:** none. **VEGETATIVE CHARACTERISTICS: Growth Habit:** loosely tufted to solitary, rhizomatous. **Culms:** weakly decumbent, slender, 5–15 dm tall, simple to branched above, nodes pubescent. **Sheaths:** glabrous to retrorsely scabrous, striate, generally puberulent at summit. **Ligules:** firm,

Figure 118. *Leersia oryzoides.*

217

minutely erose-ciliolate, truncate, to 1 mm long. **Blades:** flat to folded, up to 30 cm long, 6–15 mm wide, surfaces strongly retrorsely scabrous, margins scabrous. **HABITAT:** Found in wet areas, along streams, and in standing water. **COMMENTS:** This species supplies food for waterfowl. The leaf margins can cause cuts.

VI. PANICOIDEAE

Annuals or perennials, cespitose or occasionally rhizomatous and more rarely stoloniferous. Sheaths open; auricles absent; ligule membranous and ciliate, a ring or hairs, or rarely absent. Inflorescence bractless (Paniceae) or with bracts (Andropogoneae), panicles, racemes, spikes, or a complex series of rames (Andropogoneae); usually bisexual; disarticulation usually below the glumes or in the upper axes. Spikelets bisexual, rarely unisexual, paired, or in triplets. Glumes usually 2, equal or subequal, shorter or greater than fertile floret. Spikelets with 2 florets, the lower sterile or represented only by a sterile lemma, the upper fertile; lemmas often terminally awned. Palea often as long as fertile lemma.

7. ANDROPOGONEAE

Plants perennial (in ours), cespitose, occasionally rhizomatous. Sheaths open; auricles absent or sometimes present in *Sorghastrum;* ligule membranous, ciliate or not. Inflorescence terminal and axillary, usually of 1 to many spicate branches, these usually in digitate clusters on a peduncle, often enclosed by a subtending leaf sheath; disarticulation usually in branch axes beneath sessile florets; the dispersal unit consists of a sessile floret, internode to the next sessile floret, pedicel, and pedicellate spikelet (rames). Spikelets usually in pairs, one sessile, the other pedicellate. Glumes usually 2, enclosing florets, tougher than lemmas, rounded or dorsally compressed; lower floret usually sterile and reduced to a scale, upper floret bisexual or pistillate, lemmas often hyaline and extending into an awn.

32. *Andropogon* L.

Cespitose perennials with usually stiffly erect culms, rounded or flattened and keeled sheaths, and flat or folded blades. Ligule membranous. Flowering culm much branched and "broomlike" in some species, unbranched or little branched above the base in others. Each culm or branch terminated by an inflorescence of 2 to several racemose branches. Sessile spikelet of a pair fertile. Pediceled spikelet well developed, rudimentary, or absent. Disarticulation in the rachis, sessile spikelets falling attached to the associated pedicel and section of rachis. Glumes large, firm, awnless. Lemmas of sterile and fertile florets membranous, lemma of fertile floret awned or awnless. Basic chromosome numbers, $x = 9$ and 10. Photosynthetic pathway, C_4.

Represented in Colorado by 2 species.

1. Sessile spikelets with awns 8–25 mm long; ligules 0.4–2.5 mm long; hairs of rame internodes 2.2–4.2 mm long; rhizomes sometimes present, internodes usually less than 2 cm . . . . 97. *A. gerardii*
1. Sessile spikelets awnless or with an awn less than 11 mm long; ligules 2.5–4.5 mm long; hairs of rame internodes 3.7–6.6 mm long; rhizomes always present, internodes usually more than 2 cm long . . . . 98. *A. hallii*

97. *Andropogon gerardii* Vitman (Fig. 119).

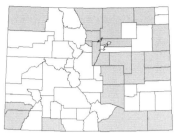

Synonyms: *Andropogon chrysocomus* Nash, *Andropogon furcatus* Muhl. *ex* Willd., *Andropogon gerardii* Vitman var. *chrysocomus* (Nash) Fern., *Andropogon provincialis* Lam. **Vernacular Name:** Big bluestem. **Life Span:** Perennial. **Origin:** Native. **Season:** C$_4$, warm season. **FLORAL CHARACTERISTICS: Inflorescence:** panicle compound, both terminal and axillary, 2–6 digitate, spicate branches; branches 4–10 cm long, internodes sparsely to densely pubescent with whitish to grayish or rarely yellowish hairs 2.2–4.2 mm long. **Spikelets:** in pairs of 1 sessile (6–11 mm long) and 1 pediceled (6–10 mm long); disarticulating as a pair with a portion of rachis; fertile spikelet 2-flowered; lower floret reduced; upper floret perfect. **Glumes:** glaucous, nerves scabrous above; of sessile spikelet subequal, margins of lower enrolled and clasping the upper; lower 4- to 6-nerved; upper 3-nerved; of pedicellate spikelet subequal, lower 9-nerved; upper 3-nerved. **Lemmas:** of sessile spikelet membranous, transparent, deeply cleft, awned. **Awns:** fertile lemma of lower (sessile) spikelet 8–25 mm long, geniculate, tightly twisted below, loosely twisted above. **VEGETATIVE CHARACTERISTICS: Growth Habit:** clumped, appearing cespitose occasionally with many short rhizomes; when present internode is usually less than 2 cm long. **Culms:** erect, 0.5–1.5 m long, stout, glaucous. **Sheaths:** open, compressed, occasionally purplish, glabrous to villous or papillose-pilose, basal sheaths occasionally sparsely hairy. **Ligules:** membranous, 0.4–2.5 mm long, ciliate. **Blades:** flat, 3–10 mm wide, 5–50 cm long, sometimes pilose near throat, occasionally lower ones villous. **HABITAT:** Plains, mesas, and foothills; a dominant species of heavier soils in the Great Plains region. **COMMENTS:** Stubbendieck, Hatch, and Landholt (2003) listed this species as excellent and palatable forage for all kinds and classes of animals. Some authors combine this taxon with *A. hallii* (Allred 2005), but *A. gerardii* has less-developed rhizomes, grows in heavier soils, and has whitish or grayish hairs on rachis joints and pedicels.

98. *Andropogon hallii* Hack. (Fig. 120).

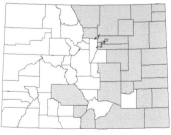

Synonyms: *Andropogon gerardii* Vitman var. *incanescens* (Hack.) Boivin, *Andropogon gerardii* Vitman var. *paucipilus* (Nash) Fern., *Andropogon hallii* Hack. var. *incanescens* Hack., *Andropogon paucipilus* Nash. **Vernacular Names:** Sand bluestem, turkey-foot. **Life Span:** Perennial. **Origin:** Native. **Season:** C$_4$, warm season. **FLORAL CHARACTERISTICS: Inflorescence:** panicle compound, both terminal and axillary, 2–6 digitate, spicate branches; branches 4–10 cm long, internodes mostly pubescent with yellowish or golden hairs 3.7–6.6 mm long. **Spikelets:** in pairs of 1 sessile (7–11 mm long) and 1 pediceled (7–12 mm long); disarticulating as a pair with a portion of rachis; fertile spikelet 2-flowered; lower floret reduced; upper floret perfect. **Glumes:** of sessile spikelet subequal, margins of lower enrolled and clasping the second; lower 4-nerved, grooved; upper 3-nerved, ciliate on margins and keel; of pedicellate

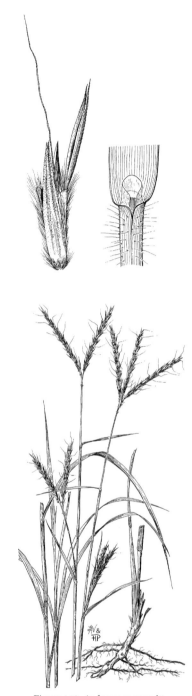

Figure 119. *Andropogon gerardii.*

Figure 120. *Andropogon hallii*.

spikelet subequal; lower 5- to 9-nerved; upper 3-nerved. **Lemmas:** of sessile spikelet membranous, transparent, deeply cleft, sometimes awned. **Awns:** fertile lemma of lower (sessile) spikelet sometimes delicately awned, usually less than 11 mm long. **VEGETATIVE CHARACTERISTICS: Growth Habit:** with well-developed and creeping rhizomes, internodes usually more than 2 cm long. **Culms:** erect, 0.5–2 m tall, glabrous to glaucous. **Sheaths:** open, compressed, smooth, glabrous to glaucous or with short papillose-based hairs, occasionally with auricles. **Ligules:** membranous, puberulent, 2.5–4.5 mm long, ciliate. **Blades:** glaucous, flat or loosely rolled, 3–10 mm wide, 5–50 cm long, scabrous. **HABITAT:** Dry sandy plains or dunes, common in sandy soils throughout the Great Plains. **COMMENTS:** Internode length is not seen very well on those portions of rhizomes near the base of the plant. Some authors include this species with *A. gerardii* (Allred 2005), but *A. hallii* has more extensively developed rhizomes, occurs only in sandy soils, and has yellowish or golden hairs on rachis joints and pedicels.

33. *Bothriochloa* Kuntze

Cespitose perennials with erect or decumbent-spreading culms and flat blades. Inflorescence a terminal panicle, spikelets on few to several spicate primary branches, these sparingly rebranched in a few species. Pedicels and upper rachis internodes with a central groove or membranous area. Sessile spikelets of a pair fertile and awned, more or less triangular in outline, lower glume dorsally flattened, upper glume with a median keel. Pediceled spikelet staminate or sterile, usually well developed. Disarticulation in rachis, sessile spikelet falling attached to a pedicel and section of rachis. Basic chromosome number, $x = 10$. Photosynthetic pathway, C_4.

Represented in Colorado by 5 species and 1 subspecies.

1. Pediceled spikelets smaller and narrower than sessile ones
    2. Sessile spikelets less than 4.5 mm long; awn of lemma less than 18 mm long . . . .
    . . . . 102. *B. laguroides*
    2. Sessile spikelets more than 4.5 mm long; awn of lemma more than 20 mm long
        3. Culm nodes with long, silky, spreading white hairs; panicle axis less than 5 cm long; panicle branches mostly 2–7, rarely more than 8 . . . . 103. *B. springfieldii*
        3. Culm nodes, if bearded, with appressed hairs less than 3 mm long; panicle axis usually more than 5 cm long; panicle branches mostly 9–30 or more . . . . 99. *B. barbinodis*
1. Pediceled spikelets about as large and broad as sessile one
    4. Panicle axis longer than branches, branches numerous . . . . 100. *B. bladhii*
    4. Panicle axis shorter than branches, branches 2–8 . . . . 101. *B. ischaemum*

---

99. *Bothriochloa barbinodis* (Lag.) Herter (Fig. 121).

Synonyms: *Andropogon barbinodis* Lag., *Andropogon perforatus* Trin. *ex* Fourn., *Bothriochloa barbinodis* (Lag.) Herter var. *palmeri* (Hack.) de Wet, *Bothriochloa barbinodis* (Lag.) Herter var. *perforata* (Trin. *ex* Fourn.) Gould, *Bothriochloa palmeri* (Hack.) Gould, **Vernacular Name:** Cane bluestem. **Life Span:** Perennial. **Origin:** Native. **Season:** C_4, warm season. **FLORAL CHARACTERISTICS: Inflorescence:** panicle terminal, 5–14 cm long, of 9–30 clustered racemes (subdigitate), fan-shaped, silvery-white; panicle axis (rachis) 5–10 cm long, straight; branches spicate, 4–9 cm long; pedicels and upper rachis internodes with a translucent

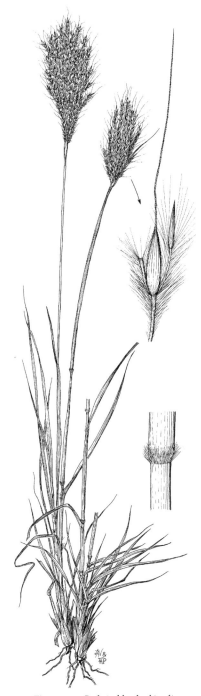

Figure 121. *Bothriochloa barbinodis.*

central groove, margins often villous. **Spikelets:** in pairs of 1 sessile (fertile, 4–7 mm long) and 1 pediceled (sterile, 3–4 mm long); pediceled one rudimentary, narrower than sessile spikelet, often deciduous; disarticulating as a pair with a portion of rachis; fertile spikelet 2-flowered; lower floret reduced; upper floret perfect. **Glumes:** of sessile spikelet subequal, margins of lower enrolled and clasping the upper; lower 7- to 11-nerved, short pilose; upper 3-nerved;

of pedicellate spikelet rudimentary. **Lemmas:** of sessile spikelet membranous, transparent, deeply cleft, awned. **Awns:** fertile lemma of sessile spikelet awned, awn (16–) 20–35 mm long, geniculate, twisted. **VEGETATIVE CHARACTERISTICS: Growth Habit:** tufted. **Culms:** erect to geniculate at base, 6–12 dm tall, generally branched below; nodes with appressed hairs less than 3 mm long, ascending to erect, off-white to tan. **Sheaths:** rounded, open, glabrous. **Ligules:** erose-lacerate membrane, 1–2.5 mm long. **Blades:** flat, 2–7 mm wide, 20–30 cm long, glabrous to scaberulous, pilose at base behind ligule. **HABITAT:** Found in southwestern Colorado on talus slopes in canyons in the Dolores River Valley (Weber and Wittmann 2001b; Wingate 1994) and along roadsides and drainageways (Barkworth et al. 2003). **COMMENTS:** Infrequent in Colorado.

100. *Bothriochloa bladhii* (Retz.) S. T. Blake (Fig. 122).

Synonyms: *Andropogon bladhii* Retz., *Andropogon caucasicus* Trin., *Andropogon intermedius* R. Br., *Bothriochloa caucasica* (Trin.) C. E. Hubbard, *Bothriochloa intermedia* (R. Br.) A. Camus. **Vernacular Names:** Caucasian bluestem, Australian bluestem. **Life Span:** Perennial. **Origin:** Introduced. **Season:** C$_4$, warm season. **FLORAL CHARACTERISTICS: Inflorescence:** panicle, 5–15 cm long, terminal, of numerous clustered racemes (racemose), elliptic to lanceolate, red-

dish at maturity; branches spicate, 3–7 cm long, shorter than panicle axis; pedicels and upper rachis internodes with a translucent central groove, margins with sparse hairs. **Spikelets:** in pairs of 1 sessile (fertile, 3.5–4 mm long) and 1 pediceled (staminate or sterile, same size and shape as sessile ones or to half their length); disarticulating as a pair with a portion of rachis; fertile spikelet 2-flowered; lower floret reduced; upper floret perfect. **Glumes:** of sessile spikelet subequal, margins of lower enrolled and clasping the upper, glabrous to scabrous; upper 3-nerved; of pedicellate spikelet subequal. **Lemmas:** of sessile spikelet membranous, transparent, deeply cleft, awned. **Awns:** fertile lemma of sessile spikelet 10–17 mm long, geniculate, twisted. **VEGETATIVE CHARACTERISTICS: Growth Habit:** cespitose. **Culms:** generally erect, 4–9 dm tall, nodes glabrous to short hispid, generally appressed hairs. **Sheaths:** rounded, open. **Ligules:** membranous, 0.5–1.5 mm long. **Blades:** 20–35 cm long, 1–4 mm wide, generally glabrous. **HABITAT:** Found along roadsides, in waste areas, and in pastures on the eastern plains. **COMMENTS:** Weber and Wittmann (2001a) state that *B. bladhii* has been introduced along roadsides in some of the easternmost counties, but the only documented collections are from El Paso and Kit Carson counties.

Figure 122. *Bothriochloa bladhii*.

101. *Bothriochloa ischaemum* (L.) Keng (Fig. 123).

Synonym: *Dichanthium ischaemum* (L.) Roberty. **Vernacular Name:** Yellow bluestem. **Life Span:** Perennial. **Origin:** Introduced. **Season:** C_4, warm season. **FLORAL CHARACTERISTICS: Inflorescence:** panicle, 3–9 cm long, terminal, of 2–8 clustered racemes (subdigitate), fan-shaped, silvery reddish purple; branches spicate, 3–9 cm long, longer than panicle axis; pedicels and upper rachis internodes with a central groove, margins ciliate (hairs 1–3 mm long).

Spikelets: in pairs of 1 sessile (fertile, 3–4.5 mm long) and 1 pediceled (staminate, 3–4.5 mm long); disarticulating as a pair with a portion of rachis; fertile spikelet 2-flowered; lower floret reduced; upper floret perfect. **Glumes:** of sessile spikelet subequal, margins of lower enrolled and clasping the upper; lower generally strongly 4-nerved, hirsute below; upper 3-nerved; of pedicellate spikelet subequal. **Lemmas:** of sessile spikelet membranous, transparent, deeply cleft, awned. **Awns:** fertile lemma of sessile spikelet 9–17 mm long, geniculate, twisted. **VEGETATIVE CHARACTERISTICS: Growth Habit:** generally cespitose but occasionally stoloniferous and almost rhizomatous; leaves generally basal. **Culms:** erect, occasionally decumbent to prostrate at base, 3–8 dm tall, nodes glabrous to short hirsute. **Sheaths:** rounded, open, glabrous. **Ligules:** membranous, minutely ciliate, 0.5–1.5 mm long. **Blades:** flat to folded, 5–25 cm long, 2–4.5 mm wide, glabrous to sparsely hairy along lower margins with papillose-based hairs. **HABITAT:** Found on the eastern plains and foothills along roadsides and in improved pastures. **COMMENTS:** This species was introduced for roadside stabilization and as a pasture grass and is now naturalized.

102. *Bothriochloa laguroides* (DC.) Herter (Fig. 124).

Synonyms: *Andropogon saccharoides* Sw. var. *torreyanus* (Steud.) Hack., *Bothriochloa saccharoides* (Sw.) Rydb. var. *torreyana* (Steud.) Gould, *Dichanthium saccharoides* (Sw.) Roberty subvar. *torreyanum* (Steud.) Roberty. **Vernacular Name:** Silver bluestem. **Life Span:** Perennial. **Origin:** Native. **Season:** C_4, warm season. **FLORAL CHARACTERISTICS: Inflorescence:** panicle, terminal, of 6 to numerous clustered racemes (racemose), 2–4 cm long, fan-shaped,

with more than 10 branches, silvery-white to light tan; branches spicate, pedicels and upper rachis internodes with a translucent central groove, margins densely hairy (hairs 3–9 mm long). **Spikelets:** in pairs of 1 sessile (fertile, 2.5–4.5 mm long) and 1 pediceled (sterile, 1.5–2.5 mm long), smaller and narrower than sessile spikelet; disarticulating as a pair with a portion of rachis; fertile spikelet 2-flowered; lower floret reduced; upper floret perfect. **Glumes:** of sessile spikelet unequal (2.5–4.5 mm long), glaucous, margins of lower enrolled and clasping the upper; lower 7- to 9-nerved, glabrous to hirtellous laterally above; upper 3-nerved; of pedicellate spikelet narrow, obscurely 7- to 9-nerved. **Lemmas:** of sessile spikelet membranous, transparent, deeply cleft, awned. **Awns:** fertile lemma

Figure 123. *Bothriochloa ischaemum*.

Figure 124. *Bothriochloa laguroides.*

of sessile spikelet 8–16 mm long, geniculate, twisted. **VEGETATIVE CHARACTERISTICS: Growth Habit:** tufted, leaves generally basal, occasionally cauline on robust plants. **Culms:** erect to geniculate at base, slender, 3–12 dm tall, glaucous, nodes glabrous to shortly hirsute to pilose with erect hairs. **Sheaths:** rounded, open. **Ligules:** membranous, 1–3 mm long. **Blades:** flat to folded, 5–25 cm long, 2–7 mm wide, generally glaucous, midrib prominent, margins whitish. **HABITAT:** Found in southeastern Colorado on the plains and mesas. **COMMENTS:** There has been some confusion between this taxon and *B. saccharoides*. According to Barkworth and colleagues (2003), *B. saccharoides* is a more southern species and is not found in the United States. The only subspecies found in Colorado is subsp. *torreyana* (Steud.) Allred & Gould.

103. *Bothriochloa springfieldii* (Gould) Parodi (Fig. 125).

Synonym: *Andropogon springfieldii* Gould. **Vernacular Names:** Springfield's beard grass, Springfield bluestem. **Life Span:** Perennial. **Origin:** Native. **Season:** C_4, warm season. **FLORAL CHARACTERISTICS: Inflorescence:** panicle, terminal, of numerous clustered racemes (subdigitate), 4–7 cm long, oblong to fan-shaped, densely white-hairy; panicle axis (rachis) 1–5 cm long, with 2–7 branches; branches spicate, pedicels and upper rachis internodes with a translucent central groove, margins densely pubescent with white hairs (5–10 mm long). **Spikelets:** in pairs of 1 sessile (fertile, 5.5–8.5 mm long) and 1 pediceled (staminate or sterile 3.5–5.5 mm long), smaller and narrower than sessile spikelet; disarticulating as a pair with a portion of rachis; fertile spikelet 2-flowered; lower floret reduced; upper floret perfect. **Glumes:** of sessile spikelet subequal, margins of lower enrolled and clasping the upper; lower generally strongly 4-nerved; upper 3-nerved. **Lemmas:** of sessile spikelet membranous, transparent, deeply cleft, awned. **Awns:** fertile lemma of sessile spikelet 15–30 mm long, geniculate, twisted. **VEGETATIVE CHARACTERISTICS: Growth Habit:** cespitose, leaves generally basal. **Culms:** erect, unbranched, 3–8 dm tall, nodes densely bearded with long, silky, spreading white hairs. **Sheaths:** rounded, open, margins often densely hairy below summit. **Ligules:** membranous, 1–2.5 mm long. **Blades:** flat to folded, 5–30 cm long, 2–3 mm wide, upper surface occasionally sparsely hairy, base of leaf pilose. **HABITAT:** Found near Mesa de Maya and the Comanche National Grassland on grassy benches and canyons (Weber and Wittmann 2001a). **COMMENTS:** Uncommon. Named for H. W. Springfield, a range scientist for the U.S. Department of Agriculture (Allred 2005), not for Springfield, Colorado.

34. *Schizachyrium* Nees

Cespitose perennials, some with short, stout rhizomes. Leaves with rounded or compressed and keeled sheaths and flat or folded, infrequently terete blades. Flowering culms much branched, each leafy branch terminating in a single pedunculate raceme. Spikelets appressed to rachis or somewhat divergent at maturity. Disarticulation in the rachis, sessile spikelet falling attached to a pedicel and a section of the rachis. Sessile spikelet fertile, awned, pediceled spikelet reduced but usually present. Glumes of sessile spikelet large

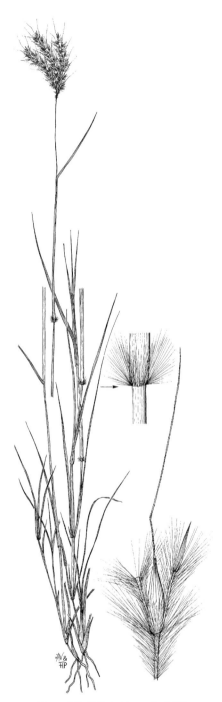

Figure 125. *Bothriochloa springfieldii*.

and firm. Lemmas of sterile and fertile florets thin and membranous, the latter usually with a geniculate arm. Basic chromosome number, $x = 10$. Photosynthetic pathway, C_4.

Represented in Colorado by a single species and a single variety.

104. *Schizachyrium scoparium* (Michx.) Nash (Fig. 126).

Synonyms: *Andropogon praematurus* Fern., *Andropogon scoparius* Michx., *Andropogon scoparius* Michx. var. *ducis* Fern. & Grisc., *Andropogon scoparius* Michx. var. *frequens* F. T. Hubbard, *Andropogon scoparius* Michx. var. *neomexicanus* (Nash) Hitchc., *Andropogon scoparius* Michx. var. *polycladus* Scribn. & Ball, *Andropogon scoparius* Michx. var. *septentrionalis* Fern. & Grisc., *Schizachyrium praematurum* (Fern.) C. F. Reed, *Schizachyrium scoparium* (Michx.) Nash subsp. *neomex-*
icanum (Nash) Gandhi & Smeins, *Schizachyrium scoparium* (Michx.) Nash var. *frequens* (F. T. Hubbard) Gould, *Schizachyrium scoparium* (Michx.) Nash var. *neomexicanum* (Nash) Gould, *Schizachyrium scoparium* (Michx.) Nash var. *polycladum* (Scribn. & Ball) C. F. Reed. **Vernacular Names:** Little bluestem, broom beard grass, prairie beard grass. **Life Span:** Perennial. **Origin:** Native. **Season:** C_4, warm season. **FLORAL CHARACTERISTICS: Inflorescence:** terminal and axillary spicate racemes, 2.5–5 cm long, usually several per culm; peduncles included in sheath with spikelets exserted. **Spikelets:** paired, sessile spikelet perfect, 6–8 mm long, 2-flowered (lower one reduced and sterile, while upper one is perfect), narrow-lanceolate; pedicellate spikelet sterile or staminate, 4–5 mm long, subulate; disarticulation below sessile spikelet and including pedicellate spikelet and internode to base of next sessile spikelet. **Glumes:** of sessile spikelet generally subequal, 2.5–4 mm long, glabrous to scabrous; lower glume 5- to 7-nerved, flat to grooved across back, margins infolded and clasping upper glume; upper 3-nerved, keeled. **Lemmas:** of sessile spikelet 4.2–6.5 mm long, membranous, cleft and bearing a geniculate awn; of pedicellate spikelet generally absent when spikelet sterile. **Awns:** lemma of sessile spikelet with geniculate awn 2.5–17 mm long, generally twisted below; lemmas of pedicellate spikelets, when present, unawned to awned, awn to 4 mm long. **VEGETATIVE CHARACTERISTICS: Growth Habit:** cespitose, sometimes with short rhizomes. **Culms:** decumbent to erect, 7–10 dm tall, no branching below. **Sheaths:** open, rounded to keeled, generally glabrous. **Ligules:** minutely puberulent, ciliolate membrane, 0.5–2.7 mm long, generally truncate. **Blades:** flat to folded, 7–25 or more cm long, 1.5–9 mm wide, lower surface and margin scabrous, upper surface glabrous to scaberulous. **HABITAT:** Found from low elevations to montane along outwash mesas, piedmont valleys, and canyon bottoms over the state. **COMMENTS:** Common. This is a widespread prairie grass and one of the main grasses of the tallgrass prairies (Barkworth et al. 2003). The typical variety, var. *scoparium,* is the only one reported from Colorado.

35. *Sorghastrum* Nash

Slender, rather tall perennials with flat blades and narrow, contracted panicles. Spikelets in pairs of 1 sessile and perfect and 1 pediceled and rudimentary, but the pediceled spikelets completely reduced and represented only by the slender, hairy pedicel. Disarticulation

JRJ

Figure 126. *Schizachyrium scoparium*.

beneath sessile spikelets. Glumes firm, subequal, awnless. Lemma of lower (sterile) floret absent or membranous and rudimentary. Lemma of upper (fertile) floret membranous, with a stout, geniculate, and twisted awn. Basic chromosome number, $x = 10$. Photosynthetic pathway, C_4.

Represented in Colorado by a single species.

105. *Sorghastrum nutans* (L.) Nash (Fig. 127).

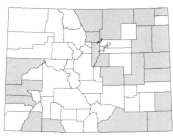

Synonyms: *Andropogon nutans* L., *Sorghastrum avenaceum* (Michx.) Nash. **Vernacular Names:** Indiangrass, yellow Indian grass. **Life Span:** Perennial. **Origin:** Native. **Season:** C_4, warm season. **FLORAL CHARACTERISTICS: Inflorescence:** loosely contracted panicle, 15–30 cm long, dense, color brownish to yellowish; branchlets villous. **Spikelets:** paired; pedicellate spikelet generally absent, leaving a pubescent pedicel 3–6 mm long; sessile spikelet perfect, 2-flowered (lower one reduced and sterile, while upper one is perfect), 5–8.7 mm long; disarticulation below sessile spikelet and including pedicellate spikelet and internode to base of next sessile spikelet. **Glumes:** subequal, 5–8 mm long; lower 5- to 7-nerved above and smooth and shiny below, flat to rounded across back, apex narrowly truncate to acuminate, margins infolded and clasping upper glume; upper 5-nerved, generally keeled, glabrous to ciliate, apex acuminate. **Lemmas:** hyaline to transparent, ciliate; sterile lemma 4–5 mm long; fertile lemma 4.5–6.5 mm long, apex bifid and awned. **Awns:** fertile lemma of sessile spikelet with geniculate awn, 12–18 mm long, twisted tightly below, loosely twisted above. **VEGETATIVE CHARACTERISTICS: Growth Habit:** rhizomatous with short, stout rhizomes. **Culms:** erect, 0.5–2.5 m tall, slender to stout, nodes with ascending hairs, internodes glabrous. **Sheaths:** glabrous to hirsute, round to occasionally flattened. **Auricles:** prominent, erect, thick and pointed, 2–7 mm long. **Ligules:** membranous, 1.5–6.5 mm long. **Blades:** flat to slightly keeled, 10–70 cm long, 1–4 mm wide, generally glabrous, constricted at base. **HABITAT:** Found in the eastern plains and foothills in pockets of remaining tallgrass prairie and infrequently along roadsides and riverbanks on the Western Slope. **COMMENTS:** Weber and Wittmann (2001a) call this species *S. avenaceum* (Michx.) Nash. The species is popular as an ornamental, and several cultivars are available.

36. *Sorghum* Moench

Annuals and perennials, many with tall, stout culms. Blades long, flat, narrow or broad. Inflorescence a large open or contracted panicle, spikelets clustered on short racemose branchlets. Spikelets in threes at branchlet tips, 1 sessile and fertile and 2 pediceled and staminate or neuter, and below the tips in pairs of 1 sessile and perfect and 1 pediceled and reduced. Disarticulation below sessile spikelet, rachis section and pedicel or pedicels falling attached to sessile spikelet. Glumes coriaceous, awnless, about equal in length. Lemma of sterile floret and lemma and palea of fertile floret membranous, lemma of fertile floret usually with a geniculate and twisted awn, this readily deciduous in *Sorghum halepense*. Basic chromosome number, $x = 5$. Photosynthetic pathway, C_4.

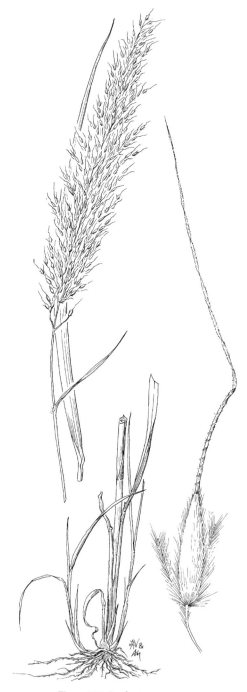

Figure 127. *Sorghastrum nutans.*

Represented in Colorado by 2 species.

1. Plant annual, without rhizomes . . . . 106. *S. bicolor*
1. Plant perennial, rhizomes present . . . . 107. *S. halepense*

106. *Sorghum bicolor* (L.) Moench (Fig. 128).

Synonyms: *Sorghum caudatum* (Hack.) Stapf, *Sorghum guineense* Stapf, *Sorghum nervosum* Besser *ex* Schultes. **Vernacular Name:** Sorghum. **Life Span:** Annual. **Origin:** Introduced. **Season:** C$_4$, warm season. **FLORAL CHARACTERISTICS: Inflorescence:** open to contracted, dense panicle, 5–60 cm long, 3–30 cm wide; primary branches rebranching, scabrous. **Spikelets:** paired with terminal spikelets as a triplet (1 sessile with 2 pediceled), sessile spikelet perfect, 3–9 mm long, 2-flowered (lower one reduced and sterile, while upper one is perfect), lanceolate to ovate; pedicellate spikelet sterile or staminate, 3–6 mm long, narrow-lanceolate, pedicel 0.7–2 mm long, pubescent; disarticulation below sessile spikelet and including pedicellate spikelet and internode to base of next sessile spikelet. **Glumes:** of sessile spikelet generally subequal; lower broadly ovate, maturing to shiny black, glabrous to pubescent, visibly 9- to 11-nerved at apex; upper narrower, visibly 3-nerved at apex; of pedicellate spikelet subequal; lower 9- to 11-nerved, flat to grooved on back, margins infolded and clasping the upper; upper narrow, margins with appressed hirsute hairs, 3- to 7-nerved. **Lemmas:** of sessile spikelet sterile lemma generally hyaline, 3.8–5 mm long; fertile lemma 3.9–4.3 mm long, unawned to awned (awn sometimes deciduous); of pedicellate spikelet short-awned. **Awns:** fertile lemma of sessile spikelet unawned to awned, awn geniculate, 5–8.5 mm long or occasionally longer, twisted below; of pedicellate spikelet occasionally short-awned to 5 mm long. **VEGETATIVE CHARACTERISTICS: Growth Habit:** cespitose, sometimes tillering at base. **Culms:** erect, 1–3 m tall, robust, nodes glabrous to appressed pubescent, internodes glabrous, occasionally glaucous. **Sheaths:** open, glabrous, short appressed hairs present at junction with blade. **Ligules:** membranous, 1–5.5 mm long. **Blades:** flat, 5–100 cm long, 5–100 mm wide, occasionally glabrous. **HABITAT:** Found in cultivation and sometimes escaping in warmer climates. **COMMENTS:** Weber and Wittmann (2001a) call this taxon *S. vulgare* Pers. *Sorghum bicolor*, however, is the conserved name that includes several cultivated strains as well as several wild strains (Welsh et al. 1993). Cultivated races include grain sorghum, sweet sorghum, pop sorghum, and broomcorn used for making whisk brooms. Species also can be used to brew beer (Allred 2005).

107. *Sorghum halepense* (L.) Pers. (Fig. 129).

Synonyms: *Holcus halepensis* L., *Sorghum miliaceum* (Roxb.) Snowden. **Vernacular Name:** Johnson grass. **Life Span:** Perennial. **Origin:** Introduced. **Season:** C$_4$, warm season. **FLORAL CHARACTERISTICS: Inflorescence:** open panicle, 10–50 cm long, 5–25 cm wide; branches smooth to scabrous, branch bases puberulent. **Spikelets:** paired, terminal spikelets as a triplet (1 sessile with 2 pediceled); sessile spikelet perfect, 3.8–6.5 mm long, 2-flowered (lower one reduced and sterile, while upper one is perfect), ovate; pedicellate spikelet staminate,

Figure 128. *Sorghum bicolor.*

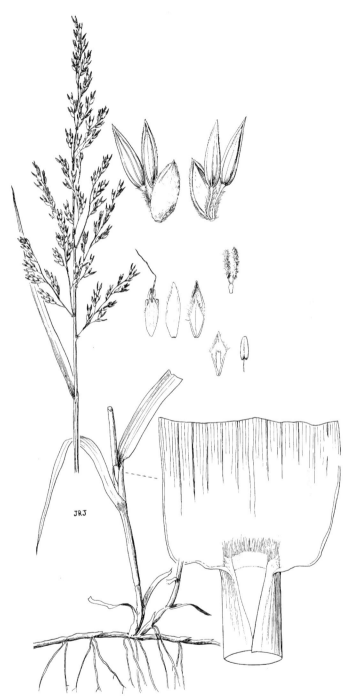

Figure 129. *Sorghum halepense.*

3.6–5.6 mm long, broadly lanceolate; pedicel 1.8–3.3 mm long, pubescent; disarticulation generally below sessile spikelets, occasionally below pedicellate spikelets. **Glumes:** of sessile spikelet generally subequal, indurate, darkening on maturity, shiny, appressed pubescent; lower obscurely 5- to 7-nerved; upper 3-nerved; of pedicellate spikelet subequal; lower 9-nerved, flat to grooved on back, lateral nerves have scabrous margins infolded and clasping the fertile 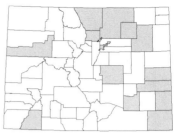 lemma; upper back keeled to rounded, glabrous to scabrous along nerves, 7-nerved. **Lemmas:** sessile spikelet with sterile lemma 4–4.5 mm long; fertile lemma greatly reduced, hyaline, 3-nerved, awned; of pediceled spikelet hyaline, reduced. **Awns:** fertile lemma of sessile spikelet awned, awn geniculate, 9–12 mm long, terminal segment 4–7 mm long, awns often early deciduous. **VEGETATIVE CHARACTERISTICS: Growth Habit:** rhizomatous, rhizomes stout, scaly, and extensive. **Culms:** erect, 0.5–2 m tall, stout, nodes occasionally appressed pubescent, internodes glabrous. **Sheaths:** glabrous, occasionally puberulent at collar. **Ligules:** densely ciliate membrane, 1.5–6 mm long. **Blades:** flat, 10–90 cm long, 8–40 mm wide, midnerve prominent, margins generally scabrous. **HABITAT:** Weedy, found on roadsides, along ditches, and in disturbed areas; escaped from cultivation in eastern Colorado. **COMMENTS:** Awns are easily broken off and/or early deciduous, and they are frequently missing from specimens. Under stress, this species may accumulate cyanide and become toxic to livestock (Allred 2005).

37. Zea L.

Monoecious plants with tall, thick, usually succulent culms and broad, flat blades. Staminate spikelets in unequally pediceled pairs on spikelike branches, these forming large panicles at culm apex. Glumes of staminate spikelet broad, thin, several-nerved. Lemma and palea hyaline. Pistillate spikelets sessile in pairs on a thickened woody or corky axis (cob). Glumes of pistillate spikelets broad, thin, rounded at apex, much shorter than mature caryopsis. Lower floret sterile or occasionally fertile. Lemma of lower floret and lemma and palea of upper (fertile) floret membranous and hyaline. Basic chromosome number, $x = 10$. Photosynthetic pathway, C_4.

Represented in Colorado by a single species commonly known as cultivated corn or maize.

108. *Zea mays* L. (Fig. 130).

Synonyms: None. **Vernacular Names:** Corn, maize. **Life Span:** Annual. **Origin:** Introduced. **Season:** C_4, warm season. **FLORAL CHARACTERISTICS: Inflorescence:** monoecious, of 2 types; staminate: terminal panicle of 1 to many branches, 10–25 cm or more; pistillate: spike in leaf axils, shortly pedunculate to sessile, 4–30 cm long, 1–10 cm wide, with 2 or more paired spikelets (4 or more rows of spikelets). **Spikelets:** 2-flowered; staminate: paired in sessile-pedicellate pairs (both flowers staminate) 9–14 cm long, 2.5–5 mm wide; pistillate: paired, subsessile pairs, each with 1 fertile floret. **Glumes:** staminate: lower membranous to papery, broad, apex acute, several-nerved, flexible and enclosing the upper; pistillate:

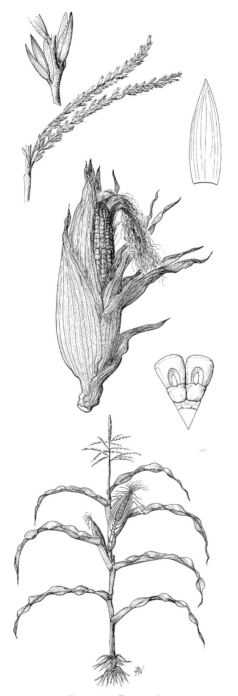

Figure 130. *Zea mays* L.

reduced; below indurate; above hyaline. **Lemmas:** staminate and pistillate; hyaline, unawned. **Paleas:** of both staminate and pistillate hyaline. **Awns:** none. **VEGETATIVE CHARACTERISTICS: Growth Habit:** solitary, lower nodes with prop roots. **Culms:** erect, 1–3 m tall, 1–5 cm thick, robust, succulent when young, becoming woody with age, internodes pith-filled. **Sheaths:** open, occasionally with auricles. **Ligules:** ciliate membrane. **Blades:** flat, distichous,  30–90 cm long, 2–12 cm wide, with well-developed midrib. **HABITAT:** Cultivated as food, fodder, or fuel; widespread throughout the state; occasionally escaping and found along roadsides, field margins, and fallow fields. **COMMENTS:** Pistillate spikes are commonly called ears, while the staminate panicles are tassels. The ears terminate reduced branches and are enclosed in several, generally bladeless leaf sheaths and prophylls called husks. The styles of the pistillate inflorescence are known as silks. Rarely collected, and obviously much more abundant than indicated by the distribution map.

8. PANICEAE

Annuals or perennials, cespitose, rhizomatous or rarely stoloniferous. Sheaths open; auricles absent; ligules membranous, a fringe of hairs, or rarely absent (in *Echinochloa*). Inflorescence a panicle; spikelets with 2 florets, lower sterile, upper perfect. Disarticulation below the glumes, in *Cenchrus* spikelets enclosed in, and falling with, bristly or spiny involucres. Lower glume usually small or absent. Upper glume and lemma of lower (sterile) floret similar in size, shape, and texture. Lemma of upper floret usually firm or indurate, shiny and smooth, indistinctly nerved, awnless (short-awned in *Eriochloa*), margins overlapping, inrolled and tightly clasping palea (except in *Digitaria*).

38. *Cenchrus* L.

Annuals and perennials, mostly with weak, geniculate-decumbent culms and soft, flat blades, but a few with tall, coarse, stiffly erect culms. Ligule a fringe of hairs. Spikelets enclosed in fascicles, these sessile or subsessile on a short, stout rachis. Fascicles of bristles, flattened spines (modified branches), or both fused together at least at base. Bristles and spines usually retrorsely barbed. Spikelets several in each fascicle, this readily disarticulating at maturity. Glumes thin, membranous, unequal. Lower glume 1- or 3-nerved, upper 1- to 7-nerved. Lemma of sterile floret thin, 1- to 7-nerved, equaling or exceeding upper glume. Palea of sterile floret about equaling lemma. Lemma of fertile floret thin, membranous, 5- or 7-nerved, tapering to a slender, usually acuminate tip, margins not inrolled. Caryopsis elliptic to ovoid, dorsally flattened. Basic chromosome number, $x = 17$. Photosynthetic pathway, C_4.

Represented in Colorado by 2 species.

1. Inner bristles 0.5–1 mm broad at base; fascicles with more than 45 bristles, outer bristles usually terete . . . . 109. *C. longispinus*
1. Inner bristles 1–3 mm broad at base; fascicles with less than 45 bristles, outer bristles usually flattened . . . . 110. *C. spinifex*

109. *Cenchrus longispinus* (Hack.) Fernald (Fig. 131).

Synonym: *Cenchrus carolinianus* Walt. **Vernacular Names:** Mat sandbur, longspine sandbur, field sandbur, innocent weed. **Life Span:** Annual. **Origin:** Native. **Season:** C_4, warm season. **FLORAL CHARACTERISTICS: Inflorescence:** spikelike panicle, 3–10 cm long, 1.2–2.2 cm wide, bearing 4–12 fascicles (burrs) or more; fascicle encloses spikelets in a subtending involucre of basally fused bristles, 45–75 in number, generally globose to urceolate; outer bristles many, terete, generally shorter and thinner than inner ones; inner bristles 3.5–7 mm long, 0.5–1

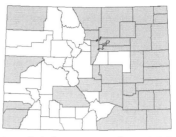

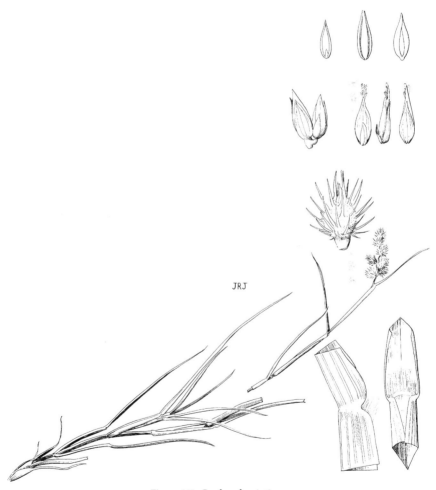

Figure 131. *Cenchrus longispinus.*

mm broad at base, located at irregular intervals and fused for half their length, forming a cupule, purplish, short-pubescent, containing 2–3 sessile spikelets. **Spikelets:** 6–7.8 mm long, 2-flowered, lower floret often staminate, upper floret fertile; disarticulation occurs below fascicle and falls as a unit. **Glumes:** unequal, glabrous, slightly convex; lower 1.5–4 mm long, narrow lanceolate, 1-nerved; upper 4.4–6 mm long, 3- to 5-nerved, broadly lanceolate. **Lemmas:** broadly lanceolate, apex acuminate; sterile lemma 5–6.5 mm long, 3- to 7-nerved; fertile lemma 5.2–7 mm long, obscurely 3- to 5-nerved, smooth. **Paleas:** sterile palea subequal to sterile lemma, narrow lanceolate, scabrid; fertile palea subequal to fertile lemma, smooth. **Awns:** none. **VEGETATIVE CHARACTERISTICS: Growth Habit:** loosely tufted, often prostrate and mat-forming. **Culms:** decumbent to erect, occasionally geniculate and rooting at nodes, 10–55 cm long, glabrous. **Sheaths:** glabrous, rarely ciliate, compressed-keeled, summit sometimes puberulent. **Ligules:** dense fringe of hairs, 0.5–1.5 mm long. **Blades:** flat to folded, 4–14 cm long, 2.8–6.5 mm wide, scabrous. **HABITAT:** Found in sandy, usually disturbed sites at low elevations over Colorado. **COMMENTS:** This is an extremely noxious weed whose spines can inflict painful injury. The spines and bristles are also an excellent dispersal mechanism. Allred (2005) says the uppermost leaf margins of this species are conspicuously crinkled at the base. *Cenchrus* is closely related to *Pennisetum*, and they are sometimes included in the same genus.

110. *Cenchrus spinifex* Cav. (Fig. 132).

Synonyms: *Cenchrus incertus* M. A. Curtis, *Cenchrus pauciflorus* Benth. **Vernacular Names:** Coastal sandbur, common sandbur. **Life Span:** Annual or short-lived perennial. **Origin:** Native. **Season:** C$_4$, warm season. **FLORAL CHARACTERISTICS: Inflorescence:** spikelike panicle, 2–7 cm long, bearing 5–15 fascicles (burrs) or more; fascicle encloses spikelets in a subtending involucre of basally fused bristles, generally globose to urceolate and covered with 20–40 bristles; outer bristles not always present, flattened in appearance; inner bristles 2–7 mm long, 1–2 mm wide, retrorsely barbed, fused for at least half their length, forming a cupule, short-pubescent, purple-tinged; containing 2–4 sessile spikelets. **Spikelets:** 3.5–5.9 mm long, 2-flowered, lower floret sterile or sometimes staminate, upper floret fertile; disarticulation occurs below fascicle and falls as a unit. **Glumes:** unequal, glabrous, slightly convex; lower 1–3 mm long, narrow lanceolate, 1-nerved; upper 3–5 mm long, 5- to 7-nerved, broadly lanceolate. **Lemmas:** broadly lanceolate, apex acuminate; sterile lemma 5–6.5 mm long, 3- to 7-nerved; fertile lemma 3.5–6 mm long, obscurely 5- to 7-nerved, smooth. **Paleas:** sterile palea sometimes reduced or absent; fertile palea subequal to fertile lemma, smooth. **Awns:** none. **VEGETATIVE CHARACTERISTICS: Growth Habit:** loosely tufted, sometimes sprawling. **Culms:** decumbent to erect, occasionally geniculate and rooting at nodes, 3–10 dm long, glabrous. **Sheaths:** glabrous, compressed-keeled, summit sometimes puberulent. **Ligules:** dense fringe of hairs, 0.5–1.5 mm long. **Blades:** flat to folded, 3–28 cm long, 3–7 mm wide, glabrous to sparsely pilose. **HABITAT:** Generally found in disturbed sandy sites at low elevations. Only 2 collections verified, but perhaps much more common than collections would indicate. **COMMENTS:** This is an extremely noxious weed whose spines

Figure 132. *Cenchrus spinifex.*

can inflict painful injury. The spines and bristles are also an excellent dispersal mechanism. Allred (2005) says only 1 margin of the uppermost leaf blades of this species is conspicuously crinkled at the base. *Cenchrus* is closely related to *Pennisetum,* and they are sometimes included in the same genus.

39. *Dichanthelium* (Hitchc. & Chase) Gould

Perennials, typically tufted, and usually forming early in the growing season a basal rosette of shorter, broader blades than those of the culms. Ligule usually a ring of hairs, infrequently a short membrane or absent. Secondary branching after first culm growth and elongation often resulting in densely clustered fascicles of much-reduced leaves and inflorescences. Inflorescences typically a small panicle with spreading or occasionally contracted branches, spikelets short- or long-pediceled. Disarticulation below the glumes. Both glumes present, but lowermost often greatly reduced. Lower floret usually neuter but staminate in a few species. Upper floret perfect, with a shiny, glabrous, coriaceous lemma and palea. Lemma tightly inrolled over palea. Basic chromosome number, $x = 9$. Photosynthetic pathway, C_3.

Represented in Colorado by 4 species and 4 subspecies.

1. Basal leaf blades similar in shape to cauline leaf blades
   2. Cauline blades 4–8 cm long . . . . 114. *D. wilcoxianum*
   2. Cauline blades, at least the upper, 10–20 cm long . . . . 112. *D. linearifolium*
1. Basal leaf blades different in shape from cauline leaf blades
   3. Spikelets 1.1–1.9 mm long; upper glumes without an orange or purple dot at base, nerves obscure . . . . .111. *D. acuminatum*
   3. Spikelets 2.5–4.3 mm long, upper glumes with an orange or purple dot at base, nerves prominent . . . . 113. *D. oligosanthes*

111. *Dichanthelium acuminatum* (Sw.) Gould & C. A. Clark (Fig. 133).

Synonyms: Subsp. *fasciculatum* (Torr.) Freckmann: *Dichanthelium acuminatum* (Sw.) Gould & C. A. Clark var. *implicatum* (Scribn.) Gould & C. A. Clark, *Dichanthelium lanuginosum* (Ell.) Gould, *Dichanthelium lanuginosum* (Ell.) Gould var. *fasciculatum* (Torr.) Spellenberg, *Dichanthelium subvillosum* (Ashe) Mohlenbrock, *Panicum acuminatum* Sw. var. *fasciculatum* (Torr.) Lelong, *Panicum acuminatum* Sw. var. *implicatum* (Scribn.) C. F. Reed, *Panicum brodiei* St. John, *Panicum curtifolium* Nash, *Panicum glutinoscabrum* Fern., *Panicum huachucae* Ashe, *Panicum huachucae* Ashe var. *fasciculatum* (Torr.) F. T. Hubbard, *Panicum implicatum* Scribn., *Panicum languidum* Hitchc., *Panicum lanuginosum* Ell., non Bosc *ex* Spreng., *Panicum lanuginosum* Ell. var. *fasciculatum* (Torr.) Fern., *Panicum lanuginosum* Ell. var. *huachucae* (Ashe) Hitchc., *Panicum lanuginosum* Ell. var. *implicatum* (Scribn.) Fern., *Panicum lanuginosum* Ell. var. *tennesseense* (Ashe) Gleason, *Panicum lassenianum* Schmoll, *Panicum lindheimeri* Nash var. *fasciculatum* (Torr.) Fern., *Panicum occidentale* Scribn., *Panicum pacificum* Hitchc. & Chase, *Panicum subvillosum* Ashe, *Panicum tennesseense* Ashe; subsp. *sericeum* (Schmoll) Freckmann: *Panicum ferventicola* Schmoll, *Panicum ferventicola* Schmoll var. *papillosum* Schmoll, *Panicum ferventicola* Schmoll var. *sericeum* Schmoll. **Vernacular Names:** Tapered rosette grass, hairy panicgrass. **Life Span:** Perennial. **Origin:** Native. **Season:** C_3, cool season. **FLORAL CHARACTERISTICS: Inflorescence:** primary panicle, 3–9 cm long, open, exserted; later panicles usually reduced, often not fully exserted from sheath. **Spikelets:** elliptic to ovate, generally pubescent, 1.1–1.9 mm long (in Colorado), apex subacute to obtuse; disarticulation below the glumes; lower floret sterile; upper floret fertile, 1.1–1.7 mm long. **Glumes:** unequal; lower 0.3–1.1 mm long; upper subequal to sterile lemma, slightly shorter to as long as upper

Figure 133. *Dichanthelium acuminatum.*

floret when mature, nerves generally obscure. **Lemmas:** sterile lemma to 1.7 mm long; fertile lemma 1.1–1.7 mm long, 0.6–1 mm wide, ellipsoid, apex variable, obtuse to acute to minutely apiculate. **Awns:** none. **VEGETATIVE CHARACTERISTICS: Growth Habit:** generally densely cespitose, basal rosettes present. **Culms:** decumbent to ascending to erect, 1.5–10 dm tall, initially simple, generally becoming much branched with age, stout to wiry; nodes sometimes swollen, glabrous to pubescent, often subtended by a viscid or glabrous ring; internodes variously pubescent to glabrous. **Sheaths:** generally shorter than internodes, glabrous to densely pubescent, margins glabrous to ciliate. **Ligules:** with pseudoligules 1–5 mm long, hairs. **Blades:** basal blades (rosette) ovate to lanceolate, usually well differentiated from cauline; cauline blades 4–7, glabrous to densely pubescent; leaves of primary culm 4–10 cm long, 3–9 mm wide; later leaves of branches generally shorter, narrower, and more crowded; basally at least with papillose-based cilia on margins. **HABITAT:** Found in low elevations on moist sandy sites. Weber and Wittmann (2001a, 2001b) further describe these sites as rocky at the base of rimrock cliffs, in association with seeps, along the Front Range and plains in eastern Colorado. In western Colorado it has only been found near a spring in Unaweap Canyon [subsp. *fasciculatum* (Torrey) Freckmann & Lelong] and near Box Canyon Hot Springs on steaming travertine [subsp. *sericeum* (Schmoll) Freckmann & Lelong] (Weber and Wittmann 2001a, 2001b; Barkworth et al. 2003). **COMMENTS:** Extremely variable species with numerous subspecies, 3 (the 2 listed here plus the typical subspecies, subsp. *acuminatum*) of which occur in the state and can be separated based on these characteristics:

1. Culms shorter than 30 cm; mid-culm sheaths almost as long as internodes . . . .
. . . . subsp. *sericeum*

1. Culms longer than 30 cm; mid-culm sheaths about half as long as internodes
  2. Culms and lower sheaths densely covered with spreading, villous hairs or soft papillose-based hairs with shorter hairs underneath; blades velvety abaxially . . . .
. . . . subsp. *acuminatum*
  2. Culms and lower sheaths pilose with papillose-based hairs to hispid; blades appressed pubescent, not velvety to the touch . . . . subsp. *fasciculatum*

112. *Dichanthelium linearifolium* (Scribn.) Gould (Fig. 134).

Synonyms: *Dichanthelium depauperatum* (Muhl.) Gould var. *perlongum* (Nash) Boivin, *Dichanthelium linearifolium* (Scribn. *ex* Nash) Gould var. *werneri* (Scribn.) Mohlenbrock, *Dichanthelium perlongum* (Nash) Freckmann, *Panicum linearifolium* Scribn. *ex* Nash, *Panicum linearifolium* Scribn. *ex* Nash var. *werneri* (Scribn.) Fern., *Panicum perlongum* Nash, *Panicum strictum* Pursh var. *linearifolium* (Scribn. *ex* Nash) Farw., *Panicum strictum* Pursh var. *perlongum* (Nash)

Farw., *Panicum werneri* Scribn. **Vernacular Names:** Slimleaf panicgrass, linear-leaved panicgrass. **Life Span:** Perennial. **Origin:** Native. **Season:** C$_3$, cool season. **FLORAL CHARACTERISTICS: Inflorescence:** primary panicle, 4–10 cm long, open, generally long-exserted; branches spreading with 12–70 spikelets; secondary panicles small, enclosed within leaf sheaths to slightly exserted. **Spikelets:** ellipsoid, generally glabrous, occasionally sparsely

Figure 134. *Dichanthelium linearifolium*.

pilose, 2–3.2 mm long, 0.8–1.4 mm wide; disarticulation below the glumes; lower floret sterile; upper floret fertile, 1.7–2.3 mm long, ovoid-ellipsoid. **Glumes:** unequal; lower 0.6–1.1 mm long, ovate-triangular; upper subequal to sterile lemma, longer than upper floret when flowering by 0.2 mm, becomes subequal by time fruit matures, 7- to 9-nerved. **Lemmas:** sterile lemma 2.5–2.8 mm long; fertile lemma 2.3–3.2 mm long, ovoid-ellipsoid, apex minutely umbonate. **Awns:** none. **VEGETATIVE CHARACTERISTICS: Growth Habit:** cespitose, basal rosettes present. **Culms:** drooping to erect, 1–5 dm tall, very slender; nodes bearded; internodes glabrous to hispid or puberulent. **Sheaths:** longer than internodes, densely pilose with fine, papillose-based hairs, or occasionally glabrous. **Ligules:** fringe of hairs, 0.3–1 mm long. **Blades:** basal blades (rosette) similar to cauline blades, poorly differentiated; cauline blades flat to involute on drying, stiffly ascending to erect, 5–20 cm long (upper blades generally more than 10 cm), 2–5 mm wide, glabrous to densely pilose on both surfaces, long tapering to apex. **HABITAT:** Found in eastern Colorado on mesas and in the plains-foothills transition on rocky slopes of outer hogbacks (Weber and Wittmann 2001a). **COMMENTS:** Infrequent.

113. *Dichanthelium oligosanthes* (Schult.) Gould (Fig. 135).

Synonyms: *Dichanthelium oligosanthes* (Schult.) Gould var. *helleri* (Nash) Mohlenbrock, *Panicum helleri* Nash, *Panicum macrocarpon* Le Conte *ex* Torr., *Panicum oligosanthes* Schult. var. *helleri* (Nash) Fern., *Panicum oligosanthes* Schult. var. *scribnerianum* (Nash) Fern., *Panicum scoparium* S. Wats. *ex* Nash, non Lam., *Panicum scribnerianum* Nash. **Vernacular Names:** Heller's rosette grass, few-flowered panicgrass. **Life Span:** Perennial. **Origin:** Native. **Season:** C$_3$, cool season.

FLORAL CHARACTERISTICS: Inflorescence: primary panicle, 5–9 cm long, 3–6 cm wide, open, long-exserted to partially enclosed, with 6–60 spikelets. **Spikelets:** broadly obovoid to elliptic, generally glabrous, 2.7–3.5 mm long; disarticulation below the glumes; lower floret sterile; upper floret fertile, 2–3 mm long, apex minutely umbonate. **Glumes:** unequal; lower 0.8–1.6 mm long; upper subequal to sterile lemma, 2.2–3.3 mm long, 7- to 9-nerved, nerves broad and rounded, orange to purple spot at base evident, sometimes fading with age. **Lemmas:** sterile lemma 2.2–3.3 mm long; fertile lemma 2.5–3 mm long. **Paleas:** sterile palea 1.2–2 mm long, well developed. **Awns:** none. **VEGETATIVE CHARACTERISTICS: Growth Habit:** cespitose, basal rosettes present. **Culms:** wiry, basally geniculate to distally erect, 2–8 dm tall; nodes glabrous to sparsely pubescent; internodes glabrous to hispid or puberulent, generally becoming much branched with age, producing fascicles of reduced leaves and additional smaller panicles. **Sheaths:** glabrous to puberulent to papillose-hispid, margins ciliate. **Ligules:** fringe of hairs, 1–3 mm long. **Blades:** basal blades (rosette) ovate to lanceolate, 2–6 cm long; cauline flat to involute on drying, 5–12 cm long, 4–15 mm wide, glabrous to pubescent below, ciliate at base, margins cartilaginous. **HABITAT:** Found in the plains on rocky ground and on sandy or clayey banks (Barkworth et al. 2003). **COMMENTS:** Colorado specimens belong to subsp. *scribnerianum* (Nash) Freckmann & Lelong, which has shorter and broader spikelets (2.7–3.5 mm long, 2–2.4 mm wide) and wider blades (6–15 mm wide) that are flat.

JRJ

Figure 135. *Dichanthelium oligosanthes.*

114. *Dichanthelium wilcoxianum* (Vasey) Freckmann (Fig. 136).

Synonyms: *Dichanthelium oligosanthes* (Schult.) Gould var. *wilcoxianum* (Vasey) Gould & C. A. Clark, *Panicum wilcoxianum* Vasey, *Panicum wilcoxianum* Vasey var. *breitungii* Boivin. **Vernacular Names:** Fall rosette grass, Wilcox's panicgrass. **Life Span:** Perennial. **Origin:** Native. **Season:** C₃, cool season. **FLORAL CHARACTERISTICS: Inflorescence:** open panicle, 3–5 cm long, 2–4 cm wide, ovoid, shortly exserted with 12–32 spikelets. **Spikelets:** elliptic to obovoid, pubescent,

2.4–3.2 mm long, generally reddish; disarticulation below the glumes; lower floret sterile; upper floret fertile, 1.9–2.5 mm long. **Glumes:** lower 0.6–1.4 mm long; upper subequal to sterile lemma, 2–2.8 mm long. **Lemmas:** sterile lemma 2–2.8 mm long; fertile lemma 2.4–3 mm long. **Paleas:** sterile palea 0.8–1.4 mm long. **Awns:** none. **VEGETATIVE CHARACTERISTICS: Growth Habit:** cespitose, basal rosettes present, poorly differentiated from cauline leaves. **Culms:** erect, 15–35 cm tall; nodes glabrous to sparsely pubescent; internodes sparsely pubescent, purplish gray. **Sheaths:** glabrous to hirsute, when hairs present with papillose bases. **Ligules:** fringe of hairs, 0.5–1 mm long. **Blades:** basal blades (rosette) similar but shorter, 2–4 cm long, flat, ascending to spreading; cauline blades flat, 4–8 cm long, 2–5 mm wide, sparsely pilose to hirsute, stiffly erect. **HABITAT:** Found in the plains and lower foothills in eastern Colorado on rocky ground. **COMMENTS:** Wingate (1994) includes this taxon with *D. oligosanthes*, whereas Weber and Wittmann (2001a, 2001b) treat them separately.

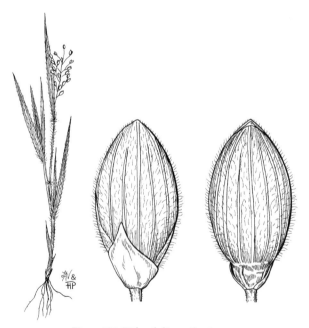

Figure 136. *Dichanthelium wilcoxianum*.

40. *Digitaria* Haller

Annuals and perennials, with erect or decumbent-spreading, stoloniferous culms. Blades mostly thin and flat. Inflorescence a panicle with few to numerous slender spikelike racemose branches, these unbranched or sparingly branched near base. Spikelets slightly plano-convex; solitary, paired, or in groups of 3–5; subsessile or short-pediceled in 2 rows on a 3-angled, often winged rachis. Disarticulation below the glumes. Lower glume minute or absent. Upper glume well developed but usually shorter than lemma of sterile floret. Nerves of the glumes and lemma of sterile floret glabrous, puberulent, or long-ciliate. Lemma of fertile floret relatively narrow, acute or acuminate, firm and cartilaginous, but not hard; margins thin, flat, not inrolled over palea. Basic chromosome number, $x = 9$. Photosynthetic pathway, C_4.

Represented in Colorado by 3 species and 1 variety.

1. Spikelets in groups of 3 on middle portion of primary branches; longer pedicels usually adnate to branch axes for part of their length . . . . 116. *D. ischaemum*
1. Spikelets paired on middle portion of primary branches; pedicels not adnate to branch axes.
   2. Upper lemmas brown when immature, turning dark brown when mature; perennial . . . . 115. *D. californica*
   2. Upper lemmas pale yellow, tan, or gray, sometimes purple-tinted when immature; annual . . . . 117. *D. sanguinalis*

115. *Digitaria californica* (Benth.) Henrard (Fig. 137).

Synonym: *Trichachne californica* (Benth.) Chase. **Vernacular Name:** Arizona cottontop. **Life Span:** Perennial. **Origin:** Native. **Season:** C_4, warm season. **FLORAL CHARACTERISTICS: Inflorescence:** contracted panicle, 8–20 cm long, 4–20 mm wide, with 4–10 alternating spikelike primary branches; branches 3–6 cm long, appressed to ascending; pedicels not adnate to branch axes, of 2 lengths, 0.1–0.3 mm long and 1–2 mm long. **Spikelets:** paired, 3–5.4 mm long, 2-flowered; lower floret staminate or sterile; upper floret fertile, obscured by hairs from lemmas and upper glume; disarticulation below the glumes. **Glumes:** lower 0.4–0.6 mm long, reduced to absent; upper 3-nerved, 2–5.1 mm long, densely villous, narrow, covered with silvery-white to purple hairs, 1.5–5 mm long. **Lemmas:** sterile lemma obscurely 7-nerved, 2.7–5 mm long, margins and lateral nerves covered with silvery-white to purple hairs, 1.5–5 mm long, areas between nerves glabrous, apex acuminate to awn-tipped; fertile lemmas 2.5–3.5 mm long, ovate-lanceolate, brown when immature and becoming dark brown when mature, acuminate. **Awns:** fertile lemma sometimes awn-tipped. **VEGETATIVE CHARACTERISTICS: Growth Habit:** cespitose. **Culms:** erect to occasionally geniculate at base, 4–10 dm tall, branching from base; base knotty. **Sheaths:** basal villous, upper glabrous to densely hairy, hairs papillose-based. **Ligules:** entire to lacerate membrane, 1.5–6 mm long, obtuse. **Blades:** flat to folded, 2–18 cm long, 2–5 mm wide, both surfaces glabrous to upper surface sparsely to densely pubescent. **HABITAT:** Plains or open ground. **COMMENTS:** Rare. The only variety found in Colorado is the typical one (var. *californica*).

Figure 137. *Digitaria californica.*

116. *Digitaria ischaemum* (Schreb.) Muhl. (Fig. 138).

Synonyms: *Digitaria ischaemum* (Schreb.) Schreb. *ex* Muhl. var. *mississippiensis* (Gattinger) Fern., *Panicum ischaemum* Schreb., *Syntherisma ischaemum* (Schreb.) Nash. **Vernacular Name:** Smooth crabgrass. **Life Span:** Annual. **Origin:** Introduced. **Season:** C$_4$, warm season. **FLORAL CHARACTERISTICS: Inflorescence:** panicle with spreading subdigitate to racemose branches; branches 1–6, 6–15.5 cm long, axes wing-margined; pedicels unequal, longer ones usually adnate for a portion below. **Spikelets:** in threes, 1.7–2.3 mm long, 2-flowered; lower floret staminate or sterile; upper floret fertile; disarticulation below the glumes. **Glumes:** lower reduced to membranous rim or absent; upper 3-nerved, 1.7–2.3 mm long, appressed pubescent. **Lemmas:** sterile lemma 5-nerved, 1.7–2.3 mm long, subglabrous to appressed pubescent; fertile lemma smooth, dark when mature. **Awns:** none. **VEGETATIVE CHARACTERISTICS: Growth Habit:** tufted. **Culms:** geniculate to decumbent, 2–6 dm tall, rooting at lower nodes. **Sheaths:** glabrous to slightly pubescent, keeled. **Ligules:** truncate, entire to lacerate, 0.6–2.5 mm long. **Blades:** flat, 1.5–9 cm long, 3–5 mm wide, glabrous except for a few papillose-based hairs near collar. **HABITAT:** Found in Colorado as a lawn weed. **COMMENTS:** Common and probably much more abundant than the distribution map indicates.

117. *Digitaria sanguinalis* (L.) Scop. (Fig. 139).

Synonyms: *Panicum sanguinale* L., *Syntherisma sanguinalis* (L.) Dulac. **Vernacular Names:** Hairy crabgrass, northern crabgrass. **Life Span:** Annual. **Origin:** Native. **Season:** C$_4$, warm season. **FLORAL CHARACTERISTICS: Inflorescence:** panicle with spreading subdigitate to racemose branches; branches 4–13, 3–30 cm long, axes wing-margined; pedicels not adnate to branch axes, unequal. **Spikelets:** paired, 2.5–3.4 mm long, 2-flowered; lower floret staminate or sterile; upper floret fertile; disarticulation below the glumes. **Glumes:** lower 0.2–0.4 mm long, nerveless; upper 0.9–2 mm long, 3-nerved, margins pubescent. **Lemmas:** sterile lemma 7-nerved, 1.7–2.3 mm long, glabrous; fertile lemma smooth, cartilaginous, 2.5–3.4 mm long, obscurely 1-nerved. **Awns:** none. **VEGETATIVE CHARACTERISTICS: Growth Habit:** tufted. **Culms:** geniculate to decumbent, 2–7 dm tall, rooting at lower nodes. **Sheaths:** sparsely pubescent with papillose-based hairs, keeled. **Ligules:** thin membrane with irregular teeth, 0.5–2.6 mm long. **Blades:** flat, 2–11 cm long, 3–8 mm wide, both surfaces generally with papillose-based hairs, occasionally glabrous. **HABITAT:** Found throughout Colorado as a lawn weed; probably much more common than the distribution map indicates. **COMMENTS:** A European species now established as a global weed.

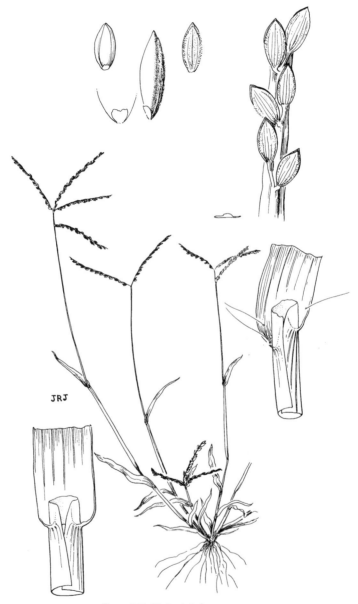

JRJ

Figure 138. *Digitaria ischaemum*.

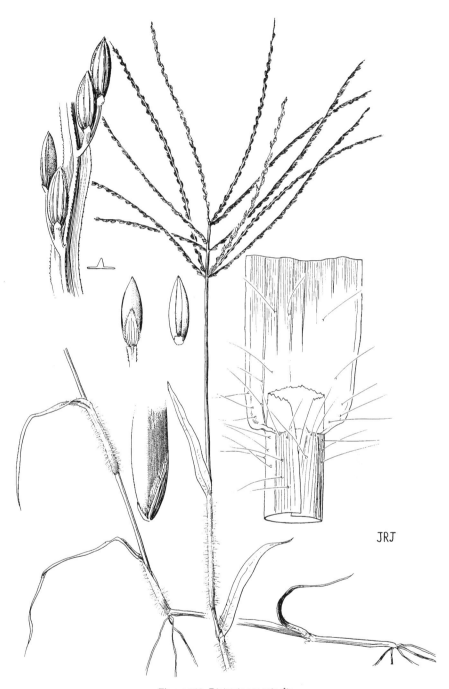

Figure 139. *Digitaria sanguinalis.*

41. *Echinochloa* P. Beauv.

Coarse annuals and perennials, usually with weak, succulent culms and broad, flat blades. Ligule a ring of hairs or absent. Inflorescence a contracted or moderately open panicle, with few to numerous, simple or rebranched, densely flowered branches. Spikelets subsessile, in irregular fascicles or regular rows, disarticulating below the glumes. Glumes and lemma of sterile floret usually with stout spicules and long or short hairs, these often glandular-pustulate at base. Lower glume well developed but much shorter than upper, acute or slightly awned. Upper glume and lemma of sterile floret about equal, acute, short-awned or with a long, flexuous awn. Lemma of fertile floret indurate, smooth and shiny, with inrolled margins and usually an abruptly pointed apex. Palea of fertile floret similar to lemma in texture, broad but narrowing to a pointed tip free from lemma margins. Basic chromosome number, $x = 9$. Photosynthetic pathway, C_4.

Represented in Colorado by 4 species and 1 variety.

1. Spikelets, particularly those at or near the bottom of panicle branches, not disarticulating at maturity; upper lemmas wider and longer than glumes and exposed at maturity . . . . 120. *E. frumentacea*
1. Spikelets all disarticulating at maturity; upper lemmas scarcely as wide and long as glumes
   2. Panicle branches 1.5–2 (rarely up to 4) cm long, secondary branches lacking; spikelets 2–3 mm long, unawned . . . . 118. *E. colona*
   2. Panicle branches 2.5–14 cm long, secondary branches present, often short and inconspicuous; spikelets 2.5–5 mm long, awned or less frequently unawned
      3. Upper lemmas with rounded or broadly acute coriaceous apices that pass abruptly into a membranous tip, a line of minute hairs present at base of tip . . . . 119. *E. crus-galli*
      3. Upper lemmas with acute or acuminate coriaceous apices that extend into membranous tip, without hairs at base of tip . . . . 121. *E. muricata*

118. *Echinochloa colona* (L.) Link (Fig. 140).

Synonym: *Panicum colonum* L. **Vernacular Names:** Jungle rice, awnless barnyard grass. **Life Span:** Annual. **Origin:** Introduced. **Season:** C_4, warm season. **FLORAL CHARACTERISTICS: Inflorescence:** erect panicle of 5–10 spikelike branches, 2–12 cm long; branches ascending to erect, 0.7–2 cm long. **Spikelets:** 2-flowered, 2–3 mm long, hispid to pubescent, apex acute to cuspidate; disarticulation below the glumes, all disarticulating at maturity; lower floret sterile or staminate; upper floret fertile, dorsally compressed. **Glumes:** membranous; lower unawned to minutely awn-tipped; upper unawned, subequal to equal to length of spikelet. **Lemmas:** sterile lemma generally similar to upper glume, unawned; fertile lemma rounded dorsally, 2.6–2.9 mm long, elliptic, generally coriaceous, scarcely as wide and as long as the glumes, distally becoming membranous to tip, this tip soon withering. **Awns:** lower glume sometimes awn-tipped. **VEGETATIVE CHARACTERISTICS: Growth Habit:**

Figure 140. *Echinochloa colona.*

cespitose to spreading, rooting from lower nodes. **Culms:** decumbent to erect, 1–7 dm tall; nodes below glabrous to hispid, those above glabrous. **Sheaths:** open, compressed, glabrous. **Ligules:** none, area generally brownish purple. **Blades:** 8–22 cm long, generally glabrous but occasionally hispid, with papillose-based hairs on or near margins. **HABITAT:** Is considered a weed and found in damp to wet disturbed areas at low elevations (Barkworth et al. 2003). **COMMENTS:** Rare in Colorado, only collected in Fremont County and perhaps not a constant member of our flora.

119. *Echinochloa crus-galli* (L.) P. Beauv. (Fig. 141).

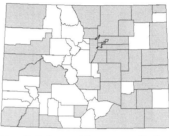

Synonyms: *Echinochloa crus-galli* (L.) P. Beauv. subsp. *spiralis* (Vasinger) Tzvelev, *Panicum crus-galli* L. **Vernacular Names:** Barnyard grass, Japanese millet. **Life Span:** Annual. **Origin:** Introduced. **Season:** C$_4$, warm season. **FLORAL CHARACTERISTICS: Inflorescence:** erect to nodding panicle of 5–12 spikelike branches, 5–25 cm long; branches appressed to spreading, 1.5–10 cm long, longer branches with secondary branches present, often inconspicuous, short. **Spikelets:** 2-flowered, 2.5–4 mm long, 1.1–2.3 mm wide, slightly turgid to generally laterally compressed; disarticulation below the glumes, all disarticulating at maturity; lower floret sterile or staminate; upper floret fertile, dorsally compressed. **Glumes:** membranous; lower unawned to short-awned, 1.2–1.6 mm long, 3-nerved, broad, scaberulous; upper 2.5–3.2 mm long, 5-nerved, lateral nerves scabrous below, pectinate when near margins, apex acuminate to awned. **Lemmas:** sterile lemma generally similar to upper glume, awned to unawned; fertile lemma rounded dorsally, 2.2–3 mm long, elliptic to broadly ovate, 5- to 9-nerved, generally coriaceous, distally becoming membranous to tip, this separation further distinguished by a line of very short hairs. **Awns:** lower glume sometimes awned, to 0.5 mm long; upper glume sometimes awned, 0.7–1.5 mm long; sterile lemma, when awned, with awns 0.5–50 mm long; fertile lemma short-awned to 0.5 mm long. **VEGETATIVE CHARACTERISTICS: Growth Habit:** cespitose to spreading. **Culms:** decumbent to erect, 0.3–2 m tall; nodes generally glabrous, lower nodes sometimes puberulent. **Sheaths:** open, slightly compressed, glabrous. **Ligules:** none, area occasionally pubescent. **Blades:** 8–65 cm long, 5–35 mm wide, generally glabrous but occasionally sparsely hirsute. **HABITAT:** Found on moist disturbed sites (gardens, irrigation ditches, and farmyards, to name a few) at low elevations to the lower foothills over Colorado. **COMMENTS:** A common weed throughout most of the state.

120. *Echinochloa frumentacea* Link (Fig. 142).

Synonyms: *Echinochloa crus-galli* (L.) P. Beauv. subsp. *edulis* Hitchc., *Echinochloa crus-galli* (L.) P. Beauv. var. *frumentacea* (Link) W. Wight, *Panicum frumentaceum* Roxb., non Salisb. **Vernacular Names:** White panic, Siberian millet. **Life Span:** Annual. **Origin:** Introduced. **Season:** C$_4$, warm season. **FLORAL CHARACTERISTICS: Inflorescence:** erect to nodding panicle of numerous spikelike branches, 7–18 cm long; branches appressed to ascending, sparsely hispid to glabrous, hairs papillose-based, 1.5–4 cm long. **Spikelets:** generally 3-

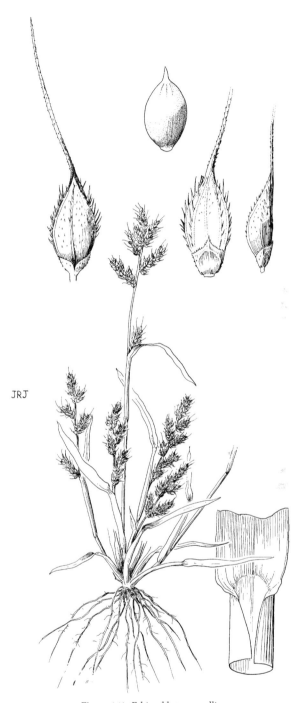

JRJ

Figure 141. *Echinochloa crus-galli*.

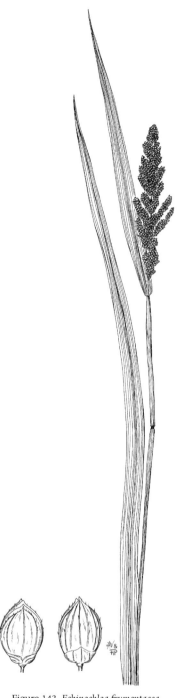

Figure 142. *Echinochloa frumentacea.*

flowered, 3–3.5 mm long, scabrous to short-hispid, hairs not papillose-based, green and pale as mature, apex generally obtuse to acute; no disarticulation at maturity, particularly those near or at base of panicle branches; lower floret sterile; upper florets fertile, dorsally compressed. **Glumes:** purplish, membranous; lower unawned; upper unawned, narrower and shorter than fertile lemma. **Lemmas:** sterile lemma generally similar to upper glume, unawned, purplish; fertile lemma 2.5–3 mm long, elliptic to ovate, 5- to 9-nerved, coriaceous, pale, portion terminating abruptly at juncture to membranous tip, this tip exposed at maturity. **Awns:** none. **VEGETATIVE CHARACTERISTICS: Growth Habit:** cespitose, usually reddish purple at maturity. **Culms:** erect, 7–15 dm tall. **Sheaths:** glabrous. **Ligules:** none. **Blades:** 8–35 cm long, 3–20 mm wide, generally glabrous. **HABITAT:** Cultivated for grain, fodder (forage grass), and beer, but occasionally escapes. Found growing on roadsides, edges of plowed ground, and in fallow fields. **COMMENTS:** Strict rules exist for planting this taxon because it can become a noxious weed.

121. *Echinochloa muricata* (P. Beauv) Fernald. (Fig. 143).

Synonyms: Var. *microstachya* Wieg.: *Echinochloa muricata* (Beauv.) Fern. subsp. *microstachya* (Wieg.) Jauzein, *Echinochloa muricata* (Beauv.) Fern. var. *occidentalis* Wieg., *Echinochloa muricata* (Beauv.) Fern. var. *wiegandii* Fassett. **Vernacular Name:** Rough barnyard grass. **Life Span:** Annual. **Origin:** Native. **Season:** C_4, warm season. **FLORAL CHARACTERISTICS: Inflorescence:** erect to nodding panicle of spikelike branches, 7–35 cm long; branches generally spreading, 2–8 cm long, often with secondary branches. **Spikelets:** 2-flowered, 2.5–3.8 mm long, 1.1–2.3 mm wide, purple to purple-streaked, generally hispid with papillose-based hairs; disarticulation below the glumes, all disarticulating at maturity; lower floret sterile or staminate; upper floret fertile, dorsally compressed. **Glumes:** membranous; lower acuminate to acute to obtuse, 0.9–1.6 mm long, 3-nerved, broad, scaberulous; upper 2.5–3.8 mm long, 5-nerved, pustular-based setae scattered over surface, apex acuminate to awned. **Lemmas:** sterile lemma with pustular-based setae scattered over surface, awned to unawned; fertile lemma 2.2–3.8 mm long, less than or scarcely as wide and long as glumes, orbicular to broadly obovate, 5- to 9-nerved, coriaceous, narrowing distally and extending into membranous tip, short hairs absent. **Awns:** lower glume sometimes awned, to 0.5 mm long; upper glume sometimes awned, to 2.5 mm long; sterile lemma, when awned, with awns to 10 mm long; fertile lemma short-awned, to 0.5 mm long. **VEGETATIVE CHARACTERISTICS: Growth Habit:** cespitose to spreading. **Culms:** spreading to erect, 0.3–2 m long; nodes generally glabrous, lower nodes glabrous to puberulent, upper glabrous, sometimes rooting at lower nodes. **Sheaths:** open, slightly compressed, glabrous. **Ligules:** none. **Blades:** 1–27 cm long, 0.8–30 mm wide, generally glabrous. **HABITAT:** Found growing in moist, generally disturbed sites in eastern Colorado (Barkworth et al. 2003). **COMMENTS:** Colorado specimens belong to

Figure 143. *Echinochloa muricata*.

var. *microstachya* Wiegand, which has spikelets 2.5–3.8 mm long, lower lemmas unawned to awned, and awns, when present, to 10 mm long.

42. *Eriochloa* Kunth

Cespitose annuals (ours) and perennials. Sheaths open; ligule a ciliated membrane; auricles absent. Inflorescence a loosely contracted panicle, spikelets subsessile or short-pediceled on unbranched or sparingly rebranched primary branches. Disarticulation below the glumes. Lower glume reduced and fused with rachis node to form a cup or disk. Upper glume and lemma of sterile floret about equal, usually scabrous, hispid or hirsute, acute or more commonly acuminate at apex. Lemma of fertile floret indurate, glabrous, finely rugose, with slightly inrolled margins, apiculate or short-awned at apex. Basic chromosome number, $x = 9$. Photosynthetic pathway, C_4.

Represented in Colorado by a single species.

122. *Eriochloa contracta* Hitchc. (Fig. 144).

Synonyms: None. **Vernacular Name:** Prairie cupgrass. **Life Span:** Annual. **Origin:** Native. **Season:** C_4, warm season. **FLORAL CHARACTERISTICS: Inflorescence:** loosely contracted panicle, 6–20 cm long, 0.3–1.2 cm wide; branches 10–28, 1.5–4.5 cm long, with 8–16 loosely appressed spikelets on short pedicels. **Spikelets:** lanceolate, 3.5–4.5 mm long, 1.2–1.7 mm wide, 2-flowered; lower floret sterile, consisting only of sterile lemma; upper floret fertile; disarticulation below the glumes. **Glumes:** lower reduced and fused with rachis node to form cuplike structure; upper 3–4.3 mm long, sparsely appressed pubescent below, glabrous to scabrous distally, 3- to 9-nerved, apex acuminate to awned. **Lemmas:** sterile lemma lanceolate, 3–4.3 mm long, 1.2–1.7 mm wide, 3- to 7-nerved, apex acuminate to mucronate; fertile lemma indurate, finely rugose, elliptic, 2.1–2.5 mm long, 5- to 7-nerved, margins inrolled over margins of fertile palea, apex awned. **Awns:** upper glume sometimes awned, awn 0.4–1 mm long; upper lemma awned, awn 0.4–1.1 mm long, easily broken off. **VEGETATIVE CHARACTERISTICS: Growth Habit:** cespitose. **Culms:** decumbent to erect, 0.2–2 m long, occasional rooting at lower nodes, glabrous to short-pubescent. **Sheaths:** variously pubescent. **Ligules:** short membrane with dense fringe of hairs, 0.4–1.1 mm long. **Blades:** flat to folded, 6–12 cm long, 2–8 mm wide, generally short-pubescent on both surfaces. **HABITAT:** Found as a weed in Bent, Otero, and Prowers counties. **COMMENTS:** Infrequent in Colorado, but it is a troublesome weed in Kansas, New Mexico, and Arizona.

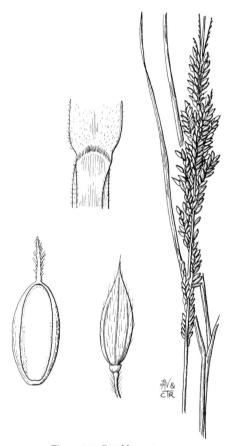

Figure 144. *Eriochloa contracta.*

43. *Panicum* L.

Annuals and perennials of extremely diverse habit. Ligule a membrane or a ring of hairs. Inflorescence an open or contracted panicle. Disarticulation below the glumes. Glumes usually both present, the lower commonly short. Lowermost floret sterile or occasionally staminate, with a lemma similar to the glumes in texture and usually as long as, or slightly longer than, the upper glume. Lemma of upper floret shiny and glabrous in our species, smooth, firm or indurate, tightly clasping palea with thick, inrolled margins. Palea of upper floret like the lemma in texture. Basic chromosome numbers, $x = 9$ and 10. Photosynthetic pathway, C_3 and C_4.

Represented in Colorado by 6 species and 5 subspecies.

1. Plant annual
    2. Lower glume to one-third as long as spikelet, obtuse or rounded; mature plant with coarse trailing culms to 1 m or more in length . . . . 124. *P. dichotomiflorum*
    2. Lower glume more than one-third as long as spikelet, acute or acuminate

3. Spikelets mostly 4–6 mm long . . . . 126. *P. miliaceum*

3. Spikelets less than 4 mm long . . . . 123. *P. capillare*

1. Plant perennial

    4. Lower glume as long as upper; long stolons present . . . . 127. *P. obtusum*

    4. Lower glume shorter than upper; stolons absent

        5. Rhizomes present . . . . 128. *P. virgatum*

        5. Rhizomes absent . . . . 125. *P. hallii*

123. *Panicum capillare* L. (Fig. 145).

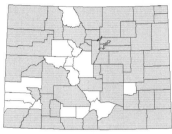

Synonyms: *Panicum barbipulvinatum* Nash, *Panicum capillare* L. subsp. *barbipulvinatum* (Nash) Tzvelev, *Panicum capillare* L. var. *agreste* Gattinger, *Panicum capillare* L. var. *barbipulvinatum* (Nash) R. L. McGregor, *Panicum capillare* L. var. *brevifolium* Vasey *ex* Rydb. & Shear, *Panicum capillare* L. var. *occidentale* Rydb. **Vernacular Name:** Witchgrass. **Life Span:** Annual. **Origin:** Native. **Season:** C$_4$, warm season. **FLORAL CHARACTERISTICS: Inflorescence:** open, diffuse panicles, 13–50 cm long, 7–24 cm wide, base included to exserted at maturity; branches spreading, densely flowered. **Spikelets:** acuminate, ellipsoid to lanceolate, 1.9–4 mm long, 2-flowered; disarticulating below the glumes; lower floret sterile and lacking palea; upper floret fertile. **Glumes:** lower 1–2 mm long, 1- to 3-nerved, one-third to half as long as spikelet; upper lanceolate, 1.8–3.1 mm long, 7- to 9-nerved, nerves scabrous, internerve spaces glabrous. **Lemmas:** sterile lemma 1.9–3 mm long, similar to upper glume, nerves prominent distally; fertile lemma 1.5–2.3 mm long, indurate, glabrous, shiny, pale to brownish. **Awns:** none. **VEGETATIVE CHARACTERISTICS: Growth Habit:** cespitose. **Culms:** decumbent to erect, 2–13 dm tall, rooting at lower nodes, stout to slender, culms hollow; nodes sparsely to densely pubescent. **Sheaths:** papillose-hirsute. **Ligules:** ciliate membrane, hairs 0.5–1.5 mm long. **Blades:** flat to folded, 5–40 cm long, 3–18 mm wide, linear, ciliate and sometimes sparsely papillose-pilose. **HABITAT:** Found as a common weed in disturbed areas over Colorado at low elevations and in the foothills. **COMMENTS:** Common. At maturity the inflorescence disarticulates at the base and becomes a "tumbleweed" (Barkworth et al. 2003). Two subspecies found in Colorado are subsp. *capillare* and subsp. *hillmanii* (Chase) Freckmann & Lelong. They can be separated by these characteristics:

1. Palea of lower floret absent; pedicels and panicle branches strongly divergent . . . .
. . . . subsp. *capillare*

1. Palea of lower floret present; pedicels and panicle branches appressed to narrowly divergent . . . . subsp. *hillmanii*

124. *Panicum dichotomiflorum* Michx. (Fig. 146).

Synonym: *Panicum dichotomiflorum* Michx. var. *geniculatum* (Wood) Fern. **Vernacular Names:** Fall panicgrass, fall panicum. **Life Span:** Annual. **Origin:** Native. **Season:** C$_4$, warm season. **FLORAL CHARACTERISTICS: Inflorescence:** diffuse panicle, 4–40 cm long, often included in upper sheath; branches alternate to opposite to occasionally whorled, spreading to ascending, pedicels generally less than 3 mm long, scabrous. **Spikelets:** elliptic to narrowly ovoid, 2.2–3.8 mm long, generally widest above the middle, apex acute to acuminate; disarticulation below the glumes; 2-flowered; lower floret sterile; upper floret fertile, 1.4–2.5 mm long, 0.7–1.1 mm wide, narrowly elliptic, smooth, shiny. **Glumes:** glabrous; lower broadly triangular, a quarter to one-third as long as spikelet, 1- to 3-nerved, some nerves obscure; upper

1.8–3.2 mm long, rounded to obtuse, obscurely 7-nerved, subcoriaceous. **Lemmas:** sterile lemma subcoriaceous, similar to upper glume in size and texture, exceeding upper floret

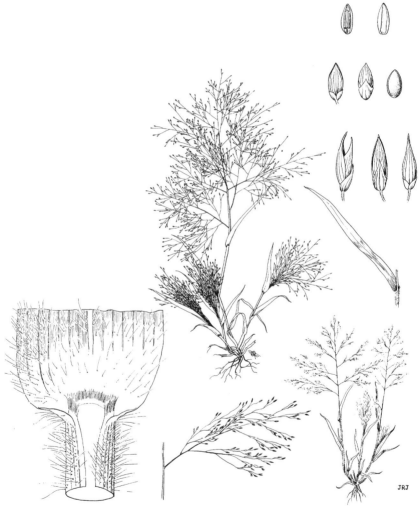

Figure 145. *Panicum capillare.*

by 0.3–0.6 mm; fertile lemma 1.8–2.3 mm long. **Awns:** none. **VEGETATIVE CHARACTERIS-TICS: Growth Habit:** tufted, terrestrial or sometimes aquatic. **Culms:** decumbent to erect to geniculate at base, rooting at nodes when in water, 0.5–2 m tall, glabrous, generally succulent; nodes generally swollen, branching occurring from lower and middle nodes, often resulting in coarse trailing culms 1 m or more in length at maturity. **Sheaths:** compressed, glabrous to sparsely pilose, hairs not papillose-based. **Ligules:** hairs on a membranous base, 0.5–2 mm long. **Blades:** flat, mostly glabrous, 10–65 cm long, 3–25 mm wide, midrib obvious, white. **HABITAT:** Found around Denver and Larimer counties and presumably spreading (Weber and Wittmann 2001a; Wingate 1994; Harrington 1954). **COMMENTS:** The only subspecies found in Colorado is the typical one (subsp. *dichotomiflorum*).

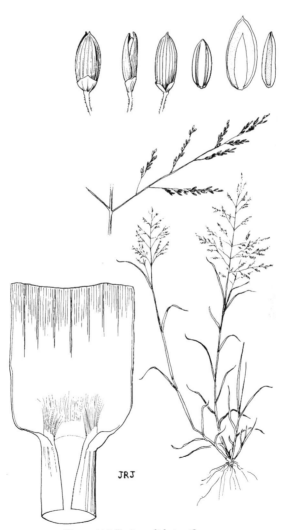

Figure 146. *Panicum dichotomiflorum.*

125. *Panicum hallii* Vasey (Fig. 147).

Synonym: *Panicum lepidulum* A. S. Hitchc. & Chase. **Vernacular Names:** Hall's panicgrass, Hall's witchgrass. **Life Span:** Perennial. **Origin:** Native. **Season:** C₄, warm season. **FLORAL CHARACTERISTICS: Inflorescence:** open, diffuse panicles, 13–50 cm long, 7–24 cm wide, base included to exserted at maturity; branches spreading, densely flowered. **Spikelets:** acuminate, ellipsoid to lanceolate, 1.9–4 mm long, 2-flowered; sometimes disarticulating below glumes;

lower floret sterile and palea present; upper floret fertile, sometimes disarticulating above the glumes. **Glumes:** lower 1–2 mm long, 1- to 3-nerved, one-third to half as long as spikelet; upper lanceolate, 1.8–3.1 mm long, 7- to 9-nerved, nerves scabrous, internerve spaces glabrous. **Lemmas:** sterile lemma 1.9–3 mm long, similar to upper glume, nerves prominent distally; fertile lemma 1.5–2.3 mm long, indurate, glabrous, shiny, brownish. **Awns:** none. **VEGETATIVE CHARACTERISTICS: Growth Habit:** cespitose. **Culms:** decumbent to erect, 1.5–13 dm tall, rooting at lower nodes, stout to slender, hollow; nodes sparsely to densely pubescent. **Sheaths:** papillose-hirsute, round. **Ligules:** ciliate membrane, hairs 0.5–1.5 mm long, truncate. **Blades:** flat to folded, ascending, 5–40 cm long, 3–18 mm wide, linear ciliate and sometimes sparsely papillose-pilose, curling as they age. **HABITAT:** Scattered in eastern Colorado in prairies, roadsides, and open ground. **COMMENTS:** The only subspecies found in Colorado is the typical one (subsp. *hallii*).

126. *Panicum miliaceum* L. (Fig. 148).

Synonyms: None. **Vernacular Names:** Broomcorn millet, hog millet, panic millet, proso millet. **Life Span:** Annual. **Origin:** Introduced. **Season:** C₄, warm season. **FLORAL CHARACTERISTICS: Inflorescence:** dense panicle, often drooping at tips when mature, 6–20 cm long, 4–11 cm wide, generally included at base at maturity; branches spreading to appressed; spikelets solitary and short-pedicellate, located distally on branches. **Spikelets:** ovoid, 4–6 mm long;

apex acuminate, nerves prominent, 2-flowered; not disarticulating at maturity; lower floret sterile; upper floret fertile, 3–3.8 mm long, 2–2.5 mm wide, smooth to striate, stramineous to orange. **Glumes:** glabrous, nerves scabridulous distally; lower broadly ovate, 2.8–3.6 mm long, half to three-quarters as long as spikelet, 5- to 7-nerved; upper ovate, 4–5.1 mm long, 11- to 13-nerved, apex acuminate to attenuate, slightly exceeding fertile floret. **Lemmas:** sterile lemma 4–4.8 mm long, 9- to 13-nerved, nerves scabrous distally; fertile lemma 3–3.8 mm long, broadly ovate, indurate, and occasionally obscurely 3-nerved. **Awns:** none. **VEGETATIVE CHARACTERISTICS: Growth Habit:** cespitose. **Culms:** decumbent to ascending to erect, stout, 2–12 dm tall; internodes glabrous to pubescent, hairs papillose-based; nodes puberulent. **Sheaths:** densely pilose with papillose-based hairs (hairs break easily and bases remain). **Ligules:** ciliate membrane, hairs 1–3 mm long. **Blades:** flat, 15–40 cm long, 7–25

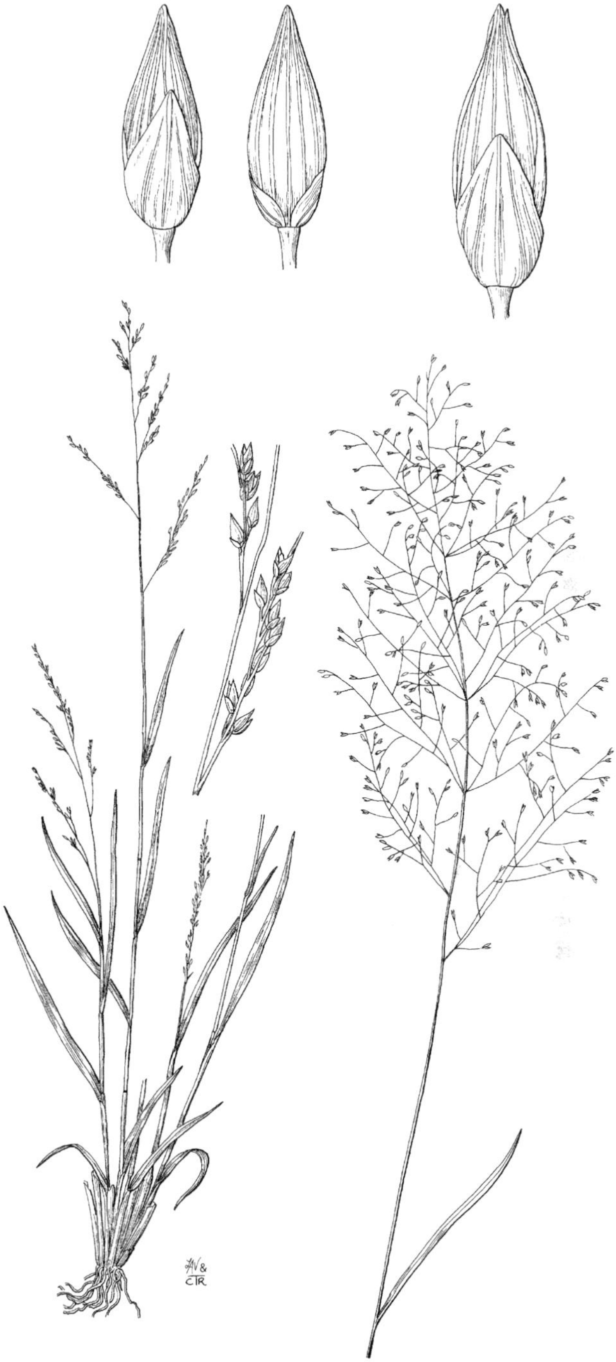

Figure 147. *Panicum hallii*.

269

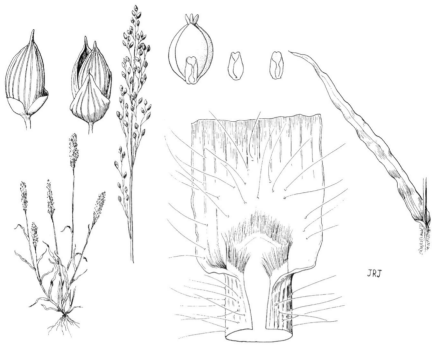

Figure 148. *Panicum miliaceum.*

mm wide, generally pubescent with papillose-based pilose hairs, rarely glabrate, margins crisped. **HABITAT:** Escapes from cultivation, and found as a weed along roadsides and field margins at low elevations in eastern Colorado. **COMMENTS:** The only subspecies found in Colorado is the typical one (subsp. *miliaceum*).

127. *Panicum obtusum* Kunth (Fig. 149).

Synonyms: None. **Vernacular Name:** Vine mesquite. **Life Span:** Perennial. **Origin:** Native. **Season:** C$_4$, warm season. **FLORAL CHARACTERISTICS: Inflorescence:** contracted panicle, 5–15 cm long, 0.8–1.5 cm wide; branches 2–6, erect to slightly spreading. **Spikelets:** oblong to obovate, 2.8–4.4 mm long, glabrous; disarticulating below the glumes; 2-flowered; lower floret staminate; upper floret fertile. **Glumes:** subequal, glabrous, 2.8–4.4 mm long, ovate to lanceolate-

ovate, apex rounded; lower three-quarters to as long as spikelet, 5- to 7-nerved; upper 5–9-nerved. **Lemmas:** sterile lemma similar to glumes, 5- to 9-nerved; fertile lemma ovate to elliptic, indurate, shiny, margins enfolded over margins of palea at base only. **Awns:** none. **VEGETATIVE CHARACTERISTICS: Growth Habit:** cespitose, stoloniferous with long stolons. **Culms:** decumbent to erect; base knotty, 0.2–2 m long; nodes pubescent; inter-

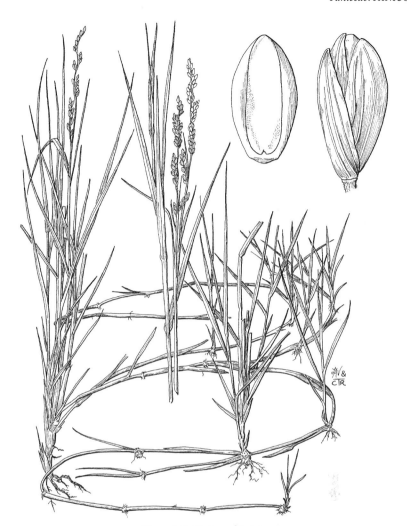

Figure 149. *Panicum obtusum.*

nodes glabrous, sprawling or climbing over shrubs. **Sheaths:** basal sheaths generally villous, upper sheaths glabrous. **Ligules:** membranous, truncate, entire to lacerate, 0.2–2 mm long. **Blades:** flat to involute near distal end, 3–26 cm long, 2–7 mm wide, base sparsely pilose, margins occasionally scabrous, midnerve whitish, prominent below. **HABITAT:** Found on heavy clay soils and moist depressions. **COMMENTS:** A fair to good forage species that withstands heavy grazing (Stubbendieck, Hatch, and Landholt 2003). Sometimes planted for erosion control. While this book was in publication, Zuloaga et al. (2007) placed this species into a monotypic genus [*Hopia obtusa* (Kunth) Zuloaga & Morrone] based on micromorphological characteristics of the upper floret, differing subtype of photosynthetic pathway, base chromosome number of 10, and genetic data. I agree with their circumscription.

128. *Panicum virgatum* L. (Fig. 150).

Synonyms: None. **Vernacular Name:** Switchgrass.
Life Span: Perennial. **Origin:** Native. **Season:** C_4,
warm season. **FLORAL CHARACTERISTICS: Inflorescence:** open panicle, 10–55 cm long, 2–15 mm wide;
branches ascending to spreading, solitary, paired or
whorled, spikelets located distally. **Spikelets:** elliptic-
ovate, glabrous, 2.5–8 mm long, 1.2–2.5 mm wide;
disarticulating below the glumes; 2-flowered; lower
floret staminate; upper floret fertile, 2.3–3 mm long,
0.8–1.1 mm wide, indurate, narrowly ovoid, glabrous, shiny. **Glumes:** unequal; lower 1.8–3.2
mm long, three-quarters as long as the upper, 5- to 9-nerved, lanceolate-ovate, apex acuminate, clasping base of the upper; upper lanceolate-ovate to ovate, 3.3–8 mm long, 7- to 11-
nerved, marginal nerves sometimes obscure, apex acuminate. **Lemmas:** sterile lemma 3.3–8
mm long, 7- to 11-nerved, large space at apex between upper glume and sterile lemma; fertile lemma indurate, shiny, clasping palea only at base. **Paleas:** sterile palea 3–3.5 mm long,
nearly as long as sterile lemma, ovate-hastate with lobes initially folded over anthers. **Awns:**
none. **VEGETATIVE CHARACTERISTICS: Growth Habit:** solitary to clumped, rhizomatous,
generally loosely interwoven, occasionally short, forming a knotty crown. **Culms:** decumbent to erect, 0.4–3 m tall, glabrous, robust. **Sheaths:** glabrous to pilose, especially near
the summit, margins generally ciliate, base often purplish to reddish. **Ligules:** ciliate membrane, hairs of various lengths, 2–6 mm long, obtuse. **Blades:** flat, 10–60 cm long, 2–15 mm
wide, glabrous to pubescent, upper surface has triangular patch of hair present at base of
leaf. **HABITAT:** Found on the plains and in the foothills of eastern Colorado on moist soils
and in the Escalante Canyon of the Western Slope. **COMMENTS:** A codominant of the
tallgrass prairie and a good forage species (Stubbendieck, Hatch, and Landholt 2003). Commonly used as a native ornamental and planted for roadside stabilization.

44. *Paspalum* L.

Annuals and perennials, many with rhizomes or stolons. Blades usually flat, often thin
and broad. Inflorescence with 1 to many unilateral spikelike branches, these scattered or,
in a few species, paired at the culm apex. Spikelets subsessile or short-pediceled, solitary or
in pairs on a flattened, occasionally broadly winged rachis, with rounded back of lemma
of fertile floret turned toward the rachis. Disarticulation at base of spikelet. Lower glume
typically absent but irregularly present in a few species. Upper glume and lemma of sterile floret usually about equal, broad and rounded at apex, infrequently acute. Lemma of
fertile floret firm or indurate, rounded on back, usually obtuse, with inrolled margins. Palea broad, flat or slightly convex, margins entirely enfolded by lemma. Basic chromosome
number, $x = 10$. Photosynthetic pathway, C_4.

Represented in Colorado by 4 species and 1 variety.

1. Spikelet margins ciliate with long hairs . . . . 129. *P. dilatatum*
1. Spikelet margins without long hairs
    2. Rachis broad and winged; plant annual; lemma of lower floret transversely wrinkled . . .
    . . . . 131. *P. racemosum*

Figure 150. *Panicum virgatum.*

2. Rachis narrow and narrowly winged; plant perennial; lemma of lower floret not transversely wrinkled

    3. Spikelets elliptic or obovate; inflorescence of 3–7 branches, axillary inflorescence absent . . . . 130. *P. pubiflorum*

    3. Spikelets suborbicular, broadly ovate or broadly obovate; inflorescence of usually 1–3 branches, axillary inflorescence usually present . . . . 132. *P. setaceum*

129. *Paspalum dilatatum* Poir. (Fig. 151).

Synonyms: None. **Vernacular Name:** Dallisgrass. **Life Span:** Perennial. **Origin:** Introduced. **Season:** C$_4$, warm season. **FLORAL CHARACTERISTICS: Inflorescence:** panicle with 2–7 unilateral spikelike branches; branches 1.5–12 cm long, ascending to drooping; rachis 0.7–1.4 mm wide, narrowly winged, glabrous, margins scabrous. **Spikelets:** paired, short-pediceled and overlapping, 2.3–4 mm long, 1.7–2.5 mm wide, ovate, apex acute, generally stramineous, margins ciliate with long hairs; disarticulating below the glumes; 2-flowered; lower floret sterile; upper floret fertile. **Glumes:** lower absent; upper 2.3–4 mm long, 5- to 7-nerved, apex abruptly pointed, margin long-pilose, hairs white. **Lemmas:** sterile lemma 2.3–4 mm long, 5- to 7-nerved, apex abruptly pointed, margin long-pilose, hairs white; fertile lemma indurate, minutely papillose, convex, broadly elliptic, margins involute clasping palea. **Awns:** none. **VEGETATIVE CHARACTERISTICS: Growth Habit:** cespitose, short-rhizomatous, forms a knotty base. **Culms:** erect to ascending to decumbent at base, 0.5–1.75 m tall; nodes glabrous to pubescent. **Sheaths:** glabrous to pubescent, lower sheaths generally pubescent, flattened, summit ciliate. **Ligules:** membranous, 1.5–3.8 mm long. **Blades:** flat, generally glabrous, to 35 cm long, 2–16.5 mm wide, base of upper surface often with a few long hairs. **HABITAT:** Weedy areas and open, disturbed ground. **COMMENTS:** Uncommon. This species is introduced from South America and has become a common weed in the southeastern United States. Sometimes used as a turf grass.

130. *Paspalum pubiflorum* Rupr. *ex* E. Fourn. (Fig. 152).

Synonyms: *Paspalum geminum* Nash, *Paspalum laeviglume* Scribn. *ex* Nash, *Paspalum pubiflorum* Rupr. *ex* Fourn. var. *glabrum* Vasey *ex* Scribn. **Vernacular Name:** Hairyseed paspalum. **Life Span:** Perennial. **Origin:** Native. **Season:** C$_4$, warm season. **FLORAL CHARACTERISTICS: Inflorescence:** panicle terminal, with 3–7 unilateral spikelike branches; branches 2.2–7.9 cm long, divergent to spreading; rachis 1.1–2.3 mm wide, narrowly winged, glabrous, margins scabrous. **Spikelets:** paired, short-pediceled, imbricate, elliptic to obovate, 2.8–3.6 mm long, 1.5–2 mm wide, glabrous to pubescent, ovate, apex acute, generally stramineous to light brown; disarticulating below the glumes; 2-flowered; lower floret sterile; upper floret fer-

Figure 151. *Paspalum dilatatum*.

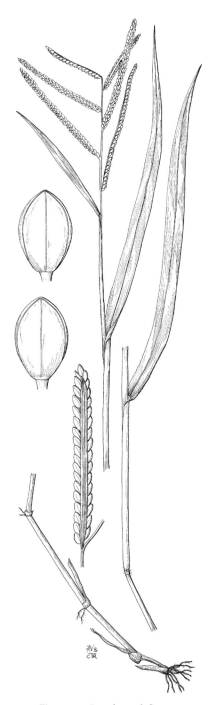

Figure 152. *Paspalum pubiflorum*.

tile. **Glumes:** lower absent; upper glume 2.3–4 mm long, glabrous to sparsely pubescent, 3-nerved, apex abruptly pointed, margin entire. **Lemmas:** sterile lemma 2.3–4 mm long, glabrous to sparsely pubescent, margin entire, smooth, 3-nerved, apex abruptly pointed; fertile lemma indurate, smooth, convex, broadly elliptic, stramineous, margins involute clasping palea. **Awns:** none. **VEGETATIVE CHARACTERISTICS: Growth Habit:** cespitose. **Culms:** erect to ascending to decumbent at base, rooting at lower nodes, 30–130 cm tall; nodes glabrous to pubescent. **Sheaths:** glabrous to pubescent, keeled, lower sheaths generally with some papillose-based hairs. **Ligules:** membranous, 1–3.2 mm long. **Blades:** flat, generally glabrous, to 31 cm long, 4–18 mm wide, base of upper surface often with a few long hairs, margins ciliate basally. **HABITAT:** Found as a lawn weed in and around Boulder. **COMMENTS:** Infrequent.

131. *Paspalum racemosum* Lam. (Fig. 153).

Synonym: *Paspalum stoloniferum* Bosc. **Vernacular Name:** Peruvian paspalum. **Life Span:** Annual. **Origin:** Introduced. **Season:** C$_4$, warm season. **FLORAL CHARACTERISTICS: Inflorescence:** panicle terminal, with 40–75 unilateral spikelike branches; branches 1–2.5 cm long, divergent to erect; rachis 1–1.5 mm wide, widely winged (foliaceous). **Spikelets:** paired, appressed to divergent, 2.5–2.9 mm long, 0.8–1.2 mm wide, elliptic to linear-elliptic, pubescent, generally stramineous or purplish, apex acute; disarticulating below the glumes; 2-flowered; lower floret sterile; upper floret fertile, 1.3–1.6 mm long. **Glumes:** lower absent; upper 2.5–2.9 mm long, thin, rugose below, fluted or wrinkled near margins, short-ciliate, apex acute, margin short ciliate. **Lemmas:** sterile lemma 2.5–2.9 mm long, thin, rugose below, fluted or transversely wrinkled on either side of midnerve, short-ciliate, apex acute, margin short-ciliate; fertile lemma indurate, 1.5–1.8 mm long, oblong, smooth, pale, shiny, margins involute clasping palea. **Paleas:** Fertile paleas indurate. **Awns:** none. **VEGETATIVE CHARACTERISTICS: Growth Habit:** cespitose. **Culms:** erect to widely spreading to sprawling, 4–9 dm tall; nodes purple to brown. **Sheaths:** glabrous, loose. **Ligules:** membranous, 0.1–0.3 mm long. **Blades:** flat, glabrous, 4–13 cm long, 10–22 mm wide, base rounded to subcordate, margins glabrous to scabrous. **HABITAT:** Found as an adventive in eastern Colorado (Weber and Wittmann 1992). **COMMENTS:** Infrequent. A recent introduction from South America.

132. *Paspalum setaceum* Michx. (Fig. 154).

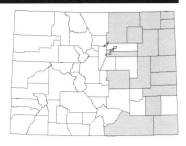

Synonyms: *Paspalum bushii* Nash, *Paspalum ciliatifolium* Michx. var. *stramineum* (Nash) Fern., *Paspalum stramineum* Nash. **Vernacular Names:** Thin paspalum, hairy bead grass. **Life Span:** Perennial. **Origin:** Native. **Season:** C$_4$, warm season. **FLORAL CHARACTERISTICS: Inflorescence:** panicle, terminal and axillary, with 1–3 unilateral spikelike branches; axillary panicles generally partially to completely enclosed by

Figure 153. *Paspalum racemosum*.

Figure 154. *Paspalum setaceum.*

subtending leaf sheath; branches 4–12 cm long, ascending to spreading; rachis 0.6–1.1 mm wide, not winged to slightly winged, glabrous, rarely scabrous. **Spikelets:** paired, short-pediceled, appressed to rachis and imbricate, 1.7–2.4 mm long, 1.5–2.1 mm wide, suborbicular to broadly obovate, generally stramineous, apex acute, pubescent to occasionally glabrous; disarticulating below the glumes; 2-flowered; lower floret sterile; upper floret fertile, 1.7–2.1 mm long. **Glumes:** lower absent; upper 1.7–2.4 mm long, 3-nerved, glabrous to shortly glandular-pubescent. **Lemmas:** sterile lemma 2.3–4 mm long, 2-nerved, glabrous to shortly glandular-pubescent; fertile lemma indurate, minutely papillose, convex, broadly elliptic, margins involute clasping palea. **Awns:** none. **VEGETATIVE CHARACTERISTICS: Growth Habit:** cespitose to short-rhizomatous. **Culms:** erect to spreading to prostrate, 2–11 dm tall; nodes glabrous to pubescent. **Sheaths:** glabrous to pubescent, lower sheaths generally pubescent, flattened, summit and margins pilose. **Ligules:** membranous, 0.2–0.5 mm long, hairs behind it to 2 mm long. **Blades:** flat, generally glabrous to occasionally pubescent, 5–25 cm long, 3–13.5 mm wide, both surfaces often with a few long hairs along midnerve, margins scabrous and with papillose-ciliate hairs. **HABITAT:** Found on the eastern plains on sandy soils. **COMMENTS:** The only variety found in Colorado is var. *stramineum* (Nash) D. Banks. This variety differs from the typical variety in the generally glabrate appearance and color of the leaf blades (yellowish green to dark green).

45. Pennisetum

Annuals or perennials with generally tall, erect culms, usually tightly cespitose (in ours). Leaf blades usually thin and flat, occasionally folded. Inflorescence a dense, bristly, tightly contracted, spicate panicle. Spikelets solitary to 2 to several in fascicles of numerous bristles. Bristles usually free at base (in ours) and in three distinct series: outer, inner, and primary. Fascicle disarticulating from rachis and containing spikelets. Lower glume small or vestigial. Upper glume and lemma of sterile floret about equal, or glume slightly shorter. Lemma and palea of fertile floret thin, smooth, and shiny; margins of lemma thin and flat. Basic chromosome numbers, $x = 5, 7, 8$, and 9. Photosynthetic pathway, C_4.

Represented in Colorado by a single species.

133. *Pennisetum alopecuroides* (L.) Spreng. (Fig. 155).

Synonym: *Panicum alopecuroides* L. **Vernacular Name:** Foxtail fountaingrass. **Life Span:** Perennial. **Origin:** Introduced. **Season:** C_4, warm season. **FLORAL CHARACTERISTICS: Inflorescence:** panicle, terminal, exserted, spicate, erect, densely flowered, 6–20 cm long, 20–50 mm wide; green, brown, deep purple, or creamy white; rachis terete with hairs; fascicles 9–16 per cm of rachis; outer bristles 10–20, 1–15 mm long; inner bristles 7–10, 11–30 mm long,

primary bristle 25–35 mm long, all bristles antrorsely barbed. **Spikelets:** glabrous, 5–8.5 mm long; disarticulation below the fascicle; 2-flowered; lower floret sterile; upper floret fertile. **Glumes:** lower 0.2–1.4 mm long, nerveless; upper 2–4.9 mm long, about half as long as spikelet, 1- to 5-nerved. **Lemmas:** sterile lemma ovate, 5–8 mm long, 7- to 9-nerved;

Figure 155. *Pennisetum alopecuroides.*

fertile lemma 5–7.5 mm long, acuminate, 5- to 7-nerved. **Paleas:** lower palea absent. **Awns:** none. **VEGETATIVE CHARACTERISTICS: Growth Habit:** strongly cespitose. **Culms:** ascending to erect; nodes glabrous, 3–10 dm tall, scabrous to base of inflorescence. **Sheaths:** generally glabrous but occasionally ciliate along margins. **Ligules:** row of hairs with a short membranaceous base, 0.2–0.5 mm long. **Blades:** flat to folded, 30–60 cm long, 2–12 mm wide. **HABITAT:** A single specimen reported as a weed growing in margin of bluegrass lawn. **COMMENTS:** A native of Southeast Asia, introduced into our area as a very showy and common ornamental. *Pennisetum* is closely related to *Cenchrus*, and they are sometimes combined into a single genus.

46. *Setaria* P. Beauv.

Cespitose annuals and perennials, with erect or geniculate culms, these often branching at base. Leaf blades typically flat and thin. Inflorescence a slender, usually contracted, densely flowered, bristly panicle, spikelets subsessile on main axis and short branches. Some or all of spikelets subtended by 1 to several persistent bristles (reduced branches), spikelets disarticulating above bristles. Glumes and lemma of sterile floret typically glabrous, prominently nerved, acute or obtuse; lower glume short, upper glume and lemma of sterile floret equal or, more frequently, upper glume half to two-thirds as long. Lemma and palea of fertile floret indurate, rounded at apex, usually finely or coarsely transverse-rugose. Basic chromosome number, $x = 9$. Photosynthetic pathway, C_4.

Represented in Colorado by 5 species, 1 subspecies, and 1 variety.

1. Plant perennial . . . . 135. *S. leucopila*
1. Plant annual
   2. Bristles retrorsely barbed . . . . 137. *S. verticillata*
   2. Bristles antrorsely barbed
      3. Inflorescence over 10 cm long
         4. Spikelet more than 2.5 mm long, upper floret smooth, shiny, rarely transversely rugose; disarticulation between upper and lower florets . . . . 134. *S. italica*
         4. Spikelet less than 2.5 mm long, upper floret distinctly finely transversely rugose; disarticulation below glumes . . . . 138. *S. viridis*
      3. Inflorescence less than 10 cm long . . . . 136. *S. pumila*

134. *Setaria italica* (L.) P. Beauv. (Fig. 156).

Synonyms: *Chaetochloa italica* (L.) Scribn., *Panicum italicum* L., *Setaria italica* (L.) P. Beauv. var. *metzgeri* (Koern.) Jáv., *Setaria italica* (L.) P. Beauv. var. *stramineofructa* (F. T. Hubbard) Bailey, *Setaria italica* (L.) P. Beauv. subvar. *metzgeri* (Koern.) F. T. Hubbard. **Vernacular Names:** Foxtail bristlegrass, foxtail millet. **Life Span:** Annual. **Origin:** Introduced. **Season:** C_4, warm season. **FLORAL CHARACTERISTICS: Inflorescence:** cylindrical panicle, short-lobed, densely

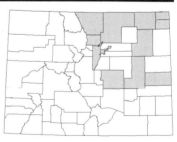

flowered, 8–30 cm long, to 30 mm wide, nodding, lobed, yellowish to greenish to purplish; rachis hispid to villous; 1–3 bristles subtending spikelets, purplish, antrorsely sca-

Figure 156. *Setaria italica.*

brid, lengths various. **Spikelets:** glabrous, 2.6–3 mm long; disarticulation between upper (fertile) and lower (staminate) florets; 2-flowered; lower floret staminate; upper floret fertile, smooth, shiny, rarely transversely rugose. **Glumes:** lower triangular-ovate, 0.8–1.4 mm long, about one-third the length of spikelet, 3-nerved; upper 2–2.6 mm long, about three-quarters the length of spikelet, 5- to 7-nerved. **Lemmas:** sterile lemma ovate, 2.2–2.8 mm long, equal or subequal to fertile lemma, 5- to 7-nerved; fertile lemma suborbicular at maturity, 2.2–2.8 mm long. **Paleas:** sterile palea absent to half as long as sterile lemma; fertile palea very finely transversely rugose to smooth, shiny. **Awns:** none. **VEGETATIVE CHARACTERISTICS: Growth Habit:** cespitose. **Culms:** ascending to erect; nodes appressed pubescent, 1–10 dm tall, scabrous to base of inflorescence. **Sheaths:** generally glabrous but occasionally scabrous to pubescent, margins sparsely ciliate. **Ligules:** row of hairs with a short membranaceous base, 1–2 mm long. **Blades:** flat, 10–40 cm long, 10–30 mm wide; both surfaces scabrous. **HABITAT:** Found on the eastern plains of Colorado as a weed in moist areas, roadsides, and abandoned fields. **COMMENTS:** Infrequent. A cultivated species in China and Europe. It is now used mostly as a hay and pasture grass. Generally a weed in our area. Closely related to *S. viridis.*

135. *Setaria leucopila* (Scribn. & Merr.) K. Schum. (Fig. 157).

Synonym: *Chaetochloa leucopila* Scribn. & Merr. **Vernacular Name:** Streambed bristlegrass. **Life Span:** Perennial. **Origin:** Native. **Season:** C$_4$, warm season. **FLORAL CHARACTERISTICS: Inflorescence:** cylindrical panicle, densely flowered, often interrupted, 6–15 cm long, 7–17 mm wide, erect, pale green; rachis scabrous to villous; bristles subtending; spikelets generally solitary, antrorsely scabrid, 4–15 mm long. **Spikelets:** elliptic, 2.2–2.8 mm long; disarticulation below the glumes; 2-flowered; lower floret staminate; upper floret fertile, 2.2–2.8 mm long. **Glumes:** lower 1.1–1.8 mm long, about half the length of spikelet, 3-nerved; upper 1.7–2.4 mm long, about three-quarters the length of spikelet, 5-nerved. **Lemmas:** sterile lemma ovate, 2.2–2.8 mm long, equal or subequal to fertile lemma, 5-nerved; fertile lemma 2.3–3.2 mm long, finely papillate with minute cross-ridges. **Paleas:** sterile palea lanceolate, 0.9–1.8 mm long; fertile palea flat, finely papillate with minute cross-ridges. **Awns:** none. **VEGETATIVE CHARACTERISTICS: Growth Habit:** cespitose. **Culms:** erect to geniculate at base, slightly flattened and grooved on 1 side, 2–10 dm tall; nodes appressed pubescent. **Sheaths:** glabrous, compressed-keeled, upper margins ciliate to villous. **Ligules:** dense fringe of hairs on basal membrane, 1–2.5 mm long. **Blades:** flat to folded at maturity, 8–25 cm long, 2–5 mm wide; both surfaces scabrous. **HABITAT:** Found on rocky slopes around Mesa de Maya and the Arkansas Valley near Cañon City (Weber and Wittmann 2001a). **COMMENTS:** A fair to good forage species.

136. *Setaria pumila* (Poir.) Roem. & Schult. (Fig. 158).

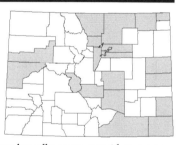

Synonyms: *Chaetochloa glauca* (L.) Scribn., *Chaetochloa lutescens* (Weigel) Stuntz, *Panicum glaucum* L., *Setaria glauca* (L.) P. Beauv., *Setaria lutescens* (Weigel) F. T. Hubbard. **Vernacular Name:** Yellow bristlegrass. **Life Span:** Annual. **Origin:** Introduced. **Season:** C$_4$, warm season. **FLORAL CHARACTERISTICS: Inflorescence:** cylindrical panicle, densely flowered, 3–10 cm long, to 10 mm wide, erect, yellowish, densely spicate; rachis hispid; 4–12 bristles subtending spikelets, antrorsely scabrid, lengths various, yellowish. **Spikelets:** broadly ovate, turgid, 3–3.4 mm long; disarticulation below the glumes; 2-flowered; lower floret staminate; upper floret fertile. **Glumes:** lower broadly ovate, 1–1.7 mm long, about one-third the length of spikelet, 3-nerved, apex acute; upper ovate, 1.4–2.2 mm long, about half the length of spikelet, 5-nerved, apex acute to apiculate, margins scarious. **Lemmas:** sterile lemma ovate, 2.2–3 mm long, equal or subequal to fertile lemma, apex apiculate, 5-nerved; fertile lemma ovate, 2.3–3.2 mm long, apex acute, coarsely transversely rugose. **Paleas:** sterile palea broad, equal to sterile lemma; fertile palea coarsely transversely rugose. **Awns:** none. **VEGETATIVE CHARACTERISTICS: Growth Habit:** cespitose. **Culms:** ascending to prostrate to geniculate at base, generally rooting at lower nodes, 3–13 cm tall. **Sheaths:** glabrous, compressed-

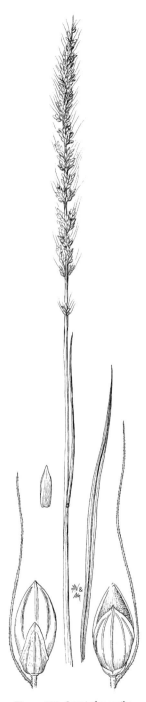

Figure 157. *Setaria leucopila*.

Figure 158. *Setaria pumila*.

keeled, margins scarious. **Ligules:** ciliate membrane, 0.2–1.2 mm long. **Blades:** flat to occasionally twisted, yellowish green, 5–20 cm long, 4–10 mm wide, upper surface glabrous to occasionally scabrous or pilose, base generally papillose-pilose near throat, lower surface glabrous. **HABITAT:** Found as a weed along roadsides, lawns, and fields in eastern Colorado. **COMMENTS:** There has been much confusion over the correct name for this taxon. Weber and Wittmann (2001a, 2001b) and Wingate (1994) call this species *S. glauca* (L.) P. Beauvois, and Harrington (1954) calls it *S. lutescens* (Weigel) F. T. Hubbard. The only subspecies found in Colorado is the typical one (subsp. *pumila*).

137. *Setaria verticillata* (L.) P. Beauv. (Fig. 159).

Synonyms: *Chaetochloa verticillata* (L.) Scribn., *Panicum verticillatum* L., *Setaria carnei* Hitchc. **Vernacular Names:** Hooked bristlegrass, bristly foxtail. **Life Span:** Annual. **Origin:** Introduced. **Season:** C$_4$, warm season. **FLORAL CHARACTERISTICS: Inflorescence:** cylindrical panicle, densely flowered, often interrupted near base, 5–15 cm long, erect, yellowish green to purplish; branches short, verticillate, retrorsely hispid; bristles subtending; spikelets generally solitary, retrorsely scabrid, 4–7 mm long. **Spikelets:** oblong-elliptic, glabrous, 2.2–2.5 mm long; disarticulation below the glumes; 2-flowered; lower floret staminate; upper floret fertile, 2.2–2.8 mm long. **Glumes:** lower 0.7–1.2 mm long, broadly ovate, about one-third the length of spikelet, 1-nerved; upper 1.6–2 mm long, ovate, about as long as sterile lemma, 5- to 7-nerved. **Lemmas:** sterile lemma ovate, 1.9–2.4 mm long, equal or subequal to fertile lemma, 7-nerved; fertile lemma ovate, 1.8–2.3 mm long, obscurely 3-nerved, finely transversely rugose. **Paleas:** sterile palea membranous, small; fertile palea flat, finely transversely rugose. **Awns:** none. **VEGETATIVE CHARACTERISTICS: Growth Habit:** cespitose. **Culms:** erect to ascending to geniculate at base, generally glabrous, 3–10 dm tall, grooved on 1 side. **Sheaths:** glabrous, compressed-keeled, distal margins ciliate. **Ligules:** dense fringe of hairs on basal membrane, to 1 mm long. **Blades:** flat, 8–20 cm long, 5–15 mm wide; upper surface generally scabrous. **HABITAT:** Found at low elevations in cultivated areas and along roadsides over Colorado. **COMMENTS:** A recent introduction from Europe, and now a fairly aggressive weed.

138. *Setaria viridis* (L.) P. Beauv. (Fig. 160).

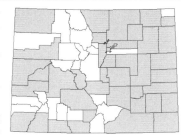

Synonyms: *Chaetochloa viridis* (L.) Scribn., *Panicum viride* L., *Setaria viridis* (L.) P. Beauv. var. *breviseta* (Doell) Hitchc., *Setaria viridis* (L.) P. Beauv. var. *weinmannii* (Roem. & Schult.) Borbás. **Vernacular Names:** Green bristlegrass, green foxtail. **Life Span:** Annual. **Origin:** Introduced. **Season:** C$_4$, warm season. **FLORAL CHARACTERISTICS: Inflorescence:** cylindrical panicle, densely flowered, apex nodding at maturity, occasionally interrupted near base, 3–20

JRJ

Figure 159. *Setaria verticillata.*

cm long, erect, greenish; branches short, verticillate, antrorsely hispid to villous; 1–3 bristles subtending spikelet, antrorsely scabrous, generally green, rarely purple. **Spikelets:** oblong-elliptic, glabrous, 1.8–2.2 mm long; disarticulation below the glumes; 2-flowered; lower floret staminate; upper floret fertile, 2.2–2.8 mm long, distinctly very finely transversely rugose. **Glumes:** lower 1–1.2 mm long, ovate, about one-third the length of spikelet, 1- to 3-nerved; upper 2–2.4 mm long, ovate, about as long as spikelet, 5- to 7-nerved. **Lemmas:** sterile lemma ovate, 2–2.5 mm long, slightly exceeding fertile lemma, 5- to 7-nerved; fertile lemma ovate, pale green, 1.8–2.5 mm long, obscurely 3- to 5-nerved. **Paleas:** sterile palea hyaline, about one-third the size of sterile lemma; fertile palea flat, very finely transversely rugose. **Awns:** none. **VEGETATIVE CHARACTERISTICS: Growth Habit:** cespitose. **Culms:** erect to ascending to geniculate at base, generally glabrous, 0.2–2.5 m tall. **Sheaths:** glabrous to occasionally scabridulous, compressed-keeled, distal margins ciliate. **Ligules:** dense fringe of hairs on basal membrane, 1–2 mm long. **Blades:** flat, 8–20 cm long, 4–25 mm wide; surfaces scabrous to glabrous. **HABITAT:** Found at low elevations in cultivated areas and along roadsides over Colorado. **COMMENTS:** The only variety found in Colorado is the typical one (var. *viridis*). Closely related and sometimes confused with *S. italica*. This is the most common annual *Setaria* in the United States (Barkworth et al. 2003).

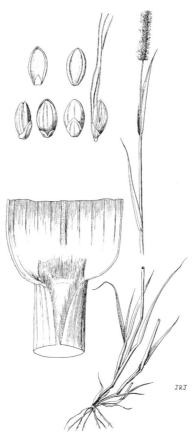

Figure 160. *Setaria viridis.*

VII. POOIDEAE

Annuals or perennials, cespitose, rhizomatous, rarely stoloniferous. Sheath margins free or connate; ligules membranous, sometimes a ring of hairs; auricles present in some. Inflorescence a panicle or bilateral spike, rarely a raceme. Disarticulation usually above the glumes. Spikelet with 1 to several florets, laterally compressed or terete; reduced florets when present usually above fertile florets (except in *Arrhenatherum, Hierochloë, Phalaris, Phalaroides*). Lemmas 5- or more-nerved; awned or awnless.

9. BRACHYPODIEAE

Annual or perennials, cespitose or rhizomatous. Sheath open, margins overlapping most of their length. Ligules membranous, sometimes short-ciliate. Auricles absent. Inflorescence terminal, a spikelike raceme. Disarticulation above the glumes and beneath florets. Spikelets all alike, laterally compressed to terete, of several to many fertile florets with imperfect florets (when present) above. Glumes persistent, unequal, lower 3- to 7-nerved, upper 5- to 9-nerved. Rachilla prolonged beyond base of distal floret. Lemmas 5- to 9-nerved, apex entire, awnless or less frequently awned. Palea 2-keeled, shorter to slightly longer than lemma.

47. *Brachypodium* P. Beauv.

Annual (in ours) with short-pediceled or subsessile spikelets borne singly at nodes; a stiffly erect, spicate raceme. Inflorescence sometimes reduced to a single terminal spikelet. Spikelets large, several- to many-flowered. Disarticulation above the glumes and between florets. Glumes unequal, stiffly pointed, 5- to 7-nerved. Lemmas firm, rounded or flattened on back, 7-nerved, midnerve extending into a short or long awn from an entire apex. Palea about as long as lemma, 2-keeled, often with stiff, pectinate-ciliate hairs on nerves. Basic chromosome number, $x = 7$. Photosynthetic pathway, C_3.

Represented in Colorado by a single species.

139. *Brachypodium distachyon* (L.) P. Beauv. (Fig. 161).

Synonyms: *Bromus dastachyos* L., *Trachynia distachya* (L.) Link. **Vernacular Name:** Purple falsebrome. **Life Span:** Annual. **Origin:** Introduced. **Season:** C_3, cool season. **FLORAL CHARACTERISTICS: Inflorescence:** raceme, 2–7 cm long, 1–7 overlapping, appressed spikelets, basal ones sometimes diverging. **Spikelets:** laterally compressed, 15–40 mm long, 1- to 15-flowered; disarticulation above the glumes and between florets. **Glumes:** unequal, glabrous; lower 5–6 mm long, 5- to 7-nerved; upper 7–8 mm long, 7- to 9-nerved. **Lemmas:** glabrous to scabrous, 7–10 mm long, lanceolate with hyaline margins, 7-nerved, apex entire, awned. **Awns:** lemmas awned, arising from entire apex, awn 8–20 mm long, usually straight. **VEGETATIVE CHARACTERISTICS: Growth Habit:** solitary to loosely tufted. **Culms:** erect to decumbent, 5–50 cm tall, internodes glabrous; nodes conspicuously pubescent. **Sheaths:** open, mostly glabrous. **Ligules:** ciliated membrane, 1–2 mm long, obtuse. **Blades:** flat, 10–40 cm long, 2–5 mm wide, glaucous, both surfaces sparsely hairy. **HABITAT:** Found along roadsides,

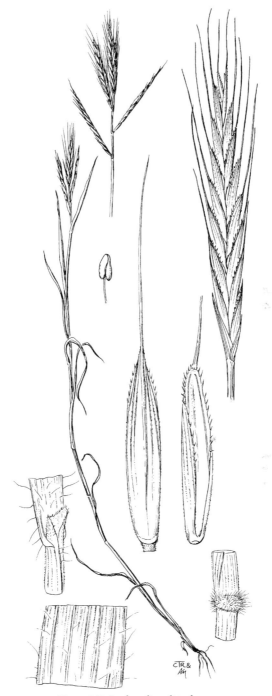

Figure 161. *Brachypodium distachyon*.

field margins, and other disturbed areas. **COMMENTS:** Rare weedy species in Colorado. Only known from a single report with no voucher specimen (Michael Piep, personal communication) and possibly never a member of our flora. This genus is sometimes included with the Poeae (Gould and Shaw 1983) or Bromeae. But recent nucleic acid data indicate that *Brachyypodium* is an isolated genus within the Pooideae and warrants distinction in a monotypic tribe (Barkworth et al. 2007).

10. BROMEAE

Annuals or perennials, cespitose, occasionally rhizomatous. Sheath margins connate for most of their length. Ligules membranous. Auricles absent. Inflorescence a panicle. Disarticulation above the glumes. Spikelets all alike, of several to many fertile florets with imperfect florets (when present) above. Glumes persistent, shorter than lowest lemma, awnless. Lemmas 5- to 13-nerved, with or without 2-tooth apex, awned or rarely awnless. Palea ciliate on the 2 keels, adhering to caryopsis.

48. *Anisantha* K. Koch

   Annuals. Sheath margins connate. Ligules membranous. Auricles absent. Inflorescence a simple panicle, sometimes racemose when immature. Disarticulation above the glumes. Lower glume 1- to 3-nerved. Lemmas long and narrow, 10–30 mm long and 0.5–1.5 mm wide, 5- or more-nerved, acute to acuminate, apical teeth usually more than 2 mm long; awned from between lemma teeth, awn usually 10–60 mm long. Basic chromosome number, $x = 7$. Photosynthetic pathway, C_3.

Represented in Colorado by 3 species. A segregate of *Bromus*.

1. Lemma 20–35 mm long . . . . 140. *A. diandra*
1. Lemma 9–20 mm or less
   2. Lemma 9–12 mm long; upper glume generally 1 cm long or less; at least some panicle branches with 4–8 spikelets . . . . 142. *A. tectorum*
   2. Lemma 13–20 mm long; upper glume 1 cm long or more; rarely panicle branches with more than 3 spikelets . . . . 141. *A. sterilis*

140. *Anisantha diandra* (Roth) Tutin *ex* Tzvelev (Fig. 162).

Synonyms: *Bromus diandrus* Roth subsp. *diandrus*, *Bromus gussonei* Parl., *Bromus rigidus* Roth. **Vernacular Names:** Great brome, ripgut brome. **Life Span:** Annual. **Origin:** Introduced. **Season:** C_3, cool season. **FLORAL CHARACTERISTICS: Inflorescence:** compact to open panicle, 7–20 cm long; branches erect to nodding. **Spikelets:** compressed, 30–50 mm long without awns, 4- to 8-flowered; disarticulation above the glumes. **Glumes:** unequal, glabrous to scabrous with

hyaline margins, keel glabrous to scabrous; lower 13–25 mm long, 1- to 3-nerved; upper 20–35 mm long, 3- to 5-nerved. **Lemmas:** glabrous to scabrous, 20–35 mm long, lanceolate with hyaline margins, 7-nerved, apex bifid, teeth 3–6 mm long. **Awns:** lemmas awned, arising from between lobes, awn 30–65 mm long. **VEGETATIVE CHARACTERISTICS: Growth**

Figure 162. *Anisantha diandra*.

Habit: solitary to loosely tufted. **Culms:** erect to decumbent, 2–9 dm tall; nodes swollen. **Sheaths:** softly pilose, margins connate. **Ligules:** membranous, 2–6 mm long, obtuse, lacerate. **Blades:** flat to involute on drying, 10–20 cm long, 3–8 mm wide, both surfaces pilose. **HABITAT:** Found along roadsides, field margins, and other disturbed areas. **COMMENTS:** Infrequent weedy species in Colorado. Excellent forage when very young, but becoming worthless at the onset of flowering (Stubbendieck, Hatch, and Landholt 2003). Stiff awns can cause serious injury to nose, eyes, and belly of grazing animals (Allred 2005).

141. *Anisantha sterilis* (L.) Nevski (Fig. 163).

Synonym: *Bromus sterilis* L. **Vernacular Names:** Barren brome, poverty brome. **Life Span:** Annual. **Origin:** Introduced. **Season:** C$_3$, cool season. **FLORAL CHARACTERISTICS: Inflorescence:** open panicle, 10–20 cm long; branches ascending to drooping, 2–10 cm long, rarely bearing more than 3 spikelets. **Spikelets:** subterete to slightly compressed, 20–35 mm long, 3- to 9-flowered, glabrous to strigulose-puberulent; disarticulation above the glumes. **Glumes:** unequal, linear to narrowly lanceolate, tapered to awnlike tips; lower 6–14 mm long, 1-nerved; upper 10–20 mm long, 3-nerved. **Lemmas:** narrowly elliptic to lanceolate, 13–20 mm long, back rounded, 5- to 7-nerved, apex bifid, teeth 1–3 mm long, very slender. **Awns:** lemmas awned, awn from between lobes, 18–26 mm long. **VEGETATIVE CHARACTERISTICS: Growth Habit:** solitary to small tufts. **Culms:** erect to occasionally geniculate below, 2–10 dm tall. **Sheaths:** densely pubescent with fine hairs. **Ligules:** erose-ciliate to lacerate membrane, 1–2.5 mm long. **Blades:** flat to loosely involute, 5–20 cm long, 2–5 mm wide, both surfaces pubescent. **HABITAT:** Found at lower elevations in the foothills and towns in waste places and disturbed ground. **COMMENTS:** Infrequent weedy species.

142. *Anisantha tectorum* (L.) Nevski (Fig. 164).

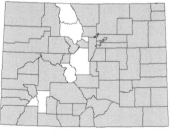

Synonyms: *Bromus tectorum* L., *Bromus tectorum* L. var. *glabratus* Spenner, *Bromus tectorum* L. var. *hirsutus* Regel, *Bromus tectorum* L. var. *nudus* Klett & Richter. **Vernacular Names:** Cheatgrass, downy brome, june grass, downy chess. **Life Span:** Winter annual. **Origin:** Introduced. **Season:** C$_3$, cool season. **FLORAL CHARACTERISTICS: Inflorescence:** open panicle, 5–20 cm long, initially erect but drooping with age; densely branched, branches 1–4 cm long, generally 1-sided, usually at least 1 branch bearing 4–8 spikelets; branches and pedicels flexuous. **Spikelets:** subterete to slightly compressed, 10–24 mm long, purplish to greenish, 4- to 8-flowered; disarticulation above the glumes. **Glumes:** unequal, narrowly lanceolate to lanceolate; lower 4–8 mm long, 1-nerved; upper 8–11 mm long, 3-nerved. **Lemmas:** lanceolate, back rounded to slightly carinate, margins hyaline, 9–12 mm long, apex bidentate, teeth membranous, 1–3 mm long. **Awns:** lemmas awned from between lobes, awn straight, 10–18 mm long.

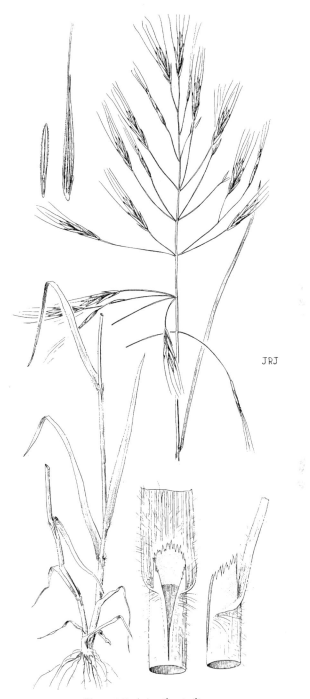

JRJ

Figure 163. *Anisantha sterilis*.

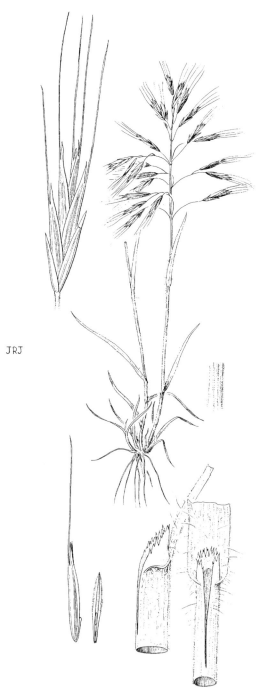

JRJ

Figure 164. *Anisantha tectorum.*

VEGETATIVE CHARACTERISTICS: Growth Habit: solitary to tufted. **Culms:** decumbent at base to erect, 0.5–9 dm tall, softly pubescent throughout. **Sheaths:** retrorsely softly pubescent, round, margins connate. **Ligules:** erose to lacerate membrane, 1–3.5 mm long. **Blades:** flat to loosely involute, narrow, 4–16 cm long, 2–7 mm wide, generally softly pubescent on both surfaces but occasionally glabrous. **HABITAT:** Found over Colorado from grassland and waste places to montane communities. **COMMENTS:** Very common weedy species that increases fire frequency where it has invaded. The awns can cause injury to grazing animals. Fair forage for domestic animals just after germination, but becoming worthless once flowering begins (Stubbendieck, Hatch, and Landholt 2003).

49. *Bromopsis* (Dumort.) Fourr.

Perennials, cespitose or rhizomatous. Sheath margins connate. Auricles absent. Ligule membranous. Inflorescence a large, open panicle. Disarticulation above the glumes and between florets. Spikelets large, with numerous florets, slightly laterally compressed to rounded on back, florets closely overlapping. Glumes unequal, acute or blunt, rarely awned, shorter than lowest floret, lower with fewer nerves than upper. Lemmas 5- to 9-nerved, bifid at apex, awned from between teeth. Palea adnate to caryopsis. Basic chromosome number, $x = 7$. Photosynthetic pathway, C_3.

Represented in Colorado by 9 species and 1 subspecies. A segregate of *Bromus*.

1. Rhizomes present
   2. Lemma usually glabrous, occasionally sparsely pubescent along margins and at base; usually awnless or with an awn up to 3.5 mm long . . . . 145 *B. inermis*
   2. Lemmas pubescent across back, or on lower portion and margins, or along marginal nerves and keel; with an awn up to 7.5 mm long or rarely awnless . . . .
   . . . . 150. *B. pumpelliana*
1. Rhizomes absent
   3. Lower glumes within a panicle generally 3-nerved, some 1-nerved
      4. Leaf blades often glaucous; glumes usually glabrous . . . . 144. *B. frondosa*
      4. Leaf blades not glaucous; glumes usually pubescent . . . . 148. *B. porteri*
   3. Lower glumes within a panicle generally 1-nerved, some 3-nerved
      5. Glumes usually pubescent, rarely glabrous
         6. Upper glume mucronate . . . . 147. *B. mucroglumis*
         6. Upper glume not mucronate
            7. Awn 1–3.5 mm long; blades 2–5 mm wide . . . . 148. *B. porteri*
            7. Awn 3–8 mm long; blades 6–19 mm wide . . . . 149. *B. pubescens*
      5. Glumes usually glabrous, rarely pubescent
         8. Lemma backs and margins usually pubescent, sometimes nearly glabrous; awns 2–4 mm long; anthers 1.8–4 mm long . . . . 146. *B. lanatipes*
         8. Lemma margins conspicuously hirsute or densely pilose along at least the lower half; awns 3–5 mm long; anthers 1–2.7 mm long
            9. Backs of all lemmas glabrous or sometimes scabrous; anthers 1–1.4 mm long; upper glume 7.1–8.5 mm long . . . . 143. *B. ciliata*
            9. Backs of upper lemmas in a spikelet hairy; anthers 1.6–2.7 mm long; upper glume 8.9–11.3 mm long . . . . 151. *B. richardsonii*

143. *Bromopsis ciliata* (L.) Holub. (Fig. 165).

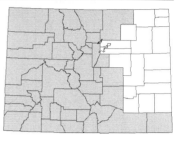

Synonyms: *Bromopsis canadensis* (Michx.) Holub, *Bromus canadensis* Michx., *Bromus ciliatus* L. var. *genuinus* Fern., *Bromus ciliatus* L. var. *intonsus* Fern., *Bromus dudleyi* Fern., *Bromus richardsonii* Link var. *pallidus* (Hook.) Shear. **Vernacular Name:** Fringed brome. **Life Span:** Perennial. **Origin:** Native. **Season:** C₃, cool season. **FLORAL CHARACTERISTICS: Inflorescence:** open panicle, 7–25 cm long; branches generally spreading or nodding. **Spikelets:** terete to compressed, 15–30 mm long, 4- to 9-flowered; disarticulation above the glumes. **Glumes:** unequal, glabrous to scabrous, narrow lanceolate to lanceolate, apex obtuse to acute; lower 5–8 mm long, generally 1-nerved; upper 7.1–8.5 mm long, 3-nerved. **Lemmas:** back glabrous, margins conspicuously hirsute on lower half to third, apex entire to emarginate, 8–15 mm long, 5- to 7-nerved. **Awns:** lemmas awned, awn 3–5 mm long. **Anthers:** 1–1.4 mm long. **VEGETATIVE CHARACTERISTICS: Growth Habit:** loosely tufted. **Culms:** erect, 4.5–12 dm tall; nodes generally pubescent; internodes glabrous. **Sheaths:** generally glabrous above, basal ones occasionally slightly retrorsely pilose, margins often ciliate near collar. **Ligules:** erose-ciliate membrane, truncate, 0.3–2 mm long. **Blades:** flat, lax, 13–25 cm long, 4–10 mm wide, both surfaces glabrous to pilose. **HABITAT:** Found along moist roadsides, trails, and stream banks from foothills to subalpine elevations in the mountains and mesas. **COMMENTS:** Species is widespread across North America. Weber and Wittmann (2001a, 2001b) call this taxon *B. canadensis* (Michx.) Holub. This species is sometimes combined with *B. richardsonii*.

144. *Bromopsis frondosa* (Shear) Holub. (Fig. 166).

Synonyms: *Bromus frondosus* (Shear) Wooton & Standl., *Bromus porteri* (J. M. Coult.) Nash var. *frondosus* Shear. **Vernacular Name:** Weeping brome. **Life Span:** Perennial. **Origin:** Native. **Season:** C₃, cool season. **FLORAL CHARACTERISTICS: Inflorescence:** open panicle, 10–20 cm long; branches (depending on period of growing season) appressed, ascending, spreading to drooping. **Spikelets:** terete to slightly compressed, 15–30 mm long, 4- to 10-flowered; disarticulation above the glumes. **Glumes:** usually glabrous to rarely puberulent, 3-nerved; lower 5–8 mm long, rarely 1-nerved; upper 6–9 mm long, frequently with a mucronate tip. **Lemmas:** glabrous to puberulent, elliptic to lanceolate, back pubescent (rarely glabrous), margins with longer hairs, apex entire, 8–12 mm long, **Awns:** lemma awn originates less than 1.5 mm below apex, 1.5–4 mm long. **VEGETATIVE CHARACTERISTICS: Growth Habit:** loosely tufted, not rhizomatous. **Culms:** erect to spreading, 4–11 dm tall; nodes usually glabrous; internodes glabrous. **Sheaths:** generally glabrous, rarely pubescent; midrib narrowed just below collar; auricles absent. **Ligules:** laciniate membrane, truncate to obtuse, 1–3 mm long. **Blades:** flat, 10–20 cm long, 3–6 mm wide, usually glabrous, often glaucous, basal ones often pubescent. **HABITAT:** Commonly found in open forests and on rocky

Figure 165. *Bromopsis ciliata*.

Figure 166. *Bromopsis frondosa.*

slopes from mid- to upper elevations. **COMMENTS:** Closely related to *B. porteri* but differs in having glabrous glumes and pedicels, lax or spreading leaf blades, and midrib of culm leaves narrowed just below collar (Barkworth et al. 2007).

145. *Bromopsis inermis* (Leyss.) Holub. (Fig. 167).

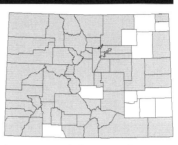

Synonym: *Bromus inermis* Leyss. subsp. *inermis* var. *inermis*. **Vernacular Names:** Smooth brome, Hungarian brome. **Life Span:** Perennial. **Origin:** Introduced. **Season:** C$_3$, cool season. **FLORAL CHARACTERISTICS: Inflorescence:** narrow to open panicle, 5–20 cm long; branches appressed to ascending or spreading. **Spikelets:** terete to slightly compressed, 15–40 mm long, 5- to 13-flowered; disarticulation above the glumes. **Glumes:** unequal, glabrous to puberulent along margins and near base, narrow lanceolate to elliptic, apex rounded to acute; lower 5–8 mm long, generally 1-nerved; upper 6–10 mm long, 3-nerved. **Lemmas:** generally glabrous,

Figure 167. *Bromopsis inermis.*

occasionally puberulent, elliptic, back rounded, apex entire to emarginate, 9–13 mm long, prominently 3-nerved, occasionally with additional obscure nerves. **Awns:** lemmas awnless to awned, when present awn up to 3.5 mm long. **VEGETATIVE CHARACTERISTICS: Growth Habit:** solitary or loosely tufted with extensive creeping rhizomes. **Culms:** erect to spreading, 5–13 dm tall, glabrous. **Sheaths:** generally glabrous, rarely pubescent. **Auricles**: sometimes present, small. **Ligules:** erose to ciliolate membrane, truncate, 0.5–3 mm long. **Blades:** flat to rarely involute, 15–40 cm long, 4–15 mm wide, both surfaces glabrous to pilose; "W" or "M" imprint noticeable on fresh specimens (less noticeable on dried specimens). **HABITAT:** Commonly found in disturbed areas from low to montane elevations over Colorado. **COMMENTS:** This species has been used extensively along roadsides and in fields for soil stabilization and forage. It is one of the most common roadside plants in the mountains. It is also planted extensively as a pasture grass and used for hay. Excellent forage early in the growing season (Stubbbendieck, Hatch, and Landholt 2003).

146. *Bromopsis lanatipes* (Shear) Holub. (Fig. 168).

Synonyms: *Bromus lanatipes* (Shear) Rydb., *Bromus anomalus* Rupr. *ex* Fourn. var. *lanatipes* (Shear) Hitchc., *Bromus porteri* (J. M. Coult.) Nash var. *lanatipes* Shear. **Vernacular Name:** Woolly brome. **Life Span:** Perennial. **Origin:** Native. **Season:** C_3, cool season. **FLORAL CHARACTERISTICS: Inflorescence:** open panicle, 10–25 cm long; branches ascending to spreading, becoming loose and nodding. **Spikelets:** subterete to compressed, 10–30 mm long, 5- to 11-flowered; disar-

ticulation above the glumes. **Glumes:** unequal, generally glabrous but occasionally pubescent, elliptic to lanceolate; lower 5–8 mm long, generally 1-nerved; upper 6–11 mm long, 3- to 5-nerved. **Lemmas:** pubescent, back rounded, apex acute to emarginate, 8–11 mm long, 5- to 7-nerved. **Awns:** lemmas awned, awn 2–4 mm long. **Anthers:** 1.8–4 mm long. **VEGETATIVE CHARACTERISTICS: Growth Habit:** solitary to small tufts. **Culms:** erect, 4–9 dm tall; nodes generally pubescent; internodes glabrous. **Sheaths:** lower sheaths densely lanate, upper sheaths nearly glabrous. **Ligules:** truncate to obtuse, 1–2 mm long. **Blades:** flat to loosely involute, 5–20 cm long, 3–6 mm wide, glabrous to scabrous to pilose. **HABITAT:** Found along roadsides and trails from foothills to subalpine elevations over the state, except for the far eastern counties (Weber and Wittmann 2001a, 2001b; Wingate 1994). **COMMENTS:** Common.

147. *Bromopsis mucroglumis* (Wagnon) Holub (Fig. 169).

Synonym: *Bromus mucroglumis* Wagnon. **Vernacular Name:** Sharpglume brome. **Life Span:** Perennial. **Origin:** native. **Season:** C_3, cool season. **FLORAL CHARACTERISTICS: Inflorescence:** open panicle, 10–20 cm long; branches ascending to spreading, becoming loose and nodding. **Spikelets:** subterete to compressed, 20–30 mm long, 5- to 10-flowered; disarticulation above the glume and between florets. **Glumes:** unequal, generally pilose to pubescent, rarely glabrous; lower 6–8 mm long, generally 1-nerved; upper 8–8.5 mm long, 3-nerved, mucronate. **Lemmas:** back rounded, apices acute to obtuse, 10–12 mm long, 5- to 7-nerved.

Figure 168. *Bromopsis lanatipes.*

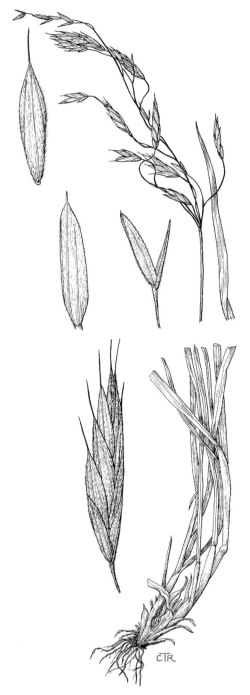

Figure 169. *Bromopsis mucroglumis.*

Awns: lemmas awned, 3–5 mm long, straight, arising less than 1.5 mm below lemma apices. **VEGETATIVE CHARACTERISTICS: Growth Habit:** cespitose. **Culms:** erect, 5–10 dm tall; nodes pilose to pubescent; internodes glabrous. **Sheaths:** pubescent to glabrous. **Ligules:** membranous, 1–2 mm long, glabrous, truncate to obtuse. **Blades:** flat, 20–30 cm long, 4–11 mm wide, glabrous to scabrous to pilose. **HABITAT:** Found at 5,000–10,000 ft (1,500–3,000 m) elevation in New Mexico and Arizona and into Mexico. The Colorado specimen was collected at 8,953 ft (2,713 m) elevation in the Medicine Bow Mountains. **COMMENTS:** Only known in Colorado from a single collection in Jackson County made in 2001 by Ronald L. Hartman.

148. *Bromopsis porteri* (J. M. Coult.) Holub (Fig. 170).

Synonyms: *Bromus porteri* (J. M. Coult.) Nash., *Bromus kalmii* var. *porteri* J. M. Coult., *Bromus ciliatus* var. *porteri* (J. M. Coult.) Rydb. **Vernacular Name:** Nodding brome. **Life Span:** Perennial. **Origin:** Native. **Season:** C$_3$, cool season. **FLORAL CHARACTERISTICS: Inflorescence:** open panicle, 5–20 cm long; branches ascending to spreading, becoming loose and nodding. **Spikelets:** subterete to compressed, 15–30 mm long, 5- to 11-flowered; disarticulation above the glumes. **Glumes:** unequal, generally pubescent but occasionally glabrous, elliptic to lanceolate; lower 5–8 mm long, generally 3-nerved but sometimes 1-nerved; upper 6–11 mm long, 3- to 5-nerved, not mucronate. **Lemmas:** pubescent, back rounded, apex acute to emarginate, 7–12 mm long, 5- to 7-nerved. **Awns:** lemmas awned, awn 1–3.5 mm long. **VEGETATIVE CHARACTERISTICS: Growth Habit:** solitary to small tufts. **Culms:** erect, 4–9 dm tall; nodes glabrous to pubescent; internodes glabrous. **Sheaths:** glabrous to variously pubescent. **Ligules:** truncate, 0.2–1 mm long, slightly erose. **Blades:** flat to loosely involute, 2–5 mm wide, glabrous to scabrous to pilose. **HABITAT:** This species occurs in mountain meadows, grassy slopes, mesic steppes, forest edges, and open forests. **COMMENTS:** Closely related to *B. frondosa* but differs in having pubescent glumes and pedicels and erect blades, and midrib does not narrow just below the collar (Allred 2005).

149. *Bromopsis pubescens* (Muhl. *ex* Willd.) Holub. (Fig. 171).

Synonyms: *Bromus pubescens* Muhl. *ex* Willd., *Bromus ciliatus* L. var. *laeviglumis* Scribn. *ex* Shear, *Bromus pubescens* Muhl. *ex* Willd. var. *laeviglumis* (Scribn. *ex* Shear) Swallen, *Bromus purgans* L., *Bromus purgans* L. var. *laeviglumis* (Scribn. *ex* Shear) Swallen. **Vernacular Names:** Hairy woodland brome, Canadian brome. **Life Span:** Perennial. **Origin:** Native. **Season:** C$_3$, cool season. **FLORAL CHARACTERISTICS: Inflorescence:**

Figure 170. *Bromopsis porteri.*

Figure 171. *Bromopsis pubescens.*

panicle very open, generally nodding, 10–25 cm long; branches erect to widely spreading, often curved to drooping. **Spikelets:** 15–30 mm long, 4- to 8-flowered. **Glumes:** generally pubescent but rarely glabrous; lower 4–8 mm long, 1-nerved; upper 5–10 mm long, usually 3-nerved, occasionally 5-nerved, not mucronate. **Lemmas:** generally pubescent with hirsute margins, rarely glabrous to scabrous, 8–10.5 mm long, usually 5- to 7-nerved. **Awns:** lemmas awned, awn 3–8 mm long. **VEGETATIVE CHARACTERISTICS: Growth Habit:** tufted. **Culms:** erect, 6–12 dm tall; nodes generally pubescent, rarely glabrous; internodes glabrous to pubescent. **Sheaths:** retrorsely pilose, summit near collar hirsute to villous to glabrous. **Ligules:** glabrous membrane, erose, 0.5–2.2 mm long, obtuse to truncate. **Blades:** flat, 12–32 cm long, 6–15 mm wide, generally pilose or pubescent on 1 or both surfaces. **HABITAT:** Found in wet to mesic areas from mid- to upper elevations. **COMMENTS:** Scattered in distribution but never abundant.

150. *Bromopsis pumpelliana* (Scribn.) Holub. (Fig. 172).

Synonyms: *Bromus inermis* Leyss. subsp. *pumpellianus* (Scribn.) Wagnon var. *pumpellianus* (Scribn.) C. L. Hitchc., *Bromopsis dicksonii* (Mitchell & Wilton) Á. & D. Löve, *Bromopsis inermis* (Leyss.) Holub. subsp. *pumpelliana* (Scribn.) W. A. Weber, *Bromus ciliatus* L. var. *coloradensis* Vasey *ex* Beal, *Bromus inermis* Leyss. var. *purpurascens* (Hook.) Wagnon, *Bromus inermis* Leyss. var. *tweedyi* (Scribn. *ex* Beal) C. L. Hitchc., *Bromus pumpellianus* Scribn., *Bromus pumpellianus* Scribn.

subsp. *dicksonii* Mitchell & Wilton, *Bromus pumpellianus* Scribn. var. *tweedyi* Scribn. *ex* Beal, *Bromus pumpellianus* Scribn. var. *villosissimus* Hultén. **Vernacular Name:** Pumpelly's brome. **Life Span:** Perennial. **Origin:** Native. **Season:** C$_3$, cool season. **FLORAL CHARACTERISTICS: Inflorescence:** narrow to open panicle, 10–20 cm long; branches erect to nodding, greenish to purplish. **Spikelets:** terete to slightly compressed, 20–30 mm long, 7- to 11-flowered; disarticulation above the glumes. **Glumes:** subequal to unequal, glabrous to pubescent; lower 6–8 mm long, 1-nerved; upper 7–9 mm long, 3-nerved. **Lemmas:** densely sericeous to pubescent on back or along margins and below, elliptic, back rounded, apex entire to emarginate, 9–14 mm long, 5- to 7-nerved. **Awns:** lemmas awned, when present awn 1–7.5 mm long, rarely awnless. **VEGETATIVE CHARACTERISTICS: Growth Habit:** with creeping rhizomes. **Culms:** erect, 5–12 dm tall; nodes and internodes glabrous to pubescent. **Sheaths:** generally glabrous to sparsely pubescent. **Auricles:** sometimes present, small. **Ligules:** truncate to obtuse, to 3 mm long. **Blades:** flat to rarely involute, 15–40 cm long, 4–15 mm wide, upper surface pilose to rarely glabrous, lower surface glabrous to pilose. **HABITAT:** Found at montane to subalpine elevations over Colorado. **COMMENTS:** It is considered the native equivalent of *B. inermis*. Colorado plants belong to the typical subspecies (subsp. *pumpelliana*). Wingate (1994) places this species under *Bromus inermis* Leyss. as a subspecies.

Figure 172. *Bromopsis pumpelliana.*

151. *Bromopsis richardsonii* (Link) Holub. (Fig. 173).

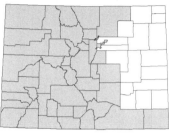

Synonyms: *Bromus ciliatus* L. var. *richardsonii* (Link) Boivin, *Bromopsis canadensis* (Michx.) Holub subsp. *richardsonii* (Link) Tzvelev, *Bromus richardsonii* Link. **Vernacular Name:** Fringed brome. **Life Span:** Perennial. **Origin:** Native. **Season:** C_3, cool season. **FLORAL CHARACTERISTICS: Inflorescence:** open panicle with nodding branches, 10–20 to rarely 25 cm long; branches filiform, ascending, spreading to drooping. **Spikelets:** lanceolate to elliptic, ranging from terete to moderately laterally compressed, 15–25 to occasionally 40 mm long, generally 6- to 10-flowered. **Glumes:** unequal, glabrous to occasionally pubescent; lower 7.5–12.5 mm long, 1- to occasionally 3-nerved; upper 8.9–11.3 mm long, 3-nerved, often with a mucronate tip. **Lemmas:** elliptic, rounded rather than keeled over midnerve, 9–14 mm long, 5- to 7- to 9-nerved, varying in pubescence between lowermost and upper lemmas; lowermost glabrous across the back, while upper ones with appressed pubescence across the back, margins on lower half to three-quarters with varying degrees of pilose hairs to densely pilose, apical portion obtuse, entire. **Awns:** upper glumes often with mucronate tip; lemmas with straight awn 3–5 mm long, arising just below apices. **Anthers:** 1.6–2.7 mm long. **VEGETATIVE CHARACTERISTICS: Growth Habit:** tufted. **Culms:** spreading to erect, 5–11 dm tall, generally glabrous, although occasionally pubescent. **Sheaths:** generally retrorsely pubescent on basal sheaths, upper sheaths glabrous; lacks auricles but instead tufts of pilose hairs present in those areas; culm leaf midrib not abruptly narrowed below collar. **Ligules:** erose-ciliolate, rounded, 0.4–2 mm long. **Blades:** flat, glabrous, 10–35 cm long, 3–12 mm wide. **HABITAT:** Found in montane to subalpine zones in meadows to open woods, sometimes growing lower. **COMMENTS:** Some authors have merged this species with *B. ciliata* in past treatments. This and other native perennial species of brome are an excellent source of forage in open forests and montane regions.

50. *Bromus* L.

Annuals. Leaf blades and sheaths usually densely pubescent. Sheaths connate for most of their length; ligules membranous, erose; auricles absent. Inflorescence a terminal, simple panicle, rarely a raceme (most common in underdeveloped plants); branches short and appressed to long and flexuous. Disarticulation above the glumes and between florets. Spikelets large, usually with 5 to many florets, terete to very slightly laterally compressed. Glumes subequal, shorter than lowest lemma, lower usually shorter and narrower than upper. Lemmas broad, only slightly laterally compressed, if at all, 5- or more-nerved, awnless or awned from an entire or bifid apex. Awns straight or only slightly geniculate at maturity. Palea nearly as long as lemma, attracted to caryopsis. Basic chromosome number, $x = 7$. Photosynthetic pathway, C_3.

Represented in Colorado by 7 species and 1 subspecies.

1. Lemma awnless or only minutely awned, inflated, 6–8 mm wide in side view; florets widely spreading . . . . 152. *B. briziformis*
1. Lemmas with long awns, not inflated, 1–7 mm wide in side view; florets not spreading

Figure 173. *Bromopsis richardsonii*.

2. Lemma margins inrolled; florets distant, exposing rachilla
   3. Lemma margins evenly rolled; awns straight or flexuous; lower leaf sheaths glabrous or loosely pubescent and glabrate . . . . 157. *B. secalinus*
   3. Lemma margins bluntly angled; awns straight; lower leaf sheaths evenly covered with stiff hairs . . . . 153. *B. commutatus*
2. Lemmas more or less keeled; florets imbricate, rachilla hidden
   4. Awns arising less than 1.5 mm below lemma apices, erect or weakly divaricate, not twisted at base
      5. Panicle compact; pedicels usually shorter than spikelets . . . . 154. *B. hordeaceus*
      5. Panicle open; at least some pedicels longer than spikelets
         6. Anthers 0.7–1.7 mm long; lemmas 8–11.5 mm long, margins bluntly angled . . .
          . . . . 153. *B. commutatus*
         6. Anthers 1.5–3 mm long; lemmas 6.5–8 mm long, margins rounded . . . .
          . . . . 156. *B. racemosus*
   4. Awns arising 1.5 mm or more below lemma apices, erect to strongly divaricate, often twisted at base
      7. Lemma margins hyaline 0.6–0.9 mm wide; branches usually with 1 spikelet . . . . .
       . . . . 158. *B. squarrosus*
      7. Lemma margins hyaline 0.3–0.6 mm wide; branches usually with more than 1 spikelet . . . . 155. *B. japonicus*

152. *Bromus briziformis* Fisch. & C. A. Mey. (Fig. 174).

Synonym: None. **Vernacular Names:** Rattlesnake brome, rattlesnake chess. **Life Span:** Annual. **Origin:** Introduced. **Season:** C$_3$, cool season. **FLORAL CHARACTERISTICS: Inflorescence:** open, nodding panicle, secund, 3–15 cm long. **Spikelets:** strongly compressed, 15–27 mm long, to 10 mm wide, 7- to 15-flowered; florets widely spreading; disarticulation above the glumes. **Glumes:** oblong-lanceolate, broad, rounded to U-shaped on back, glabrous to scabrous;  lower 4.5–6 mm long, 3- to 5-nerved; upper 5–9 mm long, 7- to 9-nerved. **Lemmas:** glabrous to scabridulous, rhombic-ovate, 6.5–10 mm long, 6–8 mm broad, strongly inflated at maturity, nerves several and obscure, apex entire to emarginate, unawned to short-awned. **Awns:** lemmas when awned, to 1 mm long. **VEGETATIVE CHARACTERISTICS: Growth Habit:** solitary or in small tufts. **Culms:** erect, 1–6 dm tall. **Sheaths:** generally densely retrorsely hispid to softly pilose, sheath closed to near summit. **Ligules:** erose, obtuse, hairy, 0.5–2 mm long. **Blades:** flat to occasionally involute, ascending to lax, 1–6 mm wide, scabrous to pilose to pubescent on both surfaces. **HABITAT:** Found at low elevations, on mesas in lower foothills in dry open sites. **COMMENTS:** Weber and Wittmann (2001a, 2001b) call this taxon an alien weed of ruderal sites. Plant sometimes used as an ornamental in gardens and floral arrangements.

Figure 174. *Bromus briziformis.*

153. *Bromus commutatus* Schrad. (Fig. 175).

Synonyms: *Bromus racemosus* L., *Bromus commutatus* Schrad. var. *apricorum* Simonkai, *Bromus popovii* Drobow. **Vernacular Names:** Meadow brome, European brome, hairy chess. **Life Span:** Annual. **Origin:** Introduced. **Season:** C_3, cool season. **FLORAL CHARACTERISTICS: Inflorescence:** open panicle, oval, 7–16 cm long; branches stiffly ascending to spreading; pedicels generally longer than spikelets. **Spikelets:** oblong-lanceolate, slightly compressed, 10–25 mm

long, 4- to 10-flowered; florets not spreading, distant, exposing rachilla; disarticulation above the glumes. **Glumes:** unequal, scabrous; lower lanceolate, 5–6 mm long, 3- to 5-nerved; upper oblong-lanceolate, 6–7 mm long, 5- to 9-nerved. **Lemmas:** scabrous, 8–11.5 mm long, 1.7–2.5 mm wide, margins inrolled, bluntly angled, nerves obscure, 7- to 9-nerved, apex bifid. **Awns:** lemmas awned, 7–9 mm long, straight. **Anthers:** 0.7–1.7 mm long. **VEGETATIVE CHARACTERISTICS: Growth Habit:** tufted. **Culms:** erect to decumbent at base, 3–10 dm tall; nodes puberulent, occasionally overall. **Sheaths:** generally retrorsely pilose with stiff hairs, especially lower ones, closed to near summit. **Ligules:** erose-ciliolate, occasion-

Figure 175. *Bromus commutatus.*

ally pubescent, 0.5–2.5 mm long. **Blades:** flat, 9–18 cm long, 1–6 mm wide, pilose on both surfaces. **HABITAT:** Found in dry, open waste places at lower elevations in the foothills and plateaus of Colorado. **COMMENTS:** This species is an introduced weed of disturbed sites.

154. *Bromus hordeaceus* L. (Fig. 176).

Synonym: *Bromus mollis* L. **Vernacular Names:** Soft brome, soft chess. **Life Span:** Annual. **Origin:** Introduced. **Season:** C$_3$, cool season. **FLORAL CHARACTERISTICS: Inflorescence:** dense panicle, when depauperate may be reduced to 1–2 spikelets, 3–10 cm long, ascending to erect; pedicels usually shorter than spikelets. **Spikelets:** soft-pubescent to glabrous, 14–20 mm long, 5- to 10-flowered; florets imbricate, not spreading, rachilla hidden; disarticulation above the glumes. **Glumes:** subequal, pilose to rarely glabrous; lower lanceolate, 5–7.5 mm long, 3- to 5-nerved; upper ovate-lanceolate, 6–8.5 mm long, 5- to 7-nerved. **Lemmas:** slightly compressed, slightly keeled, pilose to scabrous, 7–10 mm long, 3–5 mm wide, prominently 7- to 9-nerved, apices minutely bifid, to 1 mm, awned. **Awns:** lemmas awn generally arising from less than 1.5 mm below apices, awns straight to recurved at maturity, 5–9 mm long. **VEGETATIVE CHARACTERISTICS: Growth Habit:** solitary to tufted. **Culms:** ascending to erect with geniculate base, 1–8 dm tall, glabrous to retrorsely pubescent, occasionally only on nodes. **Sheaths:** lower densely, often retrorsely pilose, round, closed to near summit. **Ligules:** erose-ciliate membrane, 0.5–2 mm long. **Blades:** flat, 3–15 cm long, 1.5–5 mm wide, generally hairy to glabrous on both surfaces. **HABITAT:** Found in dry fields and waste places in counties along the Utah border. **COMMENTS:** This species is considered a weed of disturbed sites. *B. mollis* L. is included here, and Wingate (1994) places *B. racemosus* L. here as well. Colorado specimens belong to the typical subspecies (subsp. *hordeaceus*).

155. *Bromus japonicus* Thunb. (Fig. 177).

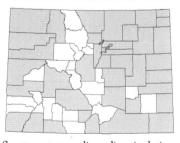

Synonyms: *Bromus arvensis* L. var. japonicus (Thunb.) Fiori, *Bromus japonicus* Thunb. *ex* Murr. var. *porrectus* Hack., *Bromus patulus* Mert. & Koch. **Vernacular Names:** Japanese brome, Japanese chess. **Life Span:** Annual. **Origin:** Introduced. **Season:** C$_3$, cool season. **FLORAL CHARACTERISTICS: Inflorescence:** open, nodding panicle, 8–22 cm long; branches slender, flexuous, ascending to spreading, generally with more than 1 spikelet. **Spikelets:** glabrous to rarely scabrous, narrow, 15–24 mm long, 5- to 12-flowered; florets not spreading; disarticulation above the glumes. **Glumes:** unequal; lower lanceolate, 4–6 mm long, 3- to 5-nerved; upper oblong-lanceolate, 5.5–8.5 mm long, 5- to 7-nerved. **Lemmas:** coriaceous, broadly oblong-lanceolate, 7–9 mm long, slightly compressed, 7- to 9-nerved, glabrous to scaberulent, margin hyaline, 0.3–0.6 mm wide, awned. **Awns:** lemmas awned, awn generally slightly geniculate and frequently slightly twisted at base, 7–13 mm long, arising below apex, generally 1.5

Figure 176. *Bromus hordeaceus.*

Figure 177. *Bromus japonicus.*

mm or more. **VEGETATIVE CHARACTERISTICS: Growth Habit:** solitary to tufted. **Culms:** ascending to erect to geniculate at base, 2–7 dm tall. **Sheaths:** densely retrorsely pilose to lanate, closed to near summit. **Ligules:** erose-ciliate to lacerate, 0.5–2 mm long, pilose. **Blades:** flat, 10–20 cm long, 1.5–7 mm wide, generally softly pilose on both surfaces. **HABITAT:** Found in disturbed areas from low elevations to the foothills over Colorado. **COMMENTS:** A very common weed of disturbed sites. Another weedy species, *B. arenarius* Labill, could occur in the Four Corners area of the state because it occurs just across the border in San Juan County, New Mexico. It differs from *B. japonicus* in having a longer lower (7–10 mm) and upper (8–12 mm) glume, sinuously curved panicle branches, and rolled lemma margins (Barkworth et al. 2007).

156. *Bromus racemosus* L. (Fig. 178).

Synonyms: *Bromus arvensis* L. var. *racemosus* (L.) Neilr., *Bromus hordeaceus* L. var. *racemosus* (L.) Fiori, *Bromus mollis* L. var. *racemosus* (L.) Fiori, *Bromus popovii* Drobow. **Vernacular Name:** Smooth brome. **Life Span:** Annual. **Origin:** Introduced. **Season:** C$_3$, cool season. **FLORAL CHARACTERISTICS: Inflorescence:** erect, open, 4–16 cm long, 2–3 cm wide; branches sometimes longer than spikelets, slender, usually ascending, curved or straight. **Spikelets:** 10–20 mm

long, 5- to 6-flowered, moderately laterally compressed; florets about 1–1.5 mm apart; disarticulation above the glumes; florets not spreading, rachilla concealed at maturity. **Glumes:** short, faintly nerved, glabrous; lower 4–6 mm long, 3- to 5-nerved; upper 4–7 mm long, 7-nerved. **Lemmas:** glabrous to scabrous, 6.5–8 mm long, 3–4.5 mm wide, 7- to 9-nerved, strongly rounded on back, margins not inrolled, apices acute to obtuse, minutely bifid with teeth less than 1 mm long. **Awns:** lemmas awned, awns 5–9 mm long, straight, arising less than 1.5 mm from lemma apex. **Anthers:** 1.5–3 mm long. **VEGETATIVE CHARACTERISTICS: Growth Habit:** solitary to tufted. **Culms:** erect, to over 1 m tall, glabrous. **Sheaths:** lower with dense, stiff, retrorse hairs; upper glabrous or hairy. **Ligules:** erose, glabrous to hairy, 1–2 mm long. **Blades:** flat, pilose on both surfaces, 1–4 mm wide, 7–18 cm long. **HABITAT:** Grows in fields, waste areas, and roadsides. **COMMENTS:** A weed of disturbed sites. Wingate (1994) included this in *B. hordeaceus* L., while Barkworth and colleagues (2007) remarked that it is closest to *B. commutatus* Schrad.

157. *Bromus secalinus* L. (Fig. 179).

Synonyms: None. **Vernacular Names:** Chess, cheat. **Life Span:** Annual. **Origin:** Introduced. **Season:** C$_3$, cool season. **FLORAL CHARACTERISTICS: Inflorescence:** loose, nodding panicle, secund, 5–20 cm long. **Spikelets:** 10–20 mm long, 4- to 10-flowered, strongly compressed; florets spreading and exposing rachilla joints; disarticulation above the glumes. **Glumes:** short, faintly nerved, glabrous; lower 4–6 mm long,

Figure 178. *Bromus racemosus.*

Figure 179. *Bromus secalinus.*

3- to 5-nerved; upper 5–7 mm long, 5- to 7-nerved. **Lemmas:** glabrous to scabrous, 6–9 mm long, 1.7–2.5 mm wide, 7- to 9-nerved, strongly rounded on back, margins scabrous, strongly and evenly rounded, inrolled, apex bifid. **Awns:** lemmas awned, awns 3–9.5 mm long, straight or flexuous, base located less than 1.5 mm below lemma apices. **VEGETATIVE CHARACTERISTICS: Growth Habit:** solitary to tufted. **Culms:** erect, 3–10 dm tall, glabrous and pubescent below nodes. **Sheaths:** glabrous to loosely pubescent, becoming glabrate with age. **Ligules:** erose, glabrous to hairy, 0.5–3 mm long. **Blades:** flat, pilose or rarely glabrous, 2–9 mm wide. **HABITAT:** A weed of disturbed sites. **COMMENTS:** Scattered across the state. Occasionally grown for hay in the Pacific Northwest (Barkworth et al. 2007).

158. *Bromus squarrosus* L. (Fig. 180).

Synonyms: None. **Vernacular Names:** Corn brome, downy brome, nodding brome. **Life Span:** Annual. **Origin:** Introduced. **Season:** C$_3$, cool season. **FLORAL CHARACTERISTICS: Inflorescence:** open panicle, occasionally racemose in appearance, secund, 6.5–20 cm long; branches few, usually with only 1 spikelet. **Spikelets:** broadly oblong to ovate-lanceolate, appear slightly inflated, 15–70 mm long, 7–15 mm wide, 8- to 30-flowered; florets not spreading, imbricate, rachilla hidden; disarticulation above the glumes. **Glumes:** unequal, glabrous to scabrous; lower 4.5–7 mm long, 3- to 5- and occasionally 7-nerved; upper 6–9 mm long, 7- to 9-nerved. **Lemmas:** chartaceous, broadly rhombic in shape, rounded over midvein and strongly angled above middle, glabrous to scabridulous, 8–11 mm long, 2–2.4 mm wide, 7- to 9-nerved, margins hyaline, 0.6–0.9 mm wide, awned. **Awns:** lemmas awned, awns becoming divaricate as mature, 8–10 mm long, flattened, occasionally twisted basally, arising 1.5 mm or more below lemma apices. **VEGETATIVE CHARACTERISTICS: Growth Habit:** solitary to tufted. **Culms:** ascending to erect, occasionally geniculate at base; nodes glabrous to pubescent. **Sheaths:** below densely pilose to lanate, above glabrous to pilose, sheath closed to near summit. **Ligules:** erose-ciliolate, obtuse, 1–2.6 mm long. **Blades:** flat, 3–15 cm long, glabrous to densely pilose on both surfaces. **HABITAT:** Found in waste places over northwestern Colorado. **COMMENTS:** Another introduced weed of disturbed sites.

51. *Ceratochloa* P. Beauv.

Annual or biennial, cespitose. Sheaths connate for most of their length; ligules prominent, membranous, erose; auricles absent. Inflorescence a simple terminal panicle (rarely a raceme in very immature plants). Disarticulation above the glumes and between florets. Spikelets relatively large, strongly laterally compressed (strongly flattened), with 3–9 florets. Glumes subequal, strongly keeled, shorter than lowermost lemma. Lemmas strongly flattened, many-nerved, awned from a bifid or entire apex or awnless. Palea nearly as long as lemma. Basic chromosome number, $x = 7$. Photosynthetic pathway, C$_3$.

Represented in Colorado by a single species. A segregate of *Bromus*.

Figure 180. *Bromus squarrosus*.

159. *Ceratochloa carinata* (Hook. & Arn.) Tutin (Fig. 181).

Synonyms: *Bromus carinatus* Hook. & Arn., *Bromus carinatus* Hook. & Arn. var. *californicus* (Nutt. *ex* Buckley) Shear, *Bromus carinatus* Hook. & Arn. var. *carinatus* Hook. & Arn., *Bromus carinatus* Hook. & Arn. var. *hookerianus* (Thurb.) Shear, *Bromus laciniatus* Beal. **Vernacular Names:** California brome, mountain brome. **Life Span:** Annual or biennial. **Origin:** Native. **Season:** C$_3$, cool season. **FLORAL CHARACTERISTICS: Inflorescence:** lax panicle, loosely contracted and nodding, 15–40 cm long; lower branches usually open to declining. **Spikelets:** laterally compressed, 20–40 mm long, 4- to 7-flowered; rachilla scabrous; disarticulation above the glumes. **Glumes:** unequal, lanceolate, evidently keeled; lower 7–11 mm long, 3- to 5-nerved; upper 9–13 mm long, 5- to 7-nerved. **Lemmas:** scabrous over entire surface or only along veins, 11–15 mm long, 7- to 11-nerved, margin hyaline, callus glabrous, apices entire to slightly bifid at maturity. **Awns:** lemmas awned, awn stout, 4–8 mm long. **VEGETATIVE CHARACTERISTICS: Growth Habit:** tufted. **Culms:** erect, 5–10 dm tall, glabrous to pubescent. **Sheaths:** retrorsely soft-pilose to glabrous-scabrous, at least at summit, closed to

Figure 181. *Ceratochloa carinata.*

near summit. **Auricles**: if present, small. **Ligules**: obtuse to lacerate to erose, 1–4 mm long, glabrous to pilose. **Blades**: flat to rarely involute, 10–30 cm long, 2.5–6 mm wide, generally both surfaces sparsely pilose but occasionally glabrous. **HABITAT:** Roadsides and old fields, escaped from cultivation. **COMMENTS:** Collections of *C. catharticus* Vahl [= *C. unioloides* (Willd.) P. Beauv.], a closely related species to *C. carinata,* from Colorado are of cultivated plants, and it is not known to grow outside of cultivation.

11. MELICEAE

Perennial, cespitose or sometimes rhizomatous; often lower culms swollen and forming a corm or bulbous base. Sheaths connate for most of their length; ligules membranous; auricles absent. Inflorescence a panicle, rarely racemose. Spikelets with 2 to several florets; reduced floret, if present, above fertile florets. Rachilla extended beyond uppermost perfect floret. Reduced florets merely underdeveloped, as a group of enveloped lemmas or as a swollen rachilla extension. Disarticulation above the glumes and between florets or below the glumes. Glumes often papery or membranous, shorter than lowermost lemma. Lemmas 3- to several-nerved, awned or awnless.

52. *Bromelica* (Thurb.) Farw.

Perennial, cespitose or frequently rhizomatous; lower culms swollen and forming a corm or bulbous base. Sheaths connate; ligules prominent, membranous, apex erose-lacerate, truncate, or acute; auricles absent. Inflorescence a simple terminal panicle with short branches. Spikelets with 3–9 florets. Rudimentary florets, empty lemmas as well as an extension of the rachilla. Disarticulation above the glumes and between florets. Glumes unequal, shorter than or nearly as long as lowermost lemma. Lower 1- to 3-nerved, upper 7- to 9-nerved. Lemmas broadly ovate to obtuse, emarginate or bifid at apex, often purplish, 9- to 15-nerved. Palea usually with ciliated nerves. Basic chromosome number, $x = 9$. Photosynthetic pathway, C_3.

Represented in Colorado by 3 species. A segregate of *Melica.*

1. Spikelets narrow; lemmas acuminate . . . . 162. *B. subulata*
1. Spikelets broad; lemmas obtuse to emarginated to rounded
    2. Bulb of culm attached to rhizome by a slender stem; glumes less than half as long as spikelet . . . . 161. *B. spectabilis*
    2. Bulb of culm sessile on rhizome; glumes more than half as long as spikelet . . . .
    . . . . 160. *B. bulbosa*

160. *Bromelica bulbosa* (Geyer *ex* Porter & Coult.) W. A. Weber (Fig. 182).

Synonyms: *Melica bulbosa* Geyer *ex* Porter & Coult., *Melica bella* Piper, *Melica bella* Piper subsp. *intonsa* Piper, *Melica bulbosa* Geyer *ex* Porter & Coult. var. *inflata* (Boland.) Boyle, *Melica bulbosa* Geyer *ex* Porter & Coult. var. *intonsa* (Piper) M. E. Peck, *Melica inflata* (Boland.) Vasey. **Vernacular Name:** Oniongrass. **Life Span:** Perennial. **Origin:** Native. **Season:** C_3, cool season. **FLORAL CHARACTERISTICS: Inflorescence:** narrow panicle, 7–30 cm long, dense; branches appressed to ascending, generally straight with 4–8 spikelets. **Spikelets:** broad, 6–24 mm long, 2- to 9-flowered with the lower 4–7 fertile; upper florets sterile; disarticulation above the glumes. **Glumes:** unequal, scabrous, ovate, more than half as long as

Figure 182. *Bromelica bulbosa*.

spikelet, apex membranous, obtuse; lower 5.5–10.5 mm long, 2–3 mm wide, 3- to 5-nerved; upper 6–14 mm long, 2.3–3.5 mm wide, 5- to 7-nerved. **Lemmas:** glabrous to scabrous, lowermost 6–12 mm long, 7- to 11-nerved, apex obtuse to emarginate, purplish band present near apex. **Paleas:** about three-fourths lemma length, nerves ciliate. **Awns:** none. **VEGETA-TIVE CHARACTERISTICS: Growth Habit:** cespitose with short rhizomes. **Culms:** erect, 3–10 dm tall, base bulbous, clustered directly on rhizomes, scabrous at base of internode. **Sheaths:** closed nearly full length, glabrous to scabrous to rarely sparsely pilose. **Ligules:** erose to deeply lacerate membrane, 2–6 mm long, acute. **Blades:** flat to involute, 10–30 cm long, 2–5 mm wide, glabrous to scabrous. **HABITAT:** Generally found along streams, on well-drained dry slopes, and in open woods (Barkworth et al. 2007). **COMMENTS:** Bulbous swellings at base of culm and disarticulation above the glumes distinguish this genus from *Melica*.

161. *Bromelica spectabilis* (Scribn.) W. A. Weber (Fig. 183).

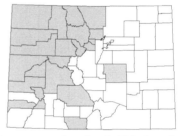

Synonyms: *Melica spectabilis* Scribn., *Melica bulbosa* Geyer *ex* Porter & Coult. var. *spectabilis* (Scribn.) Boivin. **Vernacular Name:** Purple oniongrass. **Life Span:** Perennial. **Origin:** Native. **Season:** C_3, cool season. **FLORAL CHARACTERISTICS: Inflorescence:** panicle, generally narrow but occasionally open, 5–26 cm long; branches often flexuous. **Spikelets:** purple-tinged, turgid, broad, 7–19 mm long, 4- to 8-flowered with rudiment above; disarticulation above the glumes. **Glumes:** subequal to unequal, broadly lanceolate to ovate, less than half the length of spikelet, commonly membranous and brownish to purple throughout, apex abruptly acute to acuminate; lower 3.5–6.4 mm long, 1.5–3 mm wide, 1- to 3-nerved; upper 5–7 mm long, 2.3–3.5 mm wide, 5- to 7-nerved. **Lemmas:** obtuse to broadly ovate, 7–8.5 mm long, obscurely 5- to 11-nerved, margins pilose, apex rounded to acute. **Paleas:** two-thirds the size of lemma, nerves ciliate. **Awns:** none. **VEGETATIVE CHARACTERISTICS: Growth Habit:** loosely cespitose, rhizomatous. **Culms:** base bulbous, connected to rhizome by a slender stem 10–30 mm long, erect, 4–10 dm tall. **Sheaths:** closed, occasionally open at top, glabrous to pilose, throat often pilose. **Ligules:** erose-lacerate, 1–3.2 mm long, truncate to acute. **Blades:** flat to involute 5–18 cm long, 2–4 mm wide. **HABITAT:** Found in moist areas in aspen woodlands and in montane and subalpine communities in northern and central Colorado. **COMMENTS:** Bulbous swellings at base of culm and disarticulation above the glumes distinguish this genus from *Melica*.

162. *Bromelica subulata* (Griseb.) Farw. (Fig. 184).

Synonym: *Melica subulata* (Griseb.) Scribn. var. *subulata*. **Vernacular Names:** Alaska oniongrass, tapered oniongrass. **Life Span:** Perennial. **Origin:** Native. **Season:** C_3, cool season. **FLORAL CHARACTERISTICS: Inflorescence:** lax, narrow panicle, 8–25 cm long; branches

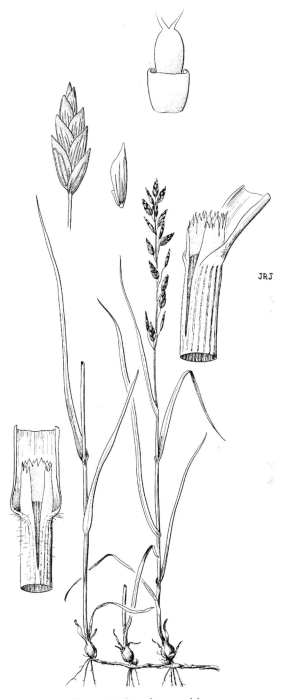

JRJ

Figure 183. *Bromelica spectabilis*.

Figure 184. *Bromelica subulata*.

short, generally ascending-erect to appressed but occasionally spreading, with 1–5 spikelets. **Spikelets:** narrow, 10–28 mm long, loosely 2- to 5-flowered, with rudiment above; disarticulation above the glumes. **Glumes:** unequal, narrow, apex acute to acuminate; lower 4–8 mm long, 1.3–2.2 mm wide, 1- to 3-nerved; upper 5.5–11 mm long, 2–3 mm wide, 3- to 5-nerved. **Lemmas:** apex tapering to an acuminate point, lowest lemma 5.5–18 mm long, 7- to 9-nerved; nerves generally strigose, prominent. **Paleas:** to half the length of lemma. **Awns:** none. **VEGETATIVE CHARACTERISTICS: Growth Habit:** cespitose, rhizomatous, rhizomes 3–10 mm long. **Culms:** ascending to erect, 5–12 dm tall, base bulbous, attached in bunches directly to short, thick rhizomes. **Sheaths:** closed nearly to or to top but with age often split, glabrous to sparsely to copiously pubescent. **Ligules:** glabrous, erose-lacerate, 0.4–5 mm long, 1.5 mm long on lower leaves to 5 mm on upper leaves. **Blades:** generally flat, 2–10 mm wide, glabrous to scabrous to sparsely hirsute. **HABITAT:** Collected in 1992 on a hillside seep in an aspen-spruce forest on silty clay loam soil at McClure Pass on the Western Slope (Weber and Wittmann 2001b). **COMMENTS:** Rare or perhaps not a constant member of the Colorado grass flora. Bulbous swellings at base of culm and disarticulation above the glumes distinguish this genus from *Melica*.

53. *Glyceria* R. Br.

Perennial, strongly rhizomatous. Sheaths connate nearly to the top; ligules prominent, membranous, erose to deeply lacerate, sometime fused in front; auricles absent. Inflorescence, in ours, usually a large open panicle, occasionally with flexuous branches. Spikelets with 3 to many florets, often purplish. Disarticulation above the glumes and between florets. Glumes shorter than lowermost lemmas, 1-nerved, awnless. Lemmas broad, firm, often erose, prominently 5- or more-nerved. Palea large, broad, sometimes exceeding lemma. Basic chromosome number, $x = 10$. Photosynthetic pathway, C_3.

Represented in Colorado by 4 species.

1. Spikelets linear and terete, except at anthesis when slightly laterally compressed, 9–22 mm long . . . . 163. *G. borealis*
1. Spikelets laterally compressed, ovate to oblong, 2–10 mm long
   2. Anthers 3; veins on glumes extending to apices; lemma apices flat . . . . . 165. *G. grandis*
   2. Anthers 2; veins on glumes terminating below apices; lemma apices prow-shaped
      3. Leaf blades narrow, 2–6 mm wide; culms usually less than 1 m long and 1.5–3.5 mm thick, base not or slightly spongy, sometimes rooting at lower nodes . . . . 166. *G. striata*
      3. Leaf blades broad, 6–15 mm wide; culms mostly over 1 m long and 2.5–8 mm thick, base spongy, rooting at lower nodes . . . . 164. *G. elata*

163. *Glyceria borealis* (Nash) Batch. (Fig. 185).

Synonym: *Panicularia borealis* Nash. **Vernacular Names:** Small floating mannagrass, northern mannagrass. **Life Span:** Perennial. **Origin:** Native. **Season:** C_3, cool season. **FLORAL**

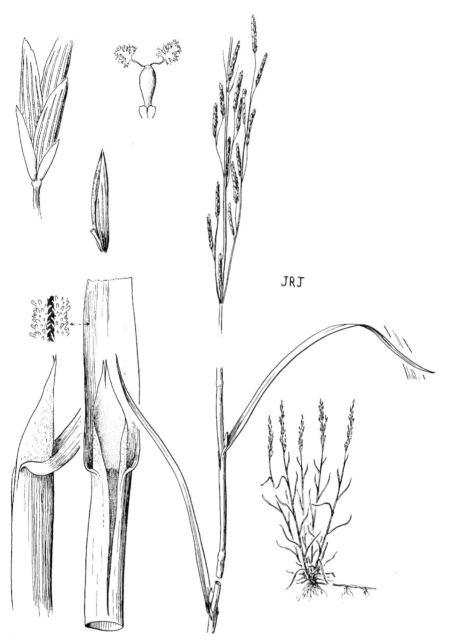

JRJ

Figure 185. *Glyceria borealis.*

CHARACTERISTICS: Inflorescence: narrow panicle, erect, 18–40 cm long; branches appressed to erect. **Spikelets:** linear, cylindrical and terete, except at anthesis when slightly laterally compressed, 9–22 mm long, 8- to 12-flowered; disarticulation above the glumes. **Glumes:** unequal, glabrous, lanceolate-elliptic to oblanceolate, 1-nerved, occasionally nerve obscure; lower 2–2.5 mm long; upper 3–4 mm long. **Lemmas:** glabrous between nerves, nerves scaberulous, 3.5–4.5 mm long, 7-nerved, apex scarious, obtuse. **Paleas:** subequal to lemma, 2-nerved, nerves winged. **Awns:** none. **Anthers:** 3. **VEGETATIVE CHARACTERISTICS: Growth Habit:** rhizomatous. **Culms:** erect to decumbent at base, 5–10 dm tall, hollow, rooting freely at nodes. **Sheaths:** glabrous, upper 1–4 cm of sheath open, slightly keeled. **Ligules:** membranous, erose to deeply lacerate, 4–12 mm long, apices acute. **Blades:** flat to folded, 8–20 cm long, 2–8 mm wide, upper surface glabrous to scaberulous to papillose. **HABITAT:** Found along pond and lake margins or in shallow water in montane and subalpine communities. **COMMENTS:** Widespread species throughout much of temperate North America.

164. *Glyceria elata* (Nash *ex* Rydb.) M. E. Jones. (Fig. 186).

Synonym: *Panicularis elata* Nash *ex* Rydb. **Vernacular Name:** Tall mannagrass. **Life Span:** Perennial. **Origin:** Native. **Season:** C_3, cool season. **FLORAL CHARACTERISTICS: Inflorescence:** open panicle, 15–30 cm long; branches lax and spreading. **Spikelets:** ovate to oblong, laterally compressed, 3–6 mm long, 5- to 8-flowered; rachilla joints 0.5–0.6 mm long; disarticulation above the glumes. **Glumes:** unequal, membranous, slightly erose-ciliolate, 1-nerved, midnerve terminating below apices; lower 1–1.5 mm long, lanceolate to ovate; upper 1.5–2.5 mm long, ovate. **Lemmas:** ovate to obovate to rounded, 2–2.5 mm long, 7-nerved, apices erose to toothed, with narrow hyaline band at apex, prow-shaped. **Paleas:** subequal to slightly longer than lemma, 2-nerved, nerves not winged. **Awns:** none. **Anthers:** 2. **VEGETATIVE CHARACTERISTICS: Growth Habit:** rhizomatous. **Culms:** decumbent, 0.75–1.5 m tall, 2.5–8 mm wide, hollow, spongy, and rooting freely at nodes. **Sheaths:** closed to top, flattened, keeled, glabrous to slightly scaberulous. **Ligules:** membranous, entire to lacerate, truncate, finely puberulent, 3–4 mm long on upper leaves. **Blades:** flat, lax, 20–40 cm long, 6–15 mm wide, upper surface scabrous. **HABITAT:** Found in wet and moist areas, aspen groves, and pond borders in communities from the foothills to the subalpine. **COMMENTS:** Closely related to *G. grandis* and *G. striata* and sometimes included within the latter. Stature and growth habitat are evident in the field, and they do not hybridize (Barkworth et al. 2007).

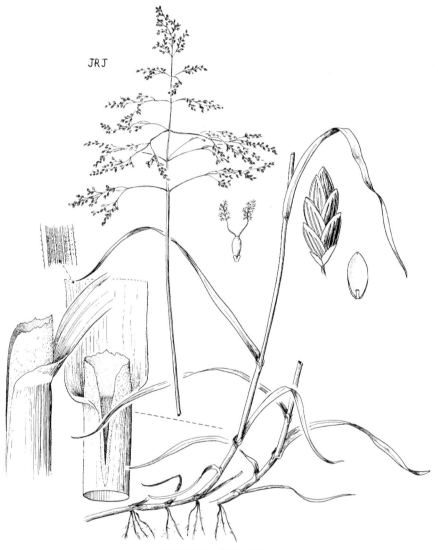

JRJ

Figure 186. *Glyceria elata*.

165. *Glyceria grandis* S. Watson (Fig. 187).

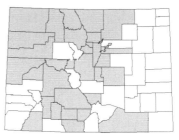

Synonyms: *Glyceria maxima* (Hartman) Holmb. subsp. *grandis* (S. Wats.) Hultén, *Glyceria maxima* (Hartman) Holmb. var. *americana* (Torr.) Boivin, *Panicularia grandis* (S. Wats.) Nash. **Vernacular Names:** American mannagrass, reed mannagrass. **Life Span:** Perennial. **Origin:** Native. **Season:** C_3, cool season. **FLORAL CHARACTERISTICS: Inflorescence:** open panicle, 20–40 cm long; branches often drooping. **Spikelets:** purplish, ovate to oblong, laterally compressed, 3.2–10 mm long, 4- to 7-flowered; rachilla joints 0.5–0.8 mm long; disarticulation above the glumes. **Glumes:** unequal, membranous, slightly erose to minutely ciliolate, 1-nerved, midnerve extending to apices; lower 1–1.8 mm long, elliptic to lanceolate; upper 1.5–2.5 mm long, ovate to obovate. **Lemmas:** ovate to elliptic, 2–2.8 mm long, 7-nerved, apices erose to toothed, truncate to obtuse, with scarious band at apex, apex flat. **Paleas:** broad, equal to slightly longer than lemmas, 2-nerved, nerves not winged. **Awns:** none. **Anthers:** 3. **VEGETATIVE CHARACTERISTICS: Growth Habit:** rhizomatous. **Culms:** erect to decumbent at base, 0.9–1.5 m tall, hollow, rooting freely at nodes. **Sheaths:** closed to open to 1 cm, flattened, keeled, glabrous to slightly scaberulous. **Ligules:** membranous, entire to lacerate, 2–4 mm long. **Blades:** flat to folded to loosely involute, 15–40 cm long, 6–12 mm wide, upper surface glabrous to scaberulent. **HABITAT:** Occurs in wet and moist areas in swamps and irrigation ditches from low elevations up to montane. **COMMENTS:** The only variety found in Colorado is the typical one (var. *grandis*). Closely related to *G. elata* but differs in having acute glumes with long veins, more evenly dark florets, flatter lemma apices, and palea keel tips that do not point toward each other. *G. grandis* is also confused with *Torreyochloa pallida* but differs in having closed sheaths and 1-nerved glumes (Barkworth et al. 2007). This is the largest of the *Glyceria* in Colorado.

166. *Glyceria striata* (Lam.) Hitchc. (Fig. 188).

Synonyms: *Glyceria nervata* (Willd.) Trin., *Glyceria striata* (Lam.) Hitchc. subsp. *stricta* (Scribn.) Hultén, *Glyceria striata* (Lam.) Hitchc. var. *stricta* (Scribn.) Fern., *Panicularia nervata* (Willd.) Kuntze, *Panicularia striata* (Lam.) Hitchc. **Vernacular Names:** Fowl mannagrass, fowl meadow grass. **Life Span:** Perennial. **Origin:** Native. **Season:** C_3, cool season. **FLORAL CHARACTERISTICS: Inflorescence:** lax and open panicle, drooping at maturity, 5–20 cm long, ovoid; branches in ones to threes. **Spikelets:** ovate to oblong, laterally compressed, purplish, 2.5–4 mm long, 3- to 7-flowered; rachilla joint 0.1–0.6 mm long; disarticulation above the glumes. **Glumes:** unequal, pale to purple-tinged, ovate to obovate, 1-nerved, midnerve terminating below apices; lower 0.5–1.2 mm long; upper 1–1.5 mm long. **Lemmas:** obovate, green to purple-tinged, scabrous, 1.5–2.5 mm long, 7-nerved, back rounded, apices flat. **Paleas:** subequal to lemma, nerves not winged. **Awns:** none. **Anthers:** 2. **VEGETATIVE**

Figure 187. *Glyceria grandis.*

JRJ

Figure 188. *Glyceria striata.*

CHARACTERISTICS: Growth Habit: rhizomatous with short rhizomes, in small to large clumps. **Culms:** slender, erect to decumbent at base, 2–10 dm tall, 1.5–3.5 mm wide, not or slightly spongy and sometimes rooting at lower nodes, hollow. **Sheaths:** glabrous to scaberulous, usually closed. **Ligules:** membranous, erose-lacerate, 1–3 mm long, truncate. **Blades:** flat to loosely involute, 5–30 cm long, 2–6 mm wide, upper surface glabrous to occasionally scaberulous. **HABITAT:** Found in moist areas along streams, generally at midmontane to subalpine elevations. **COMMENTS:** The typical variety (var. *striata*) is the only one found in Colorado. This is the most common of the *Glyceria* in Colorado. Closely related to *G. elata* and frequently grows with it. They consistently differ, however, in stature and growth form (Barkworth et al. 2007).

54. *Melica* L.

Perennial, cespitose. Sheaths connate; ligules membranous, erose-lacerate; auricles absent. Inflorescence paniculate, branches short, appressed, few-flowered. Spikelets with 2–5 florets, reflexed on short, pubescent pedicels. Disarticulation below the glumes, each spikelet falling entire at maturity. Glumes subequal. Lemma with 5 prominent and several obscure nerves. Palea about half to two-thirds as long as lemma. Basic chromosome number, $x = 9$. Photosynthetic pathway, C_3.

Represented in Colorado by a single species and a single variety.

167. *Melica porteri* Scribn. (Fig. 189).

Synonyms: None. **Vernacular Name:** Porter's melic. **Life Span:** Perennial. **Origin:** Native. **Season:** C_3, cool season. **FLORAL CHARACTERISTICS: Inflorescence:** narrow panicle, 13–25 cm long; branches few, erect to appressed, 1-sided, with few spikelets. **Spikelets:** ascending to drooping, 8–16 mm long, 1.5–3.5 mm wide, 2- to 5-flowered; rachis prolonged above fertile florets and bearing several smaller sterile lemmas; disarticulation below the glumes. **Glumes:** scarious, unequal, oblanceolate to subovate, apices acute to obtuse; lower 3.5–6 mm long, 2–3 mm wide, 3- to 5-nerved; upper 5–8 mm long, 2–3 mm wide, 5-nerved. **Lemmas:** glabrous to scaberulous, scarious margin near apices, apices rounded to acute, lowermost 6–10 mm long, 5- to 11-nerved, some faint. **Paleas:** about three-quarters as long as lemma. **Awns:** none. **VEGETATIVE CHARACTERISTICS: Growth Habit:** loosely cespitose with short rhizomes. **Culms:** erect to spreading, glabrous, 5.5–10 dm tall. **Sheaths:** closed, glabrous to retrorsely scabrous on keels. **Ligules:** membranous, 2–7 mm long. **Blades:** flat, scabrous, especially distally; 10–25 cm long, 2–5 mm wide. **HABITAT:** Found only locally on rocky cliff sides in canyons in the foothills on the Front Range, the Black Canyon, and the San Juans (Weber and Wittmann 2001a, 2001b). **COMMENTS:** Differs from *Bromelica* in location of spikelet disarticulation. This species is not frequently collected but may be more common than herbarium specimens indicate. The only variety found in Colorado is the typical one (var. *porteri*).

55. *Schizachne* Hack.

Perennial, cespitose. Sheaths closed; ligules membranous and usually united in front; auricles absent. Inflorescence a simple panicle or rarely reduced to a raceme, with few spikelets. Spikelets large, with 3–7 florets, callus bearded. Disarticulation above the glumes and between florets. Glumes unequal, membranous, shorter than lowermost lemma. Lemmas 7- to 13-nerved, nerves prominent and converging at the tip; awned from below a bifid apex. Palea with very pubescent nerves, much shorter than lemma. Basic chromosome number, $x = 10$. Photosynthetic pathway, C_3.

Represented in Colorado by the only species in the genus.

Figure 189. *Melica porteri.*

168. *Schizachne purpurascens* (Torr.) Swallen (Fig. 190).

Synonyms: *Avena torreyi* Nash, *Melica purpurascens* (Torr.) Hitchc., *Schizachne purpurascens* (Torr.) Swallen var. *pubescens* Dore, *Schizachne stricta* (Michx.) Hultén, *Trisetum purpurascens* Torr. **Vernacular Name:** False melic. **Life Span:** Perennial. **Origin:** Native. **Season:** C$_3$, cool season. **FLORAL CHARACTERISTICS: Inflorescence:** open or closed panicle or raceme when depauperate, 7–13 cm long; branches drooping, flexuous, each with 1–2 spikelets. **Spike-**
lets: narrowly elliptical, purplish at base, 11.5–17 mm long, 3- to 6-flowered, callus long-pilose; disarticulation above the glumes. **Glumes:** unequal, broadly lanceolate, glabrous, apices acute; lower 4.2–6.2 mm long, obscurely 3- to 5-nerved; upper 6–9 mm long, obscurely 3- to 5-nerved. **Lemmas:** lanceolate, glabrous to scabrous, 8–10 mm long, 7- to 9-nerved, sometimes obscurely 13-nerved, apices bifid with narrow teeth 1–2 mm long, awn originating between teeth. **Paleas:** shorter than lemmas, 2-nerved, nerves strongly ciliate. **Awns:** lemma awns 8–15 mm long, divergent, somewhat twisted to geniculate. **VEGETATIVE CHARACTERISTICS: Growth Habit:** tufted, leaves crowded at base. **Culms:** erect to decumbent at base, 5–8 dm tall. **Sheaths:** closed to top, sometimes split, glabrous, striate. **Ligules:** membranous, initially united in front but often split in front and sides, which resemble auricles; 0.5–1.5 mm long. **Blades:** flat to folded to loosely involute, glabrous to scabrous, 5–15 cm long, 2–5 mm wide. **HABITAT:** Found on deeply shaded, forested slopes in foothill and montane communities (Weber and Wittmann 2001a, 2001b). **COMMENTS:** Infrequent. This species is sometimes treated as *Trisetum purpurascens* Torr.

13. POEAE

Annuals and perennials. Sheath margins typically free, but connate in some genera; ligules membranous; auricles usually absent but infrequently present. Inflorescence usually a panicle, rarely a raceme or spike. Spikelets with 1 to several florets; reduced florets, if present, above fertile florets (except in *Anthoxanthum*, *Arrhenatherum*, *Hierochloë*, *Phalaris*, and *Phalaroides*). Disarticulation generally above the glumes and below florets but occasionally below the glumes. Glumes unequal, shorter than lowermost floret. Lemmas 5- or more-nerved, awnless or awned from the tip, or a bifid apex.

56. *Agrostis* L.

Low to moderately tall annuals and perennials with slender culms, flat or involute blades, and open or contracted panicles. Spikelets small, 1-flowered, disarticulating above the glumes. Rachilla usually not hairy or extended beyond insertion of floret. Glumes thin, lanceolate, acute to acuminate, nearly equal, lower usually 1-nerved, upper 1- or 3-nerved. Lemma thin, broad, 3- or 5-nerved, acute to obtuse or truncate at apex, glabrous or with a tuft of hair at base, awnless or awned from middle or below. Palea hyaline, usually small or absent. Basic chromosome number, $x = 7$. Photosynthetic pathway, C$_3$.

Represented in Colorado by 7 species.

Figure 190. *Schizachne purpurascens.*

1. Palea absent or minute
    2. Panicle branches naked at base
        3. Lemma with bent awn that exceeds lemma . . . . 172. *A. mertensii*
        3. Lemma awnless or with a straight, enclosed awn, occasionally geniculate in *A. scabra*
            4. Panicle diffuse with long, fine panicle branches divided beyond middle . . . .
            . . . . 173. *A. scabra*
            4. Panicle not diffuse, branches divided before middle . . . . 171. *A. idahoensis*
    2. Panicle branches (at least some) bearing spikelets to base
        5. Plants typically less than 3 dm tall; leaf blades mostly less than 2 mm wide; leaves
        mostly in basal tufts . . . . 175. *A. variabilis*
        5. Plants typically more than 2 dm tall; leaf blades over 2 mm wide; leaves mostly cauline
        . . . . 169. *A. exarata*
1. Palea well developed, at least half the length of lemma
    6. Rhizomes present, panicle open at maturity . . . . 170. *A. gigantea*
    6. Rhizomes absent, stolons often present or lowest culms decumbent; panicle narrow at
    maturity . . . . 174. *A. stolonifera*

169. *Agrostis exarata* Trin. (Fig. 191).

Synonyms: *Agrostis aenea* Trin., *Agrostis alaskana* Hultén, *Agrostis ampla* Hitchc., *Agrostis asperifolia* Trin., *Agrostis exarata* Trin. subsp. *minor* (Hook.) C. L. Hitchc., *Agrostis exarata* Trin. var. *minor* Hook., *Agrostis exarata* Trin. var. *monolepis* (Torr.) Hitchc., *Agrostis exarata* Trin. var. *pacifica* Vasey, *Agrostis exarata* Trin. var. *purpurascens* Hultén, *Agrostis longiligula* Hitchc., *Agrostis longiligula* Hitchc. var. *australis* J. T. Howell, *Agrostis melaleuca* (Trin.) Hitchc., *Agrostis*  *microphylla* Steud. var. *major* Vasey. **Vernacular Names:** Spike bentgrass, spike redtop, western redtop. **Life Span:** Perennial. **Origin:** Native. **Season:** C$_3$, cool season. **FLORAL CHARACTERISTICS: Inflorescence:** compressed panicle, occasionally appearing spikelike, interrupted, 5–25 cm long; branches commonly in dense whorls, some bearing spikelets to base. **Spikelets:** 1-flowered; rachilla not prolonged; disarticulation above the glumes. **Glumes:** about 2–4 mm long; lower sometimes tapering into a very short awn less than 0.5 mm long. **Callus:** sparse to abundant hairs to 0.3 mm long. **Lemmas:** lemma 1–3 mm long, 5-nerved, awned. **Paleas:** absent or to 0.5 mm long. **Awns:** lemmas occasionally awned from middle or above, 1–5 mm long, straight or geniculate if longer. **VEGETATIVE CHARACTERISTICS: Growth Habit:** cespitose, sometimes developing short rhizomes, leaves mostly cauline. **Culms:** hollow, 2–12 dm tall, erect or occasionally decumbent and rooting at lower nodes. **Sheaths:** open, smooth to scabrous, collar sometimes oblique. **Ligules:** membranous, 2–8 mm long, truncate to obtuse, erose-ciliate and lacerate, puberulent. **Blades:** flat, 2–10 mm broad. **HABITAT:** This species is most frequent in moist sites in montane regions on both sides of the Continental Divide (Weber and Wittmann 2001a, 2001b). **COMMENTS:** Weber and Wittmann (2001a, 2001b) state that this species is apparently introduced. Considered an excellent forage for livestock (Stubbendieck, Hatch, and Butterfield 1992).

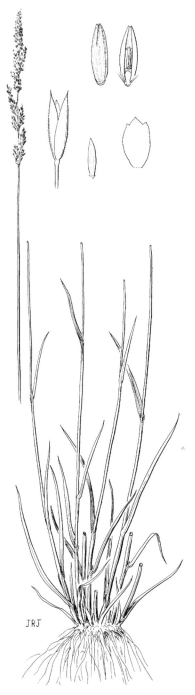

Figure 191. *Agrostis exarata*.

170. *Agrostis gigantea* Roth (Fig. 192).

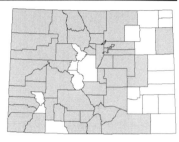

Synonyms: *Agrostis alba* L., *Agrostis gigantea* Roth var. *dispar* (Michx.) Philipson, *Agrostis nigra* With., *Agrostis stolonifera* L. subsp. *gigantea* (Roth) Schuebl. & Martens, *Agrostis stolonifera* L. var. *major* (Gaudin) Farw. **Vernacular Name:** Redtop bent. **Life Span:** Perennial. **Origin:** Introduced. **Season:** C$_3$, cool season. **FLORAL CHARACTERISTICS: Inflorescence:** open panicle, reddish, generally pyramidal-oblong, up to 20 cm long; branches spreading, especially at anthesis. **Spikelets:** 1-flowered, 2–2.5 mm long; disarticulation above the glumes. **Glumes:** nearly equal, about as long as spikelet, keel scabrous. **Callus:** sparse hairs to 0.5 mm long. **Lemmas:** 1.5–2 mm long. **Paleas:** well developed, about half the length of lemmas, 0.7–1.4 mm long. **Awns:** lemmas rarely awned. **VEGETATIVE CHARACTERISTICS: Growth Habit:** strong creeping rhizomes. **Culms:** erect, up to 1.5 m long, rarely decumbent. **Sheaths:** glabrous to scabrous. **Ligules:** 3–6 mm long. **Blades:** flat, 3–8 mm wide, 4–20 cm long. **HABITAT:** Seeded into pastures and hay meadows and used to stabilize ditches. **COMMENTS:** Very similar to *A. stolonifera* L. but differing in having rhizomes, culms erect from base, and larger stature. Weber and Wittmann (2001a, 2001b) report this species is more common than *A. stolonifera* L. Stubbendieck and colleagues (2003) list it as good forage for most kinds and classes of livestock.

171. *Agrostis idahoensis* Nash (Fig. 193).

Synonyms: *Agrostis borealis* Hartman var. *recta* (Hartman) Boivin, *Agrostis clavata* Trin., *Agrostis filicumis* M. E. Jones. **Vernacular Name:** Idaho bentgrass. **Life Span:** Perennial. **Origin:** Native. **Season:** C$_3$, cool season. **FLORAL CHARACTERISTICS: Inflorescence:** open panicle, 2–12 cm long, ovate; delicate branches ascending, forking below middle and with few spikelets, naked at base. **Spikelets:** 1-flowered, purplish; disarticulation above the glumes. **Glumes:** subequal, 1.5–2.5 mm long, lanceolate, 1-nerved, acute to acuminate; lower scabrid to minutely scabrid; upper sometimes smooth. **Callus:** sparse hairs to 0.3 mm long. **Lemmas:** 1–2 mm long, 5-nerved, membranous, awnless or rarely awned. **Paleas:** absent or to 0.1 mm long. **Awns:** lemma occasionally with a short, inconspicuous awn. **VEGETATIVE CHARACTERISTICS: Growth Habit:** cespitose, distinctly tufted. **Culms:** erect, 0.5–4 dm tall. **Sheaths:** open, glabrous to scabrous. **Ligules:** membranous, 1–3 mm long, obtuse, erose, occasionally lacerate. **Blades:** flat or involute, 0.5–2 mm wide. **HABITAT:** Moist areas in subalpine and alpine communities. **COMMENTS:** *A. idahoensis* is closely related to *A. scabra* but can be separated by the narrower, less diffuse inflorescence and panicle branches forking below the middle. Depauperate specimens of *A. idahoensis* can also be confused with *A. mertensii*, but *A. mertensii* tends to occur in better-drained habitats (Barkworth et al. 2007).

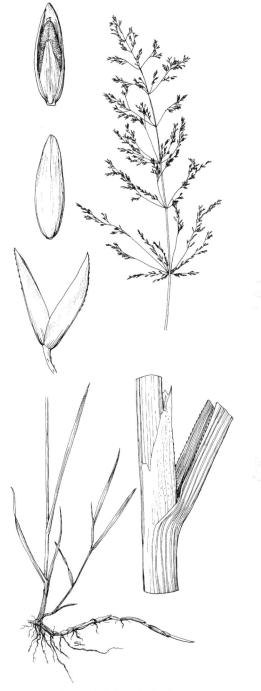

Figure 192. *Agrostis gigantea*.

Figure 193. *Agrostis idahoensis.*

172. *Agrostis mertensii* Trin. (Fig. 194).

Synonyms: *Agrostis bakeri* Rydb., *Agrostis borealis* Hartm., *Agrostis borealis* Hartm. var. *americana* (Scribn.) Fern., *Agrostis borealis* Hartm. var. *paludosa* (Scribn.) Fern., *Agrostis idahoensis* Nash var. *bakeri* (Rydb.) W. A. Weber, *Agrostis mertensii* Trin. subsp. *borealis* (Hartm.) Tzvelev, *Agrostis rupestris* All. **Vernacular Names:** Northern bentgrass, alpine bentgrass. **Life Span:** Perennial. **Origin:** Native. **Season:** C$_3$, cool season. **FLORAL CHARACTERISTICS: Inflorescence:**

panicle open, pyramidal, 5–15 cm long; lower branches whorled and spreading. **Spikelets:** 2.5–3 mm long, 1-flowered, concentrated at tips of panicle branches; disarticulation above the glumes. **Glumes:** subequal, 2–4 mm long, acute, membranous margins, reddish or purple. **Callus:** sparse hairs to 0.4 mm long. **Lemmas:** slightly shorter than the glumes, awned. **Paleas:** absent or to 0.1 mm long. **Awns:** lemma with a bent and exserted awn 3–4.4 mm long, persistent. **VEGETATIVE CHARACTERISTICS: Growth Habit:** tufted dwarf. **Culms:** 2–3 dm long. **Sheaths:** open, glabrous. **Ligules:** membranous, 1.5–2.5 mm long. **Blades:** mostly basal, flat, 5–10 cm long, 1–3 mm wide. **HABITAT:** Meadows and open grassy areas above treeline. **COMMENTS:** *Agrostis borealis* Hartm. of earlier authors. This species can be confused with *A. idahoensis*, but *A. mertensii* grows on better-drained soils. It is also mistaken for dwarfed forms of *A. scabra*, but *A. mertensii* has larger spikelets, more culm nodes, flatter leaves, and panicles less than a third of culm length (Barkworth et al. 2007).

173. *Agrostis scabra* Willd. (Fig. 195).

Synonyms: *Agrostis geminata* Trin., *Agrostis hyemalis* (Walt.) B.S.P. var. *geminata* (Trin.) Hitchc., *Agrostis hyemalis* (Walt.) B.S.P. var. *scabra* (Willd.) Blomquist, *Agrostis hyemalis* (Walt.) B.S.P. var. *tenuis* (Tuckerman) Gleason, *Agrostis scabra* Willd. subsp. *septentrionalis* (Fern.) Á. & D. Löve, *Agrostis scabra* Willd. var. *geminata* (Trin.) Swallen, *Agrostis scabra* Willd. var. *septentrionalis* Fern. **Vernacular Names:** Tickle grass, rough bentgrass. **Life Span:** Perennial. **Origin:**

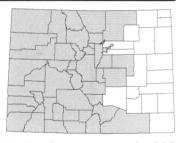

Native. **Season:** C$_3$, cool season. **FLORAL CHARACTERISTICS: Inflorescence:** panicle of diffuse branches, dividing above middle, whorled and spreading below; spikelets borne mostly near tips of panicle branches. **Spikelets:** reddish or purplish, 2.5–3.5 mm long. **Glumes:** subequal to unequal, 1.8–3.4 mm long, lanceolate, 1-nerved, nerve scabrous, apices acuminate. **Callus:** sparse hairs to 0.2 mm long. **Lemmas:** 1.4–2 mm long, 5-nerved, awnless or occasionally awned. **Paleas:** absent or up to 0.2 mm long. **Awns:** lemmas occasionally with awn from middle of back, 0.2–3 mm long, straight or geniculate, persistent when present. **VEGETATIVE CHARACTERISTICS: Growth Habit:** cespitose, tufted. **Culms:** 2–9 dm long. **Sheaths:** open, scabrous. **Ligules:** membranous, truncate, erose, sometimes lacerate. **Blades:** basal leaves mostly involute, 1–2 mm wide, scabrous, stiff. **HABITAT:** Found from foothills to subalpine, moist to dry areas, often establishes early in burned-over areas and

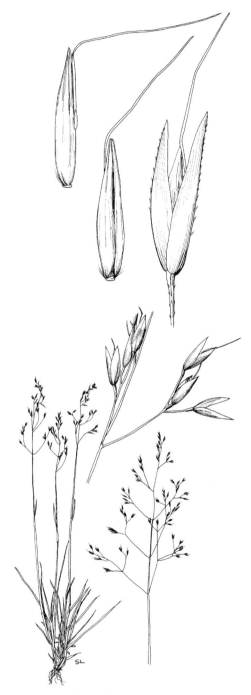

Figure 194. *Agrostis mertensii*.

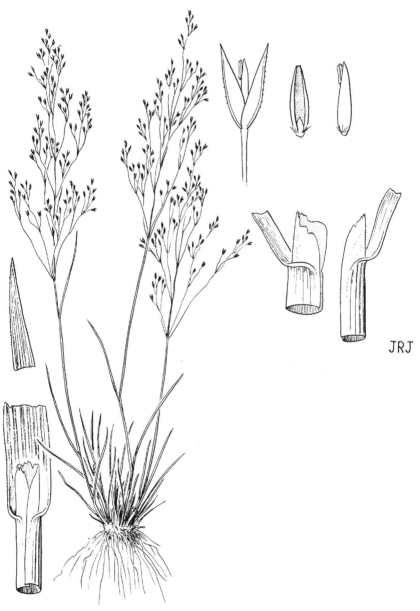

Figure 195. *Agrostis scabra*.

JRJ

disappears later. **COMMENTS:** Frequent. Closely related to *A. idahoensis* but can be separated by the more diffuse inflorescence and panicle branches forking above the middle in *A. scabra*. Dwarfed forms of *A. scabra* can be confused with *A. mertensii*, but *A. scabra* has smaller spikelets, fewer culm nodes, involute basal leaves, and panicles more than a third the length of the plant (Barkworth et al. 2007). This is one of the very first species to establish on burned-over sites; however, with competition it soon diminishes to just a few scattered plants.

174. *Agrostis stolonifera* L. (Fig. 196).

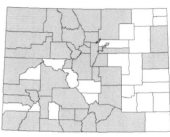

Synonyms: *Agrostis alba* L. var. *palustris* (Huds.) Pers., *Agrostis alba* L. var. *stolonifera* (L.) Sm., *Agrostis maritima* Lam., *Agrostis palustris* Huds., *Agrostis stolonifera* L. var. *compacta* Hartm., *Agrostis stolonifera* L. var. *palustris* (Huds.) Farw. **Vernacular Names:** Creeping bentgrass, creeping tickle grass, redtop bentgrass. **Life Span:** Perennial. **Origin:** Introduced. **Season:** C$_3$, cool season. **FLORAL CHARACTERISTICS: Inflorescence:** narrow panicle at maturity, 5–30 cm long, 1–5 cm wide; branches usually spreading or sometimes closing after anthesis, densely flowered, branches verticillate. **Spikelets:** 1-flowered; disarticulation above the glumes. **Glumes:** subequal to unequal, lanceolate, acute, rounded on back, 1.6–3 mm long, nerves scabrous to ciliate, purplish. **Callus:** sparse hairs to 0.5 mm long. **Lemmas:** 1.4–2 mm long, 5-nerved, membranous; usually unawned. **Paleas:** well developed, 0.7–1.4 mm long. **Awns:** lemmas usually unawned, rarely awned from below apex to 1 mm, straight. **VEGETATIVE CHARACTERISTICS: Growth Habit:** spreading from a decumbent base, rooting at lower nodes, stoloniferous. **Culms:** erect or more usually decumbent, rarely over 1 m long. **Sheaths:** round, glabrous, occasionally purplish or reddish. **Ligules:** membranous, 2–8 mm long, erose, and often lacerate. **Blades:** usually flat, up to 1 cm wide. **HABITAT:** Weber and Wittmann (2001b) report this species occurs along streams and is less common than *A. gigantea*. Also found around ponds, stock tanks, and other standing water. **COMMENTS:** Can be separated from the closely related *A. gigantea* by decumbent bases, stolons, and lower stature.

175. *Agrostis variabilis* Rydb. (Fig. 197).

Synonym: *Agrostis rossiae* Vasey. **Vernacular Name:** Mountain bentgrass. **Life Span:** Perennial. **Origin:** Native. **Season:** C$_3$, cool season. **FLORAL CHARACTERISTICS: Inflorescence:** panicle with appressed branches, 2–8 cm long; bearing spikelets nearly to base of inflorescence branches. **Spikelets:** 1-flowered, 2–3 mm long; disarticulation above the glumes. **Glumes:** 1.8–2.5 mm long, lanceolate, acute, 1-nerved, scabrous on keel, usually purple; lower slightly longer than upper. **Callus:** sparse to abundant hairs to 0.2 mm long. **Lemmas:** 1.5–2 mm long,

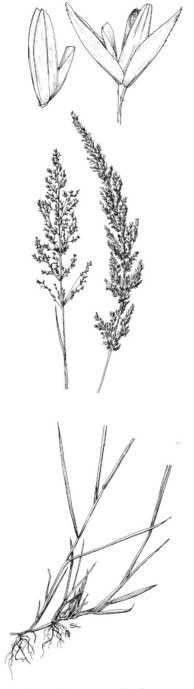

Figure 196. *Agrostis stolonifera*.

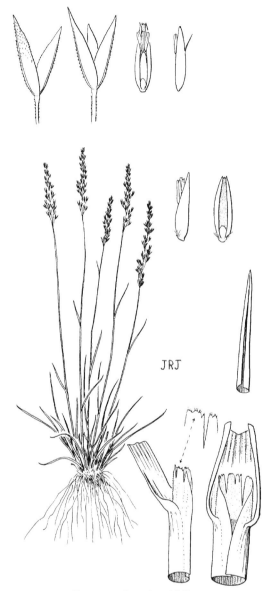

Figure 197. *Agrostis variabilis.*

membranous, 5-nerved, awnless or rarely awned. **Paleas:** minute, to 0.2 mm long. **Awns:** lemma rarely with a short awn originating from middle of the back to 1 mm long, usually not extending beyond lemmas apices. **VEGETATIVE CHARACTERISTICS: Growth Habit:** densely tufted. **Culms:** 1–3 dm long. **Sheaths:** open. **Ligules:** 1–2.8 mm long, truncate, finely erose, often lacerate. **Blades:** involute, folded or rarely flat, mostly basal. **HABITAT:** Mesic areas in subalpine to alpine regions. **COMMENTS:** Cronquist and colleagues (1977) report that *A. variabilis* is the most widespread species of the genus in western North America.

57. *Alopecurus* L.

Tufted annuals and perennials, a few rhizomatous, with flat blades and dense, cylindrical, spikelike panicles of 1-flowered spikelets. Disarticulation below the glumes, spikelets falling entire. Glumes equal, awnless, usually united at margins below, ciliate on keels. Lemma about as long as the glumes, firm, 5-nerved, obtuse, awned from below the middle. Margins of lemma usually united near base. Palea typically absent. Basic chromosome number, $x = 7$. Photosynthetic pathway, C_3.

Represented in Colorado by 6 species and 1 variety.

1. Plant annual . . . . 178. *A. carolinianus*
1. Plant perennial
   2. Glumes 1.8–3.5 mm long, apices obtuse
      3. Lemmas 1.5–2.5 mm long; lemma awns 0.7–3 mm long, straight . . . . 176. *A. aequalis*
      3. Lemmas 2.5–3 mm long; lemma awns 3.5–5 mm long, geniculate . . . .
      . . . . 179. *A. geniculatus*
   2. Glumes 3–6 mm long, apices acute
      4. Glumes densely pilose over entire surface . . . . 180. *A. magellanicus*
      4. Glumes pilose to sparsely pubescent on nerves only
         5. Lemma apices acute, glume apices parallel to convergent . . . . 181. *A. pratensis*
         5. Lemma apices truncate, occasionally obtuse, glume apices divergent . . . .
         . . . . 177. *A. arundinaceus*

176. *Alopecurus aequalis* Sobol. (Fig. 198).

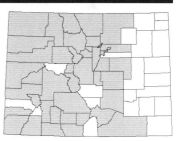

Synonyms: *Alopecurus aequalis* Sobol. var. *natans* (Wahlenb.) Fern., *Alopecurus aristulatus* Michx., *Alopecurus geniculatus* L. var. *aristulatus* (Michx.) Torr. **Vernacular Name:** Short-awn foxtail. **Life Span:** Perennial (occasionally flowering as a winter annual). **Origin:** Native. **Season:** C_3, cool season. **FLORAL CHARACTERISTICS: Inflorescence:** panicle, narrow with appressed short-pediceled spikelets, cylindrical, 1–9 cm long, 3–9 mm wide, spikelike. **Spikelets:** 1-flowered, strongly flattened, 1.8–3 mm long; disarticulation below the glumes. **Glumes:** subequal, 1.8–3 mm long, 3-nerved, connate section to 0.5 mm long, ciliate on keels and appressed-pilose on lateral nerves, apices obtuse. **Lemmas:** 1.5–2.5 mm long, lanceolate to ovate, margins connate, less than half the length, glabrous, obtuse, awned. **Paleas:** lacking. **Awns:** lemma awns short, 0.7–3 mm long, straight, often included in the glumes, attached 0.7–1 mm from base. **VEGETATIVE CHARACTERISTICS: Growth Habit:** tufted. **Culms:** erect

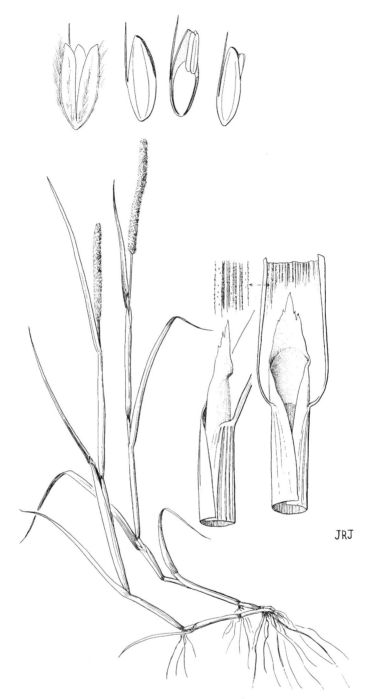

Figure 198. *Alopecurus aequalis.*

JRJ

or geniculate, occasionally rooting at lower nodes, 2–6 dm long. **Sheaths:** open to inflated below, sometimes loosely separating from culms, glabrous. **Ligules:** membranous, 2–6.5 mm long, attenuate, lacerate or entire. **Blades:** flat, 1–5 mm wide, glabrous or scabrous on upper surface. **HABITAT:** Marshes, wet meadows, margins of lakes, ponds, or streams from foothills to subalpine. Weber and Wittmann (2001a) report the plant to be widespread in muddy places, ditches, and ponds. **COMMENTS:** Cronquist and colleagues (1977) report this species to be a good forage, but it is not plentiful. The only variety found in Colorado is the typical one (var. *aequalis*).

177. *Alopecurus arundinaceus* Poir. (Fig. 199).

Synonyms: None. **Vernacular Name:** Creeping meadow foxtail. **Life Span:** Perennial. **Origin:** Introduced. **Season:** C$_3$, cool season. **FLORAL CHARACTERISTICS: Inflorescence:** panicle, contracted with short-pediceled spikelets, 3–10 cm long, 7–13 mm wide. **Spikelets:** 1-flowered, strongly flattened, 3.5–6 mm long; disarticulation below the glumes. **Glumes:** subequal, 3.6–5 mm long, connate for about 1 mm, sparsely pubescent along nerves, keel ciliate, apices acute, divergent, 3-nerved. **Lemmas:** 3.1–4.5 mm long, ovate, margins connate for about 1.5 mm, glabrous or with scattered hairs near apices, truncate to obtuse, awned. **Paleas:** lacking. **Awns:** lemma awns 1.5–7.5 mm long, geniculate, usually well exserted from spikelet. **VEGETATIVE CHARACTERISTICS: Growth Habit:** rhizomatous. **Culms:** 2–10 dm long. **Sheaths:** open to inflated, sometimes loosely separating from culms, glabrous. **Ligules:** membranous, 1.5–5 mm long, truncate, erose, becoming lacerate. **Blades:** flat, 0.3–1.5 cm wide, scabrous on both surfaces. **HABITAT:** Grows on wet soils, floodplains, along streams, bogs, and ponds. **COMMENTS:** Introduced as a pasture grass and now escaped and widely established. Sometimes used in seed mixtures for revegetation.

178. *Alopecurus carolinianus* Walter (Fig. 200).

Synonyms: *Alopecurus macounii* Vasey, *Alopecurus ramosus* Poir. **Vernacular Names:** Carolina foxtail, annual foxtail, tufted foxtail. **Life Span:** Annual. **Origin:** Native. **Season:** C$_3$, cool season. **FLORAL CHARACTERISTICS: Inflorescence:** panicle, cylindrical and tightly contracted, spikelike, 2–5 cm long. **Spikelets:** 1-flowered, strongly flattened; disarticulation below the glumes. **Glumes:** 2.1–3.1 mm long, subequal, connate only at base, membranous, appressed pubescent on sides, pale green to pale yellow. **Lemmas:** 1.9–2.7 mm long, connate to midlength, glabrous, apex obtuse, awned. **Paleas:** lacking. **Awns:** lemmas awned from slightly above base, 3–5 mm long, geniculate, exserted from spikelet. **VEGETATIVE CHARACTERISTICS: Growth Habit:** tufted. **Culms:** 1–5 dm tall, erect to decumbent at base. **Sheaths:** open to inflated at base, sometimes loosely separating from culms, glabrous. **Ligules:** membranous,

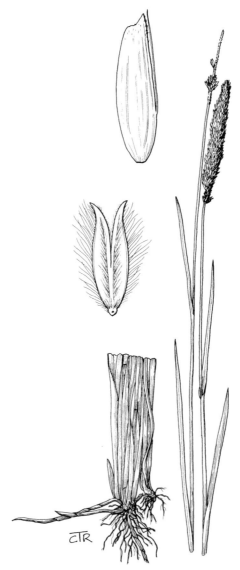

Figure 199. *Alopecurus arundinaceus.*

JRJ

Figure 200. *Alopecurus carolinianus.*

2.8–4.5 mm long, attenuate to a narrow erose tip. **Blades:** flat, scabrous above, 3–15 cm long, 0.9–3 mm wide. **HABITAT:** Wet meadows and along watercourses from the foothills to subalpine elevations. **COMMENTS:** Rare and perhaps not a constant member of our grass flora.

179. *Alopecurus geniculatus* L. (Fig. 201).

Synonym: *Alopecurus pallescens* Piper. **Vernacular Names:** Water foxtail, marsh foxtail. **Life Span:** Perennial. **Origin:** Introduced. **Season:** C$_3$, cool season. **FLORAL CHARACTERISTICS: Inflorescence:** panicle cylindrical with short-pediceled spikelets, dense, spikelike, 1.5–7 cm long, 4–8 mm wide. **Spikelets:** 1-flowered, strongly flattened, 2–4 mm long; disarticulation below the glumes. **Glumes:** subequal, 1.9–3.5 mm long, connate for less than 1 mm at base, silky pubescent on sides. **Lemmas:** 2.5–3 mm long, margins connate to near middle, glabrous, apices truncate to obtuse, awned. **Paleas:** lacking. **Awns:** lemmas awned from near base, geniculate, 3–8 mm long, exserted. **VEGETATIVE CHARACTERISTICS: Growth Habit:** weakly tufted. **Culms:** 2–6 dm long, decumbent and rooting at nodes. **Sheaths:** open to inflated, sometimes loosely separating from culms, glabrous. **Ligules:** membranous, 2–6 mm long, narrowed to a slightly erose tip. **Blades:** flat, 2–6 mm wide, scabrous above, less so below. **HABITAT:** Wet meadows and marshy areas around lakes and ponds, often growing in water. **COMMENTS:** Relatively rare, and only a few records in Colorado.

180. *Alopecurus magellanicus* Lam. (Fig. 202).

Synonyms: *Alopecurus alpinus* Sm., *Alopecurus alpinus* Sm. subsp. *glaucus* (Less.) Hultén, *Alopecurus alpinus* J. E. Sm. subsp. *stejnegeri* (Vasey) Hultén, *Alopecurus alpinus* J. E. Sm. var. *borealis* (Trin.) Griseb., *Alopecurus alpinus* Sm. var. *glaucus* (Less.) Krylov, *Alopecurus alpinus* Sm. var. *occidentalis* (Scribn. & Tweedy) Boivin, *Alopecurus alpinus* J. E. Sm. var. *stejnegeri* (Vasey) Hultén, *Alopecurus borealis* Trin., *Alopecurus glaucus* Less., *Alopecurus occidentalis* Scribn. & Tweedy. **Vernacular Name:** Alpine foxtail. **Life Span:** Perennial. **Origin:** Native. **Season:** C$_3$, cool season. **FLORAL CHARACTERISTICS: Inflorescence:** panicle, narrow, tightly contracted with short-pediceled spikelets, woolly, 1–4 cm long. **Spikelets:** 1-flowered, strongly flattened, 3–5 mm long, membranous, densely woolly; disarticulation below the glumes. **Glumes:** subequal, 3–5 mm long, connate for about 1 mm, densely villous, 3-nerved. **Lemmas:** 2.5–4.5 mm long, ovate, margins connate for about 1.5 mm, puberulent above, glabrous below; apices generally obtuse, occasionally truncate, awned. **Paleas:** lacking. **Awns:** lemma awns 2–6 (8) mm long, geniculate, attached just below middle of lemma, well exserted from spikelet. **VEGETATIVE CHARACTERISTICS: Growth Habit:** rhizomatous, stoloniferous, or both. **Culms:** 1–8 dm tall, erect to decumbent. **Sheaths:** open, inflated, sometimes loosely

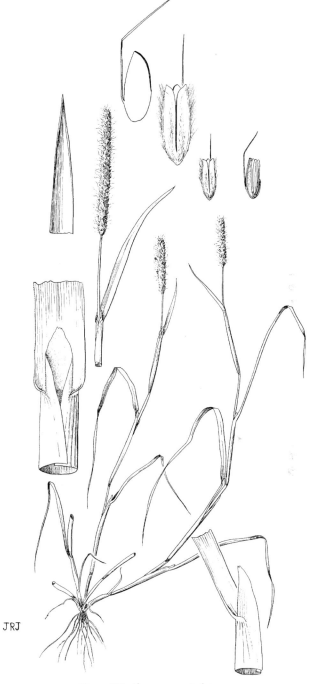

Figure 201. *Alopecurus geniculatus.*

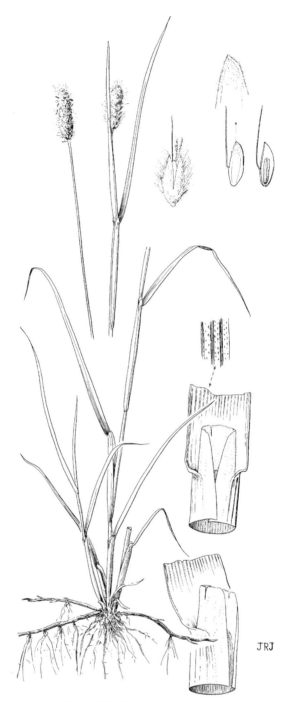

Figure 202. *Alopecurus magellanicus.*

separating from culms, glabrous. **Ligules:** membranous, 1.5–2 mm long, truncate, erose, becoming lacerate. **Blades:** flat, 4–22 cm long, 2.5–7 mm wide, scabrous on both surfaces. **HABITAT:** Along streams and in wet meadows at higher elevations. **COMMENTS:** *A. alpinus* J. E. Sm. and *A. borealis* Trin. of previous authors are included within this species.

181. *Alopecurus pratensis* L. (Fig. 203).

Synonyms: None. **Vernacular Name:** Meadow foxtail. **Life Span:** Perennial. **Origin:** Introduced. **Season:** C$_3$, cool season. **FLORAL CHARACTERISTICS: Inflorescence:** panicle cylindrical with very short-pediceled spikelets, tightly contracted, spikelike, 3.5–9 cm long. **Spikelets:** 1-flowered, strongly flattened, 4–6 mm long; disarticulation above the glumes. **Glumes:** subequal, 4–6 mm long, connate for about 1 mm, 3-nerved, pilose on lateral nerves, villous-ciliate on keel, apices acute, parallel to convergent. **Lemmas:** 4–6 mm long, margins connate for about 1 mm, 5-nerved, glabrous, occasionally keels ciliate, apices acute, awned. **Paleas:** lacking. **Awns:** lemmas awned from near base, geniculate, 5–8 mm long, well exserted. **VEGETATIVE CHARACTERISTICS: Growth Habit:** cespitose, tufted, stout. **Culms:** 3–11 dm tall, erect, sometimes rooting at lower nodes and then stoloniferous, rhizomatous, or both. **Sheaths:** open, glabrous, not as inflated as in some other members of the genus. **Ligules:** membranous, 1.5–3 mm long, obtuse to truncate, erose-ciliolate, occasionally lacerate. **Blades:** flat, 6–40 cm long, 2–10 mm wide, scabrous on both surfaces. **HABITAT:** Frequently planted in hay meadows and escaping to wet, native meadows and margins of streams and ponds. **COMMENTS:** It is commonly used to revegetate roadsides in the mountains.

58. *Anthoxanthum* L.

Annuals or perennials with fragrant herbage as a result of the presence of coumarin. Sheaths open with overlapping margins; ligules membranous, short; auricles absent. Inflorescence a contracted, often spicate panicle. Spikelets with 1 large fertile floret and 2 sterile reduced florets below. Disarticulation above the glumes, fertile and reduced florets falling together. Glumes unequal, broad, thin, acute or mucronate; lower 1-nerved and shorter than upper, upper 3-nerved and longer than florets. Lemmas of reduced florets hairy, awned from back, notched or toothed at apex. Lemmas of fertile floret broad, rounded, firm, shiny brown, awnless. Palea present and 1-nerved in fertile floret, absent from reduced florets. Basic chromosome number, $x = 5$. Photosynthetic pathway, C$_3$.

Represented in Colorado by a single species.

182. *Anthoxanthum odoratum* L. (Fig. 204).

Synonyms: None. **Vernacular Name:** Sweet vernalgrass. **Life Span:** Perennial. **Origin:** Introduced. **Season:** C$_3$, cool season. **FLORAL CHARACTERISTICS: Inflorescence:** contracted panicle, 3–8 cm long. **Spikelets:** slightly compressed laterally, 6–10 mm long with 3 florets, middle one perfect and 2 lateral ones reduced to sterile lemmas; disarticulation above the

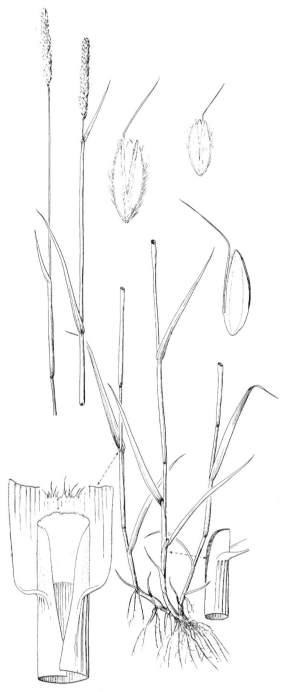

JRJ

Figure 203. *Alopecurus pratensis.*

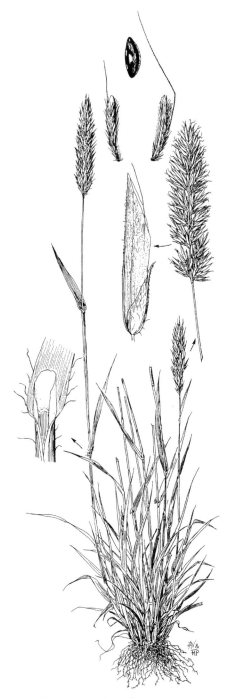

Figure 204. *Anthoxanthum odoratum*.

glumes. **Glumes:** lower about half as long as spikelet, 3–4 mm long, 1-nerved, otherwise scarious, variously pubescent or pilose; upper as long as spikelet,
8–10 mm long, 3-nerved, widely scarious-margined,
slightly pilose. **Florets:** sterile 3–4 mm long; fertile
1–2.5 mm long. **Lemmas:** center fertile, 2 mm long,
shorter than 2 lateral sterile ones, smooth, coriaceous;
first sterile 2–3 mm long, appressed pilose with yellow pubescence; second sterile similar to first with the
exception of the awn. **Awns:** first sterile lemma awned one-third of the way from bifid apex,
with a straight awn 2–4 mm long; second sterile lemma with a geniculate awn from base
exceeding spikelet in length, 4–9 mm long. **VEGETATIVE CHARACTERISTICS: Growth Habit:**
erect, tufted, without rhizomes. **Culms:** 3–6 dm tall, glabrous. **Sheaths:** glabrous. **Ligules:**
1–5 mm long. **Blades:** 2–15 cm long, 2–6 mm wide, flat, short-pilose to glabrate. **HABITAT:**
Meadows, gardens, and disturbed areas. **COMMENTS:** Rare in Colorado, and likely introduced as an ornamental. The genus *Hierochloë* is sometimes included with *Anthoxanthum*.

59. *Apera* Adans.

Tufted annuals with weak culms, flat blades, and open or contracted, densely flowered
panicles. Spikelets small, 1-flowered, disarticulating above the glumes. Rachilla prolonged
behind floret as a naked bristle. Glumes membranous, lanceolate, acuminate at tip, lower
1-nerved and usually slightly shorter than 3-nerved upper glume. Lemma firm, rounded on
back, indistinctly 5-nerved, puberulent at base, awned from near the tip with a straight or
flexuous awn. Palea well developed, 2-nerved, as long as lemma or slightly shorter. Mature
caryopsis narrowly oblong, tightly enclosed between firm lemma and palea. Basic chromosome number, $x = 7$. Photosynthetic pathway, C_3.

Represented in Colorado by a single species.

183. *Apera interrupta* (L.) P. Beauv. (Fig. 205).

Synonym: *Agrostis interrupta* L. **Vernacular Names:**
Dense silkybent, interrupted windgrass. **Life Span:**
Annual. **Origin:** Introduced. **Season:** C_3, cool season.
FLORAL CHARACTERISTICS: Inflorescence: densely
flowered, narrow panicle with ascending branches,
8–15 cm long. **Spikelets:** 1-flowered, green, occasionally tinged with purple, 2–3 mm long; disarticulating
above the glumes. **Glumes:** lower 1.5–2.2 mm long,
1-nerved; upper 2–2.8 mm long, slightly longer than
floret, 3-nerved. **Lemmas:** 1.6–2.5 mm long, subindurate at maturity. **Awns:** 6–16 mm long
from just below tip of lemmas. **VEGETATIVE CHARACTERISTICS: Growth Habit:** tufted
or rarely solitary. **Culms:** 1–7 dm tall. **Sheaths:** smooth. **Ligules:** 1.5–4.5 mm long, lacerate-erose. **Blades:** flat or involute, 1–4 mm wide. **HABITAT:** Disturbed areas on the eastern
plains, often a weed of winter wheat fields. **COMMENTS:** This genus can be mistaken for
Muhlenbergia, but the awn arises below the tip of the lemma in *Apera*.

Figure 205. *Apera interrupta*.

60. *Argillochloa* W. A. Weber

Perennial, cespitose. Culms only 2–4 dm tall. Sheaths open with overlapping margins; ligules short, membranous, auricles absent. Inflorescence simple, terminal panicle; branches divaricate, 2, 3, or 4 per node; ciliate on prominent angles. Spikelets usually with 2 florets. Disarticulation above the glumes and between florets. Glumes subequal; lower 1-nerved; upper 3-nerved. Lemmas lanceolate, green, membranous, 5-nerved, keeled, awn 2–3 mm long. Palea about as long as lemma. Basic chromosome number, $x = 7$. Photosynthetic pathway, C_3.

Represented in Colorado by a single species. A segregate of *Festuca* (*F. dasyclada*).

184. *Argillochloa dasyclada* (Hack. *ex* Beal) W. A. Weber (Fig. 206).

Synonym: *Festuca dasyclada* Hack. *ex* Beal. **Vernacular Names:** Low woollygrass, false ricegrass. **Life Span:** Perennial. **Origin:** Native. **Season:** C_4, warm season. **FLORAL CHARACTERISTICS: Inflorescence:** compact panicle, 1.5–4 cm long, 1–1.5 cm wide, light green to purple-tinged, white pubescent. **Spikelets:** typically 2, but up to 4–10 florets, 5–10 mm long, 3–6 mm wide; disarticulates above the glumes. **Glumes:** subequal, 1-nerved, glabrous, purple-tinged; lower

6–8.5 mm long; upper 6.5–9 mm long. **Lemmas:** 3-nerved, 2-lobed, 2.5–5 mm long, long-pilose on lower half, often purple-tinged. **Paleas:** 2–3.5 mm long, keels ciliate distally, long-pilose proximally. **Awns:** lemma awns 1.5–4 mm long, barely extending beyond lobes of lemma. **VEGETATIVE CHARACTERISTICS: Growth Habit:** densely tufted, stoloniferous, sometimes mat-forming. **Culms:** scabrous or puberulent, erect initially, 2–5 dm long. **Sheaths:** open, striate, margins scarious, pilose at base, bearing a tuft of hairs at apex. **Ligules:** 3–5 mm long, ciliate fringe of hairs. **Blades:** involute, 1–6 cm long, scabrid above, scabrous below. **HABITAT:** Piñon-juniper and desert steppe and on shale slopes in the Piceance Basin in western Colorado (Weber and Wittmann 2001b). **COMMENTS:** Appears similar to *Achnatherum hymenoides*. The spreading panicle of *A. dasyclada* breaks at the base when the seeds are mature and becomes a "tumbleweed" for dissemination of propagules.

61. *Arrhenatherum* P. Beauv.

Moderately tall perennials with flat blades and narrow panicles. Spikelets usually 2-flowered, upper floret perfect, lower staminate. Rachilla continued back of upper floret as a bristle. Disarticulation above the glumes, 2 florets falling together. Glumes broad, thin, acute; lower 1-nerved; upper longer, 3-nerved. Lower floret staminate, larger than upper floret and bearing a stout, geniculate awn below the middle. Upper floret perfect, lemma 5- to 7-nerved, usually awnless or with a short, straight, hairlike awn from or just below the tip. Palea 2-keeled, scabrous or hairy on keels. Caryopsis narrowly oblong, pubescent (in *A. elatius*). Basic chromosome number, $x = 7$. Photosynthetic pathway, C_3.

Represented in Colorado by a single species and 2 subspecies.

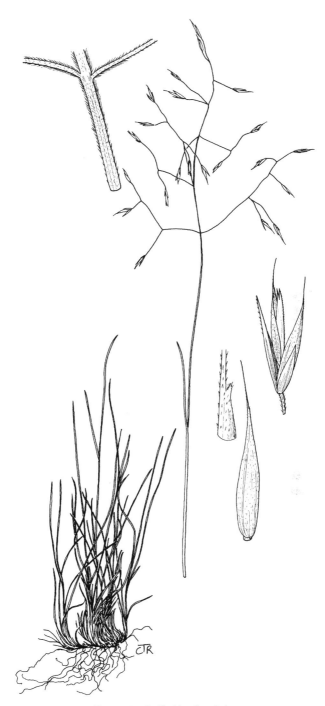

Figure 206. *Argillochloa dasyclada.*

185. *Arrhenatherum elatius* (L.) P. Beauv. *ex* J. Presl & C. Presl (Fig. 207).

Synonyms: *Arrhenatherum avenaceum* (Scop.) P. Beauv., *Arrhenatherum elatius* (L.) P. Beauv. *ex* J. & C. Presl var. *biaristatum* (Peterm.) Peterm., *Avena elatior* L. **Vernacular Name:** Tall oatgrass. **Life Span:** Perennial. **Origin:** Introduced. **Season:** C₃, cool season. **FLORAL CHARACTERISTICS: Inflorescence:** narrow panicle, 15–25 cm long, floriferous from near base; short branches spreading in anthesis. **Spikelets:** 2-flowered, 7–9 mm long; 2 florets disarticulating above the glumes as a single unit; upper floret perfect; lower floret staminate. **Glumes:** membranous, glabrous or scabrous on keel; lower 5–7 mm long, 1-nerved; upper 8–9 mm long, 3-nerved. **Lemmas:** 7–9.5 mm long, 7-nerved, scabrous or pilose, variously awned. **Awns:** lower lemma awn geniculate, from near middle, 10–14 mm long; upper lemma awn straight, up to 4 mm long or occasionally absent, attached just below apex. **VEGETATIVE CHARACTERISTICS: Growth Habit:** erect, loosely tufted, occasionally with short rhizomes. **Culms:** 7–17 dm long, glabrous or occasionally pubescent at nodes. **Sheaths:** glabrous, somewhat open. **Ligules:** 1–2 mm long, erose-ciliolate. **Blades:** flat, 3–8 mm wide, 5–25 cm long, glabrous to scabrous. **HABITAT:** Disturbed areas, sometimes planted in wet meadows for hay. **COMMENTS:** Escaped from cultivation. Currently, 2 subspecies are found in Colorado [subsp. *elatius* and subsp. *bulbosum* (Willd.) Spenner]. The subspecies can be distinguished by these characteristics:

1. Nodes glabrous; culm bases neither thickened nor bulbous. . . . . subsp. *elatius*
1. Nodes pubescent; culm bases thickened and bulbous . . . . subsp. *bulbosum*

62. *Avena* L.

In ours, annuals with moderately tall, weak culms, broad, flat blades, and panicles of large, pendulous, usually 2- to 3-flowered spikelets on slender pedicels. The inflorescence of depauperate plants may be reduced to a raceme of a few spikelets or even a single spikelet. Disarticulation above the glumes and between florets. Glumes about equal, broad, thin, several-nerved, longer than lower floret and often exceeding upper. Lemmas rounded on back, 5- to 7-nerved, tough and indurate, often hairy on callus; usually bearing a stout, geniculate awn from below the notch of a bifid apex. In the cultivated oats, *Avena fatua*, the awn is absent or reduced. Palea 2-keeled, shorter than lemma. Basic chromosome number, $x = 7$. Photosynthetic pathway, C₃.

Represented in Colorado by 3 species.

1. Florets not disarticulating from the glumes even at maturity; calluses glabrous . . . .
. . . . 187. *A. sativa*
1. Florets disarticulating at maturity, calluses bearded
   2. Florets falling from the glumes as a unit . . . . 188. *A. sterilis*
   2. Florets falling separately . . . . 186. *A. fatua*

JRJ

Figure 207. *Arrhenatherum elatius.*

186. *Avena fatua* L. (Fig. 208).

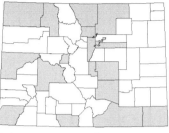

Synonyms: *Avena fatua* L. var. *glabrata* Peterm., *Avena fatua* L. var. *vilis* (Wallr.) Hausskn., *Avena hybrida* Peterm. *ex* Reichenb. **Vernacular Name:** Wild oats. **Life Span:** Annual. **Origin:** Introduced. **Season:** C$_3$, cool season. **FLORAL CHARACTERISTICS: Inflorescence:** open, loose panicle, 10–30 cm long, often partly included. **Spikelets:** 1.8–3.2 cm long, drooping, 2- to 3-flowered; florets separating easily and disarticulating above the glumes. **Glumes:** subequal, 1.8–3.2 cm long, second slightly longer, about the length of spikelet, 9- to 11-nerved. **Lemmas:** 14–22 mm long, somewhat obscurely 5- to 7-nerved, long-pilose to glabrate; calluses bearded. **Awns:** 25–45 mm long, arising from near middle of lemma, usually only on first 2 florets, geniculate, twisted below the knee. **VEGETATIVE CHARACTERISTICS: Growth Habit:** stout, erect. **Culms:** 3–9 dm long, glabrous. **Sheaths:** glabrous, lower ones sometimes pubescent. **Ligules:** 2–5.5 mm long, erose-ciliolate. **Blades:** flat, 3–10 mm wide, 10–30 cm long, scabrous to pilose. **HABITAT:** Disturbed areas, roadsides, and cultivated fields. **COMMENTS:** Hybrids between *A. fatua* and *A. sativa* are common in plantings of cultivated oats.

187. *Avena sativa* L. (Fig. 209).

Synonyms: *Avena byzantina* K. Koch, *Avena fatua* L. var. *sativa* (L.) Hausskn., *Avena hybrida* Peterm. *ex* Reichenb. In part, *Avena sativa* L. var. *orientalis* (Schreb.) Alef. **Vernacular Name:** Cultivated oats. **Life Span:** Annual. **Origin:** Introduced. **Season:** C$_3$, cool season. **FLORAL CHARACTERISTICS: Inflorescence:** open, loose panicle, 15–40 cm long. **Spikelets:** 2–3.2 cm long, drooping, mostly 2-flowered, occasionally 3; spikelets remaining on plant even when mature. **Glumes:** subequal, 2–3.2 cm long, as long as or slightly exceeding spikelet, strongly nerved. **Lemmas:** 1.4–1.8 cm long, glabrous or sparsely hirsute, indurate in lower portion and herbaceous near tip; calluses glabrous. **Awns:** absent or poorly developed from below apex of first floret only, occasionally up to 35 mm long; weakly twisted, straight, or curved. **VEGETATIVE CHARACTERISTICS: Growth Habit:** erect, tufted, stout. **Culms:** 4–10 dm tall, glabrous to scabrous. **Sheaths:** glabrous or sparsely pubescent. **Ligules:** 2–6 mm long. **Blades:** 3–20 mm wide, 10–40 cm long, flat, glabrous or scabrous, at least on the margins; rarely sparsely hairy. **HABITAT:** Cultivated, occasionally escaping to nearby roadsides and disturbed areas. **COMMENTS:** Hybrids between *A. sativa* and *A. fatua* are common in plantings of cultivated oats.

Figure 208. *Avena fatua*.

Figure 209. *Avena sativa*.

188. *Avena sterilis* L. (Fig. 210).

Synonyms: None. **Vernacular Name:** Animated oats. **Life Span:** Annual. **Origin:** Introduced. **Season:** C_3, cool season. **FLORAL CHARACTERISTICS: Inflorescence:** open, loose panicle, 10–45 cm long. **Spikelets:** 2.5–5 cm long, 2- to 5-flowered; florets disarticulating above the glumes and beneath basal floret, florets falling as a unit. **Glumes:** subequal, 2–5 cm long, strongly 9- or 11-nerved. **Lemmas:** 1.7–4 cm long, densely strigose below, apices bidentate; calluses bearded, awned. **Awns:** lemmas awned, 3–9-cm long, arising in the middle third. **VEGETATIVE CHARACTERISTICS: Growth Habit:** erect, stout. **Culms:** 4–12 dm tall, glabrous to pubescent. **Sheaths:** glabrous or sparsely pubescent. **Ligules:** 3–8 mm long, acute to truncate mucronate. **Blades:** 4–18 mm wide, 8–60 cm long, flat, often ciliate along margins. **HABITAT:** Disturbed sites. Naturalized in California and Oregon. **COMMENTS:** Collected only once in Fort Collins, Larimer County, in 1898; likely not a member of our flora.

63. *Avenula* (Dumort.) Dumort

Perennial, cespitose, infrequently stoloniferous. Sheaths open or closed; ligules membranous; auricles absent. Inflorescence a reduced panicle, many branches bearing a single spikelet. Spikelets with 2–7 florets. Disarticulation above the glumes and below florets. Glumes equaling or exceeding florets, 1–3-nerved. Lemmas 5–7-nerved, obtuse, bifid, awned from near middle. Awns geniculate, twisted below bend. Paleas with lateral wings and bifid. Basic chromosome number, $x = 7$. Photosynthetic pathway, C_3.

Represented in Colorado by a single species.

189. *Avenula hookeri* (Scribn.) Holub (Fig. 211).

Synonyms: *Avena hookeri* Scribn., *Avenochloa hookeri* (Scribn.) Holub, *Helictotrichon hookeri* (Scribn.) Henr. **Vernacular Name:** Spike oatgrass. **Life Span:** Perennial. **Origin:** Native. **Season:** C_3, cool season. **FLORAL CHARACTERISTICS: Inflorescence:** narrow panicle, 5–10 cm long or less at high elevations; branches short, erect, usually bearing just a single spikelet. **Spikelets:** 1.2–1.5 cm long, mostly 3- to 6-flowered; florets disarticulating individually above the glumes, upper ones exserted beyond the glumes. **Glumes:** subequal, thin, acute; lower 9–13 mm long, 3-nerved (the 2 lateral nerves sometimes indistinct); upper 9–14 mm long, 3- to 5 nerved; calluses bearded. **Lemmas:** 10–12 mm long, somewhat indurate, bidentate, awned from just above middle. **Awns:** lemmas awned, 10–18 mm long, geniculate, sometimes with 2 knees, twisted below lowest knee. **VEGETATIVE CHARACTERISTICS: Growth Habit:** erect, densely tufted, leaves mostly basal. **Culms:** 0.5–4.5 dm tall. **Sheaths:** keeled. **Ligules:**

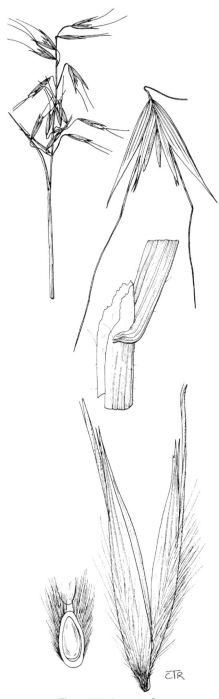

Figure 210. *Avena sterilis*.

Figure 211. *Avenula hookeri.*

2–3 mm long, membranous, acute. **Blades:** flat, 2–4 mm wide, 4–20 cm long. **HABITAT:** Open slopes, meadows, and forest openings at upper subalpine and alpine elevations. **COMMENTS:** This taxon is occasionally included in the genus *Helictotrichon,* from which it differs in having acute ligules, unribbed leaves, rachillas glabrous only on 1 side, unlobed lodicules, and liquid or semi-liquid endosperm.

64. *Beckmannia* Host

Coarse annuals with moderately thick culms and broad, flat blades. Inflorescence with 1-flowered spikelets crowded on short, spicate branches of a narrow, loosely contracted, simple or sparingly rebranched, interrupted panicle. Spikelets suborbicular, flattened, nearly sessile, closely imbricate in 2 rows on a slender rachis. Disarticulation below the glumes, spikelet falling entire. Glumes equal, broad, firm, 3-nerved, keeled, abruptly apiculate at apex. Lemma about as long as the glumes, narrow, 5-nerved, tapering to a slender, pointed tip. Palea narrow, slightly shorter than lemma. Basic chromosome number, $x = 7$. Photosynthetic pathway, C_3.

Represented in Colorado by a single species.

190. *Beckmannia syzigachne* (Steud.) Fernald (Fig. 212).

Synonyms: *Beckmannia eruciformis* (L.) Host, *Beckmannia eruciformis* (L.) Host subsp. *baicalensis* (Kusnez.) Hultén, *Beckmannia eruciformis* (L.) Host var. *uniflora* Scribn. *ex* Gray, *Beckmannia syzigachne* (Steud.) Fern. subsp. *baicalensis* (Kusnez.) Koyama & Kawano, *Beckmannia syzigachne* (Steud.) Fern. var. *uniflora* (Scribn. *ex* Gray) Boivin. **Vernacular Name:** American sloughgrass. **Life Span:** Annual. **Origin:** Native. **Season:** C_3, cool season. **FLORAL CHARAC-**

TERISTICS: Inflorescence: loosely contracted panicle with short, secund, spicate branches bearing 2 rows of closely imbricate spikelets, 6–27 cm long. **Spikelets:** 1-flowered, 2.5–3.5 mm long, flat; disarticulating below the glumes. **Glumes:** subequal, 2.5–3.5 mm long, laterally compressed, broadest above middle, keeled with 2 indistinct lateral nerves, margins scarious, apex apiculate. **Lemmas:** 2.5–3.5 mm long, acute, mucronate, or awn-pointed, tip extending slightly beyond the glumes. **Awns:** none or indistinctly from tip of lemma. **VEGETATIVE CHARACTERISTICS: Growth Habit:** erect, robust, often stoloniferous. **Culms:** 2–12 dm long, glabrous or scaberulous. **Sheaths:** open, glabrous or scabrous. **Ligules:** 5–8.5 mm long, membranous. **Blades:** 3–10 mm wide, 8–10 cm long, flat, scabrous. **HABITAT:** Sloughs, standing water, edges of ponds, streams, and rivers at lower elevations to montane. **COMMENTS:** It is considered palatable and a nutritious forage grass.

65. *Briza* L.

Tufted annuals and perennials with usually showy, open panicles of several-flowered, awnless spikelets. Florets crowded and horizontally spreading. Glumes subequal, broad, thin, and papery; rounded on back, 3- to 9-nerved. Lemmas similar to the glumes, broader than long, usually with 7–9 nerves, these distinct or indistinct, and a broadly rounded apex.

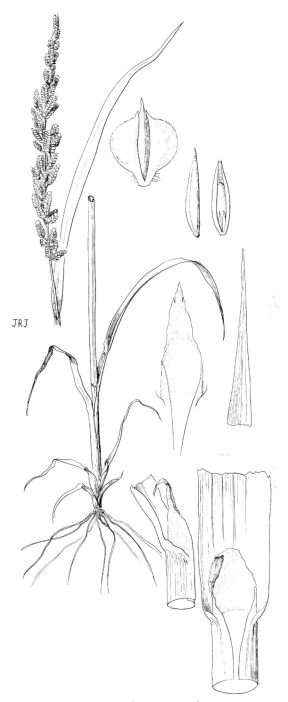

JRJ

Figure 212. *Beckmannia syzigachne.*

Palea short to nearly as long as lemma. Basic chromosome number, $x = 7$. Photosynthetic pathway, C_3.

Represented in Colorado by 2 species.

1. Panicle nodding; spikelets 10 mm wide; annual . . . . 191. *B. maxima*
1. Panicle erect; spikelet about 5 mm wide; perennial . . . . 192. *B. media*

191. *Briza maxima* L. (Fig. 213).

Synonyms: *Briza grandis* Salisb., *Macrobriza maxima* (L.) Tzvelev, *Poa maxima* (L.) Cav. **Vernacular Name:** Big quaking grass. **Life Span:** Annual. **Origin:** Introduced. **Season:** C_3, cool season. **FLORAL CHARACTERISTICS: Inflorescence:** nodding, open panicle, 3.5–10 cm long, generally 1–5 cm wide, secund; pedicels 5–20 mm long. **Spikelets:** 10–20 mm long, 10 mm wide, appear oval to elliptic-shaped, 4- to 12-flowered and rarely to 15-flowered; uppermost florets reduced; disarticulation above the glumes. **Glumes:** subequal, boat-shaped; lower 5–5.5 mm long, 5-nerved; upper 6–6.5 mm long, 7-nerved. **Lemmas:** closely imbricate, cordate, and broader than long; lowermost 7–9 mm long, obscurely 7- to 9-nerved, apices broadly obtuse, generally glabrous below, becoming villous above. **Paleas:** to 4 mm, scarious, ciliolate along margins. **Awns:** none. **VEGETATIVE CHARACTERISTICS: Growth Habit:** cespitose. **Culms:** erect, 2–8 dm tall. **Sheaths:** open to near base with overlapping margins, commonly less than half as long as internodes; internodes hollow. **Ligules:** hyaline, acute, margins entire to erose, 3–7 mm long, occasionally decurrent on lateral edges. **Blades:** flat, 2.5–20 cm long, 2–8 mm wide, margins glabrous to strigose. **HABITAT:** There is 1 historical location from roadsides in the Gunnison Valley (Weber and Wittmann 2001b). **COMMENTS:** A handsome grass frequently found in dried floral arrangements. Probably not a constant member of our flora.

192. *Briza media* L. (Fig. 214).

Synonym: *Poa media* (L.) Cav. **Vernacular Name:** Perennial quaking grass. **Life Span:** Perennial. **Origin:** Introduced. **Season:** C_3, cool season. **FLORAL CHARACTERISTICS: Inflorescence:** erect, open panicle, 8–20 cm long, lax, freely branched; branches ascending, stiff, naked below; pedicels 5–20 mm long. **Spikelets:** 4–5.5 mm long, 4–5 mm wide, oval-shaped, 3- to 6-flowered and occasionally to 10-flowered; disarticulation above the glumes. **Glumes:** subequal, obscurely 3- to 7-nerved, boat-shaped; lower 2.5–3.2 mm long; upper 2.5–4 mm long. **Lemmas:** closely imbricate, cordate, and broader than long; lowermost 3–4 mm long; uppermost florets reduced, obscurely 9- to 10-nerved, broadly obtuse. **Paleas:** to 3 mm, scarious, ciliolate-hyaline margins, V-shaped in cross-section. **Awns:** none. **VEGETATIVE**

Figure 213. *Briza maxima*.

Figure 214. *Briza media*.

CHARACTERISTICS: Growth Habit: loosely tufted to weakly rhizomatous, leaves mostly basal. **Culms:** erect, 1.5–7.5 dm tall, internodes hollow. **Sheaths:** open to half their length, shorter than internodes. **Ligules:** hyaline, truncate to occasionally erose, to 0.5 mm long. **Blades:** flat, 4–16 cm long, 2–3.2 mm wide, margins scabrid to strigose, glabrous to scabrous, blades of lower leaves longer than those above. **HABITAT:** There is 1 record for this species found in the grasslands of Boulder Open Space (Weber and Wittmann 2001a). This plant is native to Europe and is found in chalk and clay grasslands there. **COMMENTS:** Possibly not a constant member of our flora.

66. *Calamagrostis* Adans.

Moderately large, mostly rhizomatous perennials with open or, more often, contracted panicles of 1-flowered spikelets. Rachilla disarticulating above the glumes and in many species extended back of floret as a slender, hairy bristle. Glumes about equal, longer than floret, 1- or 3-nerved, acute or acuminate. Lemma 3- or 5-nerved, tapering to a narrow, notched or toothed apex, with a slender, often geniculate awn from the middle or below; usually long-hairy on callus, hairs sometimes as long as lemma. Palea about equaling lemma or slightly shorter. Basic chromosome number, $x = 7$. Photosynthetic pathway, C_3.

Represented in Colorado by 6 species, 2 subspecies, and 1 variety.

1. Awn straight
    2. Panicles open; callus hairs nearly as long as lemma . . . . 193. *C. canadensis*
    2. Panicles contracted; callus hairs half to two-thirds as long as lemma
        3. Glumes 2–4 mm long . . . . 198. *C. stricta*
        3. Glumes 4.2 mm or longer . . . . 197. *C. scopulorum*
1. Awn geniculate
    4. Awn long exserted, 6–8 mm long . . . . 195. *C. purpurascens*
    4. Awn included or only slightly exserted, 2–3.5 mm long
        5. Leaf blades involute; plants generally less than 4 dm tall . . . . 194. *C. montanensis*
        5. Leaf blades usually flat; plant mostly greater than 4 dm tall . . . . 196. *C. rubescens*

193. *Calamagrostis canadensis* (Michx.) P. Beauv. (Fig. 215).

Synonyms: *Calamagrostis anomala* Suksdorf, *Calamagrostis atropurpurea* Nash, *Calamagrostis canadensis* (Michx.) P. Beauv. var. *imberbis* (Stebbins) C. L. Hitchc., *Calamagrostis canadensis* (Michx.) P. Beauv. var. *pallida* (Vasey & Scribn.) Stebbins, *Calamagrostis canadensis* (Michx.) P. Beauv. var. *robusta* Vasey, *Calamagrostis canadensis* (Michx.) P. Beauv. var. *typica* Stebbins, *Calamagrostis expansa* (Munro *ex* Hbd.) Hitchc. var. *robusta* (Vasey) Stebbins, *Calamagrostis hyperborea* Lange, *Cala-* 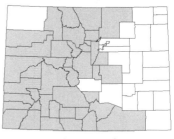 *magrostis inexpansa* Gray var. *cuprea* Kearney, *Calamagrostis inexpansa* Gray var. *robusta* (Vasey) Stebbins, *Calamagrostis scribneri* Beal. **Vernacular Name:** Bluejoint. **Life Span:** Perennial. **Origin:** Native. **Season:** C_3, cool season. **FLORAL CHARACTERISTICS: Inflorescence:** panicle, relatively open to occasionally compact, 8–25 cm long. **Spikelets:** 1- to occasionally 2-flowered, 3–4.5 mm long. **Glumes:** about the length of spikelet, subequal, lanceolate, keel

JRJ

Figure 215. *Calamagrostis canadensis.*

scabrous, tip acute or acuminate, lower slightly longer, aristate. **Calluses:** hairs nearly as long as to longer than lemmas. **Lemmas:** about as long as glumes, thin, scabrous, 2–4 teeth at apex. **Awns:** delicate straight awn arising from middle of lemma or below, 1.2–2 mm long, included and sometimes difficult to distinguish from callus hairs. **VEGETATIVE CHARACTER-ISTICS: Growth Habit:** rhizomatous, tufted. **Culms:** 6–15 dm long, stout, erect or nearly so, glabrous. **Sheaths:** glabrous to scabrous. **Ligules:** 3–8 mm long, membranous, erose-ciliate, sometimes lacerate. **Blades:** 2–8 mm wide, 10–30 cm long, flat, lax, scabrous. **HABITAT:** Wet and riparian areas of montane and subalpine communities. **COMMENTS:** Several varieties have been described but are difficult to distinguish and are not recognized here.

194. *Calamagrostis montanensis* Scribn. (Fig. 216).

Synonyms: None. **Vernacular Name:** Plains reed-grass. **Life Span:** Perennial. **Origin:** Native. **Season:** C_3, cool season. **FLORAL CHARACTERISTICS: Inflo-rescence:** narrow, spikelike panicle, 4–10 cm long, dense, interrupted; branches crowded, short, erect. **Spikelets:** 1- to occasionally 2-flowered, 3.5–5 mm long; rachilla hairs two-thirds to full length of lemma. **Glumes:** equal or subequal, as long as spikelet; lower 1-nerved; upper 3-nerved, with 2 lateral nerves indis-tinct. **Calluses:** hairs half length of lemma. **Lemmas:** 3–3.5 mm long, shorter than to as long as glumes, with 4 short teeth at tip. **Awns:** 2–3 mm long, stout, from near lemma base, geniculate, twisted below on some, protruding sideways up to 1.5 mm from the glumes, inconspicuous in inflorescence but distinguishable from callus hairs. **VEGETATIVE CHAR-ACTERISTICS: Growth Habit:** rhizomatous. **Culms:** 1.5–4 dm long, stiffly erect, scabrous below inflorescence. **Sheaths:** glabrous or scabrous above. **Ligules:** obtuse to acute, 2–3 mm long, sometimes lacerate. **Blades:** 0.5–2 mm wide, 5–15 cm long, involute. **HABITAT:** Plains and dry open areas. **COMMENTS:** Very similar to *C. purpurascens* R. Br. but differing in having less conspicuous awns.

195. *Calamagrostis purpurascens* R. Br. (Fig. 217).

Synonyms: *Calamagrostis lepageana* Louis-Marie, *Cala-magrostis maltei* (Polunin) Á. & D. Löve, *Calamagrostis purpurascens* R. Br. subsp. *maltei* (Polunin) Porsild, *Calamagrostis yukonensis* Nash, *Deschampsia congesti-formis* Booth. **Vernacular Name:** Purple reedgrass. **Life Span:** Perennial. **Origin:** Native. **Season:** C_3, cool season. **FLORAL CHARACTERISTICS: Inflorescence:** narrow, spikelike panicle, 4–12 cm long, dense, inter-rupted; branches crowded, short, erect. **Spikelets:** 1-to occasionally 2-flowered, 5–7 mm long; rachilla hairs 1–1.5 mm long. **Glumes:** subequal, as long as spikelet, keel scabrous; lower 1-nerved; upper 3-nerved, with 2 lateral nerves indis-tinct. **Calluses:** hairs 1–1.5 mm long. **Lemmas:** 4–5.5 mm long, glabrous or scabrous, tip with 4 setaceous teeth. **Awns:** 6–8 mm long, from near base of lemma, geniculate, twisted below,

Figure 216. *Calamagrostis montanensis*.

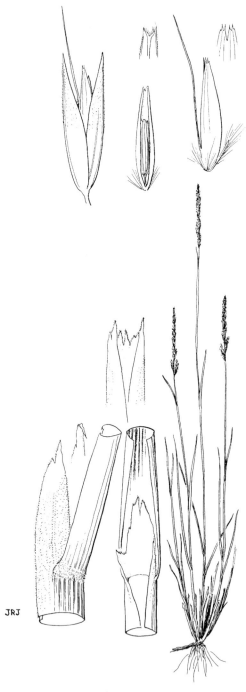

Figure 217. *Calamagrostis purpurascens.*

conspicuous in inflorescence. **VEGETATIVE CHARACTERISTICS: Growth Habit:** tufted, rhizomatous, rhizomes short and thick. **Culms:** 2–6 dm long, erect, glabrous, puberulent, or scabrous below inflorescence. **Sheaths:** usually scabrous, sometimes smooth, older ones persistent. **Ligules:** 2–5 mm long, truncate to obtuse, erose-ciliate, lacerate. **Blades:** 2–4 mm wide, 4–10 cm long, usually involute, occasionally flat. **HABITAT:** Dry rocky slopes from montane to alpine. **COMMENTS:** Similar to *C. montanensis* but differs in being conspicuously awned. The only variety found in Colorado is the typical one (var. *purpurascens*).

196. *Calamagrostis rubescens* Buckley (Fig. 218).

Synonym: *Calamagrostis fasciculata* Kearney. **Vernacular Name:** Pinegrass. **Life Span:** Perennial. **Origin:** Native. **Season:** C_3, cool season. **FLORAL CHARACTERISTICS: Inflorescence:** narrow, spikelike panicle, 6–15 cm long. **Spikelets:** 1-flowered, 3–5.5 mm long. **Glumes:** subequal, about as long as spikelet, keel scabrous; lower 1-nerved; upper 3-nerved, with 2 lateral nerves indistinct. **Calluses:** hairs one-third as long as lemma. **Lemmas:** 2.5–3.5 mm long, shorter than glumes. **Awns:** 2.8–3.5 mm long from near base of lemma, stout, twisted below, geniculate, distinguishable from callus hairs. **VEGETATIVE CHARACTERISTICS: Growth Habit:** tufted; from slender, creeping rhizomes. **Culms:** 5–11 dm long, glabrous. **Sheaths:** glabrous to pilose, at least above; pubescent on collar, sometimes obscurely so. **Ligules:** 2–6 mm long, obtuse or truncate, erose-ciliate, lacerate. **Blades:** 1.5–5 mm wide, 12–30 cm long, usually flat but occasionally involute, especially near tip; scabrous to pilose, at least above. **HABITAT:** Dry to moist sagebrush to open lodgepole pine forests. **COMMENTS:** Based on the number of collections, this is a relatively rare species in Colorado.

197. *Calamagrostis scopulorum* M. E. Jones (Fig. 219).

Synonym: *Calamagrostis scopulorum* M. E. Jones var. *bakeri* Stebbins. **Vernacular Name:** Ditch reedgrass. **Life Span:** Perennial. **Origin:** Native. **Season:** C_3, cool season. **FLORAL CHARACTERISTICS: Inflorescence:** contracted, spikelike panicle, 7–15 cm long. **Spikelets:** 1- to occasionally 2-flowered, 4.5–6 mm long; callus hairs about two-thirds length of lemma; rachilla vestige bearded. **Glumes:** equal or second slightly shorter, the length of spikelet, keel scabrous; lower 1-nerved; upper 3-nerved. **Calluses:** hairs 2–3 mm long, about half or more as long as lemma, sparse. **Lemmas:** 3.5–5 mm long, shorter than glumes, 5-nerved, lateral nerves terminating in 4 narrow apical teeth. **Awns:** arising from upper two-fifths of lemma, fine, straight, up to 2 mm long or occasionally absent. **VEGETATIVE CHARACTERISTICS: Growth Habit:** loosely tufted, short-rhizomatous. **Culms:** 5–9 dm tall, erect, glabrous, pale or glaucous. **Sheaths:** glabrous to scabrous. **Ligules:** 4–7 mm long, erose-ciliate, sometimes lacerate. **Blades:** 3–7 mm wide, 10–30 cm long, flat, scabrous, mostly on upper surface.

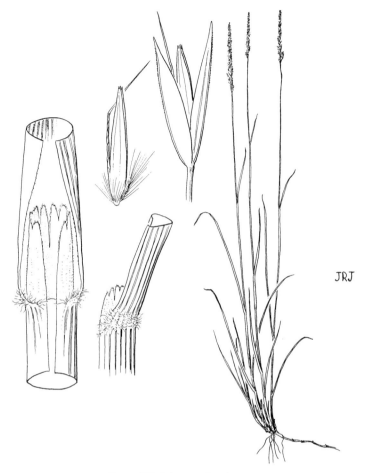

Figure 218. *Calamagrostis rubescens.*

HABITAT: Moist rocky areas, most frequently in the western part of the state. **COMMENTS:** Relatively uncommon species.

198. *Calamagrostis stricta* (Timm) Koeler (Fig. 220).

Synonyms: subsp. *inexpansa*: *Calamagrostis stricta* (Timm) Koel. subsp. *inexpansa* (Gray) C. W. Greene, *Calamagrostis californica* Kearney, *Calamagrostis canadensis* (Michx.) P. Beauv. var. *acuminata* Vasey *ex* Shear & Rydb, *Calamagrostis canadensis* (Michx.) P. Beauv. var. *arcta* Stebbins, *Calamagrostis chordorrhiza* Porsild, *Calamagrostis crassiglumis* Thurb., *Calamagrostis expansa* Rickett & Gilly, *non* (Munro *ex* Hbd.) Hitchc., *Calamagrostis fernaldii* Louis-Marie, *Calamagrostis hyperborea* Lange var. *americana* (Vasey) Kearney, *Calamagrostis hyperborea* Lange var. *elongata* Kearney, *Calamagrostis hyperborea* Lange var. *stenodes* Kearney, *Calamagrostis inexpansa* Gray, *Calamagrostis inexpansa* Gray var. *barbulata* Kearney, *Calamagrostis inexpansa*

Figure 219. *Calamagrostis scopulorum.*

Gray var. *brevior* (Vasey) Stebbins, *Calamagrostis inex-pansa* Gray var. *novae-angliae* Stebbins, *Calamagrostis labradorica* Kearney, *Calamagrostis lacustris* (Kearney) Nash, *Calamagrostis lapponica* (Wahlenb.) Hartm. var. *brevipilis* Stebbins, *Calamagrostis pickeringii* Gray var. *lacustris* (Kearney) Hitchc., *Calamagrostis robertii* Porsild, *Calamagrostis stricta* (Timm) Koel. var. *brevior* Vasey, *Calamagrostis stricta* (Timm) Koel. var. *lacustris* (Kearney) C. W. Greene; subsp. *stricta: Calamagrostis*

neglecta (Ehrh.) P. G. Gaertn., B. Mey. & Scherb, *Calamagrostis neglecta* (Ehrh.) P. G. Gaertn., B. Mey. & Scherb. subsp. *stricta* (Timm) Tzvelev, *Calamagrostis neglecta* (Ehrh.) P. G. Gaertn., B. Mey. & Scherb. var. *gracilis* (Scribn.) Scribn., *Calamagrostis neglecta* (Ehrh.) P. G. Gaertn., B. Mey. & Scherb. var. *micrantha* (Kearney) Stebbins. **Vernacular Names:** Slimstem reedgrass, slimstem small reedgrass. **Life Span:** Perennial. **Origin:** Native. **Season:** C₃, cool season. **FLORAL CHARACTERISTICS: Inflorescence:** narrow panicle, 5–15 cm long, contracted to pyramidal in shape. **Spikelets:** 1- to occasionally 2-flowered, 2–4 mm long. **Glumes:** length of spikelet, equal or subequal. **Calluses:** hairs half to as long as lemma. **Lemmas:** shorter than to as long as the glumes. **Awns:** arising from near middle of lemma, straight, reaching tip of lemma or less. **VEGETATIVE CHARACTERISTICS: Growth Habit:** cespitose with slender rhizomes. **Culms:** erect, 2.5–9 dm tall, glabrous to scabrous. **Sheaths:** glabrous to scaberulous. **Ligules:** 0.7–6 mm long, erose-ciliate. **Blades:** 1–5 mm wide, flat to involute, scabrous or sometimes glabrous. **HABITAT:** Moist areas, forests, subalpine willow carrs, and meadows. **COMMENTS:** Two subspecies are found in Colorado, the typical subspecies (subsp. *stricta*) and subsp. *inexpansa* (Gray) C. W. Greene. They can be separated by these characteristics:

1. Leaf blades stiff; awn of lemma included within spikelet; panicle appears as a tightly contracted panicle . . . . subsp. *stricta*
1. Leaf blades lax; awn of lemma slightly exceeding tip of glumes; panicle narrow but pyramidal in shape . . . . subsp. *inexpansa*

67. *Catabrosa* P. Beauv.

Aquatic perennials with decumbent-erect culms that root freely at lower nodes. Leaves with closed or partially closed sheaths and soft, flat blades. Inflorescence an open panicle of small, awnless, mostly 2-flowered spikelets. Disarticulation above the glumes and between florets. Glumes short, obtuse, unequal, nerveless or 1-nerved. Lemmas rounded on back, prominently 3-nerved, parallel nerves not converging at the broad, scarious apex. Palea broad, 2-keeled, about as long as lemma. Caryopsis flat, oval, brownish, loosely enclosed in lemma and palea. Basic chromosome number, $x = 10$. Photosynthetic pathway, C₃.

Represented in Colorado by a single species and a single variety.

199. *Catabrosa aquatica* (L.) P. Beauv. (Fig. 221).

Synonyms: *Aira aquatica* L., *Catabrosa aquatica* (L.) P. Beauv. var. *uniflora* S. F. Gray. **Vernacular Names:** Water whorlgrass, brookgrass. **Life Span:** Perennial. **Origin:** Native. **Season:** C₃, cool season. **FLORAL CHARACTERISTICS: Inflorescence:** open panicle, 7–20 cm long; erect,

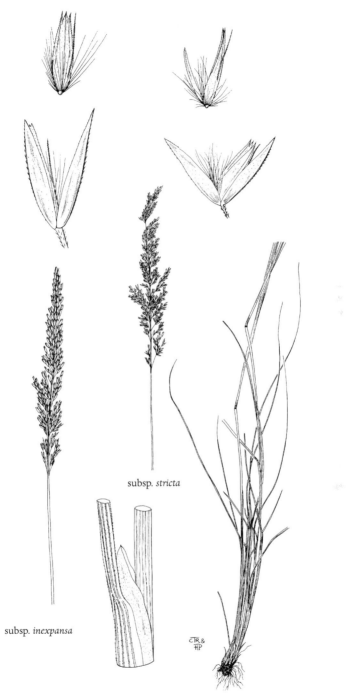

subsp. *stricta*

subsp. *inexpansa*

Figure 220. *Calamagrostis stricta.*

387

oblong, or pyramidal; branches whorled, divergent.
Spikelets: 2-flowered, occasionally 1- or 3-flowered,
2.5–3.5 mm long. **Glumes:** truncate, scarious, often
irregularly toothed at apices; lower 0.6–1.5 mm long;
upper 1.2–2.5 mm long. **Lemmas:** 2–3 mm long, with
3 prominent parallel nerves, glabrous, truncate, scarious, apices erose. **Awns:** none. **VEGETATIVE CHAR-
ACTERISTICS: Growth Habit:** stoloniferous but not
rhizomatous. **Culms:** 1–5 dm long, delicate, rooting

at nodes, decumbent at base, erect above, glabrous. **Sheaths:** glabrous or scabrous, closed at
least half their length. **Ligules:** 2–6 mm long, membranous, entire to erose, obtuse, acute,
or rounded. **Blades:** 2–8 mm wide, 3–12 cm long, flat, weak, acute, glabrous to scabrous.
HABITAT: Wet areas, usually growing in still or slow-moving water, plains to subalpine.
COMMENTS: The typical variety (var. *aquatica*) is the only one found in Colorado.

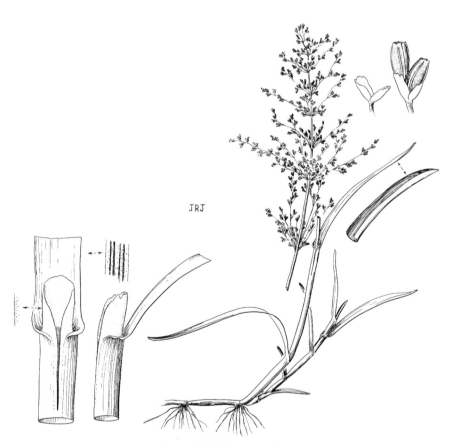

JRJ

Figure 221. *Catabrosa aquatica.*

68. Cinna L.

Tall perennials with slender culms, thin flat blades, and open or contracted panicles. Spikelets 1-flowered, disarticulating below the glumes, rachilla usually prolonged behind palea as a small glabrous or scabrous stub or bristle. Glumes about equal or lower somewhat shorter, lanceolate, 1- to 3-nerved, acute at apex. Lemma similar to the glumes, 3- to 5-nerved, midnerve usually protecting as a short, straight awn just below the acute or narrowly rounded and notched tip. Palea slightly shorter than lemma, 2- or apparently 1-nerved, with a single keel and the nerves close together. Stamens 1 or 2. Basic chromosome number, $x = 7$. Photosynthetic pathway, C_3.

Represented in Colorado by a single species.

200. *Cinna latifolia* (Trevir. *ex* Göpp.) Griseb. (Fig. 222).

Synonyms: None. **Vernacular Name:** Drooping woodreed. **Life Span:** Perennial. **Origin:** Native. **Season:** C_3, cool season. **FLORAL CHARACTERISTICS: Inflorescence:** open panicle, 10–30 cm long; branches spreading to drooping. **Spikelets:** 1-flowered, 2.5–4 mm long, strongly compressed; disarticulating below the glumes. **Glumes:** subequal, as long as spikelet, lanceolate, scaberulous, keel scabrous; lower 1-nerved; upper faintly 3-nerved, sometimes aristate. **Lemmas:** slightly shorter than glumes, lanceolate, scabrous, 3-nerved, keel scabrous, awned. **Awns:** absent or arising from just below apex of lemma, up to 1.2 mm long, straight. **VEGETATIVE CHARACTERISTICS: Growth Habit:** rhizomatous. **Culms:** 5–20 dm long, erect or decumbent, bearing most of the leaves, scabrous below nodes. **Sheaths:** open, glabrous to scabrous. **Ligules:** 3–8 mm long, membranous, lacerate. **Blades:** 4–15 mm wide, 10–20 cm long, flat, lax, scabrous. **HABITAT:** Stream banks, pond margins, bogs, sloughs, and other wet areas from montane to subalpine. **COMMENTS:** Considered a palatable forage species.

69. Dactylis L.

Perennial, erect with densely clumped culms. Sheaths distinctly flattened or keeled, open; ligules membranous, usually erose at tip; auricles absent. Inflorescence a 1-sided panicle. Spikelets with 2–5 fertile florets, laterally flattened, crowded in dense asymmetrical clusters at tips of stiff, erect or spreading inflorescence branches. Disarticulation above the glumes and between florets. Glumes unequal, keeled, hispid-ciliate on keel, 1- to 3-nerved, acute to acuminate or ending in a short awn. Lemmas keeled, awnless or short-awned, 5-nerved, hispid-ciliate on keel. Palea well developed, short-ciliate on keels. Basic chromosome number, $x = 7$. Photosynthetic pathway, C_3.

Represented in Colorado by the only species in the genus.

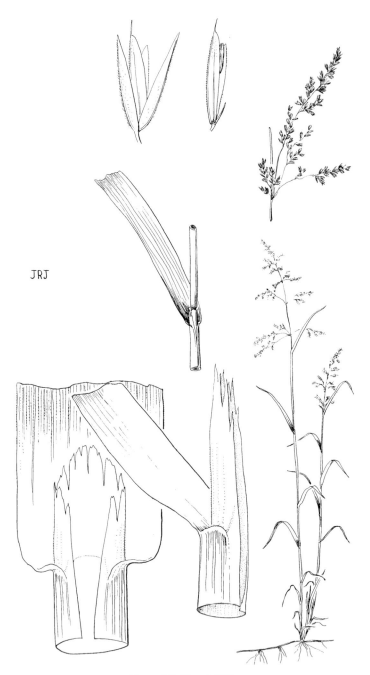

JRJ

Figure 222. *Cinna latifolia.*

201. *Dactylis glomerata* L. (Fig. 223).

Synonyms: *Dactylis glomerata* L. var. *ciliata* Peterm., *Dactylis glomerata* L. var. *detonsa* Fries, *Dactylis glomerata* L. var. *vivipara* Parl. **Vernacular Names:** Orchard grass, cocksfoot. **Life Span:** Perennial. **Origin:** Introduced. **Season:** C$_3$, cool season. **FLORAL CHARACTERISTICS: Inflorescence:** contracted panicle, 5–20 cm long; branches stiff and wiry, lower branches naked below, first internode and main branches long, spikelets clustered on 1 side of short secondary 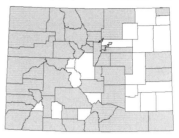 branches. **Spikelets:** 5–8 mm long, 2- to 5-flowered, laterally compressed, densely clustered on 1 side of rachis; tardily disarticulating above the glumes and between florets. **Glumes:** 3–6 mm long, often asymmetrical; rounded to ciliate-keeled, at least near tip; glabrous or scabrous, acute to awn-tipped, margins scarious. **Lemmas:** 4–6.5 mm long; keel compressed and coarsely ciliate, at least near tip; acute, mucronate, or awn-tipped. **Awns:** absent or short from tips of both glumes and lemmas. **VEGETATIVE CHARACTERISTICS: Growth Habit:** tufted, often with short, stout rhizomes. **Culms:** 3–12 dm tall, erect, glabrous. **Sheaths:** keeled, closed for most of their length, glabrous to scabrous. **Ligules:** 2–6 mm long, membranous, truncate, lacerate. **Blades:** 2–12 mm wide, 10–40 cm long, scabrous. **HABITAT:** Along roadsides and other disturbed areas, sometimes cultivated as a meadow or pasture grass. **COMMENTS:** The pollen causes hayfever (Allred 2005). A frequent pasture or meadow species that is escaped and now considered a weed. Good to excellent forage species, especially relished by deer (Stubbendieck, Hatch, and Landholt 2003).

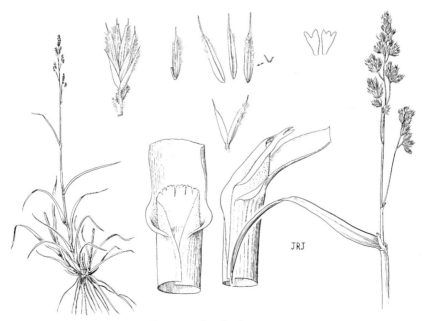

Figure 223. *Dactylis glomerata*.

70. *Deschampsia* P. Beauv

Perennial (in ours), cespitose. Sheaths open to base; ligules prominent, erose-lacerate, membranous, decurrent; auricles absent. Inflorescence an open or contracted panicle, usually with slender, wiry branches and pedicels. Disarticulation above the glumes and between florets. Spikelets with 2 fertile florets, rachilla prolonged beyond terminal floret and occasionally bearing a rudiment at the tip. Glumes 2, 1- to 3-nerved, lanceolate, nearly equal or exceeding lowermost floret. Lemma shiny, rounded on back, mostly 5- to 7-nerved but nerves often obscure, pubescent on callus, deeply cleft at apex; bearing a slender, twisted, geniculate awn from middle or near base. Palea 2-keeled, about as long as lemma. Basic chromosome number, $x = 7$. Photosynthetic pathway, C_3.

Represented in Colorado by 4 species and 1 subspecies.

1. Lemmas scabridulous or puberulent, not shiny . . . . 205. *D. flexuosa*
1. Lemmas glabrous, shiny
   2. Glumes mostly green; panicles narrow, appearing greenish . . . . 204. *D. elongata*
   2. Glumes purplish over sometimes more than half their length; panicles pyramidal to ovate to oblong-ovate
      3. Spikelets strongly imbricate, often clustered at tip of branches; glumes and lemmas purple over more than half their length . . . . 202. *D. brevifolia*
      3. Spikelets not or only moderately imbricate, not clustered at tip of branches; glumes with a green base, purple midband, and pale apex . . . . 203. *D. cespitosa*

202. *Deschampsia brevifolia* R. Br. (Fig. 224).

Synonym: *Deschampsia caespitosa* (L.) P. Beauv. subsp. *brevifolia* (R. Br.) Vasey *ex* Beal. **Vernacular Name:** Hairgrass. **Life Span:** Perennial. **Origin:** Native. **Season:** C_3, cool season. **FLORAL CHARACTERISTICS: Inflorescence:** panicle, 2–12 cm long, usually dense, narrowly cylindrical to ovate; branches usually erect to ascending; spikelets often densely clustered at branch tips, strongly imbricate. **Spikelets:** 2- (rarely 3-) flowered, 2–6 mm long; disarticulating above the  glumes. **Glumes:** nearly equal, 2.5–5.6 mm long, purplish over more than half their length, lanceolate, acuminate to acute; lower 1-nerved; upper 3-nerved, 2 lateral nerves sometimes indistinct. **Calluses:** villous, hairs 0.2–2 mm long. **Lemmas:** 2–4 mm long, smooth, shiny, glabrous, purplish over half their length, awned. **Awns:** Lemmas awned from near base to middle, 0.5–4 mm long, usually straight to weakly geniculate. **VEGETATIVE CHARACTERISTICS: Growth Habit:** strongly cespitose. **Culms:** 1–6 dm tall, erect, glabrous, leaves mostly basal. **Sheaths:** open, glabrous, somewhat keeled. **Ligules:** 1–5 mm long, acute to acuminate, entire or lacerate, membranous. **Blades:** usually less than 1 mm wide, folded or convolute. **HABITAT:** Wet places at higher elevations often in disturbed sites (Barkworth et al. 2007). **COMMENTS:** Perhaps more common than map indicates. This species is often included in *D. cespitosa* (L.) P. Beauv.

Figure 224. *Deschampsia brevifolia.*

203. *Deschampsia cespitosa* (L.) P. Beauv. (Fig. 225).

Synonyms: *Aira caespitosa* L., *Aira cespitosa* L., *Deschampsia caespitosa* (L.) P. Beauv. subsp. *genuina* (Reichenb.) Volk., *Deschampsia caespitosa* (L.) P. Beauv. subsp. *glauca* (Hartman) Hartman, *Deschampsia caespitosa* (L.) P. Beauv. subsp. *orientalis* Hultén, *Deschampsia caespitosa* (L.) P. Beauv. subsp. *paramushirensis* (Honda) Tzvelev, *Deschampsia caespitosa* (L.) P. Beauv. subsp. *parviflora* (Thuill.) Dumort., *Deschampsia caespitosa* (L.) P. Beauv. var. *abbei* Boivin, *Deschampsia*

caespitosa (L.) P. Beauv. var. *alpicola* (Rydb.) Á. & D. Löve & Kapoor, *Deschampsia caespitosa* (L.) P. Beauv. var. *arctica* Vasey, *Deschampsia caespitosa* (L.) P. Beauv. var. *glauca* (Hartman) Lindm. f., *non* Regel, *Deschampsia caespitosa* (L.) P. Beauv. var. *intercotidalis* Boivin, *Deschampsia caespitosa* (L.) P. Beauv. var. *littoralis* (Gaudin) Richter, *Deschampsia caespitosa* (L.)

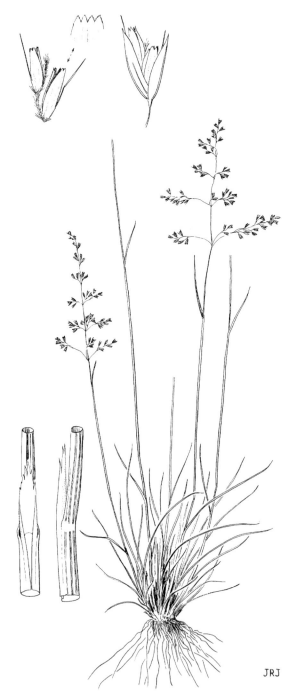

Figure 225. *Deschampsia cespitosa.*

P. Beauv. var. *longiflora* Beal, *Deschampsia caespitosa* (L.) P. Beauv. var. *mackenzieana* (Raup) Boivin, *Deschampsia caespitosa* (L.) P. Beauv. var. *maritima* Vasey, *Deschampsia caespitosa* (L.) P. Beauv. var. *parviflora* (Thuill.) Coss. & Germ., *Deschampsia cespitosa* (L.) P. Beauv. subsp. *cespitosa* (L.) P. Beauv., *Deschampsia cespitosa* (L.) P. Beauv. subsp. *genuina* (Reichenb.) Volk., *Deschampsia cespitosa* (L.) P. Beauv. var. *pallida* Gren. & Godr., *Deschampsia glauca* Hartman, *Deschampsia mackenzieana* Raup, *Deschampsia paramushirensis* Honda, *Deschampsia pumila* (Ledeb.) Ostenf., *Deschampsia sukatschewii* (Popl.) Rosh. **Vernacular Names:** Tufted hairgrass, small-flowered tickle grass. **Life Span:** Perennial. **Origin:** Native. **Season:** C$_3$, cool season. **FLORAL CHARACTERISTICS: Inflorescence:** panicle, 8–40 cm long, open, loose, spreading or nodding, often partly included, capillary branches finely villous; spikelets not clustered at branch tips, not or only moderately imbricate. **Spikelets:** 2- or 3-flowered, 2.5–7.5 mm long, shiny, usually purple; florets distant in anthesis; disarticulating above the glumes. **Glumes:** lanceolate, acute, base green with a purple midband and pale apex; lower 2.5–7 mm long, 1- to 3-nerved, entire; upper 2–7 mm long, 1- to 3-nerved, 2 lateral nerves sometimes indistinct. **Calluses:** villous. **Lemmas:** 2–5 mm long, smooth, shiny, glabrous, 5-nerved, usually purple over less than half their length, awned from near base. **Awns:** 1–8 mm long, arising from near base of lemma, exceeding glumes. **VEGETATIVE CHARACTERISTICS: Growth Habit:** strongly cespitose. **Culms:** 1–15 dm tall, ascending to erect, leaves mostly basal, hollow, glabrous. **Sheaths:** open, glabrous, somewhat keeled. **Ligules:** 2–13 mm long, acute to acuminate, entire or lacerate, membranous. **Blades:** 1–4 mm wide, 5–20 cm long, firm, usually flat or folded; scabrous, especially above. **HABITAT:** Moist areas from foothills to alpine (Weber and Wittmann 2001a, 2001b), and it is adapted to many soil types (Stubbendieck, Hatch, and Landholt 2003). **COMMENTS:** Excellent forage for all kinds and classes of livestock (Stubbendieck, Hatch, and Landholt 2003; Weber and Wittmann 2001a, 2001b). There is a large amount of variation and taxonomic confusion in this species (as illustrated by the complexity of synonyms). Three subspecies are reported for North America; the typical subspecies (subsp. *cespitosa*) is found in Colorado.

204. *Deschampsia elongata* (Hook.) Munro (Fig. 226).

Synonym: *Aira elongata* Hook. **Vernacular Name:** Slender hairgrass. **Life Span:** Perennial. **Origin:** Native. **Season:** C$_3$, cool season. **FLORAL CHARACTERISTICS: Inflorescence:** panicle, 5–30 cm long, erect or nodding, narrow or closed, greenish. **Spikelets:** 2-flowered, 3–7 mm long, appressed to branches; disarticulating above the glumes. **Glumes:** lanceolate, glabrous; usually green, occasionally purple-tipped; 3-nerved, apices acuminate; lower 3.2–5.5 mm long;  upper 3.1–5.4 mm long. **Calluses:** villous. **Lemmas:** 2–4 mm long, smooth, shiny, glabrous, apices erose or toothed, awned. **Awns:** 2–6 mm long, arising from just below to just above middle of lemma, straight to slightly geniculate. **VEGETATIVE CHARACTERISTICS: Growth Habit:** strongly cespitose. **Culms:** 1–12 dm tall, hollow, glabrous. **Sheaths:** open, glabrous, somewhat keeled. **Ligules:** 2–9 mm long, acute to acuminate, entire or lacerate, membranous. **Blades:** 0.2–2 mm wide, 5–30 cm long, firm, usually involute. **HABITAT:** Moist to wet areas, rare in Colorado and only collected in San Juan County. **COMMENTS:** Colorado plants perhaps represent an introduction (Barkworth et al. 2007).

Figure 226. *Deschampsia elongata.*

205. *Deschampsia flexuosa* (L.) Trin. (Fig. 227).

Synonym: *Aira flexuosa* L. **Vernacular Names:** Crinkled hairgrass, wavy hairgrass. **Life Span:** Perennial. **Origin:** Native. **Season:** C$_3$, cool season. **FLORAL CHARACTERISTICS: Inflorescence:** panicle, 5–15 cm long, frequently nodding, narrow to open; branches ascending to spreading, flexuous, smooth to scabridulous. **Spikelets:** 2-flowered, 4–7 mm long, oblong to cuneate; disarticulating above the glumes. **Glumes:** acute, glabrous, shiny, smaller than to subequal to adjacent floret, 1-nerved; lower 2.7–4.5 mm long; upper 3.5–5 mm long. **Calluses:** pilose, to 1 mm long. **Lemmas:** 3.3–5 mm long, scabridulous or puberulent, dull, apex acute, erose to 4-toothed, 4-nerved, awned. **Awns:** 3.7–7 mm long, arising from near base of lemma, strongly geniculate from below apex of lemma, column twisted. **VEGETATIVE CHARACTERISTICS: Growth Habit:** strongly cespitose. **Culms:** 3–8 dm tall, erect to geniculate basally, glabrous. **Sheaths:** open, glabrous, smooth. **Ligules:** 1.5–3.6 mm long, obtuse to acute, membraneous. **Blades:** 0.3–0.5 mm wide, 12–25 cm long, filiform, strongly involute, stiff, lower surfaces smooth to scabridulous, glabrous to hairy, upper surfaces scabrous. **HABITAT:** Dry, often rocky slopes, woods, thickets, and frequently in disturbed areas; rare in Colorado and only collected in Park County. **COMMENTS:** Colorado plants perhaps represent an introduction (Barkworth et al. 2007).

71. *Festuca* L.

Tufted perennials with usually thin, flat, or narrow and involute blades and 3- to several-flowered spikelets in open or contracted panicles. Disarticulation above the glumes and between florets. Glumes narrow, unequal, acute or acuminate, 1- to 3-nerved. Lemmas thin or firm, rounded on back, usually 5- to 7-nerved; awned from a narrow, entire, or minutely bifid apex or awnless. Palea free from caryopsis. Stamens 3. Basic chromosome number, $x = 7$. Photosynthetic pathway, C$_3$.

Represented in Colorado by 15 species, 3 subspecies, and 1 variety (*F. arundinacea* and *F. pratensis* are included in the genus *Schedonorus*; *F. dasyclada* is in the genus *Argillochloa*).

1. Leaf blades mostly flat and mostly greater than (2.5–) 3 mm wide (except *F. campestris* Rydb.)
    2. Lemma awns 5–20 mm long . . . . 218. *F. subulata*
    2. Lemma awns lacking or 2 mm or less long
       3. Ligules 2.5–9 mm long . . . . 219. *F. thurberi*
       3. Ligules less than 2 mm long
          4. Older sheaths not persistent, rapidly shredding into fibers; plants loosely cespitose; culm nodes usually exposed . . . . 217. *F. sororia*
          4. Older sheaths persistent; plants densely cespitose; culm nodes usually not exposed
             5. Spikelets with 2–3 (4) florets, 6.5–9.5 mm long; glumes about equaling or slightly longer than upper florets . . . . 212. *F. hallii*
             5. Spikelets with (3) 4–7 florets, 8–16 mm long; glumes shorter than upper florets
             . . . . 210. *F. campestris*

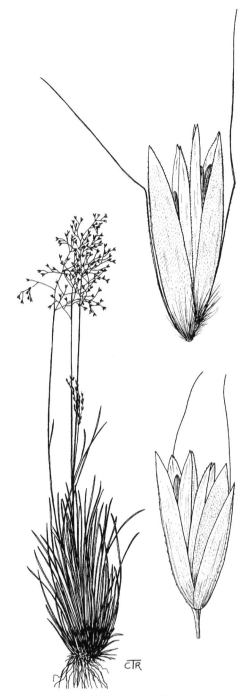

Figure 227. *Deschampsia flexuosa.*

1. Leaf blades involute or, if flat, less than 3 mm wide
  6. Ligules 2–9 mm long . . . . 219. *F. thurberi*
  6. Ligules less than 2 mm long
    7. Glumes about equaling or longer than upper florets . . . . 212. *F. hallii*
    7. Glumes distinctly shorter than upper florets
      8. Rhizomes present
        9. Ovary apex glabrous; anthers 1.8–4.5 mm long . . . . 215. *F. rubra*
        9. Ovary apex pubescent; anthers 0.6–1.4 mm long . . . . 211. *F. earlei*
      8. Rhizomes absent
        10. Culms densely scabrous or pubescent below inflorescence; ligules 0.5–2 mm
        long . . . . 206. *F. arizonica*
        10. Culms smooth and glabrous below inflorescence or, if slightly scabrous or
        pubescent, then ligule to 0.6 mm long . . . . "*F. ovina*" Complex Key

"*F. ovina*" Complex Key

The members of this difficult complex have, at one time or another, all been included in *F. ovina* L. (sheep fescue). In North American treatments, *F. ovina* has been broadly interpreted to include any fine-leaved, non-rhizomatous *Festuca*. *F. ovina*, however, is rare on the continent, and much of the information about *F. ovina* belongs to other species (Barkworth et al. 2007).

1. Anthers (1.8) 2–4.5 mm long; plants typically not of alpine habitats
  2. Lemmas 5–10 mm long, awns usually more than half as long as lemma body . . . .
  . . . . 213. *F. idahoensis*
  2. Lemmas 3.8–5 (6.5) mm long, awns usually not more than half as long as lemma
    3. Lemmas scabrous or pubescent distally, especially along margins; ovary apices glabrous . . . . 220. *F. trachyphylla*
    3. Lemmas glabrous and smooth, rarely slightly scabrous distally; ovary apices sparsely pubescent . . . . 209. *F. calligera*
1. Anthers 0.3–1.8 (2) mm long; plants mostly of alpine habitats (except *F. saximontana* Rydb.)
  4. Plant widespread, but not in alpine habitats . . . . 216. *F. saximontana*
  4. Plants restricted to alpine habitats
    5. Blades (0.2) 0.3–0.4 (0.6) mm wide; spikelet 2.5–5 mm long, with (1) 2–3 (5) florets; lemma awns 0.5–1.5 (1.7) mm long . . . . 214. *F. minutiflora*
    5. Blades (0.3) 0.4–1.2 mm wide; spikelets (3) 3.5–8.5 mm long, with 2–6 florets; lemma awns (0.2) 0.5–3.5 mm long
      6. Culms densely pubescent or pilose below inflorescence . . . . 207. *F. baffinensis*
      6. Culms usually glabrous and smooth below inflorescence, occasionally slightly scabrous or puberulent . . . . 208. *F. brachyphylla*

206. *Festuca arizonica* Vasey (Fig. 228).

Synonym: *Festuca ovina* L. var. *arizonica* (Vasey) Hack. *ex* Veal. **Vernacular Name:** Arizona fescue. **Life Span:** Perennial. **Origin:** Native. **Season:** C_3, cool season. **FLORAL CHARACTERISTICS: Inflorescence:** loosely contracted panicle, sometimes slightly secund, 6–20 cm long; branches scabrous. **Spikelets:** oblong, laterally compressed, 4- to 6-flowered, 8–12

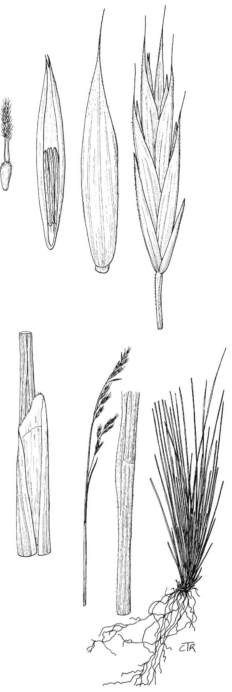

Figure 228. *Festuca arizonica*.

mm long; disarticulation above the glumes and below each fertile floret. **Glumes:** persistent, chartaceous, linear to lanceolate, exceeded by upper florets; lower 3–5 mm long, 1-nerved, sometimes with 2 indistinct lateral nerves, apices acute; upper 4.5–7 mm long, 3-nerved, surface scabrous, apices acute. **Lemmas:** lanceolate, chartaceous, 5.5–7 mm long, indistinctly 5-nerved, glabrous to scabrous, sometimes awned, apices acute. **Paleas:** slightly shorter than lemmas.

Awns: absent or to 3 mm long, from tip of lemma. **Anthers:** 0.3–0.4 mm long. **Ovary Apices:** densely pubescent. **VEGETATIVE CHARACTERISTICS: Growth Habit:** tufted, rhizomes absent. **Culms:** erect, glaucous, 2.5–9 dm long, pale blue-green, leaves mostly basal, generally densely scabrous or pubescent below inflorescence. **Sheaths:** 5–10 cm long, smooth to scabrous, firm, entire, wider than blade at collar. **Ligules:** up to 2 mm long, ciliolate membrane. **Blades:** conduplicate to involute, filiform, scabrous, 15–40 cm long, 1–1.5 mm wide, glaucous, usually greater than half the length of culms. **HABITAT:** Dry areas, foothills to subalpine, most common in southern counties. **COMMENTS:** This species is closely related to *F. idahoensis* but has much shorter awns and scabrous panicle branches.

207. *Festuca baffinensis* Polunin (Fig. 229).

Synonym: *Festuca ovina* L. **Vernacular Name:** Baffin Island fescue. **Life Span:** Perennial. **Origin:** Native. **Season:** C$_3$, cool season. **FLORAL CHARACTERISTICS: Inflorescence:** panicle, peduncle pubescent above. **Spikelets:** oblong, laterally compressed, 3- to 4-flowered, 4–6 mm long, reddish brown; disarticulation above the glumes and below each fertile floret. **Glumes:** persistent, lanceolate, chartaceous, exceeded by upper florets, 1-nerved; lower 2.5–3.5 mm long,

apices acute; upper 3.3–4.6 mm long, apices acute. **Lemmas:** lanceolate, chartaceous, 4.3–4.6 mm long, 5-nerved, apices acuminate, awned. **Paleas:** 2-nerved, equal to lemma in size. **Awns:** 0.5–3 mm long from tip of lemma. **Anthers:** 0.3–0.7 mm long. **Ovary apices:** occasionally hairy to glabrous. **VEGETATIVE CHARACTERISTICS: Growth Habit:** tufted, rhizomes lacking. **Culms:** 3–18 cm long, densely short-pubescent below inflorescence; basal sheaths persistent, new shoots intravaginal. **Sheaths:** open for about half their length, glabrous. **Ligules:** 0.1–0.4 mm long, truncate, ciliolate, glabrous, membranous. **Blades:** conduplicate to involute, 4.5–7 cm long, less than 1 mm wide, ventrally puberulent, dorsally glabrous. **HABITAT:** Found in alpine grasslands. **COMMENTS:** Infrequent. Plants are apparently cleistogamous.

208. *Festuca brachyphylla* Schult. & Schult. f. (Fig. 230).

Synonyms: *Festuca brachyphylla* Schult. *ex* Schult. & Schult. f. subsp. *breviculmis* Frederiksen, *Festuca brachyphylla* Schult. *ex* Schult. & Schult. f. var. *coloradensis* (Frederiksen) Dorn, *Festuca ovina* L. var. *brachyphylla* (Schult.) Piper, *Festuca ovina* L. var. *brevifolia* S. Wats in

Figure 229. *Festuca baffinensis.*

Figure 230. *Festuca brachyphylla.*

King. **Vernacular Name:** Shortleaf fescue. **Life Span:** Perennial. **Origin:** Native. **Season:** C₃, cool season. **FLORAL CHARACTERISTICS: Inflorescence:** narrow, somewhat spikelike panicle, sometimes secund, 1–5 cm long. **Spikelets:** oblong, laterally compressed, 3- to 5-flowered, 3–7 mm long; disarticulation above the glumes and below each fertile floret. **Glumes:** persistent, lanceolate, exceeded by upper florets; lower 2–3 mm long, 1-nerved, apices acute; upper 3–4 mm

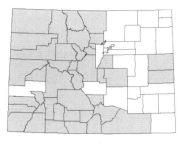

long, 3-nerved, apex acuminate. **Lemmas:** glaucous, chartaceous, 3–5 mm long, indistinctly 5-nerved, firm, puberulent above, apices acute, awned. **Paleas:** nearly the length of lemma. **Awns:** 1.5–2.5 mm long from tip of lemma. **Anthers:** 0.5–1.1 (1.3) mm long. **Ovary apices:** glabrous. **VEGETATIVE CHARACTERISTICS: Growth Habit:** densely tufted, rhizomes lacking. **Culms:** up to 1.5 dm tall, well exserted above leaves, generally smooth and glabrous, occasionally sparsely scabrous or puberulent below inflorescence. **Sheaths:** tubular, glabrous to scabrous, persistent, new basal shoots intravaginal. **Ligules:** ciliolate membrane about 0.5 mm long. **Blades:** conduplicate to involute, filiform, 5–10 cm long, about 1 mm wide, stiff and erect, glabrous to scabrous. **HABITAT:** Alpine grasslands and meadows. **COMMENTS:** Very common. The only subspecies of *F. brachyphylla* found in Colorado is subsp. *coloradensis* Frederiksen.

209. *Festuca calligera* (Piper) Rydb. (Fig. 231).

Synonyms: *Festuca. ovina* L. subsp. *calligera* Piper, *Festuca amethystine* L. var. *asperrima* Hack. *ex* Beal. **Vernacular Name:** Callused fescue. **Life Span:** Perennial. **Origin:** Native. **Season:** C₃, cool season. **FLORAL CHARACTERISTICS: Inflorescence:** loosely contracted panicle, 5–15 cm long; branches appressed. **Spikelets:** oblong, laterally compressed, 2- to 6-flowered, 6–11 mm long; disarticulation above the glumes and below each fertile floret. **Glumes:**

persistent, subequal, lanceolate, scabrous distally, exceeded by upper florets; lower 2.5–4 mm long; upper 3–5 mm long. **Lemmas:** lanceolate, chartaceous, 4–6 mm long, glabrous or scabrous distally, apices awned. **Paleas:** 2-nerved, slightly shorter than lemma. **Awns:** lemma awned from apex, 1–2.5 mm long. **Anthers:** 2.2–3.5 mm long. **Ovary apices:** sparsely pubescent. **VEGETATIVE CHARACTERISTICS: Growth Habit:** densely cespitose, rhizomes lacking. **Culms:** erect, 1.5–6.5 dm tall, smooth, glabrous. **Sheaths:** closed for less than half their length, glabrous or retrorsely hirsute. **Ligules:** ciliolate membrane, 0.3–0.5 mm long. **Blades:** conduplicate or rarely convolute, stiff, 0.4–0.8 mm wide. **HABITAT:** It is found in grasslands, open montane subalpine forests, and lower alpine regions. **COMMENTS:** The species is poorly known and collected and perhaps frequently overlooked. It is not included in Weber and Wittmann (2001a, 2001b) or Wingate (1994). It is often included in *Festuca ovina* L. based on the North American interpretation of sheep fescue.

Figure 231. *Festuca calligera*.

210. *Festuca campestris* Rydb. (Fig. 232).

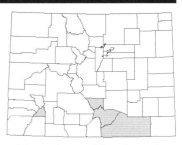

Synonyms: *Festuca altaica* Trin. var. *major* (Vasey) Gleason, *Festuca scabrella* Torr. *ex* Hook. var. *major* Vasey. **Vernacular Name:** Rough fescue. **Life Span:** Perennial. **Origin:** Native. **Season:** C$_3$, cool season. **FLORAL CHARACTERISTICS: Inflorescence:** loosely contracted to open panicle, 5–20 cm long; branches appressed, ascending, naked below in lower orders. **Spikelets:** oblong, laterally compressed, 3- to 6-flowered, 8–13 mm long; disarticulation above the glumes and below each fertile floret. **Glumes:** persistent, subequal, lanceolate to oblong-lanceolate, glabrous or scabrous, apices acute, shorter than uppermost florets; lower 5–8 mm long, 1-nerved; upper 5–9 mm long, 3-nerved. **Lemmas:** lanceolate, chartaceous, 7–10 mm long, 5-nerved, scabrous; apices acute, apiculate, or awned. **Paleas:** 2-nerved, keels scabrous, apices scaberulous. **Awns:** absent or to 1.5 mm long from tip of lemmas. **Anthers:** 4.5–6 mm long. **Ovary apices:** pubescent. **VEGETATIVE CHARACTERISTICS: Growth Habit:** densely tufted, usually not rhizomatous but occasionally with short rhizomes. **Culms:** erect, 3–14 dm tall, leaves mostly basal; scabrous below inflorescence; nodes not usually exposed. **Sheaths:** closed for less than one-third their length, glabrous or scabrous, persistent. **Ligules:** ciliolate membrane, 0.1–0.5 mm long. **Blades:** conduplicate or rarely convolute, stiff, 0.8–2 mm wide, surface scabrous, sharp-pointed. **HABITAT:** Known from Apishapa (Cordova) Pass (Weber and Wittmann 2001a; Wingate 1994) in Huerfano County and from Las Animas, Custer, and San Juan counties. **COMMENTS:** Rare; a segregate of *F. scabrella*. Not listed for Colorado in Barkworth and colleagues (2007). A common species of the Pacific Northwest.

211. *Festuca earlei* Rydb. (Fig. 233).

Synonyms: *Festuca brevifolia* R. Br. var. *utahensis* St.-Yves, *Festuca brachyphylla* Schult. var. *utahensis* Litard., *Festuca minutifolia* Rydb., *Festuca rubra* L. **Vernacular Name:** Earle's fescue. **Life Span:** Perennial. **Origin:** Native. **Season:** C$_3$, cool season. **FLORAL CHARACTERISTICS: Inflorescence:** contracted, spikelike panicle, generally less than 7 cm long, sometimes appears secund; branches slender and erect, 2 at lowest node. **Spikelets:** oblong, laterally compressed, 4–7 mm long, 1.8–3 mm wide, 2- to 5-flowered; disarticulation above the glumes and below each fertile floret. **Glumes:** shorter than upper florets, persistent, chartaceous, unequal to subequal, pubescent basally to slightly scabrous at apices with ciliate margins, apices acute; lower narrowly lanceolate, 1.5–3 mm long, 1- to sometimes obscurely 3-nerved; upper ovate-lanceolate, 2.7–3.7 mm long, 3-nerved. **Lemmas:** lanceolate, chartaceous, scabrous, 3–4.5 mm long, 5-nerved, lateral nerves obscure, entire, acute, mucronate or short-awned. **Paleas:** about as long as to slightly shorter than lemma, 2-keeled. **Awns:** lemmas sometimes awned, awn 0.4–1.6 mm long from entire apices. **Anthers:** 0.6–0.9 (1.4) mm long.

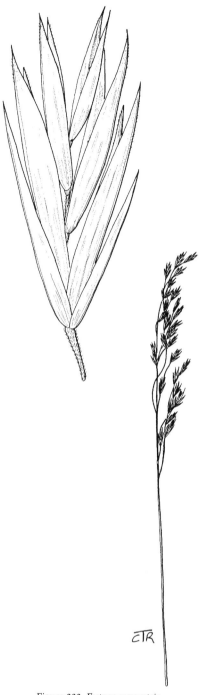

Figure 232. *Festuca campestris.*

Figure 233. *Festuca earlei*.

Ovary apices: densely and noticeably pubescent. **VEGETATIVE CHARACTERISTICS: Growth Habit:** loosely cespitose, rhizomatous. **Culms:** short, striate, 0.5–4.5 dm tall, generally less than twice as high as basal leaves, yellowish green, occasionally purplish at base, internodes glabrous; new shoots extravaginal. **Sheaths:** striate, glabrous, closed about half their length. **Auricles:** present as erect swellings that are generally distinctive. **Ligules:** 0.1–1 mm long, ciliate. **Blades:** filiform, involute, 5–12 cm long, to 1 mm wide, erect, lax to stiff, upper surface with sparse pubescence, lower glabrous. **HABITAT:** alpine grasslands, meadows, and subalpine forest openings. **COMMENTS:** This and some of the other native fescues previously lumped into sheep fescue (*F. ovina* L.) are important sources of forage for livestock and thrive in areas with 20–36 cm of annual precipitation. This has been confused with the non-rhizomatous *F. brachyphylla* Schult. and *F. minutiflora* Rydb. *F. earlei* can be distinguished from *F. rubra* L. by its pubescent ovary apices and shorter anthers (Barkworth et al. 2007).

212. *Festuca hallii* (Vasey) Piper (Fig. 234).

Synonyms: *Festuca altaica* Trin. subsp. *hallii* (Vasey) Harms, *Festuca altaica* Trin. subvar. *hallii* (Vasey) St.-Yves, *Festuca scabrella* Torr. *ex* Hook. subsp. *hallii* (Vasey) W. A. Weber, *Melica hallii* Vasey. **Vernacular Name:** Plains rough fescue. **Life Span:** Perennial. **Origin:** Native. **Season:** C$_3$, cool season. **FLORAL CHARACTERISTICS: Inflorescence:** more or less contracted panicle, open at anthesis, 6–16 cm long; primary branches appressed or ascending. **Spike-lets:** oblong, laterally compressed, 7–9.5 mm long, usually 2- to 3-flowered but occasionally 4-flowered; disarticulation above the glumes and below each fertile floret. **Glumes:** persistent, oblong-lanceolate to lanceolate, chartaceous, about equal in length to slightly exceeding uppermost florets; lower to 9.5 mm long, 1-nerved, apices acute; upper 6–9 mm long, 3-nerved, acute. **Lemmas:** lanceolate, generally 5–8 mm long, scaberulous, rounded except near apices, usually acute but rarely awn-tipped, 0.5–1.3 mm long. **Paleas:** 2-nerved, somewhat shorter than lemma, keels pubescent. **Awns:** lemma rarely awn-tipped. **Anthers:** 4–6 mm long. **Ovary apices:** sparsely pubescent. **VEGETATIVE CHARACTERISTICS: Growth Habit:** densely cespitose, short rhizomatous. **Culms:** 3–4.5 dm tall, internodes smooth to scaberulous; nodes not usually exposed. **Sheaths:** antrorsely scabrous, persistent, evident as mats at base of plant. **Ligules:** ciliolate membrane, 0.3–0.6 mm long. **Blades:** involute, 10–50 cm long, 1–2 mm wide, surface scabrous, acute, pungent. **HABITAT:** High montane meadows and alpine grasslands. **COMMENTS:** Rare; currently known only from Cameron Pass, although originally documented farther south (Weber and Wittmann 2001a, 2001b; Wingate 1994).

213. *Festuca idahoensis* Elmer (Fig. 235).

Synonyms: *Festuca idahoensis* Elmer var. *oregona* (Hack. *ex* Beal) C. L. Hitchc., *Festuca ingrata* (Hack. *ex* Beal) Rydb., *Festuca occidentalis* Hook. var. *ingrata* (Hack. *ex* Beal) Boivin, *Festuca occidentalis* Hook. var. *oregona* (Hack.) Boivin, *Festuca ovina* L. var. *columbiana* Beal, *Festuca ovina* L. var. *ingrata* Hack. *ex* Beal, *Festuca ovina* L. var. *oregona* Hack. *ex* Beal. **Vernacular**

Figure 234. *Festuca hallii*.

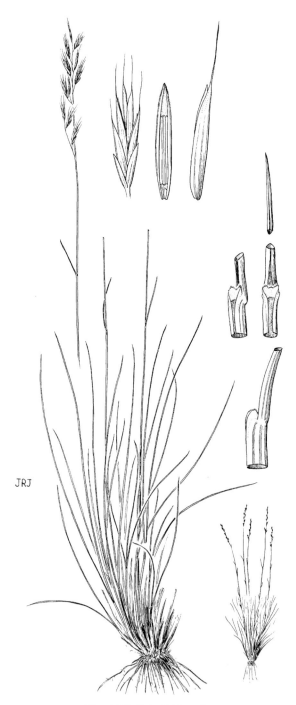

JRJ

Figure 235. *Festuca idahoensis.*

Names: Idaho fescue, blue bunchgrass, bluebunch fescue. **Life Span:** Perennial. **Origin:** Native. **Season:** C₃, cool season. **FLORAL CHARACTERISTICS: Inflorescence:** dense, narrow panicle, sometimes loose and open, 5–20 cm long. **Spikelets:** oblong, laterally compressed, 4- to 7 (9-) flowered, 6–15 mm long; rachilla joints usually visible; disarticulation above the glumes and below each fertile floret. **Glumes:** shorter than upper florets, persistent, membranous,

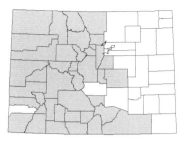

acute; lower lanceolate 3–5 (6) mm long, 1-nerved; upper lanceolate to elliptic, 2.5–5.5 mm long, 3-nerved. **Lemmas:** elliptic, chartaceous, 5–10 mm long, 5-nerved, glabrous to scabrous, acute, awned. **Paleas:** often pubescent, 2-nerved, keels scabrous. **Awns:** 2–6 mm long from tip of lemmas, usually more than half as long as lemmas. **Anthers:** 2.4–4.5 mm long. **Ovary apices:** glabrous. **VEGETATIVE CHARACTERISTICS: Growth Habit:** densely tufted, not rhizomatous, leaves mostly basal. **Culms:** erect, 3–9 dm long, generally glabrous, smooth, sometimes scabrous below inflorescence, blue-green. **Sheaths:** firm, entire, open, glabrous to scabrous, blue-green, closed for less than half their length. **Ligules:** ciliolate membrane, glabrous, 0.2–0.6 mm long. **Blades:** conduplicate to involute, filiform, 0.5–0.9 mm wide, greater than half the length of culm, blue-green, scabrous. **HABITAT:** Montane to subalpine grasslands, meadows, and other open areas. **COMMENTS:** A common plant of montane regions, and closely related to members of the *F. ovina* Complex (sheep fescues) and *F. arizonica*. Excellent forage species.

214. *Festuca minutiflora* Rydb. (Fig. 236).

Synonyms: *Festuca brachyphylla* Schult. *ex* Schult. & Schult. f. var. *endotera* (St.-Yves) Litard., *Festuca ovina* L. var. *minutiflora* (Rydb.) J. T. Howell. **Vernacular Names:** Smallflower fescue, little fescue. **Life Span:** Perennial. **Origin:** Native. **Season:** C₃, cool season. **FLORAL CHARACTERISTICS: Inflorescence:** narrow, somewhat spikelike panicle, sometimes secund, up to 5 cm long; branches 1–2 per node, erect. **Spikelets:** oblong, laterally compressed, (1) 2- to 3 (5-)

flowered, 2.5–5 mm long; disarticulation above the glumes and below each fertile floret. **Glumes:** shorter than uppermost florets, persistent, chartaceous, acute; lower lanceolate, 1.3–2.5 mm long, 1-nerved; upper lanceolate to ovate, 2–3.5 mm long, 3-nerved. **Lemmas:** lanceolate to oblong, chartaceous, 2–5 mm long, indistinctly 5-nerved, firm, puberulent above, purplish near apices, acute, awned. **Paleas:** shorter than to almost as long as lemmas. **Awns:** 0.5–1.5 (–1.7) mm long from tip of lemmas. **Anthers:** 0.3–0.7 mm long. **Ovary apices:** generally with few sparse hairs, rarely glabrous. **VEGETATIVE CHARACTERISTICS: Growth Habit:** densely tufted, rhizomes lacking. **Culms:** up to 3 dm tall, hardly exceeding leaves, new shoots intravaginal. **Sheaths:** glabrous, smooth, closed for about half their length. **Ligules:** eciliate membrane about 0.1–0.3 mm long, obtuse. **Blades:** conduplicate to involute, 5–10 cm long, 0.3–0.4 mm wide, lax, tuft usually less than 10 cm high, spreading, glabrous to scabrous. **HABITAT:** Grasslands and meadows above timberline. **COMMENTS:**

Figure 236. *Festuca minutiflora.*

This species is often overlooked or confused with *F. brachyphylla*, from which it differs by having laxer leaves, a looser panicle, smaller spikelets, and scattered hairs on the ovary (Barkworth et al. 2007).

215. *Festuca rubra* L. (Fig. 237).

Synonyms: *Festuca arenaria* Osbeck, *non* Lam., *Festuca rubra* L. var. *arenaria* (Osbeck) Fries, *Festuca rubra* L. var. *lanuginosa* Mert. & Koch., *Festuca prolifera* (Piper) Fern., *Festuca prolifera* (Piper) Fern. var. *lasiolepis* Fern., *Festuca rubra* L. subsp. *arctica* (Hack.) Govor., *Festuca rubra* L. subsp. *juncea* (Hack.) Soó, *Festuca rubra* L. subsp. *vulgaris* (Gaudin) Hayek, *Festuca rubra* L. var. *juncea* (Hack.) Richter, *Festuca rubra* L. var. *prolifera* (Piper) Piper. **Vernacular Name:** Red fescue. **Life Span:** Perennial. **Origin:** Native. **Season:** C$_3$, cool season. **FLORAL CHARACTERISTICS: Inflorescence:** narrowly contracted or often open panicle, oblong, 3–30 cm long; branches erect, ascending or spreading, angular, usually scabrous or pubescent, rarely glabrous. **Spikelets:** elliptic to oblong, 3- to 10-flowered, 7–17 mm long, blue-green to red to purplish; disarticulation above the glumes and below each fertile floret. **Glumes:** shorter than upper florets, persistent, chartaceous, acute; lower lanceolate, 2–7 mm long, 1-nerved; upper oblong, 3–8 mm long, 3-nerved. **Lemmas:** ovate, chartaceous, 4–9 mm long, 5-nerved, surface scabrous, acute, awned. **Paleas:** equal to lemmas in length, 2-nerved, keels scaberulous. **Awns:** 0.5–5 mm long from tip of lemmas. **Anthers:** 1.8–4.5 mm long. **Ovary apices:** glabrous. **VEGETATIVE CHARACTERISTICS: Growth Habit:** generally rhizomatous in loose to sandy soil, rhizomes range from long and slender to occasionally very short, loosely clumped, rarely stoloniferous. **Culms:** erect to geniculate at base, 3–13 dm tall, new shoots extravaginal or intravaginal. **Sheaths:** closed for about three-quarters of their length when young, splitting with age, glabrous to soft pubescent, lower ones reddish brown, nerves shredding into fibers. **Ligules:** ciliate membrane, 0.1–0.5 mm long. **Blades:** folded, conduplicate or convolute, filiform, stiff, 0.3–2.5 mm wide, surface smooth to scabrous, glabrous, obtuse to acute. **HABITAT:** Moist areas up to subalpine. **COMMENTS:** *F. brachyphylla* and *F. earlei* sometimes share the characteristic of basal sheath nerves separating into fibers. *F. rubra* has glabrous ovary apices, while *F. earlei* has pubescent ovary apices. *F. rubra* is usually rhizomatous, while *F. brachyphylla* lacks rhizomes. *F. rubra* is often cultivated in grass mixes and occasionally planted in pastures or lawns. Numerous subspecies are found in North America; however, further study is needed to determine which occur in Colorado.

216. *Festuca saximontana* Rydb. (Fig. 238).

Synonyms: *Festuca brachyphylla* Schult. *ex* J. A. & J. H. Schultes var. *rydbergii* (St.-Yves) Cronq., *Festuca brachyphylla* Schult. *ex* J. A. & J. H. Schultes subsp. *saximontana* (Rydb.) Hultén, *Festuca ovina* L. var. *rydbergii* St.-Yves, *Festuca ovina* L. var. *saximontana* (Rydb.) Gleason, *Festuca saximontana* Rydb. var. *robertsiana* Pavlick. **Vernacular Name:** Rocky Mountain fescue. **Life Span:** Perennial. **Origin:** Native. **Season:** C$_3$, cool season. **FLORAL CHARACTERISTICS: Inflorescence:** narrow, somewhat spikelike panicle, sometimes secund, 3–10 cm

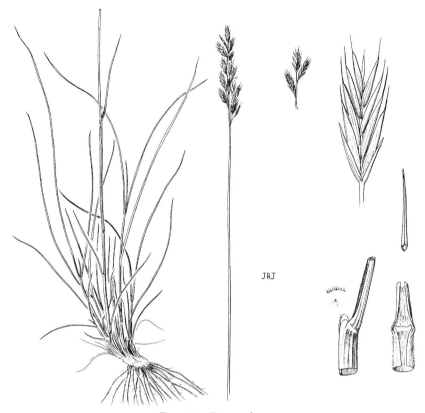

JRJ

Figure 237. *Festuca rubra.*

long; branches erect, spreading at anthesis. **Spike-lets:** oblong, laterally compressed, 3- to 5-flowered, 3–8 mm long; disarticulation above the glumes and below each fertile floret. **Glumes:** shorter than upper florets, lanceolate, chartaceous, acute; lower 2–3 mm long, 1-nerved; upper 3–4 mm long, 3-nerved. **Lemmas:** lanceolate, chartaceous, 3–5 mm long, indistinctly 5-nerved, firm, puberulent above, acute, awned. **Paleas:** as long as to slightly longer than lemmas. **Awns:** 1.5–2.5 mm long from tip of lemmas. **Anthers:** (0.8) 1.2–1.7 (2) mm long. **Ovary apices:** glabrous. **VEGETATIVE CHARACTERISTICS: Growth Habit:** densely tufted, rhizomes lacking. **Culms:** slender, 0.5–5 dm tall, new shoots intravaginal. **Sheaths:** closed for about half their length, glabrous to scabrous. **Ligules:** ciliate membrane, 0.1–0.5 mm long, obtuse. **Blades:** conduplicate, 5–10 cm long, 0.5–1.2 mm wide, stiff and erect, glabrous to scabrous. **HABITAT:** Usually below timberline in grasslands and forest openings, widespread. **COMMENTS:** A good forage species for livestock and wildlife. The typical variety is found in Colorado (var. *saximontana*).

Figure 238. *Festuca saximontana*.

217. *Festuca sororia* Piper (Fig. 239).

Synonym: *Festuca subulata* Bong. var. *sororia* St.-Yves. **Vernacular Name:** Ravine fescue. **Life Span:** Perennial. **Origin:** Native. **Season:** C₃, cool season. **FLORAL CHARACTERISTICS: Inflorescence:** open, somewhat drooping panicle, 10–20 cm long; capillary branches mostly solitary, naked below, lower ones reflexed or drooping. **Spikelets:** ovate, laterally compressed, (2) 3- to 5-flowered, 7–12 mm long; disarticulation above the glumes and below each fertile floret. **Glumes:** persistent, lanceolate, chartaceous, acute; lower 3–4 mm long, 1-nerved; upper 4.5–6 mm long, 3-nerved, scabrous on keel. **Lemmas:** lanceolate, chartaceous, 6–9 mm long, obscurely 5-nerved, smooth to scabrous, narrowly acuminate or short-awned. **Paleas:** similar in length to lemmas, 2-nerved, keels scabrous. **Awns:** absent or up to 2 mm long from tip of lemma. **Anthers:** 1.6–2.5 mm long. **Ovary apices:** pubescent. **VEGETATIVE CHARACTERISTICS: Growth Habit:** loosely cespitose, with hard, knotty base with subsurface shoots that often root and appear rhizomatous. **Culms:** 6–13 dm long, glabrous; nodes usually exposed. **Sheaths:** closed for less than one-third their length, glabrous; older sheaths not persistent, shredding rapidly into fibers. **Ligules:** eciliate membrane, 0.3–1.5 mm long. **Blades:** glabrous, 10–25 cm long, 3–6 (10) mm wide, flat, lax, margins and nerves scabrous. **HABITAT:** Canyon bottoms and along streams in Hinsdale and Archuleta counties (Weber and Wittmann 2001b). **COMMENTS:** Species is poorly collected (Weber and Wittmann 2001b).

218. *Festuca subulata* Trin. (Fig. 240).

Synonyms: *Festuca jonesii* Vasey, *Festuca subulata* Vasey var. *jonesii* (Vasey) St.-Yves. **Vernacular Name:** Bearded fescue. **Life Span:** Perennial. **Origin:** Native. **Season:** C₃, cool season. **FLORAL CHARACTERISTICS: Inflorescence:** loose, open panicle, ovate, 10–40 cm long; branches slender, spreading to reflexed, spikelets generally distal on branches. **Spikelets:** oblong to cuneate, laterally compressed, 6–12 mm long, (2) 3- to 5- (6-) flowered; disarticulation above the glumes and below each fertile floret. **Glumes:** persistent, subulate to linear-lanceolate, membranous, apex acuminate, acute to awned; lower 1.5–4 mm long, 1-nerved; upper 2–6 mm long, 1- to 3-nerved. **Lemmas:** lanceolate, chartaceous, 4–9 mm long, mostly 3-nerved, surface scabrous, apex attenuate, awned. **Paleas:** almost as long as to longer than lemma. **Awns:** 5–20 mm long from tip of lemma, glumes occasionally short awned from tip. **Anthers:** 1.5–2.5 (3) mm long. **Ovary apices:** pubescent. **VEGETATIVE CHARACTERISTICS: Growth Habit:** loosely tufted, occasionally stoloniferous. **Culms:** erect or decumbent, 5–10 dm tall, often rooting at lowermost nodes, glabrous to scabrous. **Sheaths:** closed for less than one-third their length, glabrous to scabrous. **Ligules:** erose-ciliolate membrane, 0.3–0.7 mm long. **Blades:** flat, 10–25 cm long, 3–10 mm wide, smooth or scabrous.

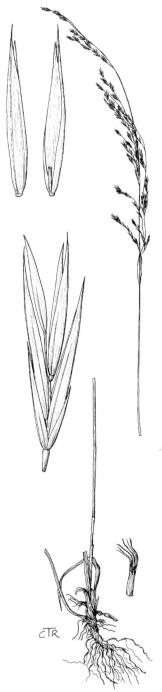

Figure 239. *Festuca sororia*.

JRJ

Figure 240. *Festuca subulata*.

HABITAT: Riparian areas (Weber and Wittmann 2001b). **COMMENTS:** Recently discovered along the White River in Rio Blanco County (Weber and Wittmann 2001b).

---

219. *Festuca thurberi* Vasey (Fig. 241).

Synonyms: *Festuca tolucensis* Kunth subsp. *thurberi* (Vasey) St.-Yves, *Poa festucoides* M. E. Jones, *Poa kaibensis* M. E. Jones. **Vernacular Name:** Thurber's fescue. **Life Span:** Perennial. **Origin:** Native. **Season:** C₃, cool season. **FLORAL CHARACTERISTICS: Inflorescence:** loose, open panicle, ovate; branches bearing spikelets in distal half. **Spikelets:** elliptic, laterally compressed, 3- to 6-flowered, 8–14 mm long; disarticulation above the glumes and below each fertile floret. **Glumes:** persistent, elliptic, chartaceous, acuminate; lower 3–5 mm long, 1-nerved; upper 4–6 mm long, 3-nerved. **Lemmas:** lanceolate, chartaceous, 6–10 mm long, obscurely 5-nerved, acute to cuspidate, sometimes minutely awned. **Paleas:** shorter than to as long as lemmas. **Awns:** absent or minute to 0.2 mm from lemma tips. **Anthers:** 3–4.5 mm long. **Ovary apices:** densely pubescent. **VEGETATIVE CHARACTERISTICS: Growth Habit:** densely tufted. **Culms:** erect, 4–10 dm tall. **Sheaths:** closed for less than one-third their length, persistent, slightly scabrous. **Ligules:** membranous, often lacerate, 2–5 (9) mm long, acute. **Blades:** 6–20 cm long, 1.5–3 mm wide when flat; when conduplicate, 0.8–1.8 mm in diameter; erect, coriaceous, surface scabrous, glabrous. **HABITAT:** Mountains, mostly of Western Slope (Weber and Wittmann 2001b). **COMMENTS:** Closely related to *F. hallii* and *F. campestris*. It differs primarily in ligule length and shorter glumes and lemmas. A good forage species for livestock (Allred 2005).

---

220. *Festuca trachyphylla* (Hack.) Krajina (Fig. 242).

Synonyms: *Festuca duriuscula* L., *Festuca longifolia* Thuill., *Festuca ovina* L., *Festuca ovina* L. var. *duriuscula* (L.) W.D.J. Koch, *Festuca stricta* Host subsp. *trachyphylla* (Hack.) Patzke. **Vernacular Name:** Hard fescue. **Life Span:** Perennial. **Origin:** Native. **Season:** C₃, cool season. **FLORAL CHARACTERISTICS: Inflorescence:** contracted panicle, with 2–3 branches per node; branches erect or sometimes spreading, lower with 2 or more spikelets. **Spikelets:** 3- to 8-flowered, 5–11 mm long; disarticulation above the glumes and below each fertile floret. **Glumes:** exceeded by upper florets, lanceolate to ovate, mostly smooth and glabrous; lower about 2–4 mm long, 1-nerved; upper glume 3–5 mm long, 3-nerved. **Lemmas:** lanceolate, chartaceous, 3.5–5 (6.5) mm long, obscurely 5-nerved, usually smooth and glabrous on lower portion and scabrous or pubescent distally, especially along margins; awned. **Paleas:** about the same length as lemmas. **Awns:** lemma awns 0.5–3 mm long, usually less than half the length of lemmas. **Anthers:** (1.8) 2.3–3.4 mm long. **Ovary apices:** glabrous. **VEGETATIVE CHARACTERISTICS: Growth Habit:** densely tufted, lacking rhizomes. **Culms:** erect, 4–9.5 dm tall, smooth, glabrous, or sometimes with sparse hairs. **Sheaths:** persistent, slightly

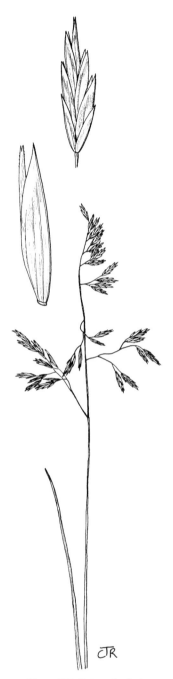

Figure 241. *Festuca thurberi.*

Figure 242. *Festuca trachyphylla*.

scabrous, closed less than one-third their length. **Ligules:** 0.1–0.5 mm long. **Blades:** 0.8–1.2 mm wide, rarely flat, mostly conduplicate, with 3–7 distinct ribs. **HABITAT:** Mountains, forest openings, and meadows. **COMMENTS:** Generally sold under the name "Hard" fescue as a turf grass and soil stabilizer (Barkworth et al. 2007).

72. *Graphephorum* Desv.

Perennial, cespitose or with short rhizomes. Sheaths open; ligules membranous; auricles absent. Inflorescence an open or contracted panicle. Spikelets with 2–3 florets, rachilla pilose. Disarticulation above the glumes and between florets. Glumes subequal, slightly shorter than spikelet. Lower glume 1-nerved, lanceolate to linear-lanceolate. Upper glume 1- or 3-nerved, lanceolate to ovate-lanceolate. Lemmas with entire apex or slightly lobed, muticous or with a short mucro. Awn attached immediately below apex, the mucro not reaching apex of lemma. Palea not gaping. Stamens 3. Basic chromosome number, $x = 7$. Photosynthetic pathway, C_3.

Represented in Colorado by a single species. A segregate of *Trisetum*.

221. *Graphephorum wolfii* (Vasey) Vasey *ex* Coult. (Fig. 243).

Synonyms: *Graphephorum brandegeei* (Scribn.) Rydb., *Graphephorum muticum* (Bol.) A. Heller, *Trisetum wolfii* Vasey, *Trisetum subspicatum* P. Beauv. var. *muticum* Bol. **Vernacular Name:** Wolf's trisetum. **Life Span:** Perennial. **Origin:** Native. **Season:** C_3, cool season. **FLORAL CHARACTERISTICS: Inflorescence:** densely erect panicle, 20–40 cm long, 1–1.5 cm wide; branches appressed. **Spikelets:** 2- or 3-flowered, 4–7 mm long; disarticulation above the glumes and between florets. **Glumes:** subequal, nearly as long as spikelet, scabrous near apices, acute; lower shorter and narrower than upper, lanceolate, 1-nerved, 4–7 mm long; upper as long as to shorter than lowermost floret, 3-nerved, 5–6.5 mm long. **Lemmas:** lanceolate, glabrous to slightly scabrous toward apices, acute to slightly bidentate, 4–7 mm long, awnless or with subapical mucro to 2 mm long, 3- to 7-nerved. **Paleas:** membranous, 2-nerved, shorter than lemmas, not gaping. **Awns:** lemmas awned, awn to 2 mm long, arising from below apices and rarely exceeding it. **VEGETATIVE CHARACTERISTICS: Growth Habit:** short rhizomes present. **Culms:** erect, 2–10 dm tall, glabrous to pilose below nodes. **Sheaths:** glabrous to pilose, open to closed basally. **Ligules:** membranous, 1.5–6 mm long, apex truncate to rounded. **Blades:** flat, to 15 cm long, 2–7 mm wide, smooth to scabrous to sparsely pilose. **HABITAT:** Found in marshy areas along swamp and pond borders in subalpine communities (Weber and Wittmann 2001b). **COMMENTS:** A segregate of *Trisetum* based primarily on the lack of apical setae on lemma, reduced or missing awn, and palea not gaping (Finot et al. 2005).

73. *Helictotrichon* Besser ex Schult. & Schult. f.

Perennial, cespitose. Sheaths open to base; ligules a short membrane with a ciliate tip; auricles absent. Inflorescence a contracted panicle. Spikelets large, with 3–7 fertile florets. Rachilla villous; disarticulation above the glumes and between florets. Glumes usually about

JRJ

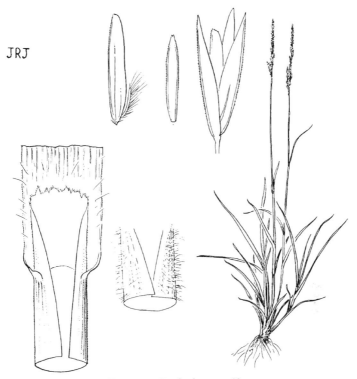

Figure 243. *Graphephorum wolfii.*

equal, larger than lemmas, 1- to 5-nerved, thin and membranous at acute apices. Lemmas rounded on back, several-nerved, firm on lower portion, membranous above and usually toothed at apices, awned from about the middle. Awn stout, geniculate, and twisted. Palea well developed, 2-keeled. Basic chromosome number, $x = 7$. Photosynthetic pathway, C_3.

Represented in Colorado by a single species.

222. *Helictotrichon mortonianum* (Scribn.) Henrard (Fig. 244).

Synonyms: *Arrhenatherum mortonianum* (Scribn.) Potztal, *Avena mortoniana* Scribn. **Vernacular Names:** Morton's alpine oatgrass, alpine oatgrass. **Life Span:** Perennial. **Origin:** Native. **Season:** C_3, cool season. **FLORAL CHARACTERISTICS: Inflorescence:** panicle, narrow, 2–5 cm long; branches short, erect, bearing a solitary spikelet. **Spikelets:** mostly 2-flowered, second floret sometimes reduced and sterile, pale or purplish. **Glumes:** subequal, margins hyaline, keel and nerves scabrous; lower 8–10.5 mm long, 1-nerved; upper 8.5–11.5 mm long, 3-nerved.

Figure 244. *Helictotrichon mortonianum.*

Calluses: hairs 1–2 mm long. **Lemmas:** 7–9.5 mm long, firm, obscurely 3-nerved, apices with 4 bristly teeth. **Awns:** 1–1.6 cm long from near middle of lemmas, geniculate, twisted below the knee; upper lemmas sometimes awnless or short-awned. **VEGETATIVE CHARAC-TERISTICS: Growth Habit:** densely tufted. **Culms:** 5–20 cm tall, glabrous. **Sheaths:** glabrous to puberulent, leaves mostly basal. **Ligules:** up to 1 mm long. **Blades:** 1–2 mm wide, 1–12 cm long, involute or convolute, becoming a pale straw color. **HABITAT:** Alpine meadows and grasslands (Weber and Wittmann 2001a, 2001b). **COMMENTS:** Fair forage species for sheep and elk grazing at higher elevations.

74. *Hierochloë* R. Br.

Perennial, rhizomatous. Herbage fragrant because of the presence of coumarin. Sheaths open; ligules membranous, fringed; auricles absent. Inflorescence a simple open or less frequently closed, few-flowered panicle. Spikelets with 3 well-developed florets, lower 2 staminate, upper fertile. Disarticulation above the glumes, and the 3 florets falling together. Glumes subequal, broad, thin, membranous, shiny, 1- to 3-nerved, awnless, as long as florets. Lemmas of staminate florets broadly elliptic, rounded on back, 5-nerved, awnless or with a short awn from a notched apex. Lemma of fertile floret about as long as lemmas of staminate florets, rounded on back, 3- to 5-nerved, awnless, short-hairy at apex, becoming indurate at maturity. Basic chromosome number, $x = 7$. Photosynthetic pathway, C_3.

Represented in Colorado by a single species and a single subspecies.

223. *Hierochloë hirta* (Schrank) Borbás (Fig. 245).

Synonyms: *Anthoxanthum hirtum* (Schrank) Y. Schouten & Veldkamp subsp. *arcticum* (J. Presl) G. Tucker, *Hierochloë odorata* (L.) P. Beauv. subsp. *arctica* (J. Presl) Tzvelev. **Vernacular Names:** Sweetgrass, holygrass. **Life Span:** Perennial. **Origin:** Native. **Season:** C_3, cool season. **FLORAL CHARACTERISTICS: Inflorescence:** panicle, open, pyramidal, 4–15 cm long. **Spikelets:** 4–6 mm long, broadly ovate, golden brown, 3-flowered, first 2 florets staminate, third perfect; disarticu-lation above the glumes with florets falling as a single unit. **Glumes:** subequal, hyaline, glabrous, shiny, acute or obtuse at apex; lower ovate-lanceolate, 3–5 mm long, 1-nerved, sometimes with 2 indistinct lateral nerves; upper 4–6 mm long, broadly ovate, 3-nerved, 2 lateral nerves indistinct. **Lemmas:** 2 staminate lemmas subequal, 3–5 mm long, ovate, shiny, 5- to 7- nerved, firm, scaberulous on back, pubescent on margins; perfect lemmas 3–3.5 mm long, apices with evenly distributed hairs to 1 mm long. **Awns:** none. **VEGETA-TIVE CHARACTERISTICS: Growth Habit:** with slender creeping rhizomes. **Culms:** 1.5–6 dm tall, erect, glabrous. **Sheaths:** glabrous to puberulent, especially on collar. **Ligules:** 2–4 mm long, membranous. **Blades:** 2–6 mm wide, mostly basal, flat, scabrous on margins. **HABI-TAT:** Moist areas, montane to lower alpine. **COMMENTS:** Fragrant plant because of the presence of coumarin, which is an anticoagulant and is the active ingredient in the prescription drug Coumadin. Coumarin, however, can be metabolized by a fungus that produces dicounarol, which is poisonous to animals (Barkworth et al. 2007). The plant is sometimes

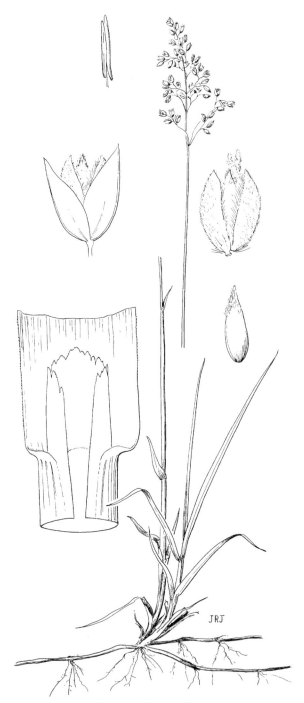

Figure 245. *Hierochloë hirta*.

used to flavor vodka in Poland or to make ale (Weber and Wittmann 2001b). Also used as incense, as an extract for hair tonic, and to make baskets (Cronquist et al. 1977). The subspecies found in Colorado is subsp. *arctica* (J. Presl) G. Weim. Some have merged *Hierochloë* with *Anthoxanthum* (Barkworth et al. 2007).

75. *Holcus* L.

Perennial (in ours), with velvety pubescent herbage. Culms weak and rather succulent. Sheaths open to base; ligules a short, fringed membrane, often puberulent; auricles absent. Inflorescence a contracted panicle. Spikelets with 2 florets, lower floret perfect, upper floret smaller, staminate or sterile. Disarticulation below the glumes. Glumes large, thin, subequal, longer than florets, lower 1-nerved, upper 3-nerved. Lemma of lower floret faintly 3- to 5-nerved, awnless, blunt at apex. Lemma of upper floret with a short, often hooked spine near apex. Palea of fertile floret as long as or slightly longer than lemmas. Basic chromosome number, $x = 7$. Photosynthetic pathway, C_3.

Represented in Colorado by a single species.

224. *Holcus lanatus* L. (Fig. 246).

Synonyms: *Aira holcus-lanata* Vill., *Aira holcul-lanatus* (L.) Vill., *Avena lanata* (L.) Cav., *Avena lanata* (L.) Koeler, *Avena pallida* Salisb., *Ginannia lanata* (L.) F. T. Hubb., *Ginannia pubescens* Bubani, *Nothoholcus lanatus* (L.) Nash *ex* Hitchc. **Vernacular Name:** Common velvetgrass. **Life Span:** Perennial. **Origin:** Introduced. **Season:** C_3, cool season. **FLORAL CHARACTERISTICS: Inflorescence:** panicle, 5–15 cm long, pale or purplish, contracted except in anthesis; branches short, ascending, pilose. **Spikelets:** 2-flowered, 3–6 mm long; upper floret staminate; lower perfect. **Glumes:** longer than florets and enclosing them, subequal; lower lanceolate, 3.5–5 mm long, 1-nerved; upper ovate, 3.8–5.5 mm long, prominently 3-nerved, sometimes with a short awn at tip. **Lemmas:** 1.7–2.5 mm long, shiny, upper cleft, hook-awned, smaller than lower. **Awns:** upper glumes sometimes with a short, straight awn at tip; fertile lemmas with short, hooked awn from below cleft apex. **VEGETATIVE CHARACTERISTICS: Growth Habit:** cespitose. **Culms:** 3–10 dm tall, erect or decumbent, grayish pubescent. **Sheaths:** velvety gray-green. **Ligules:** 1–3 mm long, truncate, erose-ciliate, puberulent. **Blades:** 3.5–10 mm wide, 2–20 cm long, flat, velvety gray-green. **HABITAT:** Moist open areas at mid-elevations. **COMMENTS:** Uncommon. Occasionally cultivated as a meadow grass.

76. *Koeleria* Pers.

Perennial, cespitose. Leaf blades frequently corkscrewing at maturity. Sheaths open to base; ligules short, membranous; auricles absent. Inflorescence usually a contracted, spicate panicle, except at anthesis when panicle branches are extended. Spikelets usually 2- to 4-flowered, shiny, strongly flattened, rachilla usually extended beyond uppermost floret as a slender bristle or bearing a vestigial floret. Disarticulation above the glumes and between florets. Glumes large, thin, acute, unequal; lower shorter and 1-nerved; upper longer,

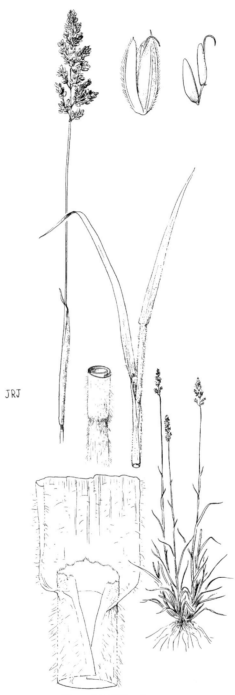

JRJ

Figure 246. *Holcus lanatus.*

broader, and 3- to 5-nerved. Lemmas thin, shiny, lowermost usually slightly longer than the glumes, awnless or rarely awned from a bifid apex. Palea large, scarious, and colorless. Basic chromosome number, $x = 7$. Photosynthetic pathway, C_3.

Represented in Colorado by a single species.

225. *Koeleria macrantha* (Ledeb.) Schult. (Fig. 247).

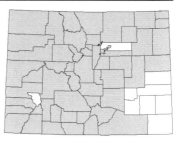

Synonyms: *Koeleria cristata* Pers. var. *longifolia* Vasey *ex* Burtt Davy, *Koeleria cristata* Pers. var. *macrantha* (Ledeb.) Griseb., *Koeleria cristata* Pers. var. *pinetorum* Abrams, *Koeleria gracilis* Pers., *Koeleria nitida* Nutt., *Koeleria yukonensis* Hultén. **Vernacular Name:** Prairie junegrass. **Life Span:** Perennial. **Origin:** Native. **Season:** C_3, cool season. **FLORAL CHARACTERISTICS: Inflorescence:** panicle, 4–5 cm long, pale green or purplish; branches short, appressed; rachis densely puberulent, interrupted at base. **Spikelets:** 3–6 mm long, 2-flowered (occasionally up to 4-flowered). **Glumes:** subequal, scabrous on keel; lower 2–4.5 mm long, 1-nerved; upper 3–5 mm long, 1- to 3-nerved, laterals indistinct. **Lemmas:** 3–5 mm long, lanceolate, 5-nerved, 4 lateral nerves faint, acute or short-awned from just below apices. **Paleas:** often as long

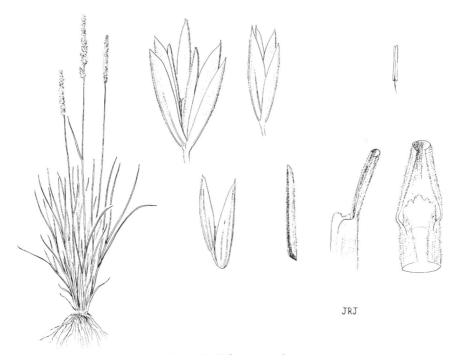

JRJ

Figure 247. *Koeleria macrantha.*

429

as lemma, colorless. **Awns:** absent or short from just below apices of lemmas. **VEGETA-TIVE CHARACTERISTICS: Growth Habit:** tufted. **Culms:** 1.5–6.5 dm tall, erect, puberulent below inflorescence. **Sheaths:** open, hispid, sometimes densely so. **Ligules:** 0.5–1.5 mm long, erose to subentire, ciliolate, sometimes puberulent. **Blades:** mostly basal, 1–3 mm wide, 5–25 cm long, flat or involute, tips prow-shaped. **HABITAT:** Scattered throughout, from eastern plains to subalpine. **COMMENTS:** Good to excellent forage grass, although not abundant enough to be of consequence (Stubbendieck, Hatch, and Landholt 2003). Twisted leaves are a good vegetative characteristic.

77. *Leucopoa* Griseb.

Our species a dioecious, rhizomatous perennial with culms in large, dense clumps. Blades firm, narrow, flat or loosely involute. Inflorescence a contracted panicle with short, mostly appressed branches, these spikelet-bearing to base. Spikelets 3- to 5-flowered, 7–12 mm long, staminate spikelets somewhat larger than pistillate. Disarticulation above the glumes and between florets. Glumes subequal, thin, lanceolate, acute, lower 1-nerved, upper 3-nerved. Lemmas awnless, acute or acuminate, 5-nerved, rounded on back. Palea as long as lemma, 2-nerved and 2-keeled, scabrous-ciliate on keels. Caryopsis beaked, bidentate at apex. Basic chromosome number, $x = 7$. Photosynthetic pathway, C_3.

Represented in Colorado by a single species.

226. *Leucopoa kingii* (S. Watson) W. A. Weber (Fig. 248).

Synonyms: *Festuca confinis* Vasey, *Festuca kingii* (S. Wats.) Cassidy, *Hesperochloa kingii* (S. Wats.) Rydb., *Poa kingii* S. Wats. **Vernacular Names:** Spike fescue, King's fescue, spikegrass, Watson's fescue-grass. **Life Span:** Perennial. **Origin:** Native. **Season:** C_3, cool season. **FLORAL CHARACTERISTICS: Inflorescence:** panicle, narrow, erect, 3–20 cm long, incompletely dioecious; branches short, appressed, floriferous to near base. **Spikelets:** 3- to 5-flowered, 6–12 mm  long, laterally compressed; disarticulating above the glumes and between florets. **Glumes:** broadly lanceolate, subcoriaceous; lower 3–5.5 mm long, 1-nerved; upper 4–6.5 mm long, 1- to faintly 3-nerved. **Lemmas:** 4.5–8 mm long, ovate, obscurely 5-nerved, scabrous, sometimes minutely awned. **Paleas:** as long as lemma. **Awns:** occasionally minute awn from tip of lemmas. **VEGETATIVE CHARACTERISTICS: Growth Habit:** single or tufted, rhizomes present. **Culms:** 3–10 dm tall, erect, glabrous, old ones persisting at base. **Sheaths:** glabrous to scabrous, old ones persisting at base. **Ligules:** 1–3.5 mm long, erose-ciliolate. **Blades:** 2–6 mm wide, 10–40 cm long, erect, flat, firm, striated, glaucous, sometimes scabrous. **HABI-TAT:** Mostly dry slopes and ponderosa pine forests (Weber and Wittmann 2001a, 2001b). **COMMENTS:** Sometimes included in *Festuca*, but it is more closely related to *Lolium* and *Schedonorus* (Barkworth et al. 2007). This species, in our area, grows predominantly under ponderosa pine in mid-elevation forests. Not a highly palatable forage species and frequently not grazed, even in heavily utilized areas.

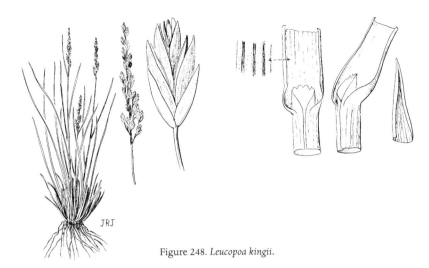

Figure 248. *Leucopoa kingii.*

78. *Lolium* L.

Annuals and short-lived perennials, usually with succulent culms and flat or folded blades. Inflorescence a spike of several-flowered spikelets, these borne solitary and oriented edgewise at nodes of a continuous rachis. Lower glume absent except on terminal spikelet. Upper glume usually large, broad, several-nerved, awnless. Lemmas 5- to 9-nerved, rounded on back, awnless or awned from a usually broad apex. Palea large. Basic chromosome number, $x = 7$. Photosynthetic pathway, C_3.

Represented in Colorado by 2 species.

1. Lemmas awnless or with awn less than 8 mm long; leaf blades narrow, 1–4 mm broad, conduplicate; long-lived perennial; 2–10 florets per spikelet . . . . 228. *L. perenne*
1. Lemmas awned; awns 2–15 mm long; leaf blades wide, 3–8 mm broad, convolute; annual, biennial, or short-lived perennial; 10–22 florets per spikelet . . . . 227. *L. multiflorum*

227. *Lolium multiflorum* Lam. (Fig. 249).

Synonyms: *Lolium perenne* L. subsp. *multiflorum* (Lam.) Husnot, *Lolium multiflorum* Lam. subsp. *italicum* (A. Braun) Volk. *ex* Schinz & R. Keller, *Lolium multiflorum* Lam. var. *diminutum* Mutel, *Lolium multiflorum* Lam. var. *muticum* DC., *Lolium perenne* L. subsp. *italicum* (A. Braun) Husnot, *Lolium perenne* L. var. *aristatum* Willd., *Lolium perenne* L. var. *multiflorum* (Lam.) Parnell. **Vernacular Names:** Annual ryegrass, Italian ryegrass. **Life Span:** Annual, short-lived perennial. **Origin:** Introduced. **Season:** C_3, cool season. **FLORAL CHARACTERISTICS: Inflorescence:** spike, distichous, erect, 3–30 cm long. **Spikelets:** 10- to 22-flowered, 9–25 mm long, laterally compressed, positioned edgewise to rachis; disarticulating above the glumes. **Glumes:** lower absent in all but terminal spikelet; upper 7–12 mm long, acute, 7- to 9-nerved, lateral

Figure 249. *Lolium multiflorum.*

nerves obscure. **Lemmas:** 3.5–9 mm long, acute, awned. **Paleas:** more or less the same size as lemmas. **Awns:** lemmas awned from 0.2–0.7 mm below apices, awns 2–15 mm long. **VEGETATIVE CHARACTERISTICS: Growth Habit:** tufted. **Culms:** ascending to erect, 5–15 dm tall, generally without basal innovations, smooth to scabrous. **Sheaths:** open, glabrous or occasionally scabrous, rounded. **Auricles:** usually well developed and up to 2 mm long. **Ligules:** 0.5–1.5 mm long, membranous, entire. **Blades:** 10–30 cm long, convolute, 3–8 mm wide, scabrous above. **HABITAT:** Disturbed areas and lawns (Weber and Wittmann 2001a, 2001b). **COMMENTS:** This species has been cultivated; it escapes and persists in disturbed areas. Sometimes considered a variety of *L. perenne* L. (var. *aristatum* Willd.) (Allred 2005). *L. multiflorum* differs from *L. perenne* in being an annual or short-lived perennial, taller, and with larger leaves that are rolled rather than folded in the bud (Barkworth et al. 2007).

228. *Lolium perenne* L. (Fig. 250).

Synonyms: *Lolium multiflorum* Lam. var. *ramosum* Guss. *ex* Arcang., *Lolium perenne* L. var. *cristatum* Pers. *ex* B. D. Jackson. **Vernacular Names:** Perennial ryegrass, English ryegrass. **Life Span:** Perennial. **Origin:** Introduced. **Season:** C$_3$, cool season. **FLORAL CHARACTERISTICS: Inflorescence:** spike, distichous, erect, 10–25 cm long. **Spikelets:** 2- to 10-flowered, 9.5–15 mm long, laterally compressed, positioned edgewise to rachis; disarticulating above the glumes.

Glumes: lower absent in all but terminal spikelet; upper 6–12 mm long, 7- to 9-nerved, only 3–5 of which are prominent. **Lemmas:** 4–9 mm long, acute, awnless or less frequently awned. **Awns:** lemmas occasionally awned from 0.2–0.7 mm below tip, awns to about 8 mm long. **VEGETATIVE CHARACTERISTICS: Growth Habit:** tufted. **Culms:** 3–10 dm tall, erect or decumbent, with basal innovations, glabrous and smooth. **Sheaths:** open, glabrous or occasionally scabrous. **Auricles:** usually well developed. **Ligules:** 0.5–1.5 mm long, membranous, entire. **Blades:** 1–6 mm wide, 5–20 cm long, flat, smooth. **HABITAT:** Disturbed areas and lawns (Weber and Wittmann 2001a, 2001b). **COMMENTS:** A common lawn and pasture grass that has escaped and persists as a weedy species. This species differs from *L. multiflorum* in being a long-lived perennial, shorter in stature, and with shorter leaves that are folded in the bud rather than rolled (Barkworth et al. 2007).

79. *Phalaris* L.

Annuals. Leaves mostly glabrous, with membranous ligules and flat blades. Inflorescence a contracted, usually spikelike panicle. Spikelets with 1 terminal perfect floret and 1–2 reduced florets below, the latter reduced to scales. Disarticulation above the glumes. Glumes about equal, large, awnless, usually laterally flattened and dorsally keeled, the keel often with a thin, membranous wing. Lemma of fertile floret awnless, coriaceous, and glossy; shorter and firmer than glumes, often more or less hairy, permanently enclosing the faintly 2-nerved palea and plump caryopsis. Basic chromosome numbers, $x = 6$ and 7. Photosynthetic pathway, C$_3$.

Represented in Colorado by 3 species.

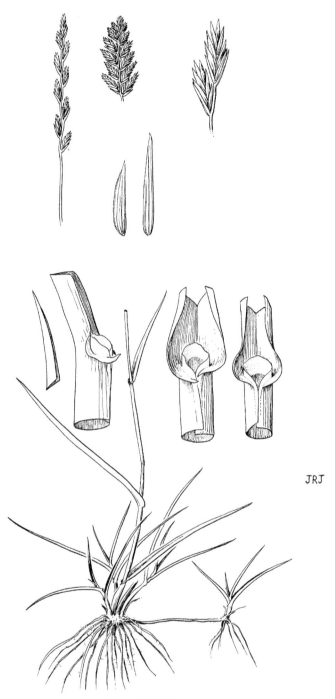

JRJ

Figure 250. *Lolium perenne.*

1. Glumes broadly winged
   2. Sterile florets 1, wings of glumes usually irregularly dentate to crenate . . . 231. *P. minor*
   2. Sterile florets 2, wings of glumes usually entire . . . . 229. *P. canariensis*
1. Glumes wingless or nearly so . . . . 230. *P. caroliniana*

229. *Phalaris canariensis* L. (Fig. 251).

Synonyms: *Phalaris avicularis* Salisb., *Phalaris ovata* Moench, *Phalaris canariensis* L. subsp. *typica* Posp. **Vernacular Names:** Annual canarygrass, common canarygrass. **Life Span:** Annual. **Origin:** Introduced. **Season:** C$_3$, cool season. **FLORAL CHARACTERISTICS: Inflorescence:** panicle, 1.5–5 cm long, ovoid to ovoid-oblong; branches obscure; spikelets borne singly, not clustered, closely imbricate. **Spikelets:** 7–9 mm long, laterally flattened, 3-flowered; terminal floret perfect; lower 2 sterile, reduced. **Glumes:** subequal, 7–9 mm long, mostly 3-nerved, these green with white between nerves; keel distinctly broadly winged, entire; the wing pale, broadened upward. **Lemmas:** fertile 4–6 mm long, lanceolate, acute, densely appressed pubescent; sterile unequal, linear, much shorter than fertile, sparsely pubescent. **Awns:** none. **VEGETATIVE CHARACTERISTICS: Growth Habit:** tufted. **Culms:** 3–6 dm tall, erect or spreading near base, robust, glabrous or scabrous. **Sheaths:** scabrous, upper ones inflated. **Auricles:** sometimes present. **Ligules:** 3–6 mm long, membranous, usually deeply lacerate. **Blades:** 4–10 mm wide, 4–20 cm long, flat. **HABITAT:** Disturbed areas, lawns, and particularly around bird feeders. **COMMENTS:** Used in commercial birdseed mixtures.

230. *Phalaris caroliniana* Walter (Fig. 252).

Synonyms: *Phalaris americana* Elliott, *Phalaris intermedia* Bosc *ex* Poir., *Phalaris microstachya* DC., *Phalaris occidentalis* Nutt., *Phalaris trivialis* Trin. **Vernacular Name:** Carolina canarygrass. **Life Span:** Annual. **Origin:** Native. **Season:** C$_3$, cool season. **FLORAL CHARACTERISTICS: Inflorescence:** panicle, dense, spikelike, 2–6 cm long. **Spikelets:** borne singly, not clustered; 4–6 mm long, laterally compressed, 3-flowered; lower 2 florets sterile, terminal floret perfect. **Glumes:** subequal, 4–6 mm long, prominently 3-nerved, keel wingless or nearly so. **Lemmas:** fertile 3–4 mm long, appressed pubescent; sterile 1.2–1.7 mm long, subulate, pubescent, appressed to fertile. **Awns:** none. **VEGETATIVE CHARACTERISTICS: Growth Habit:** tufted. **Culms:** 2.5–6 dm tall (rarely to 15 dm), erect or decumbent, glabrous or scabrous. **Sheaths:** glabrous, upper ones somewhat inflated. **Auricles:** absent to sometimes present, much reduced in size. **Ligules:** 2–6 mm long. **Blades:** 2–10 mm wide, 5–20 cm long, flat, glabrous or somewhat scabrous, occasionally glaucous, cauline blades reduced upward. **HABITAT:** Open disturbed sites or wet marshy ground in southeastern part of the state. **COMMENTS:** Documented in the state from only a few specimens.

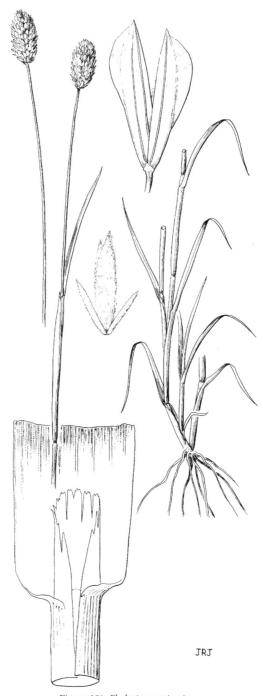

Figure 251. *Phalaris canariensis.*

Figure 252. *Phalaris caroliniana*.

231. *Phalaris minor* Retz. (Fig. 253).

Synonyms: *Phalaris ambigua* Fig. & De Not., *Phalaris aquatica* Thunb., *Phalaris brevis* Trin., *Phalaris decumbens* Moench, *Phalaris gracilis* Parl., *Phalaris haematites* Duval-Jouve & Paris, *Phalaris mauritii* Sennen, *Phalaris nepalensis* Trin., *Phalaris trivialis* Trin. **Vernacular Names:** Littleseed canarygrass, lesser canarygrass. **Life Span:** Annual. **Origin:** Introduced. **Season:** C_3, cool season. **FLORAL CHARACTERISTICS: Inflorescence:** panicle, dense, spikelike, ovate to cylindrical,

2–5 cm long. **Spikelets:** borne singly, not in clusters; closely imbricate, 4–6 mm long, laterally compressed, 2-flowered; lower floret sterile, upper floret perfect. **Glumes:** subequal, 4–6 mm long, keel distinctly winged; the wing irregularly dentate to crenate, occasionally entire. **Lemmas:** fertile, about 3 mm long, indistinctly 3- to 5-nerved, coriaceous, sparsely appressed pubescent; sterile as much as half the length of fertile, pubescent at apex. **Awns:** none. **VEGETATIVE CHARACTERISTICS: Growth Habit:** tufted. **Culms:** 3–6 dm tall, erect or occasionally spreading, glabrous. **Sheaths:** smooth or scabrous, upper ones often inflated. **Ligules:** 2–5 mm long. **Blades:** 3–6 mm wide, 5–30 cm long, flat, scabrous. **HABITAT:** Disturbed areas. **COMMENTS:** Only 1 record from Larimer County. Probably not a constant member of our flora.

80. *Phalaroides* Wolf

Perennials, rather coarse, densely cespitose or in small clusters from rhizomatous bases. Leaves mostly glabrous, with membranous ligules and flat blades. Inflorescence a contracted, usually spikelike panicle. Spikelets with 1 terminal perfect floret and 1–2 reduced florets below, the latter reduced to scales. Disarticulation above the glumes. Glumes about

Figure 253. *Phalaris minor.*

equal, large, awnless, usually laterally flattened and dorsally keeled, the keel often with a thin, membranous wing. Lemma of fertile floret awnless, coriaceous, and glossy; shorter and firmer than the glumes, often more or less hairy, permanently enclosing the faintly 2-nerved palea and plump caryopsis. Basic chromosome numbers, $x = 6$ and 7. Photosynthetic pathway, C_3.

Represented in Colorado by a single species. A segregate of *Phalaris*.

232. *Phalaroides arundinacea* (L.) Rauschert (Fig. 254).

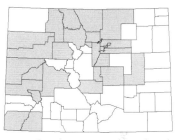

Synonyms: *Phalaris arundinacea* L., *Phalaris arundinacea* L. var. *picta* L., *Phalaroides arundinacea* (L.) Raeusch. var. *picta* (L.) Tzvelev. **Vernacular Names:** common canarygrass, reed canarygrass. **Life Span:** Perennial. **Origin:** Native. **Season:** C_3, cool season. **FLORAL CHARACTERISTICS: Inflorescence:** panicle, narrow, compact, 7–40 cm long, sometimes lobed at base, somewhat spreading in anthesis; evidently branched, at least near base. **Spikelets:** 4–8 mm long, laterally flattened, ovate, 3-flowered; terminal floret perfect, lower 2 sterile, reduced. **Glumes:** subequal, 4–6 mm long, strongly laterally compressed, 3-nerved, keel scabrous, wingless or very nearly so. **Lemmas:** fertile 3–4 mm long, ovate, coriaceous, shiny, brown to gray-brown, somewhat appressed pubescent; sterile up to 2 mm long, subulate, pubescent. **Awns:** none. **VEGETATIVE CHARACTERISTICS: Growth Habit:** creeping rhizomes present. **Culms:** 5–20 dm tall, stout, erect, glabrous. **Sheaths:** glabrous or scaberulous, open. **Ligules:** 2–8 mm long, obtuse, sometimes puberulent abaxially, entire or lacerate. **Blades:** 6–16 mm wide, 10–30 cm long, flat. **HABITAT:** Wet areas of lower elevations (Weber and Wittmann 2001a, 2001b). **COMMENTS:** The seed is a common component of birdseed mixtures. A dominant plant of many irrigation ditches.

81. *Phippsia* (Trin.) R. Br.

Low, tufted, succulent, alpine perennials with contracted, few-flowered panicles of small, awnless spikelets. Spikelets 1-flowered, disarticulating above the glumes. Glumes unequal, minute, lower sometimes absent. Lemma thin, somewhat keeled, 3-nerved, abruptly acute. Palea slightly shorter than lemma. Basic chromosome number, $x = 7$. Photosynthetic pathway, C_3.

Represented in Colorado by a single species.

233. *Phippsia algida* (Sol.) R. Br. (Fig. 255).

Synonym: *Agrostis algida* C. J. Phipps. **Vernacular Name:** Icegrass. **Life Span:** Perennial. **Origin:** Native. **Season:** C_3, cool season. **FLORAL CHARACTERISTICS: Inflorescence:** panicle, narrow, 0.5–3.5 cm long; branches short, appressed or somewhat spreading at anthesis, bearing few spikelets. **Spikelets:** 1-flowered, 1.4–1.8 mm long; disarticulating above the glumes. **Glumes:** 1 or unequal, generally colorless when present, caducous; lower obtuse, sometimes wanting, very short when present, 0.05–0.3 mm long; upper about half as long

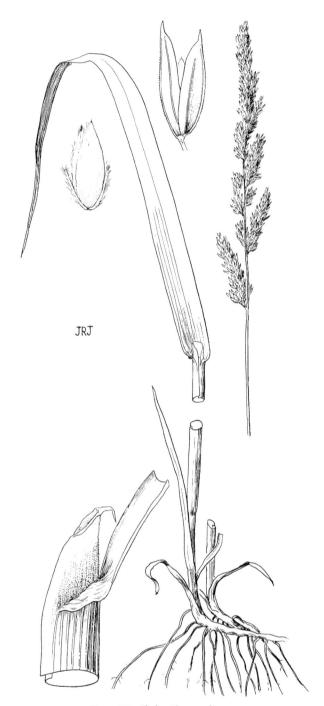

JRJ

Figure 254. *Phalaroides arundinacea.*

Figure 255. *Phippsia algida.*

as lemma, 0.3–0.6 mm long, somewhat truncate and erose at tip. **Lemmas:** 1.3–1.8 mm long, indistinctly 3-nerved, keeled, apex minutely denticulate, broadly ovate, somewhat yellowish green. **Paleas:** glabrous, 1.1–1.3 mm long. **Awns:** none. **VEGETATIVE CHAR-ACTERISTICS: Growth Habit:** densely tufted. **Culms:** less than 10 cm tall, somewhat succulent, glabrous. **Sheaths:** loose, closed, smooth, glabrous. **Ligules:** 0.3–1.6 mm long, acute, entire. **Blades:** 0.6–2.8 cm long, 1.2–3 mm wide, less than 10 cm long, folded, tip prow-shaped. **HABITAT:** Snowmelt streamlets in the alpine (Weber and Wittmann 2001a, 2001b). **COMMENTS:** Rare, perhaps an annual.

81. *Phleum* L.

Perennial, cespitose or tufted. Sheaths open, occasionally inflated and separating from culm; ligules membranous, up to 6 mm long; often with small, rounded auricles. Inflorescence a cylindrical, tightly contracted panicle. Spikelets with 1 floret, perfect. Disarticulation above or occasionally below the glumes. Glumes 2, large, subequal, laterally flattened, mostly 3-nerved, broad, abruptly narrowed at apex to a mucro or short, stout awn. Lemma membranous, 3- to 7-nerved, broad and blunt, awned, much shorter than the glumes. Palea membranous, narrow, about as long as lemma. Basic chromosome number, $x = 7$. Photosynthetic pathway, C_3.

Represented in Colorado by 2 species.

1. Culms with bulbous base; sheaths not inflated . . . . 235. *P. pratense*
1. Culms lacking bulbous base; sheaths inflated and separating from culm . . . .
. . . . 234. *P. alpinum*

234. *Phleum alpinum* L. (Fig. 256).

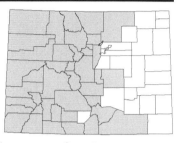

Synonyms: *Phleum alpinum* L. var. *commutatum* (Gaudin) Griseb., *Phleum commutatum* Gaudin, *Phleum commutatum* Gaudin var. *americanum* (Fourn.) Hultén. **Vernacular Name:** Alpine timothy. **Life Span:** Perennial. **Origin:** Native. **Season:** C$_3$, cool season. **FLORAL CHARACTERISTICS: Inflorescence:** panicle, dense, spikelike, ovoid to oblong, 1–5 cm long, 5–9 mm wide. **Spikelets:** 1-flowered, 3–4 mm long excluding awns. **Glumes:** subequal, 2.5–5 mm long, abruptly truncate-attenuate with horn-like awns, keels pectinate-ciliate. **Lemmas:** 1.7–2.5 mm long, erose-toothed from 5 excurrent nerves, scabrous on keel and often nerves as well. **Awns:** stout, 1.5–3 mm long from tip of glumes. **VEGETATIVE CHARACTERISTICS: Growth Habit:** tufted. **Culms:** 2–6 dm tall, erect or decumbent, bulbous base lacking. **Sheaths:** inflated and separating from culm. **Auricles:** sometimes small ones present. **Ligules:** 0.5–3 mm long, truncate or obtuse, erose or subentire. **Blades:** 2.5–8 mm wide, 2–10 cm long, flat, glabrous to scabrous. **HABITAT:** Along trails, open grasslands, and wet meadows from upper montane to alpine. **COMMENTS:** A good mountain forage grass (Stubbendieck, Hatch, and Landholt 2003).

235. *Phleum pratense* L. (Fig. 257).

Synonyms: *Phleum nodosum* L. var. *pratense* (L.) St.-Amans, *Plantinia pratensis* (L.) Bubani, *Stelephuros pratensis* (L.) Lunell. **Vernacular Name:** Timothy. **Life Span:** Perennial. **Origin:** Introduced. **Season:** C$_3$, cool season. **FLORAL CHARACTERISTICS: Inflorescence:** panicle, dense, spikelike, cylindrical, 5–15 cm long, 5–9 mm wide. **Spikelets:** 1-flowered, 2.5–3.5 mm long excluding awns. **Glumes:** subequal, 2.5–3.5 mm long, 3-nerved, abruptly truncate-attenuate with horn-like awns, keels pectinate-ciliate. **Lemmas:** 1.2–2 mm long, faintly 5-nerved, truncate to acute, erose, sometimes minutely awned. **Awns:** stout, 1–2 mm long from tip of glumes. **VEGETATIVE CHARACTERISTICS: Growth Habit:** tufted. **Culms:** 4–10 dm tall, erect, swollen and bulbous at base, glabrous or slightly scabrous. **Sheaths:** glabrous, not inflated or separating from culm. **Auricles:** sometimes small ones present. **Ligules:** 2–4 mm long, obtuse, subentire, occasionally lacerate. **Blades:** 4–8 mm wide, 5–20 cm long, flat, somewhat scabrous. **HABITAT:** Disturbed areas, cultivated hay meadows, and along roadsides and ditches. **COMMENTS:** An important meadow and hay grass that is an excellent forage; however, it does not withstand heavy grazing (Stubbendieck, Hatch, and Landholt 2003).

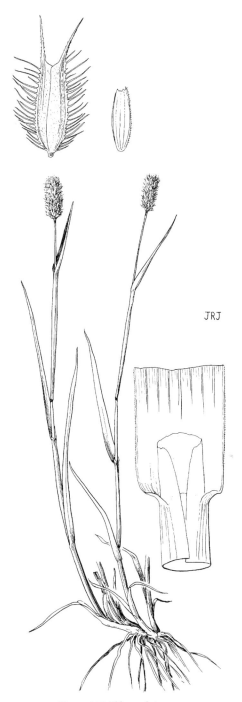

JRJ

Figure 256. *Phleum alpinum.*

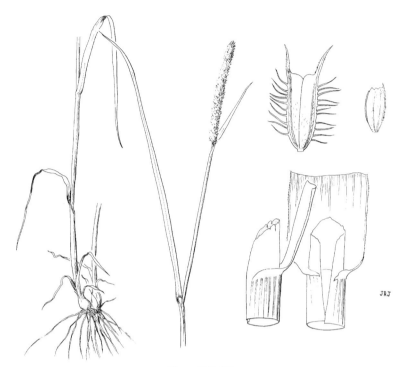

Figure 257. *Phleum pratense.*

83. *Poa* L.

Low to moderately tall annuals and perennials, many with rhizomes. Blades mostly flat or folded, with prow-shaped tips. Inflorescence an open or contracted panicle or occasionally reduced to a raceme. Spikelets mostly small, 2- to 7-flowered, awnless. Disarticulation above the glumes and between florets. Glumes relatively broad, 1- to 3-nerved. Lemmas thin, broad, usually keeled and with a membranous border, obtuse or broadly acute at apex, 5-nerved, nerves often puberulent below. Base of lemma glabrous or with long, kinky, cottony hairs. Palea glabrous. Basic chromosome number, $x = 7$. Photosynthetic pathway, C_3.

Represented in Colorado by 24 species, 15 subspecies, and 1 variety. A difficult and often confusing genus because most species are polyploid, many are apomictic, and hybrids are common (Barkworth et al. 2007).

1. Spikelets converted to bulblets
    2. Plants with bulbous bases . . . . 242. *P. bulbosa*
    2. Plants rhizomatous or tufted but not bulbous
        3. Plants rhizomatous . . . . 239. *P. arctica*
        3. Plants with tufted bases . . . . 236. *P. abbreviata*
1. Spikelets typically not converted to bulblets
    4. Plants rhizomatous
        5. Culms and nodes strongly flattened . . . . 243. *P. compressa*
        5. Culms and nodes not strongly flattened

6. Lemma with cobwebby hairs at base
  7. Sheaths closed from half to nine-tenths their length
    8. Sheath strongly keeled; keels with wings up to 0.5 mm wide .... 257. *P. tracyi*
    8. Sheath terete, not keeled
      9. Plant dioecious; florets pistillate only .... 244. *P. cusickii*
      9. Plant not dioecious; florets perfect .... 249. *P. leptocoma*
  7. Sheaths closed less than half their length, not strongly keeled; keels, if present, wingless
    10. Florets pistillate only; plant dioecious .... 244. *P. cusickii*
    10. Florets perfect; plants bisexual
      11. Panicle open at maturity; callus hairs a third to half the length of lemma
        12. Palea intercostal regions broad, distinctly hispid or softly puberulent; plants above treeline .... 239. *P. arctica*
        12. Palea intercostal regions narrow, glabrous or rarely sparsely hispid; plants generally below treeline .... 253. *P. pratensis*
      11. Panicle contracted or open only at anthesis; callus hairs absent or a quarter or less as long as lemma
        13. Spikelets laterally compressed, length 2–3.8 times width .... 240. *P. arida*
        13. Spikelets subterete or only slightly laterally compressed, length 3.8–5 times width .... 255. *P. secunda*
6. Lemma without cobwebby hairs at base .... KEY A
4. Plant not rhizomatous .... KEY B

KEY A

1. Plant dioecious, most florets pistillate
  2. Uppermost culm leaf distinctly reduced .... 245. *P. fendleriana*
  2. Uppermost culm leaf well developed .... 259. *P. wheeleri*
1. Plant not dioecious, most florets perfect
  3. Panicle narrow, branches ascending; plants of saline or alkaline riparian habitats of lower elevations .... 240. *P. arida*
  3. Panicle open, at least lower branches lax; plants of mesic forests, subalpine forests, or alpine
    4. Upper portion of lower sheaths retrorsely pubescent .... 259. *P. wheeleri*
    4. Upper portion of lower sheaths glabrous .... 239. *P. arctica*

KEY B

1. Lemmas with cobwebby hairs at base
  2. Plant annual
    3. Sheaths glabrous; panicle branches, at least the lowermost spreading .... 238. *P. annua*
    3. Sheaths scabrous; panicle branches erect .... 241. *P. bigelovii*
  2. Plant perennial
    4. Plant dioecious; florets pistillate only .... 244. *P. cusickii*
    4. Plant not dioecious; some florets perfect

    5. Lower panicle branches single or in pairs

      6. Panicle branches mostly smooth or only scabrous on angles

        7. Panicles contracted; lower portion of lemma usually covered with soft pubescence . . . . 236. *P. abbreviata*

        7. Panicles open, reflexed or loosely closed; lemmas scabrous or pubescent on nerves, usually glabrous in intercostal regions

          8. Glumes nearly equal; panicle branches reflexed at maturity . . . .

          . . . . 254. *P. reflexa*

          8. Glumes distinctly unequal; panicles loosely closed . . . . 248. *P. laxa*

      6. Panicle branches scabrous throughout . . . . 249. *P. leptocoma*

    5. Lower panicle branches usually more than 2 per node . . . . KEY D

1. Lemmas without cobwebby hairs at base

    9. Plants annual

      10. Sheaths glabrous; panicle branches, at least the lowermost spreading . . . .

      . . . . 238. *P. annua*

      10. Sheaths scabrous; panicle branches erect . . . . 241. *P. bigelovii*

    9. Plants perennial

      11. Spikelets not compressed; glumes and lemmas not keeled, rounded . . . .

      . . . . 255. *P. secunda*

      11. Spikelets noticeably laterally compressed; glumes and lemmas keeled mostly to base

        12. Lemmas pubescent on keels, marginal nerves, or both . . . . KEY E

        12. Lemmas glabrous or scabrous on keels, marginal nerves, or both

          13. Dwarf alpine plant, 0.1–1.2 dm tall; spikelets 3–4 mm long . . . .

          . . . . 250. *P. lettermanii*

          13. Erect alpine plant, up to 3 dm tall; spikelets 4–10 mm long . . . . 244. *P. cusickii*

KEY D

1. Lateral lemma nerves prominent, marginal nerves glabrous; lower glume very narrow, sickle-shaped . . . . 258. *P. trivialis*

1. Lateral lemma nerves obscure, marginal nerves pubescent; lower glume not sickle-shaped

    2. Sheaths strongly compressed-keeled; at maturity inflorescence lax, nodding, open; ligule 3–12 mm long, densely scabrous . . . . 251. *P. occidentalis*

    2. Sheaths terete; at maturity inflorescences generally not nodding, typically erect, open or contracted; ligule 0.5–6 mm long, glabrous or only sparsely scabrous

      3. Lemmas covered with short, sharp points . . . . 252. *P. palustris*

      3. Lemmas glabrous, scabrous, or sometimes villous but not covered with short, sharp points

        4. Culms, at maturity, 3.5 dm or less long; solely an alpine plant

          5. Sheaths closed up to three-quarters its length; lemmas long- or short-villous on midnerve and marginal nerves, glabrous between nerves; calluses usually webbed, rarely glabrous . . . . 248. *P. laxa*

          5. Sheaths closed up to one-third its length; lemmas long-villous along three-quarters of midnerve keel and marginal nerves, rarely glabrous, soft pubescent between nerves or sometimes glabrous; calluses without hairs or occasionally with very few webby hairs . . . . 236. *P. abbreviata*

4. Culm, at maturity, greater than 3.5 dm long; of various habitats
  6. Ligule 1.5–6 mm long . . . . 249. *P. leptocoma*
  6. Ligule 0.5–1.5 (3) mm long . . . . 247. *P. interior*

KEY E

1. Plants generally over 3 dm tall; panicle (2) 5–15 cm long; lemmas 3–7 mm long; foothills to subalpine
  2. Florets usually pistillate; plant dioecious; cauline blades reduced above, flag leaf blade absent or greatly reduced . . . . 245. *P. fendleriana*
  2. Florets perfect; plant not dioecious; cauline leaf blades not noticeably reduced above, flag leaf blade present and well developed
    3. Spikelets 4–6 mm long; panicle branches 1–3 cm long, lower ascending; inflorescence loosely contracted . . . . 248. *P. laxa*
    3. Spikelets 6–10 mm long; panicle branches 3–15 cm long, lower spreading; inflorescence open, sparse . . . . 256. *P. stenantha*
1. Plants up to 4 dm tall; panicles 1–5 (9) cm long; lemmas 2–5 mm long; subalpine to alpine
  4. Panicle branches widely spreading at maturity; spikelets ovate to subcordate, 1–2 times as long as wide; stem bases enclosed in persistent, thickened, overlapping sheaths; leaf blades 2–4 mm broad . . . . 237. *P. alpina*
  4. Panicle branches not widely spreading at maturity; spikelets elongate, generally 2–3 times as long as wide, not at all cordate at base; blades usually less than 2 mm broad
    5. Leaves curved with persistent, elongated, papery sheaths; blades green . . . .
    . . . . 236. *P. abbreviata*
    5. Leaves stiffly erect, sheaths neither elongate nor papery; blades glaucous . . . .
    . . . . 246. *P. glauca*

236. *Poa abbreviata* R. Br. (Fig. 258).

Synonyms: Subsp. *pattersonii: Poa abbreviata* R. Br. subsp. *jordalii* (Porsild) Hultén, *Poa abbreviata* R. Br. var. *jordalii* (Porsild) Boivin, *Poa jordalii* Porsild, *Poa pattersonii* Vasey. **Vernacular Name:** Patterson bluegrass. **Life Span:** Perennial. **Origin:** Native. **Season:** C_3, cool season. **FLORAL CHARACTERISTICS: Inflorescence:** narrow panicle, 2–6 cm long, dense, lax, greenish to purplish; branches usually 1–3, to 1.5 cm long. **Spikelets:** 4–6.5 mm long, 2- to 5-flowered,  generally subsessile, strongly compressed; disarticulation above the glumes. **Glumes:** subequal to slightly longer than adjacent lemma, distinctly keeled, lanceolate; lower 2.8–4 mm long, obscurely 3-nerved; upper 3.2–4 mm long, 1- to 3-nerved. **Florets:** generally perfect. **Calluses:** with cobwebby hairs sparse to absent at base. **Lemmas:** 3–4 mm long, acute, compressed-keeled, obscurely 3-nerved, pubescent on keels and marginal nerves, puberulent between nerves, scarious, usually with a purple band just below apices. **Paleas:** keels scabrous to short-villous at midlength, occasionally glabrous. **Awns:** none. **Anthers:** 0.2–1.2 (1.8) mm long. **VEGETATIVE CHARACTERISTICS: Growth Habit:** densely tufted, leaves

Figure 258. *Poa abbreviata*.

generally basal, rhizomes absent. **Culms:** erect, 0.5–2.5 dm tall, leafless above basal tuft. **Sheaths:** elongate, papery, glabrous to scabrous-puberulent, persistent, generally open for almost three-quarters of their length. **Ligules:** 0.5–5.5 mm long, truncate to obtuse to acute, lacerate-erose with a ciliolate margin, glabrous to pubescent. **Blades:** generally less than 10 cm long, 1–2 mm wide, flat to involute, generally lax to gracefully curving, typically scabrous. **HABITAT:** Moist areas in snowbed gravels and frost scars, alpine (Weber and Wittmann 2001a). **COMMENTS:** The subspecies found in Colorado is subsp. *pattersonii* (Vasey) Á. Löve & D. Löve & B. M. Kapoor.

237. *Poa alpina* L. (Fig. 259).

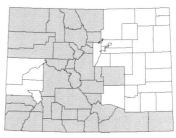

Synonyms: none. **Vernacular Name:** Alpine bluegrass. **Life Span:** Perennial. **Origin:** Native. **Season:** C$_3$, cool season. **FLORAL CHARACTERISTICS: Inflorescence:** panicle, 1–10 cm long, about as wide as long, short-pyramidal, sometimes ovoid, open; nodes with 1–2 branches; branches capillary, spreading, lower ones often reflexed. **Spikelets:** 3–5 mm long, 2- to 7-flowered, rounded to subcordate, purplish, strongly laterally compressed; disarticulation above the glumes. **Glumes:** ovate, keeled, abruptly acute, indistinctly 3-nerved: lower 2.2–3 mm long; upper 2.5–3.5 mm long. **Florets:** most perfect. **Calluses:** without cobwebby hairs at base. **Lemmas:** 3–5 mm long, broad lanceolate, compressed-keeled, indistinctly nerved, pubescent on keels and marginal nerves, pubescent between nerves below; margins scarious. **Paleas:** keels soft puberulent to short-villous, apices scabrous. **Awns:** none. **Anthers:** 1.3–2.3 mm long. **VEGETATIVE CHARACTERISTICS: Growth Habit:** densely tufted from thickly matted basal leaves, erect, without rhizomes. **Culms:** 0.4–4 dm tall, glabrous. **Sheaths:** glabrous, persisting for over a year, open to near base, closed for about a quarter of their length. **Ligules:** 1–4 mm long; truncate, obtuse, or shortly acuminate. **Blades:** short and broad, mostly 2–4 mm wide, 2–6 cm long, mostly basal, flat or folded, glabrous to somewhat scabrous on midnerve and margins, tip prow-shaped. **HABITAT:** Moist areas, montane to subalpine and alpine habitats. **COMMENTS:** The new shoots spread horizontally. Species is most common in alpine communities. The typical subspecies (subsp. *alpina*) is the only one found in Colorado.

238. *Poa annua* L. (Fig. 260).

Synonyms: *Poa annua* L. var. *aquatica* Aschers., *Poa annua* L. var. *reptans* Hausskn. **Vernacular Name:** Annual bluegrass. **Life Span:** Annual. **Origin:** Introduced. **Season:** C$_3$, cool season. **FLORAL CHARACTERISTICS: Inflorescence:** panicle, open-pyramidal, 1–8 cm long, branches spreading. **Spikelets:** 3–5 mm long, 2- to 6-flowered; disarticulation above the glumes. **Glumes:** glabrous, distinctly keeled; lower 1.5–2 mm long, 1-nerved; second 2–2.5 mm long, 1-

Figure 259. *Poa alpina.*

to 3-nerved. **Florets:** most perfect. **Calluses:** without cobwebby hairs at base. **Lemmas:** 2.5–3.5 mm long, compressed-keeled, 5-nerved, these pubescent below. **Paleas:** keels generally short to long, villous, smooth, rarely glabrous. **Awns:** none. **Anthers:** 0.6–1.1 mm long. **VEGETATIVE CHARACTERISTICS: Growth Habit:** tufted, spreading, sometimes rooting at lower nodes and forming mats, without rhizomes. **Culms:** 0.2–4.5 dm tall, strongly flattened, glabrous. **Sheaths:** open for about two-thirds their length, flattened, somewhat keeled, glabrous. **Ligules:** 0.5–3 mm long, truncate or obtuse, sometimes erose, glabrous. **Blades:** 1–4 mm wide, 2–10 cm long, lax, flat or folded, glabrous, margins scabrous, tips prow-shaped. **HABITAT:** Mostly lawns and disturbed moist areas up to subalpine. **COMMENTS:** A very common weed that invades established lawns and golf courses. Plant can produce seed heads even when closely and continually mown (Allred 2005).

JRJ

Figure 260. *Poa annua.*

239. *Poa arctica* R. Br. (Fig. 261).

Synonyms: Subsp. *aperta*: *Poa aperta* Scribn. & Merr.; subsp. *arctica*: *Poa arctica* R. Br. subsp. *longiculmis* Hultén, *Poa arctica* R. Br. subsp. *williamsii* (Nash) Hultén, *Poa arctica* R. Br. var. *glabriflora* Rosh., *Poa arctica* R. Br. var. *vivipara* Hook., *Poa cenisia* All. var. *arctica* (R. Br.) Richter, *Poa longipila* Nash, *Poa williamsii* Nash Scribn. & Merr.; subsp. *grayana*: *Poa arctica* R. Br. var. *grayana* (Vasey) Dorn, *Poa grayana* Vasey. **Vernacular Name:** Arctic bluegrass. **Life Span:** Perennial. **Origin:**

Native. **Season:** C$_3$, cool season. **FLORAL CHARACTERISTICS: Inflorescence:** panicle, open, narrowly pyramidal to oblong, 3–10 cm long; branches ascending to reflexed. **Spikelets:** 4–8 mm long, 3- to 6-flowered, mostly purplish, laterally compressed; disarticulation above the glumes. **Glumes:** lanceolate, purplish toward apex; lower 2.5–5 mm long, 3-nerved; upper 3–5.5 mm long. **Florets:** usually perfect. **Calluses:** with some cobwebby hairs at base, a third to half the length of lemma. **Lemmas:** 3–6 mm long, lanceolate, compressed-keeled, purplish toward apices, villous on keel and marginal nerves, pubescent between nerves below. **Paleas:** keels generally short- to long-villous for most of their length; intercostal regions broad, distinctly hispid to softly puberulent. **Awns:** none. **Anthers:** 1.4–2.5 mm

451

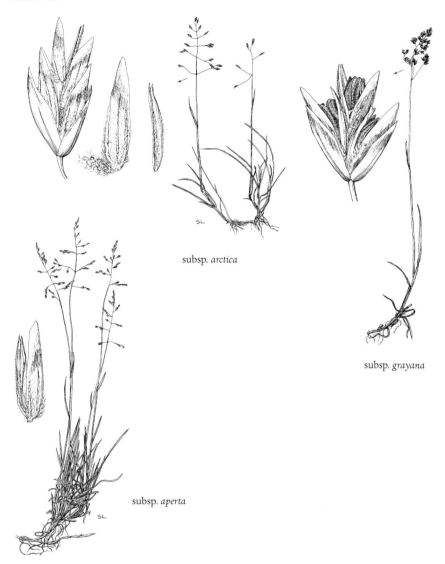

subsp. *arctica*

subsp. *grayana*

subsp. *aperta*

Figure 261. *Poa arctica.*

long, occasionally aborted at a later stage. **VEGETATIVE CHARACTERISTICS: Growth Habit:** solitary or loosely tufted from rhizomes. **Culms:** 1–3 dm tall, erect or decumbent. **Sheaths:** glabrous to short-pubescent, loose, mostly open to below midlength. **Ligules:** 1–7 mm long, obtuse or truncate, entire or erose. **Blades:** 1–4 mm wide, 2–10 cm long, flat or folded, glabrous, keeled, tip boat-shaped. **HABITAT:** Mostly above timberline in alpine meadows and grasslands. **COMMENTS:** Numerous subspecies have been described; 3 are found in Colorado, subsp. *aperta* (Scribn. & Merr.) Soreng, subsp. *arctica*, and subsp. *grayana* (Vasey) Á. & D. Löve & Kapoor. The subspecies can be distinguished by these characteristics:

1. Ligules 3–7 mm long
  2. Panicle branches up to 1/2 as long as panicle . . . . subsp. *aperta*
  2. Panicle branches fup to 2/5 as long as panicle . . . . subsp. *grayana*
1. Ligules 2–4 mm long . . . . subsp. *arctica*

240. *Poa arida* Vasey (Fig. 262).

Synonyms: *Poa fendleriana* (Steud.) Vasey var. *arida* (Vasey) M. E. Jones, *Poa glaucifolia* Scribn. & Williams *ex* Williams, *Poa pseudopratensis* Scribn. & Rydb. **Vernacular Name:** Plains bluegrass. **Life Span:** Perennial. **Origin:** Native. **Season:** C$_3$, cool season. **FLORAL CHARACTERISTICS: Inflorescence:** panicle, erect, narrow, contracted, 2–18 cm long; branches short, ascending to erect. **Spikelets:** 3–7 mm long, laterally compressed, 2- to 8-flowered; disarticulation above the glumes. **Glumes:** glabrous, green, distinctly keeled; lower 2.5–3 mm long, 3-nerved; upper slightly longer, 1- to 3-nerved. **Florets:** most perfect. **Calluses:** cobwebby hairs absent. **Lemmas:** 2–5 mm long, usually compressed-keeled, occasionally rounded, indistinctly 5-nerved, densely villous on keel and marginal nerves, sometimes slightly villous between nerves below. **Paleas:** keels scabrous, glabrous to short-villous at midpoint; intercostal regions generally glabrous, although sometimes puberulent to short-villous. **Awns:** none. **Anthers:** 1.3–2.2 mm long. **VEGETATIVE CHARACTERISTICS: Growth Habit:** tufted, rhizomatous. **Culms:** 2–8 dm tall, erect, glabrous, sometimes scabrous below panicle. **Sheaths:** glabrous or slightly scabrous above, open most of their length. **Ligules:** 1.5–5 mm long, obtuse to acute. **Blades:** 1–5 mm wide, 3–15 cm long, folded or involute, mostly basal, rather rigid and sharp-pointed. **HABITAT:** Low elevations, plains, and alkaline or saline flats (Weber and Wittmann 2001a, 2001b). **COMMENTS:** Widespread species with some grazing value because of early spring green-up. Barkworth and colleagues (2007) report it mainly on the Eastern Slope of the Rocky Mountains.

241. *Poa bigelovii* Vasey & Scribn. (Fig. 263).

Synonym: *Poa annua* L. var. *stricta* Vasey *ex* Scribn. **Vernacular Name:** Bigelow's bluegrass. **Life Span:** Annual, rarely longer-lived. **Origin:** Native. **Season:** C$_3$, cool season. **FLORAL CHARACTERISTICS: Inflorescence:** panicle, narrow, interrupted, 5–15 cm long; branches short, erect. **Spikelets:** 4–7 mm long, 3- to 7-flowered, strongly laterally compressed. **Glumes:** glabrous, somewhat gibbous at base, keel scabrous; lower 2–3 mm long, indistinctly 1- to 3-nerved; upper 2.5–4 mm long, more distinctly 3-nerved, lateral nerves sometimes faint; disarticulation above the glumes. **Florets:** most perfect. **Calluses:** base with cobwebby hairs, sometimes scant. **Lemmas:** 2–4 mm long, lanceolate, 5-nerved, compressed-keeled, keel and marginal nerves villous below, sometimes pubescent between nerves below. **Paleas:** keels at mid-

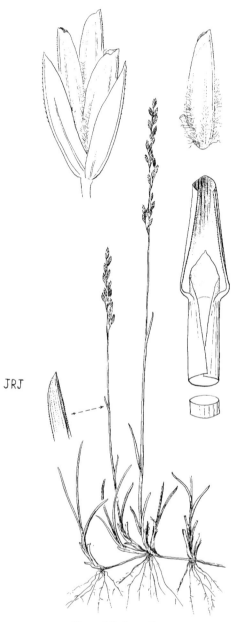

JRJ

Figure 262. *Poa arida*.

Figure 263. *Poa bigelovii*.

length softly puberulent to short-villous, scabrous near apices, generally softly puberulent on intercostal area. **Anthers:** 1–3, 0.2–1 mm long. **Awns:** none. **VEGETATIVE CHARACTER-ISTICS: Growth Habit:** more or less tufted. **Culms:** 1–5 dm tall, erect. **Sheaths:** flat, scabrous. **Ligules:** 1–3 mm long, membranous. **Blades:** 1–4 mm wide, 5–10 cm long, flat, lax. **HABITAT:** Grows in arid upland regions on rocky slopes. **COMMENTS:** Only a few records from the state.

242. *Poa bulbosa* L. (Fig. 264).

Synonyms: None. **Vernacular Name:** Bulbous bluegrass. **Life Span:** Perennial. **Origin:** Introduced. **Season:** C$_3$, cool season. **FLORAL CHARACTERISTICS: Inflorescence:** panicle, 3–12 cm long, loosely contracted to somewhat open at anthesis, narrowly oblong to ovoid; branches short, ascending to appressed, usually scabrous. **Spikelets:** mostly bulbiferous, 3–6 florets; shiny, dark purple bases and exserted, linear, foliaceous tips, strongly compressed. **Glumes:** strongly keeled; lower 2–3 mm long, indistinctly 3-nerved; upper 2.2–3.5 mm long, 3-nerved. **Florets:** forming bulblets, sometimes lower 2–3 florets are normal. **Calluses:** with cobwebby hairs at base. **Lemmas:** bulbiferous spikelets: lowest 1 or 2 usually empty or normal, next basally thickened and elongating, remaining florets modified to form bulblets; floriferous spikelets 3–5 mm long, indistinctly 5-nerved, compressed-keeled, villous on keel and marginal nerves below. **Paleas:** keels at midlength often puberulent, body scabrous. **Awns:** none. **Anthers:** 1.2–1.5 mm long, functional at times, aborted at a later stage, or not developed at all. **VEGETATIVE CHARACTERISTICS: Growth Habit:** mostly densely tufted, viviparous, without rhizomes. **Culms:** 2–6 dm tall, bases bulbous. **Sheaths:** open for about three-quarters of their length, glabrous. **Ligules:** 1–3 mm long. **Blades:** 1–2 mm wide, 5–15 cm long, flat, folded, or loosely involute, glabrous. **HABITAT:** Cultivated and otherwise disturbed areas. **COMMENTS:** Interesting species with spikelets developing into plantlets in the inflorescence that fall and produce new individuals (Allred 2005).

243. *Poa compressa* L. (Fig. 265).

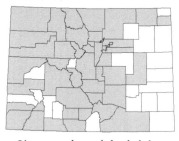

Synonyms: *Paneion compressum* (L.) Lunell, *Poa compressa* L. var. *sylvestris* Torr. **Vernacular Names:** Canada bluegrass, wiregrass. **Life Span:** Perennial. **Origin:** Introduced. **Season:** C$_3$, cool season. **FLORAL CHARACTERISTICS: Inflorescence:** panicle, 2–10 cm long, usually compact, dense, narrow; occasionally branches somewhat spreading, spikelet bearing to base, usually in twos. **Spikelets:** 3–7 mm long, 3- to 8-flowered, crowded on short branches, subsessile, laterally compressed; disarticulation above the glumes. **Glumes:** subequal, keeled; lower 2–3 mm long, acute, 3-nerved; upper slightly longer, broader, 3-nerved. **Florets:** perfect. **Calluses:** cobwebby hairs at base scant or absent. **Lemmas:** 2–4 mm long, sometimes sub-

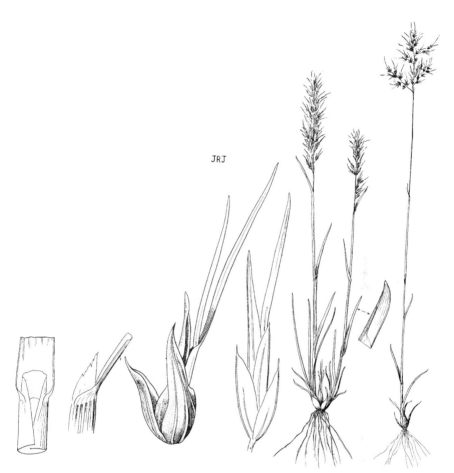

Figure 264. *Poa bulbosa.*

coriaceous, usually purplish near tip, compressed-keeled, keel and marginal nerves pubescent below, glabrous between nerves. **Paleas:** keels scabrous. **Awns:** none. **Anthers:** 1.3–1.8 mm long. **VEGETATIVE CHARACTERISTICS: Growth Habit:** mostly solitary, rhizomatous. **Culms:** 1.5–5 dm tall, decumbent, often geniculate at nodes, strongly flattened, 2-edged, glabrous; nodes strongly compressed. **Sheaths:** strongly compressed-keeled, glabrous. **Ligules:** 0.5–1.5 mm long, ciliolate, obtuse to nearly truncate. **Blades:** 1–4 mm wide, 2–10 cm long, tips prow-shaped, flat, folded, or loosely involute, bluish green, somewhat scabrous. **HABITAT:** Disturbed areas and dry hillsides. **COMMENTS:** Widespread. Culms remain green after foliage has turned. Easily identified by the very compressed culms and nodes. Sometimes seeded for soil stabilization (Barkworth et al. 2007).

JRJ

Figure 265. *Poa compressa.*

244. *Poa cusickii* Vasey (Fig. 266).

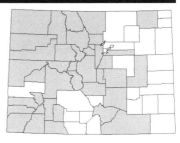

Synonyms: Subsp. *epilis*: *Poa cusickii* Vasey var. *epilis* (Scribn.) C. L. Hitchc., *Poa epilis* Scribn., subsp. *pallida*: *Poa cusickii* Vasey var. *pallida* (Soreng) Dorn. **Vernacular Name:** Cusick's bluegrass. **Life Span:** Perennial. **Origin:** Native. **Season:** C$_3$, cool season. **FLORAL CHARACTERISTICS: Inflorescence:** panicle, 2–12 cm long, erect, loosely contracted, mostly dioecious, oblong, contracted to somewhat open at anthesis, pale or tawny. **Spikelets:** 4–10 mm long, 2- to 6-flowered, strongly compressed; disarticulation above the glumes. **Glumes:** broad, keeled, acute to obtuse; lower 3.5–4 mm long, usually 1- to 3-nerved, shorter than adjacent lemma; upper 4–5 mm long, 3-nerved. **Florets:** usually pistillate. **Calluses:** glabrous or rarely slightly cobwebby at base. **Lemmas:** 4.5–7 mm long, usually distinctly 5-nerved, compressed-keeled, keel and marginal nerves smooth to slightly scabrous, glabrous to occasionally puberulent. **Paleas:** keels scabrous with intercostal areas glabrous. **Awns:** none. **Anthers:** vestigial (0.1–0.2 mm long), aborted at a later stage, or 2–3.5 mm long. **VEGETATIVE CHARACTERISTICS: Growth Habit:** densely tufted, rarely with rhizomes. **Culms:** up to 7 dm tall, erect, glabrous, often straw-colored. **Sheaths:** smooth, closed from one- to three-quarters of their length, keeled. **Ligules:** 0.5–3 mm long, entire or ciliolate, glabrous or puberulent abaxially. **Blades:** up to 2 mm wide, 5–15 cm long, folded or involute, scabrous, cauline leaves gradually reduced in size, uppermost flag leaf 0.5–5 cm long. **HABITAT:** Open plains, foothills, to alpine meadows. **COMMENTS:** Two subspecies are found in Colorado, subsp. *epilis* (Scribn.) W. A. Weber and subsp. *pallida* Soreng. Weber and Wittmann (2001b) report var. *pallida* is a dominant grass of the wetlands at the upper Colorado River near Grand Lake. The 2 subspecies can be distinguished by these characteristics:

1. Panicle branches smooth or slightly scabrous; or basal leaves more than 1.5 mm wide and flat or folded . . . . subsp. *epilis*
1. Panicle branches noticeably scabrous; leaves usually less than 1.5 mm wide, involute, rarely flat or folded . . . . subsp. *pallida*

245. *Poa fendleriana* (Steud.) Vasey (Fig. 267).

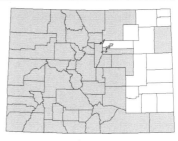

Synonyms: *Atropis fendleriana* (Steud.) Beal, *Eragrostis fendleriana* Steud. *Panicularia fendleriana* (Steud.) Kuntze, *Puccinellia fendleriana* (Steud.) Ponert, *Uralepis poaeoides* Buckley; subsp. *fendleriana*: *Poa andina* Trin. var. *major* Vasey, *Poa andina* var. *spicata* Vasey, *Poa brevipaniculata* Scribn. & T. A. Williams, *Poa eatonii* S. Watson, *Poa longepedunculata* Scribn.; subsp. *longiligula*: *Poa fendleriana* (Steud.) Vasey var. *longiligula* (Scribn. & Williams) Gould, *Poa longiligula* Scribn. & Williams. **Vernacular Names:** Muttongrass, Vasey's muttongrass, longtongue muttongrass. **Life Span:** Perennial. **Origin:** Native. **Season:** C$_3$, cool season. **FLORAL CHARACTERISTICS: Inflorescence:** erect panicle, contracted, narrowly lanceolate to ovoid, 2–30 cm

Figure 266. *Poa cusickii*.

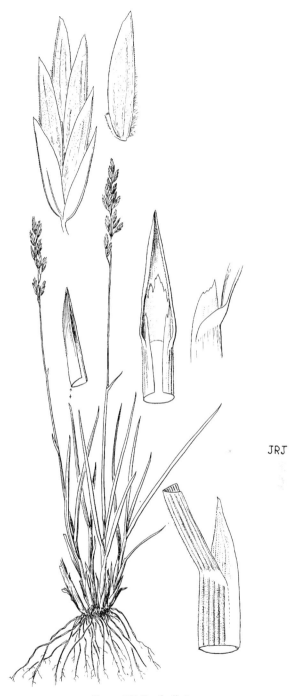

JRJ

Figure 267. *Poa fendleriana*.

long, 1–2 cm wide, usually dioecious, pale to deep purple or occasionally tawny; nodes with 1–2 branches. **Spikelets:** 4–8 (12) mm long, 2- to 7- or more-flowered, strongly compressed-keeled; disarticulation above the glumes. **Glumes:** lanceolate, distinctly keeled; lower 3–3.5 mm long, 1- to 3-nerved; upper 3–5 mm long, 1- to indistinctly 3-nerved. **Florets:** not sexually dimorphic, usually pistillate. **Calluses:** cobwebby hairs absent. **Lemmas:** 3–7 mm long, compressed-keeled, villous on keel and marginal nerves below, internerves usually glabrous. **Paleas:** keels scabrous, occasionally puberulent to long-villous with hairs to 0.4 mm long or more. **Awns:** none. **Anthers:** vestigial (0.1–0.2 mm long) or 2–3 mm long. **VEGETATIVE CHARACTERISTICS: Growth Habit:** tufted, weakly rhizomatous with short and inconspicuous rhizomes. **Culms:** generally 1.5–7 dm tall, not flattened, erect or less commonly decumbent, leaves mostly basal; usually scabrous below panicle. **Sheaths:** open for about two-thirds their length, somewhat scabrous, basal sheaths long, papery, and persistent. **Ligules:** highly variable, 0.2–18 mm long. **Blades:** mostly basal, 1–4 mm wide, 3–12 cm long, stiff, mostly folded or involute, tip pointed, prow-shaped; cauline blades reduced, flag leaf blade absent or distinctly reduced, sometimes to 1 cm. **HABITAT:** Common in the foothills to subalpine zone. **COMMENTS:** Species is dioecious, with small areas of sexual reproduction where staminate plants are common in populations. Most commonly the populations are apomictic, with only pistillate plants present (Barkworth et al. 2007). Two subspecies are found in Colorado, subsp. *fendleriana* and subsp. *longiligula* (Scribn. & Williams) Soreng. The 2 subspecies can be distinguished by:

1. Ligules 10–12 mm long . . . . subsp. *longiligula*
1. Ligules usually about 1 mm long . . . . subsp. *fendleriana*

246. *Poa glauca* Vahl (Fig. 268).

Synonyms: *Catabrosa pauciflora* (Hook.) Fries, *Poa anadyrica* Roshev., *Poa blyttii* Lindeb., *Poa bryophila* Trin.; subsp. *rupicola*: *Poa glauca* Vahl var. *rupicola* (Nash *ex* Rydb.) Boivin, *Poa rupicola* Nash *ex* Rydb. **Vernacular Name:** Glaucous bluegrass. **Life Span:** Perennial. **Origin:** Native. **Season:** C$_3$, cool season. **FLORAL CHARACTERISTICS: Inflorescence:** panicle, narrow, 1–5 cm long, purplish, branches short, stiff, ascending or appressed. **Spikelets:** 3.5–5 mm long,

length generally 2–3 times width, 2- to 4-flowered (usually 3), strongly laterally compressed; disarticulation above the glumes. **Glumes:** broad, keeled; lower 2.5–3 mm long, 1- or indistinctly 3-nerved; upper slightly longer, 3-nerved. **Florets:** mostly perfect, rarely bulb-forming. **Calluses:** cobwebby hairs absent. **Lemmas:** 2.5–3.5 mm long, compressed-keeled, villous on keel and marginal nerves below, sometimes pubescent between nerves. **Paleas:** scabrous to glabrous to puberulent, intercostal areas glabrous to puberulent. **Awns:** none. **Anthers:** (1) 1.2–2.5 mm long, rarely aborted at a later stage. **VEGETATIVE CHARACTERISTICS: Growth Habit:** densely tufted, without rhizomes. **Culms:** 1–4 dm tall, erect, rather stiff, scabrous below inflorescence, otherwise glabrous, well exserted above basal leaves. **Sheaths:** open for most of their length, terete, smooth to scabrous. **Ligules:** 0.5–4 mm long, truncate to acute. **Blades:** basal leaves elongate and narrow, 0.5–1.5 mm wide, 1–5 cm long, glaucous, stiffly erect. **HABITAT:** Dry areas, subalpine to alpine (Weber and Wittmann

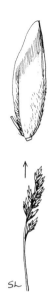

Figure 268. *Poa glauca.*

2001a, 2001b). **COMMENTS:** The Colorado plants belong to subsp. *rupicola* (Nash) W. A. Weber. *P. glauca* can be confused with *P. interior* Rydb., but *P. glauca* has a longer ligule and wider spikelets (Barkworth et al. 2007).

247. *Poa interior* Rydb. (Fig. 269).

Synonyms: *Poa nemoralis* L. var. *interior* (Rydb.) Butters & Abbe, *Poa subtrivialis* Rydb. **Vernacular Names:** Inland bluegrass, interior bluegrass. **Life Span:** Perennial. **Origin:** Native. **Season:** C$_3$, cool season. **FLORAL CHARACTERISTICS: Inflorescence:** panicle, 3–12 cm long, narrowly lanceolate to pyramidal; lower branches 2–5 at a node, branches ascending to spreading. **Spikelets:** 3–6 mm long, 2- to 5-flowered, lanceolate to ovoid, laterally compressed;

disarticulation above the glumes. **Glumes:** lanceolate, distinctly keeled, keels smooth or scabrous; lower prominently 3-nerved, 1.8–2.8 mm long; upper 2–3 mm long, 3-nerved. **Florets:** most perfect. **Calluses:** with cobwebby hairs at base, sometimes scant. **Lemmas:** 2–4 mm long, lanceolate, nerves indistinct, compressed-keeled, villous on keel and marginal nerves below. **Paleas:** keels scabrous, intercostal areas glabrous. **Awns:** none. **Anthers:** (1.1) 1.3–2.5 mm long. **VEGETATIVE CHARACTERISTICS: Growth Habit:** densely tufted, without rhizomes. **Culms:** 0.5–8 dm tall, erect, often scabrous below inflorescence, usually erect, several to many arising together. **Sheaths:** open most of their length, somewhat keeled or terete. **Ligules:** 0.5–1.5 mm long, truncate to obtuse. **Blades:** 1–3 mm wide, 2.5–10 cm

Figure 269. *Poa interior.*

long, mostly basal, crowded, flat, usually drying involute, culm leaves few, stiffly diverging from culm at an acute angle. **HABITAT:** Rocky outcrops, arid mountain slopes, and forests. **COMMENTS:** Common species. Weber and Wittmann (2001b) list this as a subspecies of *P. nemoralis* L. It can be separated from *P. glauca* by lemmas that are glabrous between the marginal nerves and keels (rarely sparsely puberulent on the lateral nerves).

248. *Poa laxa* Haenke (Fig. 270).

Synonym: Subsp. *banffiana: Poa laxa* Haenke var. *occidentalis* Vasey *ex* Rydb. & Shear. **Vernacular Name:** Lax bluegrass. **Life Span:** Perennial. **Origin:** Native. **Season:** C$_3$, cool season. **FLORAL CHARACTERISTICS: Inflorescence:** panicle, 2–8 cm long, lax, usually loosely contracted and sparse; branches mostly in twos, 1–3 cm long, capillary, subflexuous, spreading or ascending, smooth or angles sparsely scabrous. **Spikelets:** 4–6 mm long, 2- to 5-flowered, strongly compressed, mostly purplish; disarticulation above the glumes. **Glumes:** unequal, narrow, keeled, nearly equaling adjacent lemmas; lower 2–3 mm long, 1- to indistinctly 3-nerved; upper 2.5–4 mm long, 3-nerved. **Florets:** perfect. **Calluses:** usually with cobwebby hairs at base, occasionally glabrous. **Lemmas:** 3–4.5 mm long, lanceolate, compressed-keeled, pubescent on keel and marginal nerves below, or sometimes almost glabrous. **Paleas:** keels sparsely scabrous. **Awns:** none. **Anthers:** (0.6) 0.8–1.1 (1.3) mm long. **VEGETATIVE CHARACTERISTICS: Growth Habit:** densely tufted, without rhizomes. **Culms:** 1–4 dm tall, erect or decumbent, glabrous. **Sheaths:** closed no more than a third of their length, glabrous. **Ligules:** 2–4 mm long, acute, often lacerate. **Blades:** 1–3 mm wide, 4–10 cm long, usually flat, glabrous or scabrous, tip prow-shaped, not reduced above. **HABITAT:** Generally occurs in mesic alpine areas. **COMMENTS:** Not reported for Colorado by Weber and Wittmann (2001a, 2001b) or Wingate (1994). It frequently grows with *P. glauca* and they often hybridize. *P. laxa* can be distinguished from *P. glauca* by its shorter anthers and smoother branches. Colorado plants belong to the subsp. *banffiana* Soreng (Barkworth et al. 2007).

249. *Poa leptocoma* Trin. (Fig. 271).

Synonyms: *Poa crandallii* Gand., *Poa flavidula* Kom., *Poa leptocoma* Trin. var. *leptocoma, Poa leptocoma* Trin. subsp. *leptocoma, Poa nivicola* Ridl. var. *flavidula* (Kom.) Roshev., *Poa stenantha* Trin. var. *leptocoma* (Trin.) Griseb. **Vernacular Names:** Marsh bluegrass, western bog bluegrass. **Life Span:** Perennial. **Origin:** Native. **Season:** C$_3$, cool season. **FLORAL CHARACTERISTICS: Inflorescence:** panicle, 5–15 cm long, nodding; branches 1–3 (5), capillary, subflexuous, spreading or ascending, lower ones sometimes reflexed. **Spikelets:** 4–8 mm long, 2- to 5-flowered, strongly compressed, lanceolate, mostly purplish; disarticulation above the glumes. **Glumes:** unequal, narrow; lower 2–3 mm long, 1-nerved; upper 2.5–4 mm long, 3-nerved. **Florets:**

Figure 270. *Poa laxa*.

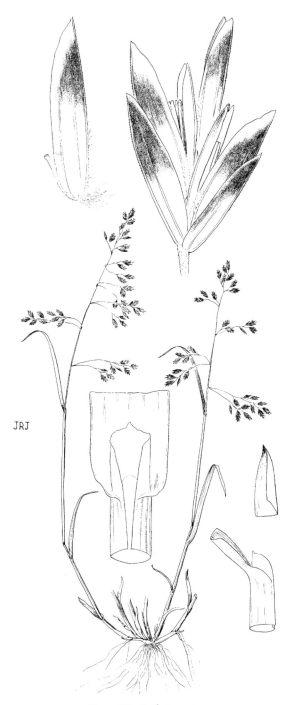

JRJ

Figure 271. *Poa leptocoma*.

generally perfect. **Calluses:** sparse cobwebby hairs at base. **Lemmas:** 3.5–4.5 mm long, acuminate, compressed-keeled, pubescent on keel and marginal nerves below, or sometimes almost glabrous. **Paleas:** keels scabrous to nearly smooth, or ciliate pectinately. **Awns:** none. **Anthers:** 0.2–1.1 mm long. **VEGETATIVE CHARACTERISTICS: Growth Habit:** solitary or few in a tuft, without rhizomes or sometimes appearing rhizomatous when rooting at lower nodes. **Culms:** 2–10 dm tall, erect or decumbent, slender, glabrous. **Sheaths:** terete, closed up to three-fifths of their length, usually scabrous, occasionally glabrous. **Ligules:** 1–5 mm long, truncate or obtuse, sometimes apiculate. **Blades:** 1–4 mm wide, 4–10 cm long, usually flat, glabrous or scabrous, tip prow-shaped. **HABITAT:** Springs, meadows, bogs, around lakes and streams and other wet ground from subalpine to alpine. **COMMENTS:** Rare. Similar to *P. reflexa* and they are sometimes treated as varieties of the same species. It can be distinguished from *P. reflexa* by its more scabrous panicle branches, shorter anthers, glabrous or pectinately ciliate palea keels, and preference for wetter sites (Barkworth et al. 2007).

250. *Poa lettermanii* Vasey (Fig. 272).

Synonyms: *Atropis lettermanii* (Vasey) Beal, *Poa brandegei* Scribn. *ex* Beal, *Poa montevansii* L. Kelso, *Puccinellia lettermanii* (Vasey) Ponert. **Vernacular Name:** Letterman's bluegrass. **Life Span:** Perennial. **Origin:** Native. **Season:** C$_3$, cool season. **FLORAL CHARACTERISTICS: Inflorescence:** panicle, narrow, contracted, erect, 1–3 cm long. **Spikelets:** 3–4 mm long, laterally compressed, 2- to 4-flowered, purplish; disarticulation above the glumes. **Glumes:** glabrous, almost as long as spikelet, subequal; lower 1-nerved; upper 1- to 3-nerved. **Florets:** most perfect. **Calluses:** cobwebby hairs absent. **Lemmas:** 2.5–3 mm long, glabrous, lanceolate, apices hyaline, broadly obtuse, somewhat erose. **Paleas:** with scabrous keels. **Awns:** none. **Anthers:** 0.2–0.8 mm long. **VEGETATIVE CHARACTERISTICS: Growth Habit:** densely tufted dwarf, without rhizomes. **Culms:** 1–12 cm tall, glabrous, shorter than to slightly longer than basal leaves. **Sheaths:** open most of their length, glabrous, strongly nerved. **Ligules:** 1–3 mm long. **Blades:** about 1 mm wide, 3–5 cm long, folded or flat, or involute. **HABITAT:** Rocky areas mostly above timberline. **COMMENTS:** Dwarf alpine species only occurs at higher elevations.

251. *Poa occidentalis* Vasey (Fig. 273).

Synonyms: *Poa platyphylla* Nash & Rydb., *Poa trivialis* L. var. *occidentalis* Vasey. **Vernacular Name:** New Mexico bluegrass. **Life Span:** Perennial. **Origin:** Native. **Season:** C$_3$, cool season. **FLORAL CHARACTERISTICS: Inflorescence:** panicle, 6–40 cm long, open, nodding or lax; branches whorled, in threes or fives, spreading or reflexed, naked below. **Spikelets:** 4–7 mm long, 3- to 7-flowered, laterally compressed; disarticulation above the glumes. **Glumes:**

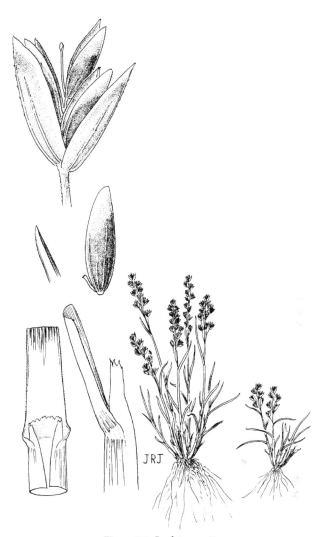

Figure 272. *Poa lettermanii.*

distinctly keeled; lower about 3 mm long, 1-nerved; upper 3.2–3.5 mm long, 3-nerved. **Florets:** most perfect. **Calluses:** cobwebby hairs present at base. **Lemmas:** 2.6–4.2 mm long, nerves obscure, distinctly keeled, villous on keel and marginal nerves below, often slightly pubescent between nerves below. **Paleas:** glabrous, scabrous keels. **Awns:** none. **Anthers:** 0.3–1 mm long. **VEGETATIVE CHARACTERISTICS: Growth Habit:** few in a tuft, without rhizomes. **Culms:** stout, erect, 4–10 dm tall, sometimes taller. **Sheaths:** closed up to half their length, somewhat retrorsely scabrous, keeled. **Ligules:** 2–12 mm long, acute to acuminate. **Blades:** 2–10 mm wide, 10–20 cm long, flat, lax, scabrous. **HABITAT:** Open woods, moist areas. **COMMENTS:** Can be confused with *P. tracyi* but *P. occidentalis* lacks rhizomes.

Figure 273. *Poa occidentalis.*

252. *Poa palustris* L. (Fig. 274).

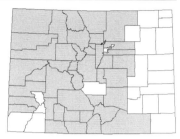

Synonyms: *Poa crocata* Michx., *Poa eyerdamii* Hultén, *Poa triflora* Gilib. **Vernacular Names:** Fowl blue-grass, marsh bluegrass, fowl meadow grass. **Life Span:** Perennial. **Origin:** Native. **Season:** C$_3$, cool season. **FLORAL CHARACTERISTICS: Inflorescence:** panicle, 13–40 cm long, loose, open, pyramidal or oblong; branches in rather distant whorls, 2–6 at a node, naked below. **Spikelets:** 3.2–5 mm long, 2- to 5-flowered, strongly laterally compressed; disarticulation above the glumes. **Glumes:** subulate to lanceolate, distinctly keeled, 1- to indistinctly 3-nerved; lower 1.7 mm long, tapering to a slender point; upper 2–3 mm long. **Florets:** most perfect. **Calluses:** cobwebby hairs present at base, sometimes scant. **Lemmas:** 2–3 mm long, lanceolate, indistinctly 5-nerved, tips bronze-colored, compressed-keeled, villous on keel and marginal nerves below. **Paleas:** keels scabrous, intercostal areas glabrous. **Awns:** none. **Anthers:** 1.3–1.8 mm long. **VEGETATIVE CHARACTERISTICS: Growth Habit:** loosely tufted; without rhizomes, although sometimes rooting at lower nodes and appearing rhizomatous. **Culms:** 2.5–12 dm tall, decumbent, bases purplish, glabrous. **Sheaths:** keeled, glabrous to sparsely retrorsely scabrous. **Ligules:** variable, 1–5 mm long. **Blades:** 1–3 mm wide, up to 20 cm long, flat, folded, or involute; tips slender, prow-shaped. **HABITAT:** Moist forest, meadows, and wet open ground. **COMMENTS:** Common species of mostly mid-elevations. Sometimes used for soil stabilization or waterfowl feed (Barkworth et al. 2007). A very variable species that is often difficult to identify.

253. *Poa pratensis* L. (Fig. 275).

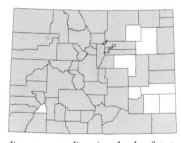

Synonyms: *Paneion pratense* (L.) Lunell, *Poa bourgeaei* E. Fourn.; subsp. *agassizensis*: *Poa agassizensis* Boivin & D. Löve, *Poa pratensis* L. subsp. *agassizensis* (Boivin & D. Löve) Taylor & MacBryde; subsp. *pratensis*: *Poa pratensis* L. var. *domestica* Laestad., *Poa pratensis* L. var. *gelida* (Roem. & Schult.) Böcher. **Vernacular Name:** Kentucky bluegrass. **Life Span:** Perennial. **Origin:** Introduced. **Season:** C$_3$, cool season. **FLORAL CHARACTERISTICS: Inflorescence:** panicle, 2–15 cm long, open, usually pyramidal; branches spreading or ascending, in whorls of 2–7. **Spikelets:** 3–7 mm long, mostly 2- to 5-flowered, strongly laterally compressed, green or purplish; disarticulation above the glumes. **Glumes:** distinctly shorter than adjacent lemmas, scabrous on keel; lower 1.5–4 mm long, 1- to 3-nerved; upper 2–4 mm long, 3-nerved. **Florets:** most perfect. **Calluses:** abundant cobwebby hairs present at base, hairs at least half the length of lemma. **Lemmas:** 2.5–6 mm long, 3- to 5-nerved, lanceolate, lateral nerves indistinct, compressed-keeled, villous on keel and marginal nerves below, glabrous between nerves. **Paleas:** scabrous, keels occasionally puberulent, intercostal areas narrow, generally glabrous but rarely sparsely hispid. **Awns:** none. **Anthers:** 1.2–2 mm long. **VEGETATIVE CHARACTERISTICS: Growth Habit:** sod-forming from extensive creeping rhizomes,

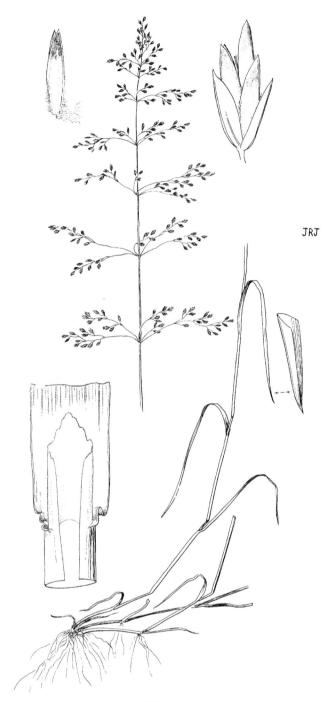

JRJ

Figure 274. *Poa palustris*.

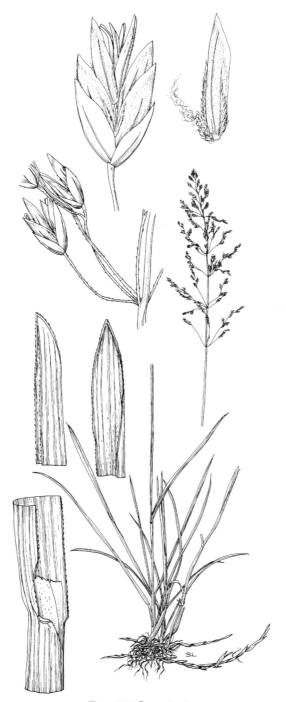

Figure 275. *Poa pratensis.*

sometimes single. **Culms:** 2–10 dm tall, erect, glabrous, somewhat compressed. **Sheaths:** compressed, closed from a quarter to half their length, mostly glabrous. **Ligules:** 0.2–3 mm long, truncate, entire or erose-ciliolate. **Blades:** basal leaves bright green, 2–3 mm wide, 5–25 cm long, abundant, soft, flat or folded, glabrous or slightly pubescent, tips strongly prow-shaped. **HABITAT:** Cultivated for lawns and pastures, escaped to moist sites generally below treeline. **COMMENTS:** The most widely used lawn grass in our area. There are 2 subspecies in Colorado: subsp. *agassizensis* (B. Boivin & D. Löve) Roy L. Taylor & MacBryde, and the typical subsp. *pratensis*. The two subspecies can be distinguished by this characteristic:

1. Panicle branches smooth or nearly so . . . . subsp. *agassizensis*
1. Panicle branches scabrous . . . . subsp. *pratensis*

254. *Poa reflexa* Vasey & Scribn. (Fig. 276).

Synonym: *Poa leptocoma* Trin. var. *reflexa* (Vasey & Scribn. *ex* Vasey) M. E. Jones. **Vernacular Name:** Nodding bluegrass. **Life Span:** Perennial. **Origin:** Native. **Season:** C_3, cool season. **FLORAL CHARACTERISTICS: Inflorescence:** panicle, 4–15 cm long, open; branches generally smooth to occasionally sparsely scabrous, mostly in remote pairs, lower ones with an abrupt geniculate-swollen base, sometimes single, usually strongly reflexed at maturity. **Spike-**

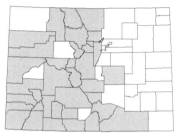

lets: 4–6 mm long, usually 2 to 5-flowered, strongly laterally compressed; disarticulation above the glumes. **Glumes:** lanceolate, distinctly keeled, subequal; lower 2.5–3 mm long, 1-nerved, narrowly lanceolate; upper 3–3.5 mm long, lanceolate, 3-nerved. **Florets:** most perfect. **Calluses:** copious cobwebby hairs present at base. **Lemmas:** 2–5 mm long, oblong-elliptic, purple-tinged, distinctly keeled, 5-nerved, villous on keel and marginal nerves below, generally glabrous between nerves. **Paleas:** keels scabrous, generally softly puberulent at midlength. **Awns:** none. **Anthers:** 0.6–1 mm long. **VEGETATIVE CHARACTERISTICS: Growth Habit:** solitary or in small tufts and rooting at nodes, without rhizomes. **Culms:** 1–6 dm tall, erect or decumbent, glabrous. **Sheaths:** glabrous to sometimes scabrous, closed about one- to two-thirds their length. **Ligules:** up to 3 mm long. **Blades:** 1–4 mm wide, 4–7 cm long, flat, bright green, glabrous, tip prow-shaped. **HABITAT:** Drier and often disturbed sites in subalpine forests, meadows, and low alpine sites. **COMMENTS:** Closely related to *P. leptocoma* and perhaps should be treated as a variety of that species. It differs from *P. leptocoma* in having hairs on the palea keels and lateral nerves of lemmas and smooth panicle branches (Barkworth et al. 2007).

255. *Poa secunda* J. Presl (Fig. 277).

Synonyms: Subsp. *juncifolia*: *Poa ampla* Merr., *Poa brachyglossa* Piper, *Poa confusa* Rydb., *Poa juncifolia* Scribn., *Poa juncifolia* Scribn. subsp. *porteri* Keck, *Poa juncifolia* Scribn. var. *ampla* (Merr.) Dorn, *Poa juncifolia* Scribn. var. *juncifolia* Scribn., *Poa nevadensis* Vasey *ex* Scribn., *Poa nevadensis* Vasey *ex* Scribn. var. *juncifolia* (Scribn.) Beetle; subsp. *secunda*: *Poa buckleyana* Nash, *Poa canbyi* (Scribn.) T. J. Howell, *Poa gracillima* Vasey, *Poa gracillima* Vasey var. *multnomae* (Piper) C. L. Hitchc., *Poa incurva* Scribn. & Williams, *Poa laevigata* Scribn.,

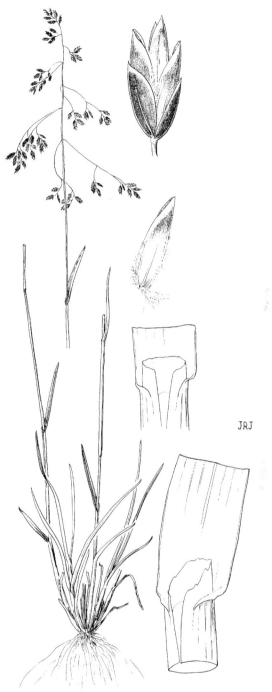

JRJ

Figure 276. *Poa reflexa*.

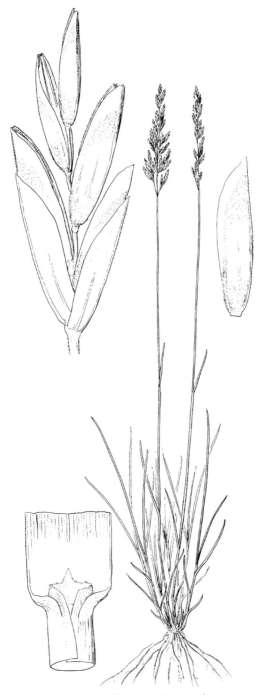

JRJ

Figure 277. *Poa secunda.*

Poa orcuttiana Vasey, *Poa sandbergii* Vasey, *Poa sca-brella* (Thurb.) Benth. *ex* Vasey, *Poa secunda* J. Presl subsp. *juncifolia* (Scribn.) Soreng, *Poa secunda* J. Presl subsp. *secunda* J. Presl, *Poa secunda* J. Presl var. *elongata* (Vasey) Dorn, *Poa secunda* J. Presl var. *incurva* (Scribn. & Williams) Beetle, *Poa secunda* J. Presl var. *stenophylla* (Vasey) Beetle, *Poa stenantha* Trin. var. *sandbergii* (Vasey) Boivin. **Vernacular Names:** Sandberg bluegrass, second bluegrass. **Life Span:** Perennial. **Origin:** Native. **Season:** C$_3$, cool season. **FLORAL CHARACTERISTICS: Inflorescence:** panicle, 2–25 cm long, erect or lax, contracted or somewhat open at anthesis; branches short, appressed. **Spikelets:** 4–10 mm long, usually 3- to 5-flowered, narrowly lanceolate, subterete to weakly laterally compressed, purple-tinged; disarticulation above the glumes. **Glumes:** keels indistinct, broadly lanceolate; lower 3–5 mm long, 1-nerved; upper 3.5–4.5 mm long, 3-nerved. **Florets:** perfect. **Calluses:** cobwebby hairs absent at base. **Lemmas:** 3.5–6 mm long, mostly rounded, weakly keeled, crisp-pubescent near base, especially on keel. **Paleas:** keels scabrous to glabrous to puberulent, to short-villous at midlength. **Awns:** none. **Anthers:** 1.5–3 mm long. **VEGETATIVE CHARACTERISTICS: Growth Habit:** densely tufted with short basal leaves, erect, rarely with rhizomes. **Culms:** 1–10 dm tall, glabrous, terete. **Sheaths:** closed no more than a quarter their length, glabrous to scabrous. **Ligules:** up to 6 mm long, truncate to long-acute. **Blades:** 1–3 mm wide, 3–10 cm long, soft, usually folded or involute, glabrous, tips prow-shaped. **HABITAT:** Very common in high deserts, dry forests, and mountain grasslands. **COMMENTS:** Previously known as *P. sandbergii*, among others. An important spring forage species (Barkworth et al. 2007). Two subspecies occur in the state (subsp. *secunda* and subsp. *juncifolia* [Scribn.] Bettle). The 2 subspecies can be distinguished by this characteristic:

1. Lemmas glabrous or minutely scabrous . . . . .subsp. *juncifolia*
1. Lemmas puberulent on the basal two-thirds . . . . subsp. *secunda*

256. *Poa stenantha* Trin. (Fig. 278).

Synonyms: *Poa chorizanthe* E. Desv., *Poa englishii* St. John & Hardin, *Poa flavicans* Griseb., *Poa macroclada* Rydb. **Vernacular Names:** Northern bluegrass, narrow-flower bluegrass. **Life Span:** Perennial. **Origin:** Native. **Season:** C$_3$, cool season. **FLORAL CHARACTERISTICS: Inflorescence:** panicle, 5–25 cm long, lax, open, sparse, pyramidal; branches 3–15 cm long, mostly spreading, distant in twos and threes, naked most of their length. **Spikelets:** 6–10 mm long, 2- to 7-flowered, lanceolate to narrowly ovate, laterally compressed; disarticulation above the glumes. **Glumes:** subequal, lanceolate, keeled; lower about 3–4 mm long, 1- to 3-nerved; upper 4–6.5 mm long, 3-nerved. **Florets:** perfect. **Calluses:** not webbed at base but crowned with hairs up to 2 mm long. **Lemmas:** 4–6 mm long, compressed-keeled, sericeous-villous on keel and marginal nerves below, glabrous to pubescent between nerves, long hairs present at base and along edge basally. **Paleas:** keels scabrous, sometimes puberulent at mid-

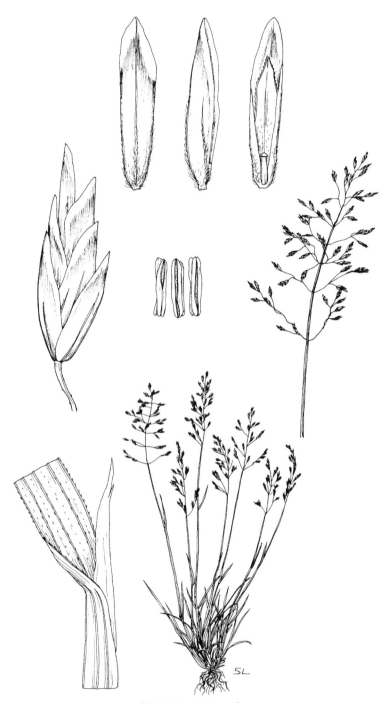

Figure 278. *Poa stenantha*.

length, intercostal areas glabrous to puberulent. **Awns:** none. **Anthers:** 1.2–2 mm long. **VEGETATIVE CHARACTERISTICS: Growth Habit:** tufted, erect or decumbent, without rhizomes. **Culms:** 2.5–6 dm tall. **Sheaths:** closed approximately a quarter of their length, smooth. **Ligules:** 2–5 mm long, acute, usually lacerate. **Blades:** 1–4 mm wide, up to 10 cm long, flat, folded, or loosely involute, lax, soft; cauline blades and flag leaf not noticeably reduced. **HABITAT:** Moist areas of middle elevations. **COMMENTS:** Weber and Wittmann (2001a, 2001b) list this species as *P. macroclada* Rydb., while Wingate (1994) does not include it. Only the typical variety (var. *stenantha*) occurs in Colorado.

257. *Poa tracyi* Vasey (Fig. 279).

Synonyms: *Poa autumnalis* Muhl. *ex* Elliott var. *robusta* (Vasey) Beal, *Poa flexuosa* Muhl. var. *occidentalis* Vasey, *Poa flexuosa* Muhl. var. *robusta* Vasey, *Poa lacustris* A. Heller, *Poa nervosa* (Hook.) Vasey var. *tracyi* (Vasey) Beal, *Poa occidentalis* (Vasey) Rydb. **Vernacular Name:** Tracy's bluegrass. **Life Span:** Perennial. **Origin:** Native. **Season:** C$_3$, cool season. **FLORAL CHARACTERISTICS: Inflorescence:** panicle, narrow, open, pyramidal, 15–30 cm long; branches spreading, becoming reflexed, not capillary; 2–5 per node, naked below middle. **Spikelets:** 4.5–7 mm long, 2- to 8-flowered, laterally compressed; disarticulation above the glumes. **Glumes:** unequal; lower generally 1-nerved, 1.5–3.5 mm long; upper 3-nerved, 2–5 mm long. **Florets:** pistillate to bisexual. **Calluses:** base sparsely cobwebby. **Lemmas:** 3–4 mm long, generally 5-nerved, keeled, keel and marginal nerves villous below. **Paleas:** keels scabrous, rarely puberulent at midlength. **Awns:** none. **Anthers:** vestigial (0.1–0.2 mm long) to (1.3) 2–3 mm long. **VEGETATIVE CHARACTERISTICS: Growth Habit:** tufted with short rhizomes. **Culms:** erect, to 4–10 dm tall. **Sheaths:** glabrous, compressed and strongly keeled with a distinct wing to 0.5 mm wide, margins fused half to nearly all their length. **Ligules:** 2–4.5 mm long, obtuse or acute. **Blades:** 3–5 mm wide, flat, lax. **HABITAT:** Coniferous forest openings in montane to subalpine communities. **COMMENTS:** Tall stature and broad blades make this species distinctive. Sometimes confused with *P. occidentalis* but differs in having rhizomes.

258. *Poa trivialis* L. (Fig. 280).

Synonyms: *Poa attica* Boiss. & Heldr., *Poa stolonifera* Hall. *ex* Muhlenb. **Vernacular Name:** Rough bluegrass. **Life Span:** Perennial. **Origin:** Introduced. **Season:** C$_3$, cool season. **FLORAL CHARACTERISTICS: Inflorescence:** panicle, 6–25 cm long, erect or lax, pyramidal, open; branches spreading or ascending, lower ones in fives. **Spikelets:** 2–4 mm long, strongly compressed, mostly 2- to 4-flowered; disarticulation above the glumes. **Glumes:** distinctly keeled, acute; lower 1.7–2.5 mm long, narrow and sickle-shaped, 1-nerved; upper 2–3 mm long, broader than lower. **Florets:** perfect. **Calluses:** cobwebby hairs present at base. **Lemmas:** 2.5–3.5 mm

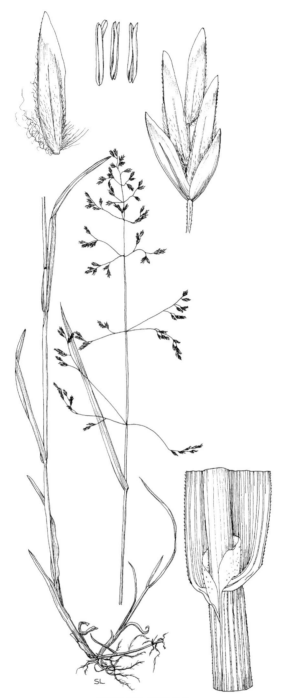

Figure 279. *Poa tracyi*.

Figure 280. *Poa trivialis*.

long, lanceolate, strongly keeled and prominently 5-nerved, glabrous except for sericeous hairs at base of keel. **Paleas:** keels smooth, muriculate, tuberculate, to scaberulous. **Awns:** none. **Anthers:** 1.3–2 mm long. **VEGETATIVE CHARACTERISTICS: Growth Habit:** rooting at nodes, sometimes stoloniferous, without rhizomes. **Culms:** 3–10 dm tall, decumbent, often strongly geniculate. **Sheaths:** retrorsely scabrous. **Ligules:** membranous, 3–10 mm long. **Blades:** 1.7–5 mm wide, up to 20 cm long, bright green, flat, scabrous. **HABITAT:** Lawn, pastures, and escaped and established in moist areas. **COMMENTS:** Uncommon. Several cultivars have been used for lawns and pastures (Barkworth et al. 2007).

259. *Poa wheeleri* Vasey (Fig. 281).

Synonym: *Poa nervosa* (Hook.) Vasey. **Vernacular Name:** Wheeler's bluegrass. **Life Span:** Perennial. **Origin:** Native. **Season:** C_3, cool season. **FLORAL CHARACTERISTICS: Inflorescence:** panicle, 4–18 cm long, open, pyramidal, apex erect or nodding; branches mostly in twos and threes, naked below, spreading, lax or reflexed. **Spikelets:** generally 6–10 mm long, 2- to 8-flowered, strongly compressed, loosely flowered with rachilla often visible, purplish; disarticulation above the glumes. **Glumes:** shorter than adjacent lemmas, lanceolate, distinctly keeled; lower 2.5–4 mm long, 1- to 3-nerved; upper 3–4.5 mm long, 3-nerved. **Florets:** pistillate, sometimes perfect. **Calluses:** cobwebby hairs absent. **Lemmas:** 3–6 mm

long, compressed-keeled, strongly nerved, glabrous or pubescent on nerves below. **Paleas:** keels scabrous, intercostal areas glabrous. **Awns:** none. **Anthers:** generally vestigial (0.1–0.2 mm long) or aborted at a later time and then up to 2 mm long, rarely normal. **VEGETATIVE CHARACTERISTICS: Growth Habit:** solitary to tufted, with creeping rhizomes. **Culms:** 3–6 dm tall, erect, terete to slightly compressed. **Sheaths:** glabrous or lower ones retrorsely pubescent and purplish. **Ligules:** up to 2 mm long. **Blades:** 2–4 mm wide, 2–7 cm long, cauline gradually reduced, sometimes middle blades are longest, flag leaf blade 1–10 cm long. **HABITAT:** Coniferous forests and meadows, montane to subalpine. **COMMENTS:** Rhizomes frequently missing from herbarium specimens. This species was *P. nervosa* (Hook.) Vasey in Weber and Wittmann (2001a, 2001b) and Wingate (1994). Barkworth and colleagues (2007) restrict *P. nervosa* to the Cascade Mountains and coastal ranges in the Pacific Northwest.

84. *Podagrostis* Scribn. & Merr.

Perennial, sometimes with short rhizomes. Sheaths open to base, smooth, glabrous, margins overlapping; ligules short, membranous; auricles absent. Inflorescence a terminal panicle, open or slightly closed. Disarticulation above the glumes and beneath floret. Spikelet with a single floret, rachilla prolonged beyond base of floret, sometimes absent. Glumes subequal or lower sometimes longer than upper, longer than floret, acute, 1-nerved, purplish. Lemma awnless or with awn up to 1.3 mm long, blunt, occasionally toothed, 3- to 5-nerved. Palea more than half as long as lemma, 2-nerved, thinner than lemma. Basic chromosome number, $x = 7$. Photosynthetic pathway, C_3.

Represented in Colorado by a single species. A segregate of *Agrostis*.

260. *Podagrostis humilis* (Vasey) Björkman (Fig. 282).

Synonyms: *Agrostis humilis* Vasey, *Agrostis thurberiana* Hitchc., *Podagrostis thurberiana* (Hitchc.) Hultén. **Vernacular Names:** Alpine bentgrass, Thurber bentgrass. **Life Span:** Perennial. **Origin:** Native. **Season:** C_3, cool season. **FLORAL CHARACTERISTICS: Inflorescence:** panicle, loosely contracted, 2–15 cm long, lax or sometimes drooping. **Spikelets:** 1-flowered, 1.5–2.5 mm long; rachilla occasionally extended behind palea as a bristle or up to 0.5 mm long, with a tuft of hairs at apex; disarticulation above the glumes. **Glumes:** subequal, 1.6–2.3 mm long, lanceolate, acute, usually purple, 1-nerved. **Lemmas:** 1.5–2.5 mm long, membranous, 5-nerved, awnless or short-awned, blunt, occasionally minutely toothed. **Paleas:** 0.9–1.6 mm long. **Awns:** lemmas occasionally with awn up to 1.3 mm long, attached subapically or near middle. **VEGETATIVE CHARACTERISTICS: Growth Habit:** low tufted, occasionally with short rhizomes. **Culms:** 0.5–6 dm long, ascending to erect, occasionally bases decumbent. **Sheaths:** open, smooth. **Ligules:** 0.5–1.5 (3) mm long, truncate to obtuse, usually lacerate.

Figure 281. *Poa wheeleri*.

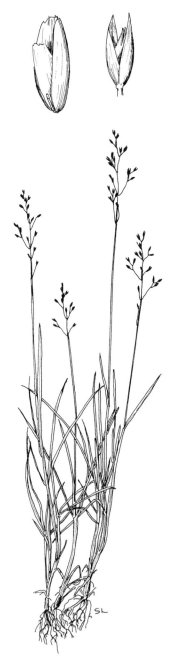

Figure 282. *Podagrostis humilis.*

Blades: mostly basal, flat, or involute. **HABITAT:** Wet meadows and bogs in alpine and subalpine. **COMMENTS:** Segregated into the genus *Podagrostis* based on the well-developed palea and extension of the rachilla.

85. *Polypogon* Desf.

Low to moderately tall annuals and perennials, with weak, decumbent-erect culms, these often rooting at lower nodes. Blades thin and flat. Inflorescence a dense, contracted panicle of small, 1-flowered spikelets, these disarticulating below the glumes and falling entire. Glumes about equal, 1-nerved, abruptly awned from an entire or notched apex. Lemma broad, smooth, and shining, mostly 5-nerved, much shorter than the glumes, awnless or with a short, delicate awn from the broad, often minutely toothed apex. Palea slightly shorter than lemma. Stamens 1–3. Basic chromosome number, $x = 7$. Photosynthetic pathway, C_3.

Represented in Colorado by 4 species.

1. Glumes with awns 3–10 mm long
2. Plants annual; glumes awn 4–10 mm long . . . . 263. *P. monspeliensis*
2. Plants perennial; glumes awn 1–3.2 mm long . . . . 262. *P. interruptus*
1. Glumes unawned to awned, awns to 3.2 mm long
3. Glumes unawned . . . . 264. *P. viridis*
3. Glumes awned, awns 1.5–3.2 mm long
4. Stipes 0.2–0.7 mm long; glumes not tapering to apex, acute to truncate, sometimes minutely lobed . . . . 262. *P. interruptus*
4. Stipes 1.5–2.5 mm long; glumes tapering to an acute, unlobed apex . . . .
. . . . 261. *P. elongatus*

261. *Polypogon elongatus* Kunth (Fig. 283).

Synonyms: *Alopecurus elongatus* (Kunth) Poir., *Chaetotropis elongata* (Kunth) Björkman, *Polypogon elongatus* Kunth. var. *strictus* E. Desv., *Polypogon intermedius* Carmich., *Vilfa acutiglumia* Steud. *ex* Lechler. **Vernacular Name:** Southern beardgrass. **Life Span:** Perennial. **Origin:** Native. **Season:** C_3, cool season. **FLORAL CHARACTERISTICS: Inflorescence:** panicle, 10–30 cm long, 0.5–3 cm wide, interrupted, dense, contracted to open, green to purple-tinged; branches  whorled, erect. **Spikelets:** laterally compressed to subterete, lanceolate, 1-flowered; disarticulation at base of stipe; stipe 1.5–2.5 mm long. **Glumes:** 3–5 mm long, subequal, hispid, tapering to an acute unlobed apices, awned. **Lemmas:** 0.8–1.5 mm long, glabrous, shiny, obtuse. **Awns:** glumes awned, 1–3 mm long; lemmas awned, 1–2 mm long, arising from midlength. **VEGETATIVE CHARACTERISTICS: Growth Habit:** tufted. **Culms:** generally decumbent or erect, to 10 dm tall. **Sheaths:** smooth, glabrous. **Ligules:** 4–8 mm long, apex truncate to obtuse, entire to lacerate. **Blades:** 10–30 cm long, 4–15 cm wide, flat. **HABITAT:** Introduced weed from Mexico and South America. **COMMENTS:** A single record from Boulder County; perhaps not a constant member of our flora.

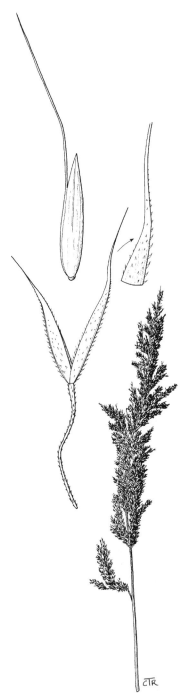

Figure 283. *Polypogon elongatus.*

262. *Polypogon interruptus* Kunth (Fig. 284).

Synonyms: *Alopecurus interruptus* (Kunth) Poir., *Polypogon brachyatherus* Phil., *Polypogon breviaristatus* Phil. **Vernacular Names:** Ditch beardgrass, ditch polypogon. **Life Span:** Perennial. **Origin:** Native. **Season:** C$_3$, cool season. **FLORAL CHARACTERISTICS: Inflorescence:** interrupted panicle, contracted to open, 3–15 cm long, 0.5–3 cm wide, green to purpletinged; branches whorled, erect. **Spikelets:** laterally compressed to subterete, lanceolate, 1-flowered; disarticulation at base of stipe; stipe 0.2–0.7 mm long. **Glumes:** 2–3 mm long, subequal, scabrous on keel, elliptic to narrowly lanceolate, not tapering toward apices, acute to truncate; apices occasionally minutely lobed to 0.1 mm. **Lemmas:** 0.8–1.5 mm long, glabrous, shiny, apices obtuse. **Awns:** glumes awned from sinus between lobes, 1.5–3.2 mm long; lemmas awned, 1–3.2 mm long. **VEGETATIVE CHARACTERISTICS: Growth Habit:** tufted. **Culms:** generally decumbent and rooting at lower nodes, occasionally erect, 2–8 dm tall. **Sheaths:** smooth, thick. **Ligules:** nerved membrane, scabrous-pubescent, 2–6 mm long, truncate to obtuse, entire to lacerate. **Blades:** flat, scabrous, 2.5–10.5 cm long, 3–6 mm wide. **HABITAT:** Found in Boulder-Denver area along streams, on sandy banks, and in Montezuma County. **COMMENTS:** This was previously misidentified as *P. elongatus* in Colorado (Weber and Wittmann 1992). Allred (2005) reports this is a sterile hybrid between *P. monspeliensis* and *Agrostis stolonifera*.

263. *Polypogon monspeliensis* (L.) Desf. (Fig. 285).

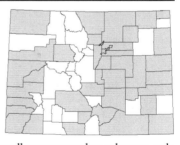

Synonyms: *Agrostis alopecuroides* Lam., *Alopecurus monspeliensis* L., *Phalaris aristata* Gouan *ex* P. Beauv., *Phalaris crinita* Forssk., *Phleum crinitum* Schreb., *Phleum monspeliense* (L.) Koeler, *Polypogon crinitus* (Schreb.) Nutt. **Vernacular Name:** Annual rabbitsfoot grass. **Life Span:** Annual. **Origin:** Introduced. **Season:** C$_3$, cool season. **FLORAL CHARACTERISTICS: Inflorescence:** compact to open panicle, occasionally lobed, narrowly elliptic, 1–17 cm long, 0.3–3 cm wide, greenish; branches appressed ascending. **Spikelets:** laterally compressed to subterete, pale green to tawny, 1-flowered; disarticulation at base of stipe; stipe 0.1–0.2 mm long. **Glumes:** 1–2.7 mm long, subequal, scabrous on keel, elliptic to narrowly lanceolate, apices rounded, bilobed, spreading upward. **Lemmas:** 0.5–1.5 mm long, glabrous, ovate, shiny, apices minutely toothed, awnless to short-awned. **Awns:** glumes awned from sinus between lobes, 4–10 mm long; lemmas unawned to short-awned, awn deciduous, 0.5–1 mm long. **VEGETATIVE CHARACTERISTICS: Growth Habit:** tufted. **Culms:** erect to ascending, geniculate basally, rooting at lower nodes, 0.5–6.5 dm tall. **Sheaths:** smooth to scabrous, uppermost sometimes inflated. **Ligules:** prominent, puberulent, 2.5–16 mm long, acute, lacerate-erose. **Blades:** generally flat, glabrous to scabrous, 1–20 cm long, 1–7 mm wide. **HABITAT:** Found at low elevations in wet, often alkaline areas, pond shores, and ditches. **COMMENTS:** Common,

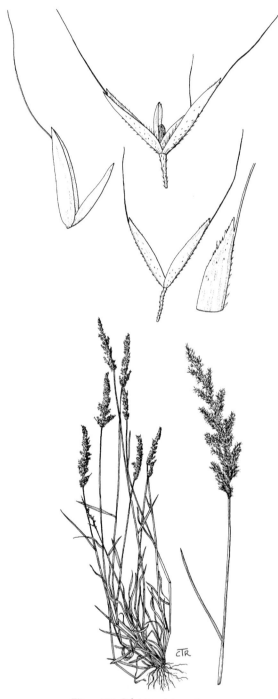

Figure 284. *Polypogon interruptus.*

JRJ

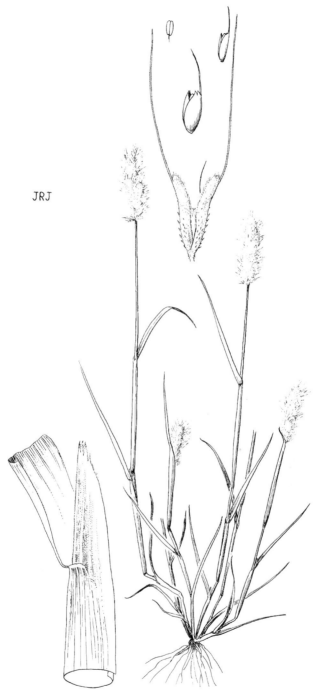

Figure 285. *Polypogon monspeliensis.*

sometimes used in dried floral arrangements. Occasionally hybridizes with *Agrostis stolonifera,* resulting in *P. interruptus,* according to Allred (2005).

264. *Polypogon viridis* (Gouan) Breistr. (Fig. 286).

Synonyms: *Agrostis semiverticillata* (Forssk.) C. Christens., *Agrostis verticillata* Vill., *Agrostis viridis* Gouan, *Polypogon semiverticillatus* (Forssk.) H. Hyl. **Vernacular Names:** Water beardgrass, beardless rabbitsfoot grass. **Life Span:** Perennial. **Origin:** Introduced. **Season:** C$_3$, cool season. **FLORAL CHARACTERISTICS: Inflorescence:** panicle, 2–10 cm long, 1–3.5 cm wide, compact to interrupted and lobed, pale green to purple-tinged; branches erect to widely spreading. **Spikelets:** laterally compressed to subterete, lanceolate, green to purple-tinged, 1-flowered; disarticulation at base of stipe; stipe 0.1–0.6 mm long. **Glumes:** 1–2.2 mm long, subequal, keel scabrous, elliptic, apices rounded to abruptly acute, unawned. **Lemmas:** about 1 mm long, glabrous, shiny, erose, unawned. **Awns:** none. **VEGETATIVE CHARACTERISTICS: Growth Habit:** tufted to stoloniferous. **Culms:** erect to decumbent basally and rooting at lower nodes, 1–9 dm tall. **Sheaths:** smooth, glabrous. **Ligules:** ciliate membrane, 1–5 mm long, erose. **Blades:** flat, lax to ascending, 2–13 cm long, 1–6 mm wide. **HABITAT:** Found in Baca and Montezuma counties in wet habitats (Weber and Wittmann 2001a, 2001b). **COMMENTS:** Rare. Sometimes confused with *Agrostis.*

86. *Puccinellia* Parl.

Low, tufted annuals or less frequently moderately tall perennials with culms decumbent and becoming stoloniferous and mat-forming or rhizomatous. Leaf sheaths free, margins overlapping. Auricles absent. Ligules membranous. Spikelets small, several-flowered, awnless, in open or contracted panicles. Disarticulation above the glumes and between florets. Glumes unequal, short, firm, lower 1-nerved (occasionally 3-nerved), upper usually 3-nerved. Lemmas firm, rounded on back, acute or obtuse and often erose at apex. Nerves of lemma usually 5, distinct or not, parallel and not converging at apex, occasionally with an additional pair of short, faint lateral nerves. Palea as long as or slightly shorter than lemma. Stamens 3. Caryopsis ovoid to oblong, usually adhering to palea. Basic chromosome number, $x = 7$. Photosynthetic pathway, C$_3$.

Represented in Colorado by 2 species and 1 subspecies.

1. Lower panicle branches reflexed at maturity; lemmas 1.4–2.2 mm long; anthers 0.4–0.8 mm long; ligules 0.8–1.2 mm long . . . . 265. *P. distans*
1. Lower panicle branch erect or occasionally descending at maturity; lemmas 2.2–3.5 mm long; anthers 0.6–2 mm long; ligule 1–3 mm long . . . . 266. *P. nuttalliana*

265. *Puccinellia distans* (Jacq.) Parl. (Fig. 287).

Synonyms: *Poa distans* Jacq., *Puccinellia distans* (Jacq.) Parl. var. *tenuis* (Uechtr.) Fern. & Weatherby, *Puccinellia retroflexa* (W. Curtis) Holmb., *Puccinellia suksdorfii* St. John. **Vernacular**

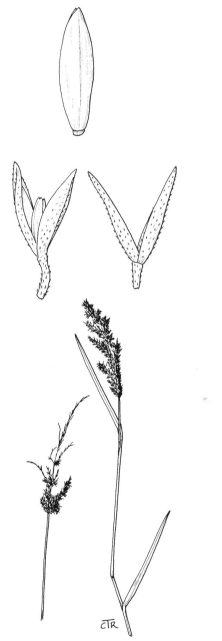

Figure 286. *Polypogon viridis.*

Figure 287. *Puccinellia distans.*

Names: Weeping alkali grass, European alkali grass.
Life Span: Perennial. **Origin:** Introduced. **Season:**
C_3, cool season. **FLORAL CHARACTERISTICS: Inflo-**
rescence: loosely pyramidal panicle, open, 5–20
cm long; branches fascicled, lower often reflexed at
maturity and spikelet-bearing above middle. **Spike-**
lets: slightly flattened, 3–7 mm long, 2- to 7-flowered;
disarticulation above the glumes and between florets. **Glumes:** ovate, unequal, erose; lower 1-nerved,

0.4–1.3 mm long; upper obscurely 3-nerved, 1.1–1.8 mm long. **Lemmas:** hyaline, broadly
ovate, greenish to purplish below, obscurely 5-nerved, 1.4–2.2 mm long, finely erose-ciliate,
rounded to truncate. **Awns:** none. **Anthers:** 0.4–0.8 mm long. **VEGETATIVE CHARACTERIS-**
TICS: Growth Habit: cespitose. **Culms:** erect to decumbent below, glabrous to scabrous, 1–5
dm tall. **Sheaths:** glabrous, prominently nerved. **Ligules:** entire, obtuse, 0.8–1.2 mm long.
Blades: flat to involute as mature, 2–10 cm long, 1–3 mm wide, glabrous to sparsely scabrous. **HABITAT:** Found in generally moist habitats in conjunction with heavy soils, some of
which are alkaline. **COMMENTS:** The only subspecies found in Colorado is the typical one
(subsp. *distans*). Barkworth and colleagues (2007) do not recognize subspecies.

266. *Puccinellia nuttalliana* (Schult.) Hitchc. (Fig. 288).

Synonyms: *Puccinellia airoides* (Nutt.) S. Wats. &
Coult., *Puccinellia cusickii* Weatherby. **Vernacular**
Name: Nuttall's alkali grass. **Life Span:** Perennial.
Origin: Native. **Season:** C_3, cool season. **FLORAL**
CHARACTERISTICS: Inflorescence: pyramidal to
elongate panicle, open, 10–25 cm long; branches
fascicled, lower often erect to occasionally descend-
ing at maturity and spikelet-bearing above middle.
Spikelets: slightly flattened, slender, 3–12 mm long,

3- to 7-flowered; disarticulation above the glumes and between florets. **Glumes:** narrowly
ovate, unequal, erose; lower 1-nerved, 0.5–1.5 mm long; upper obscurely 3-nerved, 1–
2.4 mm long. **Lemmas:** hyaline, broadly oblong, greenish to purplish below, obscurely
5-nerved, 2.2–3.5 mm long, finely erose-ciliate, obtuse. **Awns:** none. **Anthers:** 0.6–2 mm
long. **VEGETATIVE CHARACTERISTICS: Growth Habit:** cespitose. **Culms:** erect, 3–7 dm
tall. **Sheaths:** glabrous, prominently nerved. **Ligules:** entire, obtuse, 1–3 mm long. **Blades:**
generally involute, 3–10 cm long, 1–3 mm wide, glabrous to sparsely scabrous. **HABITAT:**
Found from low elevations to montane on moist, generally alkaline soils (Wingate 1994).
COMMENTS: Weber and Wittmann (2001a, 2001b) report this species as *P. airoides* (Nutt.)
S. Watson & Coult.

87. *Schedonorus* P. Beauv.

Perennial with short rhizomes, often tufted, and can grow to 2 m tall. Sheaths open,
round, smooth or scabrous. Ligules short, membranous. Auricles present, prominent, ciliate or not. Inflorescence panicles, terminal, large but simple; branches erect, glabrous,
smooth or scabrous. Disarticulation above the glumes and below florets. Spikelets large,

Figure 288. *Puccinellia nuttalliana.*

2–22 florets, terete or laterally compressed. Glumes subequal, shorter than lowermost floret, unawned. Lemmas firm, with 5–7 faint nerves, awnless or with a short awn that rarely exceeds 2 mm in length. Paleas slightly shorter and narrower than lemma, occasionally with cilia along upper margins. Basic chromosome number, $x = 7$. Photosynthetic pathway, C_3.

Represented in Colorado by 2 species. A segregate of *Festuca*, although recent genetic studies show it to be most closely related to *Lolium*.

1. Auricles ciliate, with at least 1–2 hairs along margins (check several leaves); lemmas generally scabrous to hispid, at least at apices, lemmas unawned or awned, to 4 mm long . . . . 267. *S. arundinaceus*
1. Auricles glabrous; lemmas generally smooth, occasionally slightly scabrous at apices, lemmas unawned or with a short mucro, to 0.2 mm long . . . . 268. *S. pratensis*

267. *Schedonorus arundinaceus* (Schreb.) Dumort. (Fig. 289).

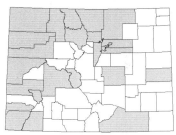

Synonyms: *Schedonorus phoenix* (Scop.) Holub, *Festuca arundinacea* Schreb., *Festuca elatior* L. subsp. *arundinacea* (Schreb.) Hack., *Festuca elatior* L. var. *arundinacea* (Schreb.) C.F.H. Wimmer, *Festuca fenas* Lag., *Lolium arundinaceum* (Schreb.) S. J. Darbyshire. **Vernacular Names:** Tall fescue, reed fescue, rye grass. **Life Span:** Perennial. **Origin:** Introduced. **Season:** C_3, cool season. **FLORAL CHARACTERISTICS: Inflorescence:** open to narrow panicle, slightly nodding, 15–30 cm long. **Spikelets:** 10–15 mm long, 3- to 9-flowered; disarticulation above the glumes. **Glumes:** unequal, lanceolate; lower 3–7 mm long, 1- to 3-nerved; upper 5–7 mm long, 3- to 5-nerved, margins membranous. **Lemmas:** 7–10 mm long, 5- to 7- or more obscure nerved, generally scabrous to hispidulous, at least at apices; apices hyaline, acute, awned or unawned. **Awns:** lemmas, when awned, to 4 mm long; awns terminal or slightly below apices. **VEGETATIVE CHARACTERISTICS: Growth Habit:** tufted, short rhizomatous. **Culms:** ascending to erect, stout, glabrous, 7–14 dm tall. **Sheaths:** smooth to glabrous, margins ciliate. **Auricles:** prominent, ciliate, appearing falcate. **Ligules:** 0.2–0.6 mm long. **Blades:** flat to slightly involute, 5–45 cm long, generally 3–8 mm wide, scabrous above. **HABITAT:** Found generally at lower elevations, occasionally in montane, in pastures, and along roadsides (Weber and Wittmann 2001a, 2001b; Wingate 1994). **COMMENTS:** Weber and Wittmann (2001a, 2001b) and Wingate (1994) include this species in the genus *Festuca*. Species is used for forage, soil stabilization, and turf. Fungi growing on the plant can sometimes cause alkaloid toxicity in livestock (Barkworth et al. 2007).

268. *Schedonorus pratensis* (Huds.) P. Beauv. (Fig. 290).

Synonyms: *Festuca elatior* L., *Festuca pratensis* Huds., *Lolium pratense* (Huds.) S. J. Darbyshire. **Vernacular Names:** Meadow ryegrass, meadow fescue, ryegrass. **Life Span:** Perennial. **Origin:** Introduced. **Season:** C_3, cool season. **FLORAL CHARACTERISTICS: Inflorescence:** narrow panicle, slightly nodding, 10–25 cm long. **Spikelets:** 10–15 mm long, 4- to 10-flowered;

Figure 289. *Schedonorus arundinaceus.*

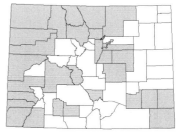

disarticulation above the glumes. **Glumes:** unequal, lanceolate; lower 2.4–4.5 mm long, 1-nerved; upper 3.5–5 mm long, 3- to 5-nerved, margins hyaline. **Lemmas:** 5–8 mm long, 5-nerved, apices hyaline, acute, generally smooth, occasionally slightly scabrous at apices, unawned to occasionally with short mucro. **Awns:** lemmas, when awned, with short mucro to 0.2 mm long. **VEGETATIVE CHARACTERISTICS: Growth Habit:** tufted, short rhizomatous. **Culms:** ascending to erect, stout, glabrous, 60–120 cm tall. **Sheaths:** smooth to glabrous. **Auricles:** prominent, glabrous, appearing falcate. **Ligules:** 0.3–0.5 mm long. **Blades:** flat to slightly involute, 5–20 cm long, generally 2–5 mm wide, generally scabrous. **HABITAT:** Found generally at lower elevations, occasionally in montane, in pastures, and along roadsides (Weber and Wittmann 2001a, 2001b; Wingate 1994). **COMMENTS:** Weber and Wittmann (2001a, 2001b) and Wingate (1994) include this species in the genus *Festuca*. Species was once a popular forage grass, but it is rarely planted now (Barkworth et al. 2007).

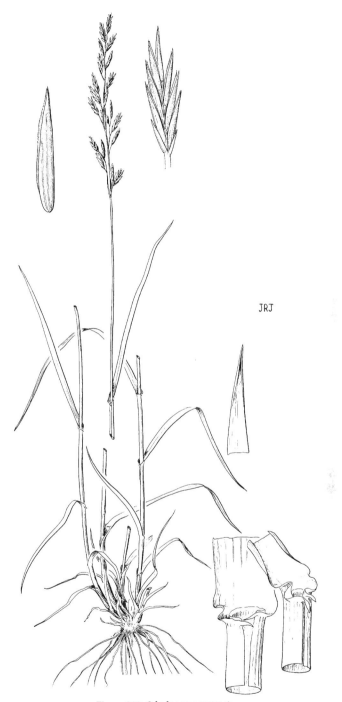

JRJ

Figure 290. *Schedonorus pratensis.*

88. *Sclerochloa* P. Beauv.

Low, tufted annual with culms mostly less than 10 cm long and inflorescences partially included in greatly enlarged upper leaf sheaths. Leaves scarcely differentiated into sheath and blade, without a ligule or thickened area at the junction. Inflorescence a spicate raceme or contracted panicle, 1–2 cm long. Spikelets awnless, 3-flowered, subsessile, and crowded on main inflorescence axis and on short branches. Disarticulation below the glumes, spikelet falling entire. Glumes broad, rather firm, the lower short, 3-nerved, the upper about half as long as spikelet, 7-nerved. Lemmas broad, rounded on back and obtuse at apex, membranous on margin above, 5-nerved. Palea hyaline. Basic chromosome number, $x = 7$. Photosynthetic pathway, C_3.

Represented in Colorado by a single species.

269. *Sclerochloa dura* (L.) P. Beauv. (Fig. 291).

Synonyms: *Cynosurus durus* L., *Eleusine dura* (L.) Lam., *Festuca dura* (L.) Vill., *Poa dura* (L.) Raspail, *Poa dura* (L.) Roth, *Poa dura* (L.) Scop., *Sesleria dura* (L.) Kunth. **Vernacular Names:** Common hardgrass, fairgrounds grass. **Life Span:** Annual. **Origin:** Introduced. **Season:** C_3, cool season. **FLORAL CHARACTERISTICS: Inflorescence:** generally racemes, occasionally reduced panicle, 1-sided, 1–4 cm long, often partially enclosed in subtending leaf sheath. **Spikelets:** laterally compressed, 3- to 4-flowered, 5–12 mm long; paleas dorsally compressed, subequal to equal to lemmas; disarticulation below the glumes. **Glumes:** unequal, glabrous, becoming indurate as mature, margins wide, hyaline, obtuse to emarginated; lower 1.4–3 mm long, 3- to 5-nerved; upper 2.6–5.4 mm long, 5- to 9-nerved. **Lemmas:** membranous, blunt, glabrous, 4.5–6 mm long, becoming successively smaller above lowermost, 7- to 9-nerved, nerves prominent and distally scabrous, rounded to emarginate. **Awns:** none. **VEGETATIVE CHARACTERISTICS: Growth Habit:** matted to solitary. **Culms:** prostrate to procumbent, 2–15 cm long, glabrous. **Sheaths:** open to closed for half of length, glabrous. **Ligules:** glabrous, acute, 0.7–2 mm long. **Blades:** flat to folded, 0.5–5 cm long, 1–4 mm wide, glabrous, strongly overlapping, usually exceeding inflorescence. **HABITAT:** Found in Garfield and Mesa counties. It is expected to be found elsewhere on the Western Slope in dryland communities. It can persist in severely compacted soils because it endures high foot and vehicular traffic (Weber and Wittmann 2001b). **COMMENTS:** Similar in appearance to *Poa annua.*

89. *Sphenopholis* Scribn.

Low, tufted annuals or short-lived perennials with soft, flat blades and usually contracted panicles of 2- (rarely 1-) to 3-flowered spikelets, these disarticulating below the glumes. Lower glume narrow, acute, 1-nerved (rarely 3-nerved); upper glume broad, obovate, 3- to 5-nerved, obtuse or broadly acute at apex, usually slightly shorter than lowermost lemma. Lemmas firm, faintly 5-nerved or nerves not visible, rounded on smooth or rugose back, awnless or less frequently with an awn from just below apex. Palea thin, membranous, colorless. Basic chromosome number, $x = 7$. Photosynthetic pathway, C_3.

JRJ

Figure 291. *Sclerochloa dura.*

Represented in Colorado by 2 species.

1. Upper glumes hooded, about half as wide as long; panicles erect, spikelike, spikelets dense . . . . 271. *S. obtusata*
1. Upper glumes not hooded, about a quarter as wide as long; panicles usually nodding, not spikelike; spikelets loosely arranged . . . . 270. *S. intermedia*

270. *Sphenopholis intermedia* (Rydb.) Rydb. (Fig. 292).

Synonyms: *Eatonia intermedia* Rydb., *Sphenopolis pallens* Scribn., *Vilfa alba* Buckley. **Vernacular Names:** Slender wedgescale, slender wedgegrass. **Life Span:** Perennial. **Origin:** Native. **Season:** C$_3$, cool season. **FLORAL CHARACTERISTICS: Inflorescence:** loose panicle, 7–20 cm long, usually nodding, not spikelike. **Spikelets:** 2–4 mm long, 1- to 4-flowered; rachilla prolonged beyond uppermost floret; disarticulation above uppermost floret and below the glumes.

Glumes: unequal; lower narrower than upper; upper 2–3 mm long, not hooded, about a quarter as wide as long, scabrous, rounded or truncate. **Lemmas:** lanceolate, smooth to

499

Figure 292. *Sphenopholis intermedia.*

scabrous, 2–3 mm long, obscurely 5-nerved. **Paleas:** colorless, shorter than lemmas. **Awns:** none. **VEGETATIVE CHARACTERISTICS: Growth Habit:** tufted to solitary. **Culms:** erect to geniculate basally, glabrous, hollow, 3.5–12 dm tall. **Sheaths:** open, glabrous to scabrous to pubescent. **Ligules:** membranous, glabrous to scabrous, finely erose-ciliate, 1.5–2.5 mm long. **Blades:** flat, 8–15 cm long, 2–6 mm wide, glabrous to scabrous, flat or involute. **HABITAT:** Found in wet to damp sites in forests, meadows, and waste areas. **COMMENTS:** This species is often confused with *Koeleria macrantha* but it differs in having a more open panicle and in disarticulating below the glumes (Barkworth et al. 2007).

271. *Sphenopholis obtusata* (Michx.) Scribn. (Fig. 293).

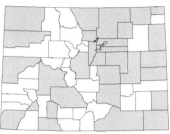

Synonyms: *Aira obtusata* Michx., *Sphenopholis obtusata* (Michx.) Scribn. var. *lobata* (Trin.) Scribn., *Sphenopholis obtusata* (Michx.) Scribn. var. *pubescens* (Scribn. & Merr.) Scribn. **Vernacular Names:** Prairie wedgescale, prairie wedgegrass. **Life Span:** Perennial or annual. **Origin:** Native. **Season:** C$_3$, cool season. **FLORAL CHARACTERISTICS: Inflorescence:** panicle, erect, generally spikelike, 5–15 cm long; spikelets dense. **Spikelets:** 1- to 2-flowered, 1.5–5 mm long; rachilla prolonged beyond uppermost floret; disarticulation above uppermost floret and below the glumes. **Glumes:** unequal; lower linear, 1.7–2.5 mm long, keel scabrous, 1-nerved; upper broadly obovate to oblanceolate, 2–2.8 mm long, hooded, about half as wide as long, apices rounded to subacute, 3-nerved, nerves prominent, scabrous. **Lemmas:** lanceolate, smooth to scabrous, 2–3 mm long, obscurely 5-nerved. **Paleas:** colorless, shorter than lemmas. **Awns:** none. **VEGETATIVE CHARACTERISTICS: Growth Habit:** tufted to solitary. **Culms:** erect to geniculate basally, glabrous, hollow, 3–9 dm tall. **Sheaths:** open, glabrous to scabrous to pubescent. **Ligules:** membranous, glabrous to scabrous, finely erose-ciliate, 1.5–3.5 mm long. **Blades:** flat, 5–20 cm long, 1.5–6 mm wide, glabrous to scabrous. **HABITAT:** Found in moist woodlands and gulches from foothill to montane communities over Colorado. **COMMENTS:** This genus looks much like *Poa* or *Koeleria* except for the very broadly obovate upper glume.

90. *Torreyochloa* G. L. Church

Perennials, strongly rhizomatous (in ours). Sheaths free but overlapping, cross-septate; ligules membranous, erose-lacerate; auricles absent. Inflorescence a terminal, open panicle. Disarticulation above the glumes and between florets. Spikelets distal on flexuous branches, with 3–7 florets. Glumes shorter than lowermost floret, unequal, lower 1-nerved, upper 3-nerved. Lemmas firm, rounded on back, 5- to 7-nerved, nerves not convergent at apex, frequently purplish, usually truncate, awnless. Palea about as long as lemma. Basic chromosome number, $x = 7$. Photosynthetic pathway, C$_3$.

Represented in Colorado by a single species.

This genus has been included in *Puccinellia* and *Glyceria*, although it is not closely related to them. It can be distinguished from *Glyceria* by its open sheaths and from *Puccinellia* by its distinct and converging nerves.

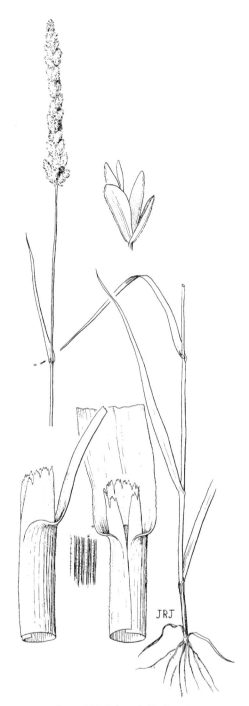

Figure 293. *Sphenopholis obtusata*.

272. *Torreyochloa pallida* (Torr.) G. L. Church (Fig. 294).

Synonyms: Var. *pauciflora*: *Glyceria pauciflora* J. Presl, *Puccinellia pauciflora* (J. Presl) Munz, *Puccinellia pauciflora* (J. Presl) Munz var. *holmii* (Beal) C. L. Hitchc., *Puccinellia pauciflora* (J. Presl) Munz var. *microtheca* (Buckley) C. L. Hitchc., *Torreyochloa pauciflora* (J. Presl) G. L. Church, *Torreyochloa pauciflora* (J. Presl) G. L. Church var. *holmii* (Beal) Taylor & MacBryde, *Torreyochloa pauciflora* (J. Presl) G. L. Church var. *microtheca* (Buckley) Taylor & MacBryde. **Vernacular Names:** Pale false mannagrass, weak mannagrass. **Life Span:** Perennial. **Origin:** Native. **Season:** C$_3$, cool season. **FLORAL CHARACTERISTICS: Inflorescence:** panicle, obovoid to ovoid to conic, 5–25 cm long, 1.8–16 cm wide; branches scabrous, stiff to flexuous, reflexed to erect. **Spikelets:** laterally compressed to terete, 3.6–6.9 mm long, 2- to 8-flowered; disarticulation above the glumes. **Glumes:** unequal, rounded to weakly keeled; lower 0.7–2.1 mm long, 1-nerved; upper 0.9–2.7 mm long, generally 3-nerved. **Lemmas:** rounded to weakly keeled, 2–3.6 mm long, occasionally pubescent, 7- to 9-nerved, generally parallel, truncate to acute. **Paleas:** subequal to lemma. **Awns:** none. **VEGETATIVE CHARACTERISTICS: Growth Habit:** rhizomatous. **Culms:** decumbent to erect to occasionally matted, 2–15 dm tall; internodes hollow. **Sheaths:** open. **Ligules:** membranous, 2–9 mm long, truncate to acute to attenuate. **Blades:** flat, 1.5–17.5 mm wide. **HABITAT:** Found in subalpine communities along pond margins. **COMMENTS:** The variety found in Colorado is var. *pauciflora* (J. Presl) J. I. Davis, the most common variety in the western United States (Barkworth et al. 2007).

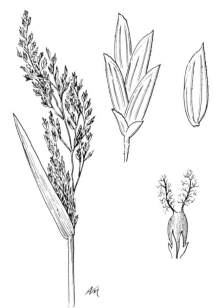

Figure 294. *Torreyochloa pallida*.

91. *Trisetum* Pers.

Tufted perennials and a few annuals, with slender culms, flat blades, and usually narrow panicles of 2-flowered (rarely 3- to 4-flowered) spikelets. Rachilla usually villous, prolonged above uppermost floret, disarticulating above the glumes and between florets or, in a few species, below the glumes. Glumes thin, nearly equal to very unequal, 1- to 3-nerved, acute, awnless, one or both usually equaling or exceeding florets in length. Lemmas 5-nerved, bifid at membranous apex, mostly with a straight or bent awn from base of notch but occasionally awnless, usually puberulent at base. Basic chromosome number, $x = 7$. Photosynthetic pathway, C_3.

Represented in Colorado by 4 species.

1. Plants annual . . . . 275. *T. interruptum*
1. Plants perennial
   2. Plants rhizomatous, culms usually single . . . . 274. *T. flavescens*
   2. Plants not rhizomatous, culms clumped
      3. Glumes generally subequal, both lanceolate, upper less than twice as wide as first . . .
      . . . . 276. *T. spicatum*
      3. Glumes generally unequal, although subequal on occasion; lower subulate to linear-lanceolate; upper at least twice as wide as lower
         4. Upper glumes equaling or longer than lowest floret; panicles yellowish brown . . .
         . . . . 274. *T. flavescens*
         4. Upper glumes shorter than lowest floret; panicles green or tan . . . .
         . . . . 273. *T. canescens*

273. *Trisetum canescens* Buckley (Fig. 295).

Synonyms: None. **Vernacular Name:** Tall trisetum. **Life Span:** Perennial. **Origin:** Native. **Season:** C_3, cool season. **FLORAL CHARACTERISTICS: Inflorescence:** panicle, narrow, erect or nodding, 10–25 cm long, green or tan or purple-tinged; branches erect to slightly divergent, spikelet bearing their entire length. **Spikelets:** 7–9 mm long, 2- to 4-flowered; disarticulation above the glumes. **Glumes:** unequal or occasionally subequal; lower 3–5 mm long, narrow, linear-  lanceolate to subulate, acute to acuminate; upper 5–9 mm long, shorter than lowermost floret, at least twice as wide as lower glume, broadly lanceolate to obovate, widest at or below middle, acute. **Lemmas:** glabrous, 5–7 mm long, 5-nerved, apices bifid with narrow teeth, setaceous, awned. **Paleas:** as long as or slightly longer than lemma. **Awns:** lemmas awned, awn geniculate, 7–14 mm long, exserted, arising from the upper third of lemmas. **VEGETATIVE CHARACTERISTICS: Growth Habit:** cespitose. **Culms:** 4–15 dm tall, clumped, ascending to erect, slender, hollow, puberulent to densely pilose or villous. **Sheaths:** open to base, frequently pubescent to pilose, less commonly smooth. **Ligules:** 1.5–6 mm long, membranous, erose, finely ciliate, rounded to truncate. **Blades:** flat, 1–3 dm long, 4–10 mm wide, flat, sometimes with scattered hairs. **HABITAT:** Usually near or along stream banks or in mesic forests. Most common in ponderosa pine and spruce-fir forests. **COMMENTS:**

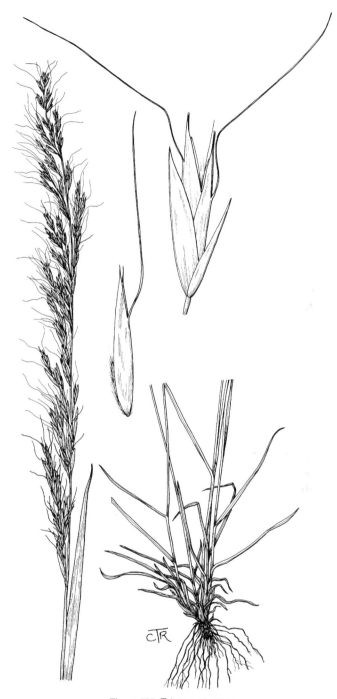

Figure 295. *Trisetum canescens.*

Rare, not reported in Weber and Wittmann (2001a, 2001b) or Wingate (1994). Much more common along the Pacific Coast into Idaho and Montana.

274. *Trisetum flavescens* (L.) P. Beauv. (Fig. 296).

Synonyms: *Avena flavescens* L., *Rebentischia flavescens* Opiz, *Trisetaria flavescens* (L.) Baumg., *Trisetaria flavescens* (L.) Maire, *Trisetum pratense* Pers. **Vernacular Name:** Yellow oatgrass. **Life Span:** Perennial. **Origin:** Introduced. **Season:** C$_3$, cool season. **FLORAL CHARACTERISTICS: Inflorescence:** panicle, 5–20 cm long, 2–7 cm wide, narrow, erect or nodding, yellowish brown and sometimes purplish. **Spikelets:** 4–8 mm long, 2- to 4-flowered; disarticulation above the glumes. **Glumes:** unequal, shiny; lower 2–5 mm long, narrowly lanceolate to subulate; upper 4–7 mm long, equaling or longer than lowermost floret, twice as wide as lower, acute. **Lemmas:** 3–6 mm long, ovate-lanceolate, minutely pubescent, 5-nerved, apices bifid or bicuspidate; teeth extending into conspicuous bristles 3–6 mm long, awned. **Paleas:** shorter than lemmas. **Awns:** lemmas awned, geniculate, tightly twisted, 3–9 mm long, exserted, arising from upper third of lemmas. **VEGETATIVE CHARACTERISTICS: Growth Habit:** generally cespitose but occasionally rhizomatous, rhizomes sometimes short. **Culms:** generally 5–10 dm or more tall, solitary or clumped, erect to decumbent, glabrous, sometimes scabrous to pubescent near upper nodes, hollow. **Sheaths:** open to base, throats usually with hairs over 2 mm long. **Ligules:** membranous, 0.5–2 mm long, obtuse, lacerate, sometimes ciliolate with hairs to 0.5 mm long. **Blades:** 5–20 cm long, 2–7 mm wide, flat or involute, pilose to slightly pubescent. **HABITAT:** Found in seeded pastures, along roadsides, and as a weed in croplands. Introduced as a forage species. **COMMENTS:** Weber and Wittmann (2001a, 2001b) and Wingate (1994) do not report this species. Barkworth and colleagues (2007) report it contains calcinogenic glycosides, which can lead to vitamin D toxicity in livestock. Only reported from Larimer County, and it may not be a constant member of our flora.

275. *Trisetum interruptum* Buckley (Fig. 297).

Synonym: *Sphenopholis interrupta* (Buckley) Scribn. **Vernacular Name:** Prairie trisetum. **Life Span:** Annual. **Origin:** Native. **Season:** C$_3$, cool season. **FLORAL CHARACTERISTICS: Inflorescence:** panicle, 2–15 cm long, 0.5–1.5 cm wide, narrow, spikelike, often interrupted in the lower third, green or tan. **Spikelets:** 2- to 3-flowered, 3–6 mm long; often paired, with 1 subsessile and 1 pediceled. **Glumes:** subequal; about as long as lowest lemma; lower 0.5–1 mm wide, 3-nerved, acuminate or sometimes apiculate; upper about twice as wide as lower, elliptical or oblanceolate, acuminate; disarticulation below the glumes. **Lemmas:** 3–5 mm long, scabrous, 5-nerved, apices bifid with narrow teeth. **Paleas:** usually two-thirds

Figure 296. *Trisetum flavescens.*

Figure 297. *Trisetum interruptum*.

as long as lemmas. **Awns:** lemmas awned, awns geniculate, 4–8 mm long, exserted, arising from midlength to just below teeth. **VEGETATIVE CHARACTERISTICS: Growth Habit:** cespitose. **Culms:** 1–5 dm tall, puberulent to densely pilose or villous, ascending to erect, slender, hollow. **Sheaths:** open to base, scabrous to pilose. **Ligules:** 1–2.5 mm long, membranous, erose, finely ciliate, truncate. **Blades:** flat, 3–12 cm long, 1–6 mm wide, pilose to scaberulous. **HABITAT:** It has been found in open arid environments from Arizona to Louisiana. It is often weedy. **COMMENTS:** Weber and Wittmann (2001a, 2001b) and Wingate (1994) did not list this species. There is only a single specimen collected from Hinsdale County, and perhaps it is not a constant member of our flora. This species is sometimes included in *Sphenopholis* because its spikelets disarticulate below the glumes (Barkworth et al. 2007).

276. *Trisetum spicatum* (L.) K. Richt. (Fig. 298).

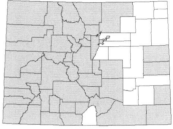

Synonyms: *Aira spicata* L., *Trisetum molle* Kunth, *Trisetum montanum* Vasey, *Trisetum montanum* Vasey var. *shearii* Louis-Marie, *Trisetum spicatum* (L.) Richter subsp. *alaskanum* (Nash) Hultén, *Trisetum spicatum* (L.) Richter subsp. *congdonii* (Scribn. & Merr.) Hultén, *Trisetum spicatum* (L.) Richter subsp. *majus* (Rydb.) Hultén, *Trisetum spicatum* (L.) Richter subsp. *molle* (Kunth) Hultén, *Trisetum spicatum* (L.) Richter subsp. *montanum* (Vasey) W. A. Weber, *Trisetum spicatum* (L.) Richter subsp. *pilosiglume* (Fern.) Hultén, *Trisetum spicatum* (L.) Richter var. *alaskanum* (Nash) Malte *ex* Louis-Marie, *Trisetum spicatum* (L.) Richter var. *congdonii* (Scribn. & Merr.) Hitchc., *Trisetum spicatum* (L.) Richter var. *maidenii* (Gandog.) Fern., *Trisetum spicatum* (L.) Richter var. *majus* (Rydb.) Farw., *Trisetum spicatum* (L.) Richter var. *molle* (Kunth) Beal, *Trisetum spicatum* (L.) Richter var. *pilosiglume* Fern., *Trisetum spicatum* (L.) Richter var. *spicatiforme* Hultén, *Trisetum spicatum* (L.) Richter var. *villosissimum* (Lange) Louis-Marie, *Trisetum subspicatum* (L.) P. Beauv., *Trisetum triflorum* (Bigelow) Á. & D. Löve, *Trisetum triflorum* (Bigelow) Á. & D. Löve subsp. *molle* (Kunth) Á. & D. Löve, *Trisetum villosissimum* (Lange) Louis-Marie. **Vernacular Names:** Spike trisetum, narrow false oats. **Life Span:** Perennial. **Origin:** Native. **Season:** C$_3$, cool season. **FLORAL CHARACTERISTICS: Inflorescence:** panicle, 5–50 cm long, 0.5–5 cm wide, narrow, spikelike or open usually at anthesis, frequently interrupted basally, silvery, shiny. **Spikelets:** 5–7.5 mm long, 2- or occasionally 3-flowered, purplish to tawny to silvery in color; disarticulation above the glumes. **Glumes:** lanceolate, generally subequal, on occasion unequal, shiny, acute to acuminate; lower 3–5 mm long, 1-nerved; upper 4–7 mm long, 3-nerved, as long as or longer than adjacent florets, less than twice as wide as lower. **Lemmas:** 3–6 mm long, 5-nerved, apices bifid with narrow teeth. **Paleas:** subequal to lemma, bifid. **Awns:** lemmas awned, awns geniculate, 3–9 mm long, exserted, arising from upper third of lemmas. **VEGETATIVE CHARACTERISTICS: Growth Habit:** cespitose. **Culms:** 1–12 dm tall, puberulent to densely pilose or villous, ascending to erect, slender, hollow. **Sheaths:** open to base, variously pubescent to occasionally glabrous. **Ligules:** 0.5–4 mm long, membranous, erose, finely ciliate, 1.5–4 mm long, truncate to rounded. **Blades:** 10–20 cm long, rarely to 40 cm long; 1–6 mm wide, flat, folded, or involute, pilose to scabrous. **HABITAT:** Found in meadows at middle elevations extending into the alpine. **COMMENTS:** An important forage species in the mountains. Weber and Wittmann

Figure 298. *Trisetum spicatum.*

(2001a, 2001b) and Wingate (1994) recognize *T. montanum* Vasey, but this is probably an extreme phase of *T. spicatum*. No subspecies are recognized here.

92. *Vahlodea* Fr.

Perennial, loosely cespitose. Sheaths open; ligules membranous; auricles absent. Inflorescence a terminal panicle, open or closed; branches often flexuous. Spikelets distal on panicle branches; usually with 2 fertile florets and sometimes with reduced florets distally; rachilla not prolonged. Disarticulation above the glumes and below florets. Glumes equaling or exceeding florets, membranous, acute or acuminate. Lemmas ovate, obscurely 5- to 7-nerved, awned. Awn from near middle of the lemma back, twisted, geniculate, visible between the glumes. Paleas almost equaling lemmas. Basic chromosome number, $x = 7$. Photosynthetic pathway, C_3.

Represented in Colorado by the only species in the genus. Sometimes included in *Deschampsia*.

277. *Vahlodea atropurpurea* (Wahlenb.) Fr. *ex* Hartm. (Fig. 299).

Synonyms: *Aira atropurpurea* Wahlenb., *Deschampsia atropurpurea* (Wahlenb.) Scheele, *Deschampsia atropurpurea* (Wahlenb.) Scheele var. *latifolia* (Hook.) Scribn. *ex* Macoun, *Deschampsia atropurpurea* (Wahlenb.) Scheele var. *paramushirensis* Kudo, *Deschampsia atropurpurea* (Wahlenb.) Scheele var. *payettii* Lepage, *Deschampsia pacifica* Tatew. & Ohwi, *Vahlodea atropurpurea* (Wahlenb.) Fr. *ex* Hartman subsp. *latifolia* (Hook.) Porsild, *Vahlodea atropurpurea* (Wahlenb.) Fr. *ex* Hartm. subsp. *paramushirensis* (Kudo) Hultén, *Vahlodea flexuosa* (Honda) Ohwi, *Vahlodea latifolia* (Hook.) Hultén. **Vernacular Name:** Mountain hairgrass. **Life Span:** Perennial. **Origin:** Native. **Season:** C_3, cool season. **FLORAL CHARACTERISTICS: Inflorescence:** closed to open panicle, 3–20 cm long; branches capillary, generally flexuous. **Spikelets:** generally 2-flowered, 4–7 mm long; disarticulation above the glumes. **Glumes:** membranous, subequal to equal, keels and marginal nerves generally glabrous to scabrous, acute to acuminate; lower 4–5 mm long, 1-nerved; upper 4–5.5 mm long, 3-nerved. **Lemmas:** 1.8–3 mm long, obscurely 5-nerved, ovate, scabrous, ciliate, awned. **Paleas:** subequal to lemmas. **Awns:** lemmas awned, attachment near middle of back, awns twisted, geniculate, 2–4 mm long, and visible between glumes. **VEGETATIVE CHARACTERISTICS: Growth Habit:** loosely cespitose. **Culms:** erect, 1.5–8 dm tall. **Sheaths:** open but closed at base, basal sheaths generally retrorsely hirsute but occasionally glabrous, upper sheaths glabrous to scabrous. **Ligules:** membranous, occasionally ciliate or lacerate, 0.8–3.5 mm long, rounded to truncate. **Blades:** flat, 1–30 cm long, 1–8.5 mm wide, glabrous to pilose. **HABITAT:** Found in subalpine meadows and in rocky gorges on moist ledges (Weber and Wittmann 2001a, 2001b). **COMMENTS:** Weber and Wittmann (2001a, 2001b) report that the Colorado plants belong to subsp. *latifolia* (Hooker) Porsild.; however, no subspecies are included here. Sometimes this species is included in *Deschampsia,* but it differs in having loosely cespitose shoots, in lacking a rachilla prolongation beyond the distal floret (or one not more than 0.5 mm long) and certain caryopsis characteristics (Barkworth et al. 2007).

Figure 299. *Vahlodea atropurpurea.*

92. *Vulpia* C. C. Gmel.

Tufted annuals with narrow blades, usually contracted, spikelike panicles, with 3 to many florets. Disarticulation above the glumes and between florets. Glumes narrow, lanceolate or acuminate, 1- to 3-nerved, lower often very short. Lemmas rounded on back, inconspicuously 5-nerved, tapering to a fine awn or merely acuminate. Caryopsis cylindrical and elongate. Anthers usually 1, infrequently 3, per flower. Basic chromosome number, $x = 7$. Photosynthetic pathway, C_3.

Represented in Colorado by a single species and 3 varieties.

278. *Vulpia octoflora* (Walter) Rydb. (Fig. 300).

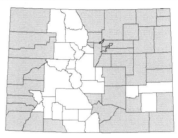

Synonyms: Var. *glauca: Festuca gracilenta* Buckley, *Festuca octoflora* Walt. var. *glauca* (Nutt.) Fern., *Vulpia octoflora* (Walter) Rydb. var. *hirtella: Festuca octoflora* Walter subsp. *hirtella* Piper, *Festuca octoflora* Walter var. *hirtella* (Piper) Piper *ex* Hitchc.; var. *octoflora: Festuca octoflora* Walter, *Festuca octoflora* Walter var. *aristulata* Torr. *ex* L. H. Dewey. **Vernacular Name:** Sixweeks fescue. **Life Span:** Annual. **Origin:** Native. **Season:** C_3, cool season. **FLORAL CHARACTERISTICS:**
Inflorescence: narrow, erect panicle or occasionally a raceme, 2–5 cm long; 1–2 branches per node. **Spikelets:** 4–13 mm long, 6- to 12-flowered, laterally compressed; disarticulation above the glumes. **Glumes:** linear-lanceolate to subulate, unequal; lower 2–4.5 mm long, 1- to 3-nerved, shorter than upper; upper 2.5–7 mm long, 3-nerved. **Lemmas:** 3–7 mm long, lanceolate, back rounded, margins inrolled, glabrous to scabrous to pubescent, obscurely 5-nerved, tapering into short awn. **Paleas:** slightly shorter than lemmas, sometimes with a bifid apices with short teeth. **Awns:** lemmas short-awned, awn 0.3–9 mm long. **VEGETATIVE CHARACTERISTICS: Growth Habit:** solitary to tufted. **Culms:** slender, erect to occasionally geniculate at lower nodes, 0.5–6 dm tall, glabrous to pubescent. **Sheaths:** smooth to finely pubescent. **Ligules:** erose-ciliate, 0.2–1 mm long at center. **Blades:** narrow, flat or involute, 1–5 cm long, to 1.2 mm wide, upper surface glabrous to finely pubescent. **HABITAT:** Found at low elevations on disturbed soils over Colorado. **COMMENTS:** Previously placed in *Festuca*. Three varieties are found in Colorado: var. *glauca* (Nutt.) Fern., var. *hirtella* (Piper) Henr., and var. *octoflora*. The varieties can be distinguished by these characteristics:

1. Spikelets 4–6.5 mm long; awns of lowermost lemmas 0.3–3 mm long . . . . var. *glauca*
1. Spikelets 5.5–13 mm long; awns of lowermost lemmas 2.5–9 mm long
   2. Lemmas glabrous to slightly scabrous, often scabrous on margins . . . . . . var. *octoflora*
   2. Lemmas scabrous to pubescent . . . . var. *hirtella*

13. STIPEAE

Perennials, usually cespitose or infrequently rhizomatous. Sheath open; ligule membranous or a fringe of hairs; auricles absent. Inflorescence a simple panicle. Spikelets with a single floret; callus well developed and often bearded. Disarticulation above the glumes. Glumes usually papery or membranous; faintly nerved. Lemma usually indurate or occasionally membranous, obscurely nerved, awned from tip.

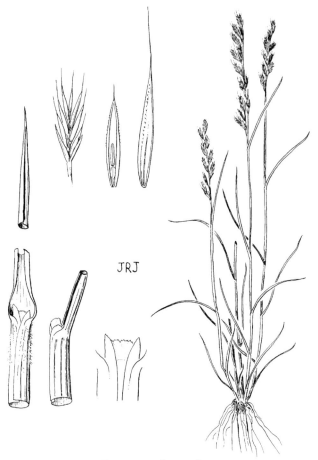

Figure 300. *Vulpia octoflora.*

94. *Achnatherum* P. Beauv.

Cespitose perennials, sometimes with short rhizomes. Inflorescence a terminal, usually contracted panicle of 1-flowered, awned spikelets. Disarticulation above the glumes. Glumes longer than florets, acute to acuminate, and usually tapering to a hairlike tip. Lemma membranous to coriaceous, usually pubescent, and terminating in a prominent awn. Palea one-third to as long as lemma, usually pubescent. Basic chromosome numbers, $x = 10$ and 11. Photosynthetic pathway, C_3.

Represented in Colorado by 11 species and 3 subspecies. At some time or another, all of the following have been included in *Stipa, Oryzopsis,* or both.

1. Awns with at least some hairs 0.5–8 mm long; callus sharp . . . . 283. *A. occidentale*
1. Awns with basal segments scabrous or with hairs less than 0.5 mm long; callus sharp or blunt

2. Lemmas evenly long-pubescent throughout, hairs 1.2–6 mm long, hairs on lemma body not evidently shorter than those of apices

   3. Awns deciduous (sometimes tardily so)

      4. Panicle branches terminating in a pair of conspicuously divaricate spikelets having subequal pedicels; florets 2.5–4.5 mm long . . . . 280. *A. hymenoides*

      4. Panicle branches terminating in a single spikelet or, if paired, spikelets not conspicuously divaricate; florets more than 4.5 mm long . . . . 289. *A. webberi*

   3. Awn persistent . . . . 285. *A. pinetorum*

2. Lemmas glabrous or with hairs 0.2–2 mm long at middle, glabrous or hairy distally, hairs at middle often shorter than those at apices

     5. Apical lemma hairs 2–3 mm long, usually 1 or more mm longer than those at middle
     . . . . 288. *A. scribneri*

     5. Apical lemma hairs absent or less than 2.2 mm long, usually less than 1 mm longer than those at middle

        6. Terminal awn segments flexuous . . . . 279. *A. aridum*

        6. Terminal awn segments straight

           7. Panicle branches flexuous, ascending to strongly divergent; spikelets divergent to pendulous . . . . 286. *A. richardsonii*

           7. Panicle branches straight, generally appressed to ascending, occasionally divergent; spikelets appressed to branches

              8. Flag leaves with a densely pubescent collar, hairs 0.5–2 mm long . . . .
              . . . . 287. *A. robustum*

              8. Flag leaves glabrous or sparsely pubescent on collar, hairs shorter than 0.5 mm long

                  9. Glumes subequal, lower glumes exceeding upper by 1 mm or less

                      10. Paleas one- to two-thirds as long as lemmas; lemma hairs (at middle) 0.5–1 mm long; awns 19–45 mm long . . . . 282. *A. nelsonii*

                      10. Paleas three-quarters to four-fifths (nine-tenths) as long as lemmas; lemma hairs (at middle) up to 0.5 mm long; awns 12–25 mm long . . . . .
                      . . . . 281. *A. lettermanii*

                  9. Glumes unequal, lower exceeding upper by 1–4 mm . . . 284. *A. perplexum*

---

279. *Achnatherum aridum* (M. E. Jones) Barkworth (Fig. 301).

Synonym: *Stipa arida* M. E. Jones. **Vernacular Name:** Mormon needlegrass. **Life Span:** Perennial. **Origin:** Native. **Season:** C_3, cool season. **FLORAL CHARACTERISTICS: Inflorescence:** panicle, 4–15 cm long, narrow, compact, pale or silvery. **Spikelets:** large, 1-flowered, 7–12 mm long; disarticulation above the glumes. **Glumes:** long-acuminate, papery; lower 8–15 mm long, 3-nerved; upper 7–10 mm long, indistinctly 5-nerved. **Florets:** 3–4.5 mm long. **Calluses:**

sharp. **Lemmas:** 4–6 mm long, appressed pubescent along margins and on lower half with short, white hairs 0.2–0.5 mm long, often glabrous above. **Paleas:** 2–3 mm long, 1/2 to 3/4 as long as lemma, pubescent. **Awns:** originating from a distinctly darkened joint, 4–8

Figure 301. *Achnatherum aridum.*

cm long, persistent, capillary, scaberulous, loosely twisted for first 1–2 cm, flexuous above. **VEGETATIVE CHARACTERISTICS: Growth Habit:** cespitose, densely tufted. **Culms:** 3–10 dm tall, erect, scaberulous below nodes. **Sheaths:** open, persistent, often densely covering base of plant, sometimes pubescent below collar; collars of basal sheaths sometimes with a tuft of hairs on sides; collars on upper leaves glabrous to sparsely pubescent. **Ligules:** membranous, truncate, lacerate at tips, decurrent, 0.2–1.5 mm long. **Blades:** mostly involute-filiform, 10–20 cm long, culm blades sometimes flat, 1–2 mm wide. **HABITAT:** Rocky shadscale and sagebrush flats and foothills up to piñon-juniper woodlands (Cronquist et al. 1977). Weber and Wittmann (2001b) report the species from the Western Slope in desert canyons south of the Colorado River Valley. **COMMENTS:** *Achnatherum* is a segregate of *Stipa*.

280. *Achnatherum hymenoides* (Roem. & Schult.) Barkworth (Fig. 302).

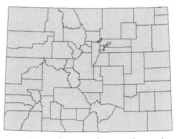

Synonyms: *Eriocoma cuspidata* Nutt., *Oryzopsis hymenoides* (Roem. & Schult.) Ricker *ex* Piper, *Stipa hymenoides* Roem. & Schult. **Vernacular Name:** Indian ricegrass. **Life Span:** Perennial. **Origin:** Native. **Season:** C_3, cool season. **FLORAL CHARACTERISTICS: Inflorescence:** open panicle, 5–20 cm or more long, 8–14 mm wide; branches in pairs; branchlets dichotomous, flexuous, divaricately spreading, terminating with a pair of conspicuously divaricate spikelets having subequal pedicels. **Spikelets:** large, 1-flowered, 4–8 mm long; disarticulation above the glumes. **Glumes:** subequal, 5–9 mm long, 3- to 5- to 7-nerved, margins hyaline, puberulent to glabrous; lower slightly longer than upper, sometimes with a long acuminate tips. **Florets:** 3–4.5 mm long. **Calluses:** sharp. **Lemmas:** 3–5 mm long; dark brown to nearly black, shiny, densely pubescent with whitish hairs nearly exceeding length of the glumes, hairs 2.5–6 mm long. **Paleas:** 3 mm long, from 3/4 to as long as lemmas, indurate, glabrous. **Awns:** lemma awns 6–10 mm long, straight, scabrous, early deciduous. **VEGETATIVE CHARACTERISTICS: Growth Habit:** cespitose, densely tufted. **Culms:** 3–9 dm tall, erect, hollow with thick walls, glabrous. **Sheaths:** glabrous to puberulent, open, persistent, and becoming papery and fibrous in older clumps; collars glabrous and occasionally with tufts of hairs on sides to 1 mm. **Ligules:** 2–8 mm long, acuminate, entire or becoming lacerate with age. **Blades:** usually convolute, 0.1–1 mm wide. **HABITAT:** Abundant and varied in its distribution. Cronquist and colleagues (1977) report it from deserts, sagebrush plains, piñon-juniper woodlands, to subalpine, typically on sandy soils where it can dominate. Weber and Wittmann (2001b) state the taxon is common of shale and clayey soils in desert steppe communities. Stubbendieck and colleagues (1992) list it as moderately salt- and alkali-tolerant. **COMMENTS:** This taxon has recently been moved to *Achnatherum* and was historically known as *Oryzopsis hymenoides* (Roem. & Schult.) Ricker. Reportedly a good forage for all kinds of animals, especially in winter because the plant cures well (Stubbendieck, Hatch, and Landholt 2003). Also, it was extensively used as a food source by Native Americans. Frequently used in reseeding efforts. This and *Elymus elymoides* are the only species reported from every county in the state.

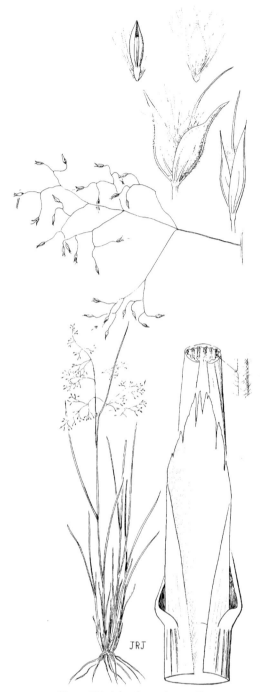

Figure 302. *Achnatherum hymenoides.*

281. *Achnatherum lettermanii* (Vasey) Barkworth (Fig. 303).

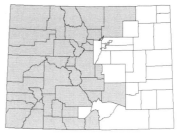

Synonym: *Stipa lettermanii* Vasey. **Vernacular Name:** Letterman's needlegrass. **Life Span:** Perennial. **Origin:** Native. **Season:** C$_3$, cool season. **FLORAL CHARACTERISTICS: Inflorescence:** contracted panicle, 6–20 cm long, 0.5–1 cm wide; branches erect, straight, with few spikelets. **Spikelets:** large, appressed to branches, 1-flowered, 5–10 mm long; disarticulation above the glumes. **Glumes:** subequal, upper slightly shorter than lower, 6–10 mm long, acuminate, glabrous or rarely scaberulous, often purple with hyaline margins. **Florets:** 4.5–6 mm long. **Calluses:** blunt. **Lemmas:** 4–7 mm long; covered with long, straight, soft hairs to 0.5 mm long, hairs at summit longer, to 1.5 (2) mm long. **Paleas:** 3–4 mm long, 3/4 to 9/10 as long as lemma, apices with hairs to 1 mm long. **Awns:** lemmas awned, persistent, 12–25 mm long, twice-geniculate; lower segment twisted and scaberulous; upper segment glabrous, straight. **VEGETATIVE CHARACTERISTICS: Growth Habit:** cespitose, densely tufted and forming large clumps. **Culms:** 2–9 dm tall, glabrous or occasionally minutely scabrous. **Sheaths:** glabrous or scabrous; collars glabrous or sparsely pubescent, even on sides, collars of flag leaves glabrous. **Ligules:** 0.2–2 mm long, truncate, decurrent. **Blades:** involute, filiform, if flat, no more than 2 mm wide, 10–20 cm long, roughly pubescent above, glabrous to scaberulous below. **HABITAT:** Cronquist and colleagues (1977) report this taxon in sagebrush and open coniferous and aspen woods at middle and subalpine levels. **COMMENTS:** Abundant, particularly on the Western Slope. Moved to *Achnatherum* from *Stipa*, and older

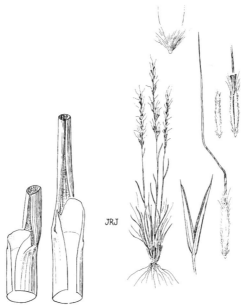

JRJ

Figure 303. *Achnatherum lettermanii*.

519

texts refer to the species as *S. lettermanii* Vasey. The long paleas in this species differentiates it from *A. nelsonii* (Scribn.) Barkworth and *A. perplexum* Hoge & Barkworth (Barkworth et al. 2007).

282. *Achnatherum nelsonii* (Scribn.) Barkworth (Fig. 304).

Synonyms: Subsp. *dorei* (Barkworth & Maze) Barkworth: *Achnatherum nelsonii* (Scribn.) Barkworth var. *dorei* (Barkworth & Maze) Dorn, *Stipa columbiana* auct. *non* Macoun [misapplied], *Stipa minor* (Vasey) Scribn., *Stipa nelsonii* Scribn. subsp. *dorei* Barkworth & Maze, *Stipa nelsonii* Scribn. var. *dorei* (Barkworth & Maze) Dorn, *Stipa occidentalis* Thurb. *ex* S. Wats. var. *minor sensu* C. L. Hitchc.: subsp. *nelsonii*: *Stipa colum-* 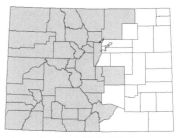 *biana* Macoun var. *nelsonii* (Scribn.) St. John, *Stipa nel-sonii* Scribn., *Stipa occidentalis* Thurb. *ex* S. Wats. var. *nelsonii* (Scribn.) C. L. Hitchc., *Stipa williamsii* Scribn. **Vernacular Names:** Columbia needlegrass, Nelson's needlegrass, Dore's needlegrass, subalpine needlegrass. **Life Span:** Perennial. **Origin:** Native. **Season:** C_3, cool season. **FLORAL CHARACTERISTICS: Inflorescence:** loosely contracted panicle, 10–35 cm long, 1–2 cm wide; branches relatively short and appressed, with numerous spikelets; lower branches sometimes removed from the rest. **Spikelets:** 1-flowered, 7–14 mm long; disarticulation above the glumes. **Glumes:** subequal, 6–14 mm long, acuminate to gradually extending into a short awn, purplish, papery, scaberulous, 3-nerved; lower exceeding upper by less than a millimeter. **Florets:** 4.5–7 mm long. **Calluses:** blunt to sharp, up to 1 mm long. **Lemmas:** 5–7 mm long, appressed pubescent (hairs 0.5–1 mm long) with longer hairs at apices (hairs to 2 mm long), awned. **Paleas:** 2–4 mm long, 1/3 to 2/3 as long as lemmas, pubescent, hairs not exceeding apices. **Awns:** 19–45 mm long, twice-geniculate, segments straight; lower segment twisted and scabrous; upper segment not twisted and scabrous to glabrous. **VEGETATIVE CHARACTERISTICS: Growth Habit:** cespitose, tufted. **Culms:** 3–18 dm tall, erect; nodes may be purple, glabrous; most leaves basal. **Sheaths:** glabrous to densely pubescent, longer hairs at apex, strongly ribbed; collar glabrous to somewhat pubescent without hair tufts on sides, collar of flag leaves glabrous to sparsely pubescent. **Ligules:** membranous, only 0.2–2 mm long, truncate, entire or erose, sometimes longer on sides, decurrent. **Blades:** typically involute but, if flat, up to 5 mm wide, 10–20 cm long, glabrous to densely pubescent. **HABITAT:** Weber and Wittmann (1992) report the species from dry montane forests on both sides of the Rockies. **COMMENTS:** This species, previously included in *Stipa* as *S. columbiana* Macoun var. *nelsonii* (Scribn.) Hitchc. *Stipa williamsii* Scribn. reported by Wingate (1994), belongs here. There has been some confusion of this taxon with *A. lemmonii* (Vasey) Barkworth in the past, but *A. nelsonii* can be separated by its broader leaves, terete florets, apical lemma lobes that are membranous and flexible, and shorter paleas. *A. lemmonii* is found in eastern Utah along the Colorado border and may occur in the Grand Junction area. Two subspecies occur in Colorado, the typical one (subsp. *nelsonii*) and subsp. *dorei* (Barkworth & Maze) Dorn. They can be separated by these characteristics:

1. Calluses blunt, awns 19–31 mm long . . . . subsp. *dorei*
1. Calluses sharp, awns 19–45 mm long . . . . subsp. *nelsonii*

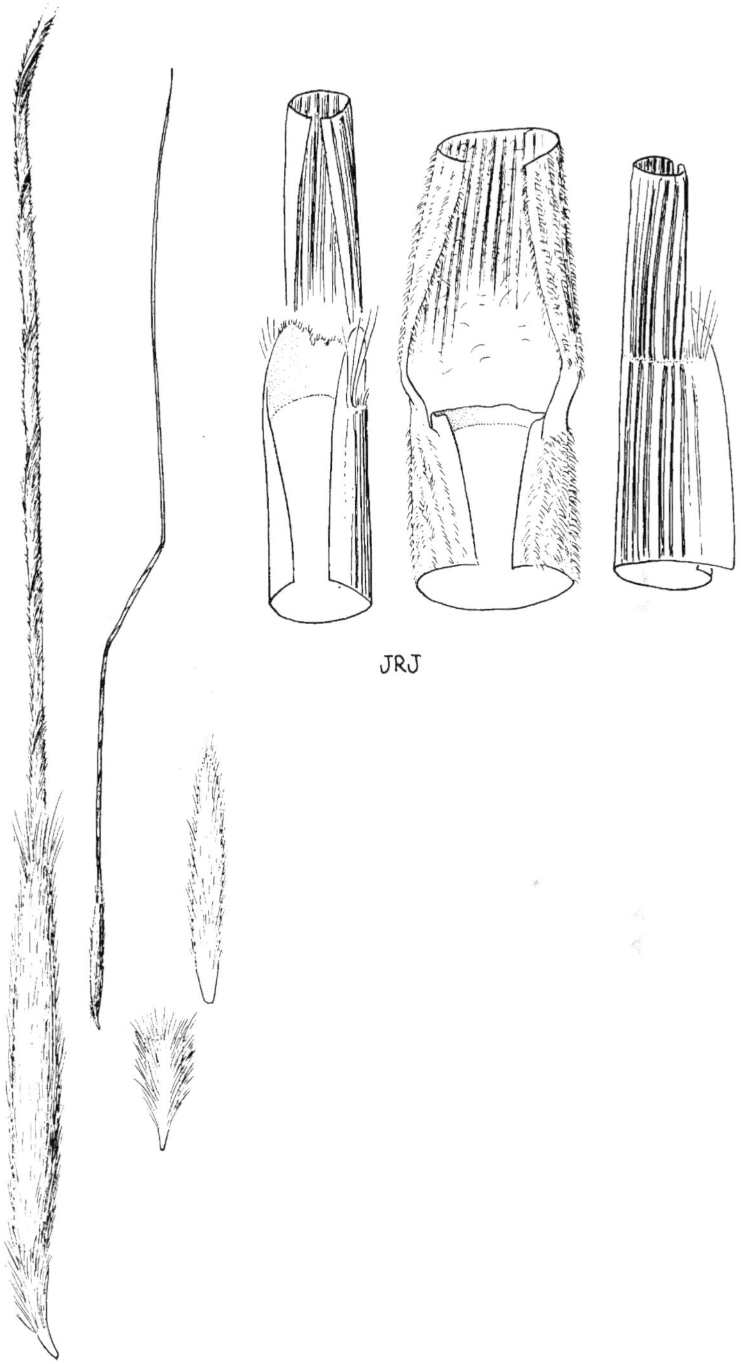

JRJ

Figure 304. *Achnatherum nelsonii*.

283. *Achnatherum occidentale* (Thurb.) Barkworth (Fig. 305).

Synonyms: *Achnatherum occidentale* (Thurb.) Barkworth subsp. *pubescens* (Vasey) Barkworth: *Stipa elmeri* Piper & Brodie *ex* Scribn., *Stipa occidentalis* Thurb. *ex* S. Wats. var. *pubescens* (Vasey) Maze, Taylor & MacBryde. **Vernacular Name:** Western needlegrass. **Life Span:** Perennial. **Origin:** Native. **Season:** C$_3$, cool season. **FLORAL CHARACTERISTICS: Inflorescence:** panicle, 5–30 cm long, 0.5–1.5 cm wide, narrow; branches appressed or ascending, short, few-flowered. **Spikelets:** 1-flowered, 7–12 mm long; disarticulation above the glumes. **Glumes:** subequal, 9–15 mm long, less than 1 mm wide. **Florets:** 5.5–7.5 mm long. **Calluses:** sharp. **Lemmas:** 5–8 mm long and less than 1 mm wide, evenly hairy, hairs 0.5–1.5 mm long at middle, apical hairs longer than those below, terete. **Paleas:** 2–3.5 mm long, 2/3 to 3/5 as long as lemma, hairs at tip extending beyond apices. **Awns:** 1.5–5.5 cm long, twice-geniculate, 2 lower segments evenly hairy, at least some hairs 0.5–8 mm long, terminal segment glabrous or partly to wholly pilose. **VEGETATIVE CHARACTERISTICS: Growth Habit:** tightly cespitose. **Culms:** 2 dm to over a meter tall, glabrous or puberulent to densely pubescent. **Sheaths:** open, glabrous or puberulent to densely pubescent, brown to gray-brown when older; collars often with hair tufts on sides. **Ligules:** membranous, 0.3–1.5 mm long. **Blades:** flat or convolute, 0.5–3 mm wide, flexuous, apices green, not sharp. **HABITAT:** Relatively widespread and common in the mountains and on the Western Slope. **COMMENTS:** The Colorado plants belong to subsp. *pubescens* (Vasey) Barkworth, which differs from the typical variety by pubescent leaves, presence of semi-plumose hairs over the first and second joints of the lemma awns, and sometimes dwarfed appearance of the plants. It is the most common, widespread, and variable subspecies. This species is not listed by either Wingate (1994) or Weber and Wittmann (2001a, 2001b).

284. *Achnatherum perplexum* Hoge & Barkworth (Fig. 306).

Synonym: *Stipa perplexa* J. Wipff & S. D. Jones. **Vernacular Name:** Perplexing needlegrass. **Life Span:** Perennial. **Origin:** Native. **Season:** C$_3$, cool season. **FLORAL CHARACTERISTICS: Inflorescence:** panicle, 10–25 cm long, 0.5–1.5 cm wide, narrow; branches ascending to appressed; spikelets appressed to branches. **Spikelets:** 10–15 mm long, 1-flowered, disarticulation above the glumes. **Glumes:** unequal, lanceolate, acuminate, glabrous, both 3- to 5-nerved; lower 10–15 mm long, exceeding upper by 1–4 mm. **Florets:** 5.5–11 mm long. **Calluses:** blunt. **Lemmas:** 5–11 mm long, uniformly hairy, hairs about 1 mm long, apical hairs 1–2 mm long, apices shallowly lobed. **Paleas:** 3–6 mm long, 1/2 to 2/3 the length of lemma, hairy, hairs not exceeding apices, rounded. **Awns:** 1–2 cm long, persistent, twice-geniculate; lower segments scabrous; upper segment straight. **VEGETATIVE CHARACTERISTICS: Growth Habit:** cespitose. **Culms:** 3–9 dm tall, puberulent below nodes, otherwise glabrous.

Figure 305. *Achnatherum occidentale.*

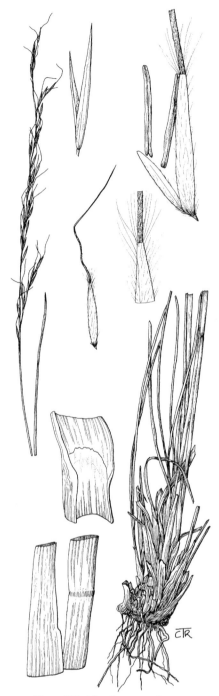

Figure 306. *Achnatherum perplexum*.

Sheaths: mostly glabrous; collars glabrous on all leaves. **Ligules:** membranous; basal very short, 0.1–0.5 mm long; upper 0.2–3.5 mm long, rounded to acute. **Blades:** linear, to 30 cm long, 1–3 mm wide. **HABITAT:** Grows in piñon-juniper communities. **COMMENTS:** Rare, or specimens have been misidentified as other species (Barkworth et al. 2007). Wingate (1994) and Weber and Wittmann (2001b) do not report this species.

285. *Achnatherum pinetorum* (M. E. Jones) Barkworth (Fig. 307).

Synonym: *Stipa pinetorum* M. E. Jones. **Vernacular Names:** Pine needlegrass, pinewoods needlegrass. **Life Span:** Perennial. **Origin:** Native. **Season:** C_3, cool season. **FLORAL CHARACTERISTICS: Inflorescence:** panicle, 4–20 cm long, 0.5–1 cm wide, narrow, contracted; branches appressed or ascending, short, few-flowered. **Spikelets:** 7–12 mm long, 1-flowered; disarticulation above the glumes. **Glumes:** subequal, 7–12 mm long, lanceolate, acuminate, glabrous, often purple, 1- to 3-nerved. **Florets:** 3.5–5.5 mm long. **Calluses:** sharp. **Lemmas:** 3–6 mm long, terete, densely and evenly pilose, at midlength of body 1.5–3.5 mm long; hairs at apices longer, to 5 mm long; apices with 0.3–2 mm long lobes. **Paleas:** 2.5–4 mm long, from 2/3 to as long as lemma, hairy. **Awns:** 1–2 cm long, persistent, twice-geniculate; lower segment twisted; upper segment not twisted; lower 2 segments scabrous, usually deep purple to nearly black. **VEGETATIVE CHARACTERISTICS: Growth Habit:** cespitose, grows in circular tufts when center dies in old bunches. **Culms:** 1–5 dm tall; puberulent below nodes, otherwise glabrous. **Sheaths:** often shiny, persistent at base of plant, glabrous. **Ligules:** membranous, very short, 0.1–0.8 mm long, truncate or rounded, sometimes decurrent. **Blades:** usually involute, 5–12 cm long, mostly basal, flexuous, scabrous to puberulent. **HABITAT:** Wingate (1994) reported the species from low foothills on both the Eastern and Western slopes of the Rockies, while Weber and Wittmann (2001b) lists it from the Gunnison Basin. Cronquist and colleagues (1977) found it on high sagebrush slopes and pine woodlands from middle elevations to nearly timberline. **COMMENTS:** *A. pinetorum* differs from the closely related *A. lettermanii* in its sharp calluses and longer lemma hairs (Barkworth et al. 2007).

286. *Achnatherum richardsonii* (Link) Barkworth (Fig. 308).

Synonym: *Stipa richardsonii* Link. **Vernacular Name:** Richardson needlegrass. **Life Span:** Perennial. **Origin:** Native. **Season:** C_3, cool season. **FLORAL CHARACTERISTICS: Inflorescence:** open panicle, 7–25 cm long, 7–15 cm wide; branches flexuous, divergent, mostly in pairs at rachis nodes, naked below, spikelets divergent, on upper 1/4 of branch. **Spikelets:** 1-flowered, 6–10 mm long; disarticulation above the glumes. **Glumes:** lower 2–3 mm longer than upper. **Florets:** 5–6 mm long. **Calluses:** blunt. **Lemmas:** about 5 mm long, dark brown, terete,

Figure 307. *Achnatherum pinetorum*.

Figure 308. *Achnatherum richardsonii.*

evenly pubescent below to 0.5 mm long, often glabrate above. **Paleas:** 2–3.5 mm long, about half as long as lemma, hairy, hairs not exceeding apices, rounded. **Awns:** 1–3 cm long, persistent, twice-geniculate; lower segments twisted and with appressed pubescence to 0.1 mm long; upper segment straight, scabrous. **VEGETATIVE CHARACTERISTICS: Growth Habit:** cespitose, tufted. **Culms:** 3–10 dm long, erect, glabrous. **Sheaths:** glabrous or lower portions sometimes puberulent, occasionally with a few pilose hairs at apex. **Ligules:** 0.3–1 mm long, truncate. **Blades:** mostly basal, involute-filiform, 15–25 cm long, scabrous. **HABITAT:** Harrington (1954) reports the species from bottomlands and rocky or wooded slopes, while Weber and Wittmann (2001a, 2001b) report it from the Eastern and Western slopes of the Rockies in lodgepole pine forests. **COMMENTS:** Less common than some other members of the genus in Colorado. It is readily recognized by its combination of flexuous panicle branches, drooping spikelets, and straight distal awn segments (Barkworth et al. 2007).

287. *Achnatherum robustum* (Vasey) Barkworth (Fig. 309).

Synonyms: *Stipa robusta* (Vasey) Scribn., *Stipa vaseyi* Scribn. **Vernacular Name:** Sleepygrass. **Life Span:** Perennial. **Origin:** Native. **Season:** C$_3$, cool season. **FLORAL CHARACTERISTICS: Inflorescence:** narrow panicle with appressed branches, 15–30 cm long, 0.8–3.5 cm wide; branches with appressed spikelets. **Spikelets:** 1-flowered, 8–11 mm long; disarticulation above the glumes. **Glumes:** subequal, 9–11 mm long, 1–1.5 mm wide, acuminate or tapering into a soft point, lower 3-nerved, upper 5-nerved. **Florets:** 5.9–8.5 mm long. **Calluses:** blunt. **Lemmas:** 6–8 mm long, dark brown at maturity, evenly hairy, hairs at midlength 0.3–0.8 mm long, hairs a little longer at apex. **Paleas:** 3.5–5.5 mm long, 2/3 to 3/4 as long as lemmas, hairy, hairs not exceeding apices, rounded. **Awns:** 2–3 cm long, persistent, twice-geniculate; lower segment twisted; upper segment straight, scabrous throughout. **VEGETATIVE CHARACTERISTICS: Growth Habit:** cespitose, large and robust. **Culms:** 1–2 m tall, erect. **Sheaths:** glabrous or with scattered pubescence, pilose on lower margins; collars densely hairy, especially those of flag leaves; hairs 0.5–2 mm long. **Ligules:** 2–4 mm long on culm leaves. **Blades:** elongate, those on culm flat and up to 1 cm wide, 20–50 cm long, glabrous but scabrous on margins. **HABITAT:** Generally occurring in foothill to montane meadows. **COMMENTS:** Reported to have a narcotic effect on animals that graze it, especially horses (Allred 2005). Often confused with *Nassella viridula* (Trin.) Barkworth, but *A. robustum* is a larger plant, the palea is about as long as the lemma and dense hairy collars on its flag leaves (Barkworth et al. 2007).

288. *Achnatherum scribneri* (Vasey) Barkworth (Fig. 310).

Synonym: *Stipa scribneri* Vasey. **Vernacular Name:** Scribner's needlegrass. **Life Span:** Perennial. **Origin:** Native. **Season:** C$_3$, cool season. **FLORAL CHARACTERISTICS: Inflorescence:** narrow panicle with loosely appressed branches, 7–20 cm long. **Spikelets:** 1-flowered, 1–2 cm long, up to 1 cm wide; disarticulation above the glumes. **Glumes:** distinctly unequal; lower 10–17 mm long, 2–5 mm longer than second, prominently 3-nerved. **Florets:** 6–9.5

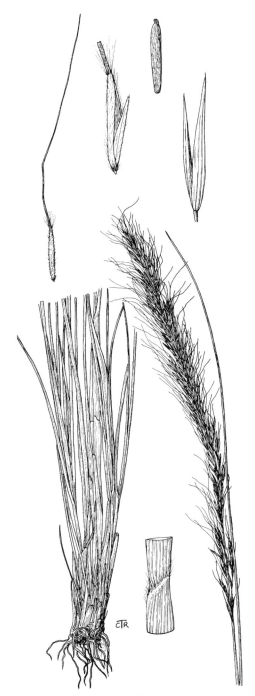

Figure 309. *Achnatherum robustum.*

Figure 310. *Achnatherum scribneri.*

mm long. **Calluses:** sharp. **Lemmas:** 6–10 mm long, villous, hairs at midlength to 1 mm long, with hairs becoming more abundant and longer at apex, to 3 mm long. **Paleas:** 2.5–3.5 mm long, up to half as long as lemma, hairy, hairs not exceeding apices, rounded. **Awns:** lemma 1–2 cm long, persistent, once- or twice-geniculate, basal segment scabrous, twisted below, straight above. **VEGETATIVE CHARACTERIS-TICS: Growth Habit:** cespitose, tufted. **Culms:** 3–9

dm long; nodes purplish and puberulent. **Sheaths:** strongly striate, persistent at base of culms; throat ciliate with long hairs that become shorter down margins. **Ligules:** 0.2–2 mm long, truncate, ciliolate, lacerate or merely a crown, abruptly longer on 1 side. **Blades:** typically involute, especially at tip; 15–30 cm long, margins scabrous to ciliate. **HABITAT:** Found on mesa and rocky foothill slopes in piñon-juniper and ponderosa pine communities on both sides of the Continental Divide. **COMMENTS:** Infrequent. It differs from the closely related *A. robustum* in its sharp lemma callus.

289. *Achnatherum webberi* (Thurb.) Barkworth (Fig. 311).

Synonyms: *Oryzopsis webberi* (Thurb.) Benth. *ex* Vasey, *Stipa webberi* (Thurb.) B. L. Johnson. **Vernacular Name:** Webber's needlegrass. **Life Span:** Perennial. **Origin:** Native. **Season:** C$_3$, cool season. **FLORAL CHARACTERISTICS: Inflorescence:** contracted panicle, 2.5–7 cm long, 0.5–2 cm wide; branches appressed, generally terminating in a single spikelet; when paired, spikelets do not appear divaricate; base often enclosed in subtending sheath. **Spikelets:**

1-flowered, 7–10 mm long; disarticulation above the glumes. **Glumes:** subequal, 6–10 mm long, acuminate, often purplish, 3- to 5-nerved. **Florets:** 4.5–6 mm long. **Calluses:** blunt. **Lemmas:** 4–6 mm long, pilose with spreading hairs 2.5–3.5 mm long, apices with 2 papery teeth. **Paleas:** 4–6 mm long, as long as lemma. **Awns:** lemma 4–11 mm long, scabrous, straight and not twisted, readily deciduous. **VEGETATIVE CHARACTERISTICS: Growth Habit:** cespitose, low, and densely tufted. **Culms:** 1–3 dm long, from a decumbent base, numerous tillers. **Sheaths:** shiny, smooth, persistent at base of culms, often enclosing lower portion of inflorescence. **Ligules:** 0.2–2 mm long, higher on edges than in middle. **Blades:** involute-filiform, 5–9 cm long, often flexuous, glabrous below and puberulent above. **HABITAT:** Cronquist and colleagues (1977) report this species from dry, open sagebrush valleys and foothills. **COMMENTS:** Rare in Colorado, with only a single reported collection from Mesa County. Cronquist and colleagues (1977) view this species as the desert counterpart to *A. pinetorum*. It has deciduous awns, while *A. pinetorum* has persistent awns.

95. *Hesperostipa* (Elias) Barkworth

Cespitose perennials. Inflorescence a terminal, open or contracted panicle of 1-flowered spikelets. Disarticulation above the glumes and below florets. Glumes long and narrow, tapering from base to a hairlike tip, longer than lemmas. Lemmas indurate, margins

Figure 311. *Achnatherum webberi*.

overlapping at maturity, upper portion fused into a ciliate crown, terminating into a stout, persistent, geniculate awn. Paleas equal to lemma in length, pubescent, coriaceous. Basic chromosome number, $x = 11$. Photosynthetic pathway, C_3.

Represented in Colorado by 3 species and 2 subspecies. A segregate of *Stipa.*

1. Terminal segment of awns plumose, hairs usually 1–3 mm long . . . . 291. *H. neomexicana*
1. Terminal segment of awns scabrous to strigose, hairs less than 1 mm long
    2. Lemmas evenly white pubescent; lower ligules scarious, often lacerate; most glumes under 3 cm long . . . . 290. *H. comata*
    2. Lemmas with brownish hairs at base and along margins; lower ligules thick, not lacerate; most glumes over 3 cm long . . . . 292. *H. spartea*

290. *Hesperostipa comata* (Trin. & Rupr.) Barkworth (Fig. 312).

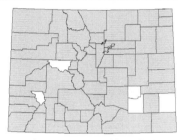

Synonyms: Subsp. *comata*: *Hesperostipa comata* (Trin. & Rupr.) Barkworth var. *comata, Hesperostipa comata* Schult. *ex* Schult. & Schult. f. var. *comata, Stipa comata* Trin. & Rupr., *Stipa comata* Trin. & Rupr. subsp. *intonsa* Piper; subsp. *intermedia*: *Hesperostipa comata* (Trin. & Rupr.) Barkworth var. *intermedia* Scribn. & Tweedy, *Hesperostipa comata* (Trin. & Rupr.) Barkworth var. *intermedia* (Scribn. & Tweedy) Dorn, *Stipa comata* Trin. & Rupr. var. *falcata* Boivin p.p., *Stipa comata* Trin. & Rupr. var. *intermedia* Scribn. & Tweedy, *Stipa comata* Trin. & Rupr. var. *suksdorfii* St. John. **Vernacular Name:** Needle and thread. **Life Span:** Perennial. **Origin:** Native. **Season:** C_3, cool season. **FLORAL CHARACTERISTICS: Inflorescence:** contracted to open panicle, 10–40 cm long, lower portion partially enclosed in uppermost subtending leaf sheath to exserted at anthesis. **Spikelets:** few, generally drooping at anthesis, 1-flowered, 18–30 mm long; disarticulation above the glumes. **Glumes:** subequal, long and narrow, 3- to 5-nerved, tapering to fine point, glabrous; lower 18–30 mm long; upper 15–27 mm long, 1–3 mm shorter than lower. **Florets:** 7–13 mm long. **Calluses:** 2–4 mm long, very sharp, sometimes glabrate. **Lemmas:** indurate, pale green to yellow- or brown-tinged, 4–12 mm long, evenly appressed pubescent with white hairs to about 1 mm long, apices often glabrate. **Paleas:** nearly as long as lemma, pubescent between nerves. **Awns:** lemma awned, 6.5–22.5 cm long, terminal segment 4–12 cm long, lower 2 segments scabrous to strigose with hairs less than 1 mm long, distal segment scabridulous; sinuous to curled at maturity. **VEGETATIVE CHARACTERISTICS: Growth Habit:** densely tufted. **Culms:** erect, 1–11 dm tall, glabrous, commonly puberulent at nodes. **Sheaths:** open, glabrous to pubescent, strongly ribbed, generally longer than internodes, nodes of lower leaves exposed or concealed by leaf sheaths. **Ligules:** lower scarious; acute, becoming lacerate; 1–6.5 mm long, decurrent; upper to 7 mm long. **Blades:** flat to involute, 5–40 cm long, 0.5–4 mm wide, scabrous above, glabrous to scabrous below. **HABITAT:** Grows from low elevations to montane throughout Colorado; common in grasslands or open forested areas. **COMMENTS:** *H. comata* is an important constituent of the mixed-grass prairies of Colorado and the West. It is a good forage in the spring and fall; however, in the summer the sharp callus on mature fruit can cause mechanical injury to grazing animals (Stubbendieck, Hatch, and

Landholt 2003). Two subspecies occur in Colorado, the typical subspecies (subsp. *comata*) and subsp. *intermedia* (Scribn. & Tweedy) Barkworth. The subspecies can be distinguished by these characteristics:

1. Distal segment of awns 4–12 cm long; lower nodes often concealed by sheaths; panicles often partially enclosed in uppermost sheath at maturity . . . . subsp. *comata*
1. Distal segment of awns 3–8 cm long; lower cauline nodes usually exposed; panicles usually completely exserted at maturity . . . . subsp. *intermedia*

JRJ

Figure 312. *Hesperostipa comata.*

291. *Hesperostipa neomexicana* (Thurb.) Barkworth (Fig. 313).

Synonyms: *Stipa neomexicana* (Thurb.) Scribn., *Stipa pennata* L. var. *neomexicana* Thurb. **Vernacular Names:** New Mexico feathergrass, New Mexican needlegrass. **Life Span:** Perennial. **Origin:** Native. **Season:** C$_3$, cool season. **FLORAL CHARACTERISTICS: Inflorescence:** narrow panicle, 10–30 cm long, lower portion included in inflated uppermost leaf sheath. **Spikelets:** 3–6 cm long, 1-flowered, papery; disarticulation above the glumes. **Glumes:** subequal, 30–60 mm long, 3- to 7-nerved, apices long and tapering. **Florets:** 15–18 mm long. **Calluses:** sharp. **Lemmas:** indurate, 15–18 mm long, 5-nerved, evenly appressed pubescent, hairs less than 1 mm long; callus 4–5 mm long; awned. **Paleas:** similar to lemma in length and texture, glabrous to sparsely hairy. **Awns:** lemma awned, 12–22 cm long; lower segment twisted and appressed pubescent; terminal segment plumose with hairs 1–3 mm long. **VEGETATIVE CHARACTERISTICS: Growth Habit:** cespitose. **Culms:** erect, 4–10 dm tall; lower nodes glabrous. **Sheaths:** glabrous to puberulent. **Ligules:** lower leaves rounded, thick, ciliate membrane, 0.5–1 mm long; upper leaves acute, scarious, to 3 mm. **Blades:** strongly involute, filiform, 10–30 cm long, 0.5–1 mm wide, abruptly narrowed at sheath-blade juncture. **HABITAT:** Found at low elevations, frequently in lower canyons south of the Colorado River on the Western Slope and in rocky places along the Front Range. **COMMENTS:** Closely related to *H. comata* but differs in the pubescent awn and shorter ligule.

292. *Hesperostipa spartea* (Trin.) Barkworth (Fig. 314).

Synonym: *Stipa spartea* Trin. **Vernacular Name:** Porcupinegrass. **Life Span:** Perennial. **Origin:** Native. **Season:** C$_3$, cool season. **FLORAL CHARACTERISTICS: Inflorescence:** narrow, nodding panicle, 10–25 cm long, base of panicle exserted from lowermost leaf sheath; branches few. **Spikelets:** 30–45 mm long, 5-nerved, 1-flowered, papery; disarticulation above the glumes. **Glumes:** subequal, 22–45 mm long, pale, glabrous. **Florets:** 15–25 mm long. **Calluses:** 7 mm long, sharp. **Lemmas:** indurate; base, margins, and lower portion pubescent; hairs beige to brown, 16–25 mm long. **Paleas:** similar to lemma in length and texture, glabrous to sparsely hairy. **Awns:** stout, twice-geniculate, 9–19 cm long; lower segment scabrid and twisted; second segment flexuous and loosely twisted; third segment straight and scabrous. **VEGETATIVE CHARACTERISTICS: Growth Habit:** cespitose. **Culms:** erect, glabrous to scabrous, 4.5–9 dm tall. **Sheaths:** lower sheaths with ciliate margins. **Ligules:** lower truncate to rounded, firm, thick, 0.3–3 mm long; upper 3–7.5 mm long, thin, acute, frequently lacerate. **Blades:** flat to involute on drying, 20–30 cm long, 1.5–4.5 mm wide, upper surface scabrous. **HABITAT:** Infrequent species, found in tallgrass prairie remnants along the base of the Front Range (Weber and Wittmann 2001a). **COMMENTS:** Native Americans used bundles of florets as combs.

Figure 313. *Hesperostipa neomexicana.*

Figure 314. *Hesperostipa spartea*.

96. *Jarava* Ruiz & Pav.

Cespitose perennials. Inflorescence a terminal, open or contracted panicle of 1-flowered spikelets. Disarticulation above the glumes and below florets. Glumes usually shorter than lemma, glabrous, awnless but pointed, lanceolate, typically very thin to almost hyaline, 1-nerved. Lemmas firmer than the glumes but less firm than in most Stipeae; brown in fruit, smooth, crown absent, hairy on upper portion and pappus-like, awned. Awn geniculate, hairless, much longer than body of lemma, persistent. Paleas conspicuous but much shorter than lemma, glabrous, tightly clasped by lemma. Basic chromosome number, $x = 11$. Photosynthetic pathway, C_3.

Represented in Colorado by a single species. A segregate of *Stipa* and *Achnatherum*.

293. *Jarava speciosa* (Trin. & Rupr.) Peñail. (Fig. 315).

Synonyms: *Achnatherum speciosum* (Trin. & Rupr.) Barkworth, *Stipa speciosa* Trin. & Rupr. **Vernacular Name:** Desert needlegrass. **Life Span:** Perennial. **Origin:** Native. **Season:** C_3, cool season. **FLORAL CHARACTERISTICS: Inflorescence:** panicle, 7–20 cm long, narrow, dense, scarcely exceeding leaves; spikelets nearly to base of branches. **Spikelets:** 1-flowered, 12–25 mm long; disarticulation above the glumes. **Glumes:** long-acuminate, terminating in a fine point, papery; lower 16–24 mm long, 1- to 3-nerved; upper 13–19 mm long, 3- to 5-nerved. **Florets:** 6–10 mm long. **Calluses:** sharp. **Lemmas:** 7–10 mm long, terete, indurate, appressed pilose below, glabrate above, awned. **Paleas:** 3–5 mm long, 2/5 to 4/5 the length of lemmas, hairy, about 0.5 mm long. **Awns:** lemma awned, awn 3–5 cm long, once-geniculate; lower segment tightly twisted and densely pubescent, with hairs up to 8 mm long; upper segment not twisted, scaberulous. **VEGETATIVE CHARACTERISTICS: Growth Habit:** cespitose, tufted, stout, bases a reddish brown. **Culms:** 3–7 dm long, erect. **Sheaths:** striate, persistent at base of culms; lowermost pilose, tuft of hairs on each side of throat. **Ligules:** 0.5–2 mm long, densely ciliolate. **Blades:** tightly involute, 10–40 cm long, firm, pungent-tipped. **HABITAT:** Found in the canyon country south of the Colorado River (Weber and Wittmann 2001b; Wingate 1994). **COMMENTS:** Common. Cronquist and colleagues (1977) suggest that the species be used as an ornamental in native desert gardens. Differs from other genera of the Stipeae by very narrow lemmas (less than 0.5 mm wide) and the upper quarter bearing strongly divergent hairs 1.5–2.5 mm long below apices.

97. *Nassella* E. Desv.

Cespitose perennials. Inflorescence a terminal contracted or occasionally an open panicle of 1-flowered spikelets. Glumes longer than floret, tapering to an acuminate tip, frequently purplish at base before drying. Lemmas strongly convolute, coriaceous, with a crown at apex, terminating into a persistent, usually twisted and geniculate awn. Paleas up to 1/3 the length of lemmas, glabrous, and without veins. Basic chromosome number, $x = 11$. Photosynthetic pathway, C_3.

Figure 315. *Jarava speciosa.*

Represented in Colorado by 2 species. A segregate of *Stipa*.

1. Mature lemmas 2–3 mm long; awns about 45–100 mm long . . . . 294. *N. tenuissima*
1. Mature lemmas at least 5 mm long; awns much less than 19–32 mm long . . . .
. . . . 295. *N. viridula*

294. *Nassella tenuissima* (Trin.) Barkworth (Fig. 316).

Synonym: *Stipa tenuissima* Trin. **Vernacular Names:** Fineleaved nassella, slender needlegrass. **Life Span:** Perennial. **Origin:** Native. **Season:** C$_3$, cool season. **FLORAL CHARACTERISTICS: Inflorescence:** panicle, 10–50 cm long, soft, to 1 cm wide, loosely contracted; branches erect to slightly nodding. **Spikelets:** 10–12 mm long, 1-flowered; disarticulation above the glumes. **Glumes:** subequal, acuminate, 5–13 mm long. **Florets:** 2.5–3 mm long. **Calluses:** blunt, bearded, with dense strigose hairs 1/4 to 1/3 the length of lemma. **Lemmas:** oblong-elliptic, 2–3 mm long, nearly glabrous, minutely papillose-roughened; awned, rounded to a crown; crown 0.2 mm long. **Paleas:** up to half as long as lemma. **Awns:** lemma awn capillary, inconspicuously twice-geniculate, 3.5–8 cm long. **VEGETATIVE CHARACTERISTICS: Growth Habit:** tufted. **Culms:** slender and wiry, erect, 3–10 dm, unbranched, smooth to scabrous above. **Sheaths:** glabrous with ciliate to long-hairy on margins, throat villous to pilose. **Ligules:** 1–5 mm long. **Blades:** 7–60 cm long, filiform, closely convolute and stiffly erect, 0.5 mm thick, strongly scabrous. **HABITAT:** Recently located growing along a trail and stream bank in Larimer County. **COMMENTS:** This is a new record for Colorado and is perhaps a disjunct population several hundred miles from its closest records in New Mexico or more likely an escaped ornamental. This species is previously known from New Mexico and Texas in dry open ground, rocky slopes, and open dry woods. It has been used as an ornamental but readily escapes and has become a noxious weed in Australia.

295. *Nassella viridula* (Trin.) Barkworth (Fig. 317).

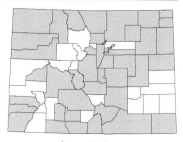

Synonym: *Stipa viridula* Trin. **Vernacular Name:** Green needlegrass. **Life Span:** Perennial. **Origin:** Native. **Season:** C$_3$, cool season. **FLORAL CHARACTERISTICS: Inflorescence:** panicle, 10–25 cm long, to 2 cm wide, oblong, greenish to tawny at maturity; branches erect to slightly spreading. **Spikelets:** 7–13 mm long, 1-flowered; disarticulation above the glumes. **Glumes:** subequal, acuminate, 7–10 mm long; 3- to rarely 5-nerved, distinct and greenish; apices long-tapered to acute with hairlike tips. **Florets:** 3.4–5.5 mm long. **Calluses:** strigose, moderately sharp. **Lemmas:** brown and leathery, 4–7 mm long, sparsely appressed pubescent, plump at maturity, awned, tapering to a crown; crown about 0.5 mm long. **Paleas:** less than half as long as lemma, rounded at apex, nerveless, membranous and glabrous. **Awns:** lemma awns obviously twice-geniculate, 19–32 mm long; lower segments scabrid.

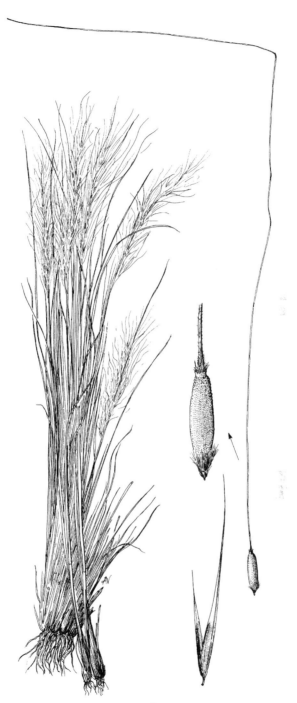

Figure 316. *Nassella tenuissima*.

Figure 317. *Nassella viridula*.

VEGETATIVE CHARACTERISTICS: Growth Habit: tufted. **Culms:** erect to geniculate basally, 3.5–12 dm long, smooth to scabrous above. **Sheaths:** glabrous with ciliate margins, throat villous to pilose. **Ligules:** truncate, 0.5–2 mm long. **Blades:** flat to involute, 6–30 cm long, 1–3 and occasionally to 6 mm wide, smooth to scabrous to puberulent, margins scabrous. **HABITAT:** Found from low elevations to the montane over Colorado. **COMMENTS:** Distinguishable from *Achnatherum robustum* by its tightly convolute leaves and in having glabrous to sparsely pubescent collars on its flag leaves. *Achnatherum nelsonii* differs in having a tuft of hairs at the leaf sheath throat.

98. *Oryzopsis* Michx.

Perennial, cespitose. Basal leaves with twisted bases placing abaxial surface uppermost, upper cauline blades reduced. Sheaths open, glabrous; auricles absent. Inflorescence a terminal, contracted panicle. Spikelets with a single floret, disarticulation above the glumes. Glumes unequal, many-nerved, mucronate. Floret terete or laterally compressed, calluses usually less than one-fifth the length of floret, blunt, distal portion pilose. Lemmas coriaceous; pubescent, at least basally, margins strongly overlapping at maturity, awned. Palea similar to lemma in length, texture, and pubescence; concealed by lemma. Basic chromosome numbers, $x = 11$ and 12. Photosynthetic pathway, C_3.

Represented in Colorado by a single species.

296. *Oryzopsis asperifolia* Michx. (Fig. 318).

Synonyms: *Oryzopsis aspera* Michx. *ex* Muhl., *Oryzopsis leucosperma* Link *ex* Walp., *Oryzopsis mutica* Link, *Urachne asperifolia* (Michx.) Trin., *Urachne leucosperma* Link, *Urachne mutica* (Link) Steud. **Vernacular Names:** Roughleaf ricegrass, winter grass. **Life Span:** Perennial. **Origin:** Native. **Season:** C_3, cool season. **FLORAL CHARACTERISTICS: Inflorescence:** contracted panicle at times reduced to a raceme, 3.5–13 cm long; branches short, appressed to slightly spread-ing at anthesis. **Spikelets:** 1-flowered; disarticulation above the glumes. **Glumes:** subequal, obovate to broadly elliptic, 5–7.5 mm long, 2.5–4 mm wide, 6- to 10-nerved, apex mucronate. **Florets:** 5–7 mm long. **Calluses:** blunt, dense collar of soft hairs on distal region. **Lemmas:** 5–7 mm long, coriaceous, pubescent below, 3- to 5- or rarely 9-nerved, awned; 0.8–2 mm long with dense ring of hairs at apices. **Paleas:** concealed by lemma. **Awns:** lemma awns generally straight, early deciduous, 7–15 mm long. **VEGETATIVE CHARACTERISTICS: Growth Habit:** tufted. **Culms:** prostrate to erect, 2–7 dm tall, glabrous, hollow. **Sheaths:** glabrous, sometimes auriculate. **Ligules:** ciliolate, truncate, 0.1–0.7 mm long. **Blades:** flat to loosely involute, 30–90 cm long, 4–9 mm wide, erect, and generally exceeding culm; upper surface glabrous; lower surface scaberulous to hirsute, margins scabrous; cauline blades much reduced; flag leaf blades 8–12 mm long. **HABITAT:** Found infrequently in shaded forests from foothills to montane over much of Colorado. **COMMENTS:** Unusual in that the flowering culms are widely spreading or prostrate and are nearly naked, with only 2–3 short leaf sheaths and very reduced or obscure blades. Considered threatened or endangered in several Midwest and Atlantic Coast states.

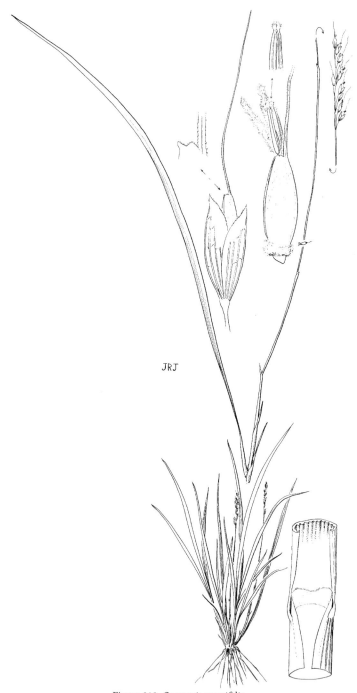

JRJ

Figure 318. *Oryzopsis asperifolia.*

99. *Piptatherum* P. Beauv.

Cespitose perennials. Inflorescence a closed or occasionally open terminal panicle of 1-flowered spikelets. Disarticulation above the glumes. Glumes about equal and as long as or longer than floret. Lemma not convolute; firm and becoming indurate, with a straight, persistent awn. Palea not covered by lemma, as long as lemma, with 2 nerves. Basic chromosome number, $x = 11$. Photosynthetic pathway, C_3.

Represented in Colorado by 3 species. All were once included in *Oryzopsis.*

1. Awn persistent, once-geniculate; floret 3–6 mm long . . . . 297. *P. exiguum*
1. Awn early deciduous, straight or almost so; floret 1.5–4 mm long
    2. Panicle 10–15 cm long; lemma generally glabrous, although occasionally sparsely pubescent; awn 4–8 mm long . . . . 298. *P. micranthum*
    2. Panicle 4–6 cm long; lemma evenly pubescent; awn 1–2 mm long, early deciduous, and may be absent on immature florets . . . . 299. *P. pungens*

297. *Piptatherum exiguum* (Thurb.) Dorn (Fig. 319).

Synonym: *Oryzopsis exigua* Thurb. **Vernacular Name:** Little ricegrass. **Life Span:** Perennial. **Origin:** Native. **Season:** C_3, cool season. **FLORAL CHARACTERIS-TICS:** spikelike panicle, 3–9 and occasionally to 11 cm long, few-flowered; branches appressed ascending; spikelets short-pediceled. **Spikelets:** pale in color, 4–6 mm long; disarticulation above the glumes. **Glumes:** subequal, broadly lanceolate, 4–5 mm long, obscurely 3- to 5- or rarely 7-nerved, glabrous to scaberulous, tips acute to rounded and erose to entire. **Florets:** 3–6 mm long. **Calluses:** blunt. **Lemmas:** leathery to indurate, terete to slightly flattened, 3–6 mm long, to 1 mm wide, evenly appressed puberulent, notched at apices, awned. **Paleas:** similar to lemmas but slightly shorter, not enclosed. **Awns:** lemma awns stout, 4–7 mm long, geniculate, somewhat twisted below, attached terminally or behind bifid apices, early deciduous. **VEGETATIVE CHARACTERISTICS: Growth Habit:** tufted, at times forming large bunches. **Culms:** erect, slender, 1–5 dm tall, hollow; leaves generally basal. **Sheaths:** glabrous to slightly scabrous, strongly nerved, persistent, ultimately becoming fibrous. **Ligules:** puberulent, 1.5–4 mm long, acute. **Blades:** involute-filiform, 5–10 cm long, to 2 mm wide, stiffly erect, scabrous. **HABITAT:** Found on rocky slopes and foothills and on canyon slopes in the northern Park Range (Weber and Wittmann 2001a, 2001b). **COMMENTS:** Locally abundant.

298. *Piptatherum micranthum* (Trin. & Rupr.) Barkworth (Fig. 320).

Synonyms: *Oryzopsis micrantha* (Trin. & Rupr.) Thurb., *Urachne micrantha* Trin. & Rupr. **Vernacular Names:** Littleseed ricegrass, small-flowered piptatherum. **Life Span:** Perennial. **Origin:** Native. **Season:** C_3, cool season. **FLORAL CHARACTERISTICS: Inflorescence:** narrow to open panicle, 5–20 cm long; branches 1–3 at node. **Spikelets:** 3–4 mm long; disarticulation above the glumes. **Glumes:** subequal, thin-membranous to hyaline, 2.7–3.5 mm

JRJ

Figure 319. *Piptatherum exiguum*.

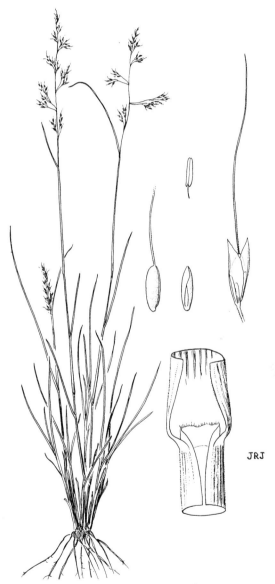

JRJ

Figure 320. *Piptatherum micranthum.*

long, ovate-acute, 3- to 5-nerved, exceeding lemma. **Lemmas:** pale to dark brown, 1.8–3 mm long, terete to slightly compressed, glabrous to sparsely appressed pilose. **Calluses:** short and blunt, glabrous to minutely hairy. **Paleas:** similar to lemma in size and texture. **Awns:** lemma awns straight to flexuous, hairlike, 5–10 mm long. **VEGETATIVE CHARACTERISTICS: Growth Habit:** tufted, leaves generally basal. **Culms:** ascending to erect, slender, glabrous to scaberulous,

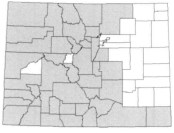

2–7 dm tall, hollow. **Sheaths:** glabrous to puberulent. **Ligules:** finely ciliolate, truncate, 0.2–2 mm long. **Blades:** flat to loosely involute, 5–15 cm long, 0.5–2 mm wide, glabrous to scabrous. **HABITAT:** Found on shaded rocky slopes on mesas and foothills. **COMMENTS:** A delicate plant frequently found growing in litter under piñon and juniper trees.

299. *Piptatherum pungens* (Torr.) Dorn (Fig. 321).

Synonyms: *Oryzopsis pungens* (Torr. *ex* Spreng.) Hitchc., *Milium pungens* Torr. **Vernacular Names:** Mountain ricegrass, sharp piptatherum. **Life Span:** Perennial. **Origin:** Native. **Season:** C_3, cool season. **FLORAL CHARACTERISTICS: Inflorescence:** panicle, 3–8 cm long; branches closely appressed to ascending or ovoid and open at anthesis; spikelets long-pediceled. **Spikelets:** 3–5 mm long; disarticulation above the glumes. **Glumes:** elliptic-obovate, 3–5 mm

long, obscurely 5-nerved, minutely scabrous above, apex obtuse. **Lemmas:** gray to pale green, variably appressed pubescent, 2.5–5 mm long. **Paleas:** equaling lemma, similar in texture and pubescence to lemmas. **Awns:** lemma awns lacking or 1–2 mm long, straight to slightly bent. **VEGETATIVE CHARACTERISTICS: Growth Habit:** densely tufted. **Culms:** slender, erect, 1–9 dm tall. **Sheaths:** glabrous to slightly pubescent. **Ligules:** obtuse, 0.5–2.5 mm long, 1.5–3 mm long on upper leaves. **Blades:** flat to convolute, at least when dry; 1–2 mm wide. **HABITAT:** Found infrequently in pine forests on both the Cochetopa Divide on the Western Slope and in the Black Forest area on the Arkansas Divide, as well as in Baca County (Weber and Wittmann 2001a, 2001b). **COMMENTS:** The presence of *P. pungens* in Colorado represents a disjunct population; other known populations occur much farther north.

100. *Ptilagrostis* Griseb.

Cespitose perennial. Inflorescence a terminal, open panicle of 1-flowered spikelets. Disarticulation above the glumes. Glumes about equal, hyaline, slightly shorter than florets, purplish. Lemma membranous, evenly pubescent, margins overlapping at maturity. Palea about equal to lemma, enclosed, 2-nerved. Basic chromosome number, $x = 11$. Photosynthetic pathway, C_3.

Represented in Colorado by a single species.

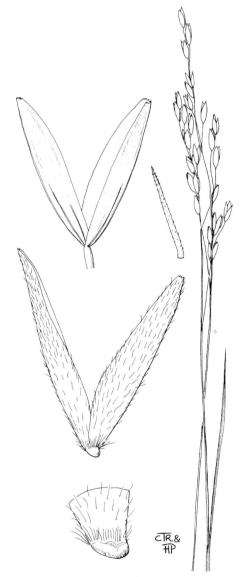

Figure 321. *Piptatherum pungens.*

300. *Ptilagrostis porteri* (Rydb.) W. A. Weber (Fig. 322).

Synonyms: *Ptilagrostis mongholica* (Turcz. *ex* Trin.) Griseb. subsp. *porteri* (Rydb.) Barkworth, *Stipa porteri* Rydb. **Vernacular Names:** Porter's false needlegrass, Porter's ptilagrostis, Rocky-Mountain ptilagrostis. **Life Span:** Perennial. **Origin:** Native. **Season:** C₃, cool season. **FLORAL CHARACTERISTICS: Inflorescence:** panicle, 5–12 cm long, open; branches flexuous, slender, spreading. **Spikelets:** 1-flowered, 4.5–6 mm long; disarticulation above the glumes. **Glumes:**

equal, hyaline, purplish, glabrous below, pubescent above, 4.5–6 mm long, obscurely 5-nerved, rounded to acute. **Lemmas:** thickly membranous, appressed pilose below, scaberulous above, 2.5–4 mm long, margins open at maturity, lobes to 0.8 mm long, awned. **Paleas:** subequal to lemmas, nerves do not extend to distal edge of paleas. **Awns:** lemma awned, awns plumose entire length, twisted at base, once- to weakly twice-geniculate, 5–25 mm long. **VEGETATIVE CHARACTERISTICS: Growth Habit:** tufted. **Culms:** erect, glabrous, 2–5 dm tall. **Sheaths:** glabrous, smooth to scabrous. **Ligules:** membranous, rounded to acute, 1.5–3 mm long. **Blades:** filiform, 2–12 cm long, 0.3–0.6 mm in diameter. **HABITAT:** Found on the northern edge of South Park in peat and willow carrs and on Hoosier Pass in peat

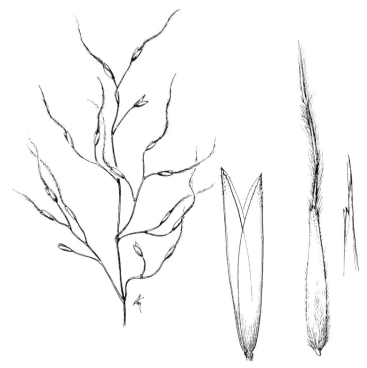

Figure 322. *Ptilagrostis porteri.*

located in calcareous fens (Weber and Wittmann 2001a, 2001b). Also recently discovered at the U.S. Air Force Academy. **COMMENTS:** Rare, but locally abundant in peaty areas.

14. TRITICEAE (HORDEAE)

Annuals or occasionally perennials. Sheaths open; ligules membranous; auricles usually present. Inflorescence a bilateral spike or spicate raceme with both sessile and pediceled spikelets at rachis nodes. Spikelets with 2 to several florets except in *Hordeum*, which typically has only a single floret per spikelet. Disarticulation above the glumes, at rachis joints, or sometimes at rachis bases. Glumes occasionally borne parallel in front of spikelets. Lemmas 5- to 7-nerved, awned or awnless.

According to some, the tribal name should be Hordeae because it evidently has priority over the more commonly recognized Triticeae (Allred 2005).

101. *Aegilops* L.

Annuals. Inflorescence a bilateral spike usually with 1–3 rudimentary spikelets below. Disarticulation either at base of spikes or in rachis, spikelets falling attached to internodes above and below. Spikelets solitary at nodes, with 2–8 florets. Glumes oblong to nearly rectangular, rounded on back, coriaceous and becoming indurate at maturity; truncate, dentate, or 1- to 5-awned. Lemmas rounded on back, toothed, 1- to 3-awned. Paleas chartaceous, 2-keeled, keels ciliate. Basic chromosome number, $x = 7$. Photosynthetic pathway, C$_3$.

Represented in Colorado by a single species.

301. *Aegilops cylindrica* Host (Fig. 323).

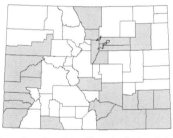

Synonyms: *Aegilops cylindrica* Host var. *rubiginosa* Popova, *Cylindropyrum cylindricum* (Host) Á. Löve, *Triticum cylindricum* (Host) Ces. Pass. & Gib. **Vernacular Name:** Jointed goatgrass. **Life Span:** Annual. **Origin:** Introduced. **Season:** C$_3$, cool season. **FLORAL CHARACTERISTICS: Inflorescence:** spike, 2–15 cm long, 0.3 cm wide, terete, entire spike frequently disarticulates at base; spikelets appressed and sunken into rachis flush with nodes, 3–12 per spike. **Spikelets:** 9–12 mm long, 2- to 5-flowered, lower 2–3 fertile, single at each node of rachis; disarticulation initially at base of inflorescence, then in rachis, spikelets remaining attached to internode (i.e., below the glumes). **Glumes:** 7–10 mm long, asymmetrical, 9- to 13-nerved, tipped by a lobe in front and an awn in back, glumes of terminal spikelet awned from between 2 short teeth. **Lemmas:** 8–11 mm long, asymmetrical, 5-nerved, awned. **Awns:** glumes with awn 2–9 mm long; lemmas awned from a bifid apex, 1–5 mm long or sometimes a mucronate; lemma of terminal spikelet with awn 20–80 mm long. **VEGETATIVE CHARACTERISTICS: Growth Habit:** weedy. **Culms:** 3–7 dm tall, erect or geniculate at base, hollow. **Sheaths:** variable, glabrous with ciliate margins to hirsute or pilose. **Auricles:** present but not obvious. **Ligules:** 0.2–0.8 mm long, truncate. **Blades:** flat, 1–6 mm wide, glabrous to hirsute. **HABITAT:** Troublesome weed in waste places and wheat fields. **COMMENTS:** Much controversy surrounds the placement of this genus. Weber and Wittmann (2001a, 2001b) include the species in the genus *Cylindropyrum* [*C. cylindricum* (Host) Á. Löve]. Others have included it

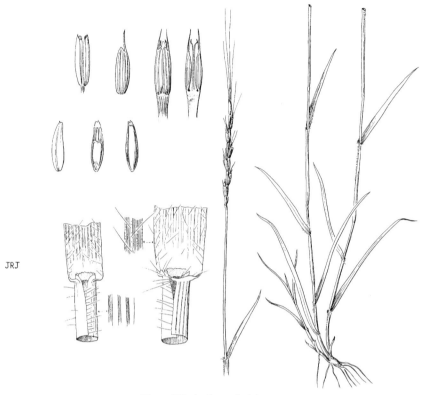

JRJ

Figure 323. *Aegilops cylindrica.*

with wheat, since it easily hybridizes with *Triticum*. Barkworth and colleagues (1983) point out that if hybridization was a good reason to merge genera within the Triticeae, the tribe would consist of a single genus.

102. *Agropyron* Gaertn.

Perennials, densely cespitose. Inflorescence a bilateral spike, with spikelets 1 per node. Spikelets divergent or pectinate on a very thick and tough rachis, with 3–10 florets; disarticulation is above the glumes below each floret. Glumes strongly keeled to base, tapering to an acuminate or short awn tip in which the nerves converge. Lemmas keeled, tipped like that of the glumes. Palea membranous, about as long as lemma, usually adhering to caryopsis. Basic chromosome number, $x = 7$. Photosynthetic pathway, C_3.

Represented in Colorado by 2 species.

1. Lemmas usually awned, awn 1–6 mm long; spikelets widely spreading at an angle greater than 30° from rachis, giving the spike a bristly appearance; spike lanceolate, rectangular, or ovate in outline . . . . 302. *A. cristatum*
1. Lemmas unawned, occasionally mucronate; spikelets spreading at an angle less than 30° from rachis; spike linear to narrowly lanceolate in outline . . . . 303. *A. fragile*

302. *Agropyron cristatum* (L.) Gaertn. (Fig. 324).

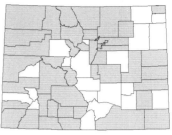

Synonyms: *Agropyron cristatiforme* Sarkar, *Agropyron desertorum* (Fisher *ex* Link) Schult., *Agropyron pectinatum* (Labill.) P. Beauv., *Agropyron pectiniforme* Roem. & Schult. **Vernacular Name:** Crested wheatgrass. **Life Span:** Perennial. **Origin:** Introduced. **Season:** C₃, cool season. **FLORAL CHARACTERISTICS: Inflorescence:** spike, 2–10 cm long, 5–25 mm wide; lanceolate, rectangular, or ovate, tapering at base; bristly in appearance; rachis internodes short, scabrous to pilose on angles. **Spikelets:** 7–16 mm long, 3- to 8-flowered, single at each node, closely imbricate and divergent from rachis at an angle of 30°–95°; disarticulation above the glumes and between florets; laterally compressed. **Glumes:** subequal, second slightly longer than first, 3–6 mm long, lanceolate, midnerve prominent and ciliate, short-awned. **Lemmas:** 5–9 mm long, 5-nerved, short-awned. **Awns:** glumes and lemmas tapering to an awn 1–6 mm long. **VEGETATIVE CHARACTERISTICS: Growth Habit:** cespitose, distinct tufts, occasionally rhizomatous. **Culms:** hollow, 3–10 dm tall, erect or sometimes geniculate. **Sheaths:** glabrous or lowermost sometimes pilose to villous, open, margins overlapping. **Auricles:** about 1 mm long. **Ligules:** membranous, short, up to 1.5 mm long, erose-ciliolate. **Blades:** 5–20 cm long, flat, 2–10 mm wide, margins weakly scabrous, glabrous to puberulent. **HABITAT:** This species has been extensively seeded in pastures and hay meadows and used for roadside stabilization and restoration. It is a good winter forage for livestock (Stubbendieck, Hatch, and Butterfield 1992). **COMMENTS:** Much confusion exists concerning the delineation of species in this complex. Weber and Wittmann (2001a, 2001b) include 2 other species (*A. mongolicum* Keng, *A. pectiniforme* Roem. & Schult.) in the state, and they believe that *A. cristatum* is purely Asiatic. *Agropyron desertorum* (Fischer *ex* Link) Schult. is included here.

303. *Agropyron fragile* (Roth) P. Candargy (Fig. 325).

Synonyms: *Agropyron cristatum* (L.) Gaertn. subsp. *fragile* (Roth) Á. Löve, *Agropyron cristatum* (L.) Gaertn. var. *fragile* (Roth) Dorn, *Agropyron fragile* (Roth) P. Candargy subsp. *mongolicum* (Keng) D. R. Dewey, *Agropyron fragile* (Roth) P. Candargy subsp. *sibiricum* (Willd.) Melderis, *Agropyron fragile* (Roth) P. Candargy var. *sibiricum* (Willd.) Tzvelev, *Agropyron mongolicum* Keng, *Agropyron sibiricum* (Willd.) P. Beauv. **Vernacular Name:** Siberian wheatgrass. **Life Span:** Perennial. **Origin:** Introduced. **Season:** C₃, cool season. **FLORAL CHARACTERISTICS: Inflorescence:** spike, 6–15 cm long, 5–13 mm wide, narrow-oblong (subcylindrical), linear to narrowly lanceolate in outline, 6–10 cm long; rachis internodes 2.5–3.5 mm long. **Spikelets:** 7–16 mm long, 5- to 7-flowered, appressed or diverging at an angle up to 30° from rachis; disarticulation above the glumes and below each floret. **Glumes:** subequal, twisted, gradually tapering into short awns, 3–5 mm long, margins narrow, 3-nerved. **Lemmas:** elliptic, 5–9 mm long, 5-nerved, acute, unawned but occasionally with a short (to 0.5 mm) mucro. **Awns:** glumes

Figure 324. *Agropyron cristatum.*

Figure 325. *Agropyron fragile.*

with awn 1–3 mm long. **VEGETATIVE CHARACTERISTICS: Growth Habit:** cespitose, not rhizomatous. **Culms:** 3–10 dm tall, ascending to erect. **Sheaths:** glabrous with lowermost ones occasionally pilose to villous. **Auricles:** about 1 mm long. **Ligules:** membranous, erose-ciliolate, to 1 mm long. **Blades:** flat, 6–15 cm long, 1–6 mm wide, generally pilose above. **HABITAT:** Cultivated, escaped from cultivation, or used in roadside stabilization and recla-mation projects. **COMMENTS:** Weber and Wittmann (2001a) call this species *A. mongolicum* Keng. This species is more drought-tolerant than *A. cristatum* (Barkworth et al. 2007).

103. *Critesion* Raf.

Annual or cespitose perennial. Inflorescence a spicate raceme with 3 spikelets at a node (triad). Rachis usually fragmenting at maturity, disarticulating above each node, short internode falling with triad of spikelets. Central spikelet usually sessile, with 1 floret and a bristle-like extension of rachilla. Glumes side by side, usually awnlike and separate to base. Lemmas firm, rounded on back, dorsally compressed, and terminating in a short awn. Palea slightly shorter than lemma and usually adnate to caryopsis. Lateral spikelets typically smaller than central one, staminate or sterile, and frequently reduced to 3 awnlike structures. Basic chromosome number, $x = 7$. Photosynthetic pathway, C_3.

Represented in Colorado by 4 species and 5 subspecies. A segregate of *Hordeum*.

1. Blades with well-developed auricles up to 8 mm long . . . . 306. *C. murinum*
1. Blades lacking auricles, or auricles less than 0.3 mm long
    2. Annual
        3. Glumes bent, strongly divergent at maturity, 15–85 mm long . . . . . . 305. *C. jubatum*
        3. Glumes straight, not divergent at maturity, 8–17 mm long . . . . 307. *C. pusillum*
    2. Perennial
        4. Central (fertile) lemma awns 1–9 cm long; inflorescence nearly as wide as long, nodding; glumes 15–85 mm long, divergent to strongly divergent at maturity . . . . 305. *C. jubatum*
        4. Central (fertile) lemma awns to 1.5 cm long; inflorescence much longer than broad, strictly erect; glumes 7–19 mm long, divergent or not at maturity . . . . 304. *C. brachyantherum*

304. *Critesion brachyantherum* (Nevski) Barkworth & D. R. Dewey (Fig. 326).

Synonyms: *Hordeum brachyantherum* Nevski subsp. *brachyantherum*, *Critesion jubatum* (L.) Nevski subsp. *breviaristatum* (Bowden) Á. & D. Löve, *Hordeum boreale* Scribn. & J. G. Sm., *Hordeum jubatum* L. subsp. *breviaristatum* Bowden, *Hordeum jubatum* L. var. *boreale* (Scribn. & J. G. Sm.) Boivin, *Hordeum nodosum* L., *Hordeum nodosum* L. var. *boreale* (Scribn. & J. G. Sm.) Hitchc. **Vernacular Name:** Meadow barley. **Life Span:** Perennial. **Origin:** Native. **Season:** C_3, cool

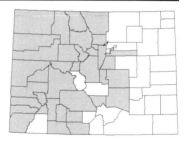

season. **FLORAL CHARACTERISTICS: Inflorescence:** erect spike, 3–10 cm long, much longer than broad; rachis disarticulating at maturity. **Spikelets:** 3 spikelets per node, 3–9 mm long, central spikelet perfect and sessile; lateral spikelets much reduced, pedicellate and staminate, pedicels curved, 0.7–1 mm long. **Glumes:** both central and lateral awnlike, generally 7–19 mm long, mostly straight, erect to slightly divergent. **Lemmas:** lemmas of lateral spikelet reduced to well developed; lemmas of central spikelet floret 5–10 mm long, 5-nerved, glabrous to scaberulous near apices, tapering to awn. **Awns:** glumes awnlike, to 19 mm long; central spikelet lemma awned, 3.5–14 mm long; lateral spikelet lemma awned, to 7.5 mm long. **VEGETATIVE CHARACTERISTICS: Growth Habit:** tufted. **Culms:** erect to occasionally spreading or geniculate at base, 3–7 dm tall. **Sheaths:** open, glabrous to puberulent to strigose. **Auricles:** absent. **Ligules:** membranous, truncate, sometimes ciliate, 0.2–0.7 mm long. **Blades:** flat, scabrous to pilose, 4–12 cm long, 2–5 mm wide. **HABITAT:** Found in valleys and mountain parks from low elevations to subalpine over central and western Colorado. **COMMENTS:** Wingate (1994) lists this taxon as *Hordeum brachyantherum* Nevski. Colorado plants belong to the typical subspecies (subsp. *brachyantherum*) (Barkworth et al. 2007).

JRJ

Figure 326. *Critesion brachyantherum.*

305. *Critesion jubatum* (L.) Nevski (Fig. 327).

Synonym: *Hordeum jubatum* L. subsp. *jubatum*. **Vernacular Names:** Foxtail barley, squirrel-tail grass. **Life Span:** Perennial, or sometimes flowering the first year and appearing annual. **Origin:** Native. **Season:** C$_3$, cool season. **FLORAL CHARACTERISTICS: Inflorescence:** nodding spike at maturity, 4–15 cm long excluding awns, 4–6 cm wide, nearly as broad as long; rachis disarticulating at maturity. **Spikelets:** 3 spikelets per node, central spikelet perfect and sessile; lateral spikelets much reduced, pedicellate and staminate or sterile, pedicels curved, 0.7–1.2 mm long. **Glumes:** subequal to equal, setaceous; central spikelet glumes awnlike, generally 15–85 mm long, scabrous, narrow, divergent at maturity; lateral glumes similar in shape but slightly shorter. **Lemmas:** central spikelet lemmas 4–8 mm long, obscurely 5-nerved, tapering into long awn; lateral spikelet lemmas 0.7–1.2 mm long, staminate or sterile, sometimes reduced to an awn. **Awns:** central spikelet lemmas awned, awns 10–90 mm long; lateral spikelet lemmas awned, awns 2–15 mm long. **VEGETATIVE CHARACTERISTICS: Growth Habit:** cespitose. **Culms:** erect to decumbent below, 2–8 dm tall, slender, soft-pubescent to glabrous. **Sheaths:** open, glabrous to pilose. **Auricles:** absent or small, to 0.5 mm long. **Ligules:** ciliolate membrane, truncate, 0.2–1 mm long. **Blades:** flat to involute, 5–15 cm long, 2–5 mm wide, scabrous to hirsute. **HABITAT:** Found from low elevations to subalpine over Colorado in wet disturbed ground, around ponds, and in alkaline areas. **COMMENTS:** Common weed. It has poor forage value but is grazed some before the inflorescence emerges. Awns have been known to cause sores on the nose, eyes, and mouth of animals (Stubbendieck, Hatch, and Butterfield 1992). Sometimes grown as an ornamental. Two subspecies can be found in Colorado, subsp. *jubatum* and subsp. *intermedium* Bowden. The 2 subspecies can be separated by these characteristics:

1. Glumes of central spikelet 35–85 mm long; lemma awns of central spikelet 35–90 mm long . . . . subsp. *jubatum*
1. Glumes of central spikelet 15–35 mm long; lemma awns of central spikelet 11–35 mm long . . . . subsp. *intermedium*

306. *Critesion murinum* (L.) Á Löve (Fig. 328).

Synonym: *Hordeum murinum* L. subsp. *murinum*. **Vernacular Names:** Smooth barley, mouse barley. **Life Span:** Annual. **Origin:** Introduced. **Season:** C$_3$, cool season. **FLORAL CHARACTERISTICS: Inflorescence:** small spike, 3–8 cm long, 7–16 mm wide, exserted to partially enclosed by inflated sheath of uppermost leaf; rachis disarticulating at maturity. **Spikelets:** spikelets 3 to a node, all 3 well developed, central spikelet sessile, fertile; lateral spikelets pedicellate, sterile to staminate. **Glumes:** 10–30 mm long, 0.8–1.8 mm wide, those of central sessile spikelet and inner ones of lateral pedicellate spikelets slightly broadened above base and

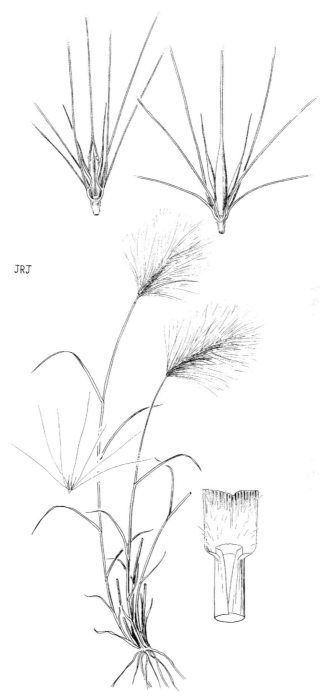

JRJ

Figure 327. *Critesion jubatum*.

Figure 328. *Critesion murinum*.

ciliate, then narrowing near apices, while the outer glumes of pedicellate spikelets remain awnlike and scabrous. **Lemmas:** of central spikelet 8–14 mm long, to 2 mm wide, glabrous to scabrous, fertile, obscurely 5-nerved; of lateral spikelets 7–18 mm long, sterile to staminate. **Paleas:** of lateral spikelet 8–15 mm long, glabrous to scabrous to hairy. **Awns:** of central and lateral lemmas stout, 15–50 mm long, characteristically reddish. **VEGETATIVE CHARACTERISTICS: Growth Habit:** solitary or loosely to densely tufted. **Culms:** generally geniculate at some of the nodes, up to a meter tall, nodes glabrous. **Sheaths:** glabrous. **Auricles:** well developed, up to 8 mm long, often completely surrounding culm. **Ligules:** truncate, entire to erose, ciliolate, 1–4 mm long. **Blades:** flat, folded, or involute; to 25 cm long, 2–10 mm wide, glabrous to scabrous to softly hairy, underside occasionally glabrous. **HABITAT:** Introduction from Eurasia; found along roadsides, along ditches, and in disturbed areas where moisture persists. **COMMENTS:** This is a highly variable species. Two subspecies are found in Colorado, subsp. *glaucum* (Steud.) Tzvelev and subsp. *leporinum* (Link) Arcang. They can be distinguished by these characteristics:

1. Lemmas of central floret much shorter than those of lateral florets; paleas of lateral florets scabrous on lower half; winter annual . . . . subsp. *leporinum*
1. Lemmas of central floret about equal to those of lateral florets; paleas of lateral florets distinctly pilose on lower half; summer annual . . . . subsp. *glaucum*

307. *Critesion pusillum* (Nutt.) Á. Löve (Fig. 329).

Synonyms: *Hordeum pusillum* Nutt., *Hordeum pusillum* Nutt. var. *pubens* Hitchc. **Vernacular Name:** Little barley. **Life Span:** Annual. **Origin:** Native. **Season:** C$_3$, cool season. **FLORAL CHARACTERISTICS: Inflorescence:** erect spike, 2–9 cm long excluding awns, 3–8 mm wide, occasionally partially enclosed in uppermost subtending leaf sheath; rachis disarticulating at maturity. **Spikelets:** 3 spikelets per node, central spikelet sessile, fertile; lateral spikelets pedicellate on  curved pedicels, 0.3–0.7 mm long, sterile with reduced, awn-tipped florets. **Glumes:** subequal to equal, 7–18 mm long, 0.5–1.5 mm wide, straight, not divergent; outermost glumes of lateral spikelets awnlike from base, scabrous to puberulent, slightly shorter; central glumes and inner awns of lateral spikelets obviously to slightly broadened above base, then narrowing into slender awns. **Lemmas:** of central spikelet scabrous, obscurely 5-nerved, 5–9 mm long, tapering into a small awn; of lateral spikelets 2.5–6 mm long, usually awned. **Awns:** central and inner glumes of lateral spikelets awned, awns 7–15 mm long; lemma of central spikelet awned, awns 2–10 mm long, lateral lemmas short-awned to 2 mm long. **VEGETATIVE CHARACTERISTICS: Growth Habit:** solitary to cespitose. **Culms:** erect, sometimes with geniculate base, 1–6 dm tall, glabrous. **Sheaths:** inflated, glabrous to pilose, round. **Auricles:** absent. **Ligules:** erose to short ciliate membrane, 0.2–0.8 mm long. **Blades:** flat, erect, glabrous to scabrous to hispidulous, 1–12 cm long, 2–5 mm wide, margin weakly scabrous. **HABITAT:** Found in dry areas at low elevations over Colorado. **COMMENTS:** One of the first weedy grass species to emerge in the spring.

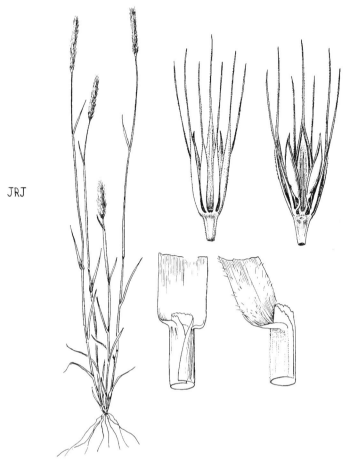

JRJ

Figure 329. *Critesion pusillum.*

104. *Elymus* L.

Perennial, occasionally with rhizomes. Inflorescence a bilateral spike or spicate raceme. Spikelets 1–4 per node, generally ascending, with 2–7 florets. Glumes lanceolate to subulate, rarely lacking. Lemmas longer than the glumes, rounded on back, acute to long-awned at tip, awn straight or curved outward. Palea well developed. Basic chromosome number, x = 7. Photosynthetic pathway, C_3.

Represented in Colorado by 14 species, 6 subspecies, and 1 variety.

Many species previously classified as *Agropyron, Hystrix,* or *Sitanion* are now placed in *Elymus.* A difficult genus with numerous hybrids to confuse identification. Hybrids exist between members of the genus as well as with members of other genera. See Barkworth and colleagues (2007) for specimens that do not fit the key or descriptions. Their work keys, illustrates, and describes many of the hybrids.

1. Spikelets solitary at all or most nodes of rachis
  2. Rhizomes present
    3. Lemmas 12–14 mm long . . . . 314. *E. glaucus*
    3. Lemmas 7–12 mm long
      4. Lemma awnless or, if short-awned, to 4 mm long, awns straight (if awn up to 10 mm, then plant is a weed)
        5. Glumes keeled distally, keels smooth and inconspicuous proximally; plants green; alien weed of cultivated areas . . . . 317. *E. repens*
        5. Glumes not keeled or keeled throughout their length; plants glaucous; native . . .
        . . . . 315. *E. lanceolatus*
      4. Lemmas with long awn, 5–12 mm long, divergent at maturity . . . . 308. *E. albicans*
  2. Rhizomes absent (although short rhizomes occasionally present in *E. glaucus*)
    6. Culms prostrate, strongly decumbent, or spreading; plants usually less than 5 dm tall, of alpine or upper subalpine communities . . . . 318. *E. scribneri*
    6. Culms ascending or erect, sometimes geniculate; plants usually taller than 5 dm, of lower communities
      7. Lemmas awned, awns 7–40 mm long
        8. Awns strongly arcuate, outcurved, or recurved
          9. Plants 4.5–10 dm tall; anthers 3–5 mm long; spike nodding or pendent . . . . . .
          . . . . 309. *E. arizonicus*
          9. Plants 3–5 dm tall; anthers 0.8–1.5 mm long; spike erect . . . . . 310. *E. bakeri*
        8. Lemma awns usually straight or flexuous
          10. Glumes 1.8–2.3 mm wide, margins 0.2–0.3 mm wide . . . .
          . . . . 319. *E. trachycaulus*
          10. Glumes 0.6–1.5 mm wide, margins 0.1–0.2 mm wide . . . . . . 314. *E. glaucus*
      7. Lemma unawned or awn up to 7 mm long
        11. Glume margins unequal, widest near the tip . . . . 320. *E violaceus*
        11. Glume margins equal, widest near middle.
          12. Glumes longer than adjacent lemma; anthers 1.2–2.5 mm long . . . .
          . . . . 319. *E. trachycaulus*
          12. Glumes shorter than adjacent lemma; anthers (2.5) 3–6 mm long . . . .
          . . . . 315. *E. lanceolatus*
1. Spikelets 2–3 or more at all or most nodes of rachis
  13. Rachis disarticulating at maturity
    14. Glume awns entire or split into 2–3 divisions; rachis internodes 3–15 mm long . . . . .
    . . . . 313. *E. elymoides*
    14. Glume awns split into 3–9 divisions; rachis internodes 3–5 mm long . . . .
    . . . . 316. *E. multisetus*
  13. Rachis not disarticulating
    15. Glumes body less than adjacent lemma; lemma awns usually flexuous to curving
      16. Glumes with more or less terete bases, awns 10–27 mm long or glume bodies indistinguishable from awns; cauline nodes usually concealed by sheaths . . . .
      . . . . 311. *E. canadensis*
      16. Glumes with more or less flat bases, awns 1–9 mm long; cauline nodes usually exposed . . . . 314. *E. glaucus*
    15. Glumes body longer than adjacent lemma; lemma awns usually straight
      17. Lemma awns 5–20 mm long . . . . 321. *E. virginicus*
      17. Lemma awns 0.5–4 mm long (rarely 5–10 mm long on distal spikelets) . . . .
      . . . . 312. *E. curvatus*

308. *Elymus albicans* (Scribn. & J. G. Sm.) Á. Löve (Fig. 330).

Synonyms: *Agropyron albicans* Scribn. & J. G. Sm., *Agropyron albicans* Scribn. & J. G. Sm. var. *griffithii* (Scribn. & J. G. Sm. *ex* Piper) Beetle, *Agropyron dasystachyum* (Hook.) Scribn. & J. G. Sm. subsp. *albicans* (Scribn. & J. G. Sm.) D. R. Dewey, *Agropyron griffithii* Scribn. & J. G. Sm. *ex* Piper, *Agropyron griffithii* Scribn. & J. G. Sm. *ex* Piper, *Elymus albicans* (Scribn. & J. G. Sm.) Á. Löve var. *griffithii* (Scribn. & J. G. Sm. *ex* Piper) Dorn, *Elymus griffithii* (Scribn. & J. G. Sm.
ex Piper) Á. Löve, *Elymus lanceolatus* (Scribn. & J. G. Sm.) Gould subsp. *albicans* (Scribn. & J. G. Sm.) Barkworth & D. R. Dewey, *Elytrigia dasystachya* (Hook.) Á. & D. Löve subsp. *albicans* (Scribn. & J. G. Sm.) D. R. Dewey, *Roegneria albicans* (Scribn. & J. G. Sm.) Beetle, *Roegneria albicans* (Scribn. & J. G. Sm.) Beetle var. *griffithii* (Scribn. & J. G. Sm. *ex* Piper) Beetle. **Vernacular Names:** Montana wheatgrass, wildrye. **Life Span:** Perennial. **Origin:** Native. **Season:** C$_3$, cool season. **FLORAL CHARACTERISTICS: Inflorescence:** erect, slender spikes, 6–14 cm long, middle internodes of spike 6–10 mm long. **Spikelets:** generally 1 spikelet per node, rarely 2 at some nodes; 10–18 mm long, 3- to 12-flowered, generally closely overlapping near base of spikelet; disarticulation above the glumes and below each floret. **Glumes:** subequal to unequal, oblong-lanceolate to lanceolate-elliptic, acute to acuminate, broadest at or above the middle, 3- to 5- to rarely 7-nerved, scabrous to glabrous, half to nearly as long as adjacent lemma; lower 4–8 mm long; upper 5.5–8 mm long. **Lemmas:** back rounded, glabrous to pubescent, 7.5–9.5 mm long, 5- to occasionally 7-nerved, but nerves may be obscure below. **Awns:** glumes generally with a short awn 0.5–3 mm long; lemma subacute to awn-tipped to awned, 5–12 mm long, divergent when mature, arcuate. **Anthers:** 3–5 mm long. **VEGETATIVE CHARACTERISTICS: Growth Habit:** rhizomatous; stems usually solitary along rhizomes, although occasionally clustered and cespitose. **Culms:** generally glaucous, 4–10 dm tall. **Sheaths:** glabrous to minutely strigulose to hirsute-pilose, rarely ciliate on margins. **Auricles:** generally present, to 1.5 mm long. **Ligules:** erose-ciliolate, to 0.5 mm long. **Blades:** firm, stiff, narrow, involute to rarely flat, 1–3 mm wide, upper surface glabrous to occasionally scabrous. **HABITAT:** Found on dry sagebrush hillsides up to wooded slopes on shallow, rocky soils in the mountains. **COMMENTS:** This species is considered by Barkworth and colleagues (2007) to be the result of hybridization between *E. lanceolatus* (Scribn. & J. M. Small) Gould and *Pseudoroegneria spicata* (Pursh) Á. Löve.

309. *Elymus arizonicus* (E. Nels.) Á. Löve (Fig. 331).

Synonyms: *Agropyron arizonicum* Scribn. & J. G. Sm., *Agropyron spicatum* Pursh var. *arizonicum* (Scribn. & J. G. Sm.) M. E. Jones, *Elytrigia arizonica* (Scribn. & J. G. Sm.) D. R. Dewey *Pseudoroegneria arizonica* (Scribn. & Merr.) Á. Löve. **Vernacular Name:** Arizona wheatgrass. **Life Span:** Perennial. **Origin:** Native. **Season:** C$_3$, cool season. **FLORAL CHARACTERISTICS: Inflorescence:** spike, 12–25 cm long, 2.5–6 cm wide with awns included, flexuous, nodding to pendent at maturity. **Spikelets:** generally 1 spikelet per node, 14–26 mm long, 6–8 mm wide, 4- to 6-flowered, appressed to divergent; disarticulation above the glumes and below

Figure 330. *Elymus albicans.*

Figure 331. *Elymus arizonicus*.

each floret. **Glumes:** subequal to unequal, narrowly lanceolate, bases flat, margins hyaline, acute to acuminate, unawned to awned; lower 5–9 mm long, 3- to 5-nerved; upper 8–10 mm long, 3- to 5-nerved. **Lemmas:** lanceolate, back rounded, scabrous, 8–15 mm long, 5- to 7-nerved. **Awns:** glumes, when awned, to 4 mm, straight; lemma awned, awn 10–25 mm long, strongly arcuate. **Anthers:** 3–5 mm long. **VEGETATIVE CHARACTERISTICS: Growth Habit:** cespitose, non-rhizomatous. **Culms:** erect to basally decumbent, 4.5–10 dm tall, glabrous. **Sheaths:** glabrous. **Auricles:** generally present, to 1 mm, falcate in shape. **Ligules:** to 1 mm long on basal sheaths, 1–3 mm on flag leaves. **Blades:** flat to conduplicate, 2.5–6 mm wide, upper surface scabrous, with scattered hairs; lower surface glabrous and smooth. **HABITAT:** Found in moist, rocky soil in mountain canyons. **COMMENTS:** When immature, this species has an erect spike that is sometimes mistaken for *Pseudoroegneria spicata*, but the culms are thicker, the ligules are longer, there are more basal leaves in general, and the leaf blades are wider. Once mature, the drooping spike and solitary spikes are very distinctive and easy to identify (Barkworth et al. 2007).

310. *Elymus bakeri* (E. Nels.) Á. Löve (Fig. 332).

Synonyms: *Agropyron bakeri* E. Nels., *Agropyron trachycaulum* (Link) Malte *ex* H. F. Lewis var. *bakeri* (E. Nels.) Boivin, *Elymus trachycaulus* (Link) Gould *ex* Shinners subsp. *bakeri* (E. Nels.) Á. Löve. **Vernacular Name:** Baker's wheatgrass. **Life Span:** Perennial. **Origin:** Native. **Season:** C_3, cool season. **FLORAL CHARACTERISTICS: Inflorescence:** slender spike, 5–12 cm long. **Spikelets:** 10–19 mm long, 3- to 6-flowered, densely to loosely imbricate on rachis;

rachilla scabrous to short-pubescent; disarticulation above the glumes and below each floret. **Glumes:** subequal to unequal, awn-tipped; lower to 9 mm long, 3- to 5-nerved; upper to 10 mm long, 5- to 7-nerved. **Lemmas:** scabrous to glabrous, 9–12 mm long, nerves obscure on back. **Awns:** glumes awned from an entire to sometimes bifid apices, 2–8 mm long, straight or divergent; lemma awned, usually from bifid apices, 6–35 mm long, divergent when dry. **Anthers:** 0.8–1.5 mm long. **VEGETATIVE CHARACTERISTICS: Growth Habit:** loosely tufted, non-rhizomatous. **Culms:** erect, 3–7 dm tall, glabrous. **Sheaths:** glabrous. **Ligules:** to 1 mm long. **Auricles:** small, rudimentary, 0.3–0.6 mm long. **Blades:** flat, 10–30 cm long, 2–5 mm wide, both surfaces scabrous. **HABITAT:** Found in high mountain meadows but not alpine, on open slopes and woods (Barkworth et al. 2007). Allred (2005), however, considers it to grow in a similar habitat to the alpine species *E. scribneri* (Vasey) M. E. Jones. **COMMENTS:** A variable species that some consider a variety of *E. trachycaulus*. *E. bakeri* can be distinguished from *E. scribneri* by its elongated spike, with the rachis easily observed (Allred 2005).

Figure 332. *Elymus bakeri.*

311. *Elymus canadensis* L. (Fig. 333).

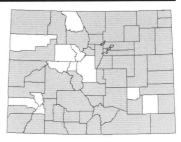

Synonyms: *Elymus brachystachys* Scribn. & Ball, *Elymus canadensis* L. var. *brachystachys* (Scribn. & Ball) Farw., *Elymus canadensis* L. var. *hirsutus* (Farw.) Dorn, *Elymus canadensis* L. var. *robustus* (Scribn. & J. G. Sm.) Mackenzie & Bush, *Elymus crescendus* L. C. Wheeler, *Elymus philadelphicus* L., *Elymus philadelphicus* L. var. *hirsutus* Farw., *Elymus robustus* Scribn. & J. G. Sm. **Vernacular Name:** Canada wildrye. **Life Span:** Perennial. **Origin:** Native. **Season:** C$_3$, cool season. **FLORAL CHARACTERISTICS: Inflorescence:** erect to nodding spike, 7–25 cm long, thick, open, bristly, occasionally interrupted below. **Spikelets:** 2–4 per node, 12–15 mm long, 2- to 5-flowered, imbricate; disarticulation generally above the glumes and below each floret. **Glumes:** subequal, body 6–13 mm long, 0.8–1.5 mm wide, 3- to 5-nerved, linear to lanceolate-linear, base generally subterete, broadest above base and tapering into a long awn, nerves scabrous to slightly ciliate. **Lemmas:** scabrous-pubescent, 8–14 mm long, 5- to 7-nerved, although sometimes obscure; broad at base and gradually tapering into a long awn. **Awns:** glume awns long, tapering, 10–27 mm long, generally straight to arcuate; lemma awn long, spreading, flexuous, divergent, 16–40 mm long. **Anthers:** 2–3.5 mm long. **VEGETATIVE CHARACTERISTICS: Growth Habit:** tufted. **Culms:** erect to decumbent at base, 80–150 cm tall; nodes mostly concealed by leaf sheaths. **Sheaths:** glabrous to occasionally pubescent. **Auricles:** well developed, 1–2 mm long, clasping stem. **Ligules:** ciliolate membrane, 0.2–2 mm long, truncate. **Blades:** flat to folded, 5–40 cm long, 7–20 mm wide, scabrous above, midnerve prominent below, margin finely toothed. **HABITAT:** Commonly found along roadsides and moist areas at lower elevations up to the foothills over Colorado. **COMMENTS:** Closely related to *E. virginicus,* with which it will hybridize where they are sympatric.

312. *Elymus curvatus* Piper. (Fig. 334).

Synonyms: *Elymus submuticus* (Hook.) Smyth, *Elymus virginicus* L. var. *jenkinsii* Bowden, *Elymus virginicus* L. var. *submuticus* Hook. **Vernacular Name:** Awnless wildrye. **Life Span:** Perennial. **Origin:** Native. **Season:** C$_3$, cool season. **FLORAL CHARACTERISTICS: Inflorescence:** erect spike, 9–15 cm long, base of inflorescence slightly included in uppermost leaf sheath to exserted; rachis flattened. **Spikelets:** 10–15 mm long, elliptic to oblong and laterally compressed, spikelets generally 2 per node, 3- to 4-flowered but occasionally with as few as 2 to as many as 5 flowers, often reddish brown in appearance; disarticulating at maturity below the glumes and each floret. **Glumes:** subequal to equal, coriaceous, base terete, firm, indurate, and strongly curved outward; body 7–15 mm long, 1.2–2.1 mm wide, linear-lanceolate, broadest above base, curved outward, 3- to 5-nerved above base, generally glabrous to scabrous, sometimes hispidulous, and rarely hirsute on nerves; apex awn-tipped to awned. **Lemmas:** lanceolate to oblong, coriaceous, glabrous to scabrous, rarely

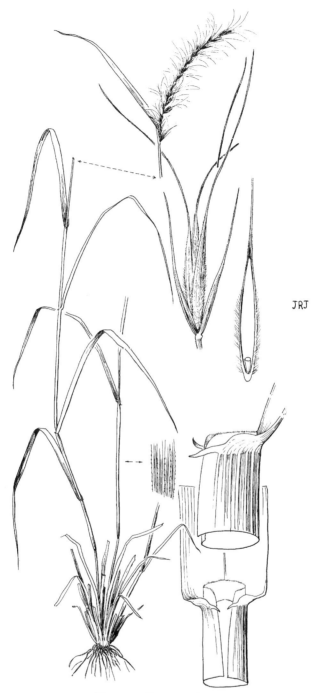

JRJ

Figure 333. *Elymus canadensis.*

Figure 334. *Elymus curvatus.*

hirsute, 6–10 mm long, 3- to 5-nerved above, apex acuminate, generally awned. **Awns:** glumes awn-tipped to awned, 0–5 mm long; lemma awns 0.5–4 mm long, rarely 5–10 mm on distal spikelets, straight. **Anthers:** 1.5–3 mm long. **VEGETATIVE CHARACTERIS-TICS: Growth Habit:** cespitose and often glaucous. **Culms:** erect to occasionally geniculate basally, 6–11 dm tall. **Sheaths:** glabrous, generally reddish brown in appearance, sheaths longer than nodes. **Auricles:** present to occasionally lacking, to 1 mm long, falcate. **Ligules:** ciliolate, less than 1 mm long. **Blades:** those near base usually drooping and dying earlier, while upper blades tend to be more involute and are ascending in appearance, 15–20 cm long, 5–15 mm wide, upper surfaces smooth to occasionally scabrous. **HABITAT:** Grows in moist or damp soils found in open forests, thickets, grasslands, around ditches and on disturbed land, especially bottomland. **COMMENTS:** Closely related to *E. virginicus* L. and at one time included as var. *submuticus* Hook. These two taxa overlap in range but can be distinguished by the shorter awns on *E. curvatus*. Additionally, in Colorado this taxon tends to have its spikes completely exserted, and anthesis tends to be 1–2 weeks behind that of *E. virginicus* (Barkworth et al. 2007).

313. *Elymus elymoides* (Raf.) Swezey (Fig. 335).

Synonyms: Subsp. *brevifolius* (J. G. Sm.) Barkworth: *Elymus elymoides* (Raf.) Swezey var. *brevifolius* (J. G. Sm.) Dorn, *Elymus longifolius* (J. G. Sm.) Gould, *Sitanion hystrix* (Nutt.) J. G. Sm. var. *brevifolium* (J. G. Sm.) C. L. Hitchc., *Sitanion longifolium* J. G. Sm., *Sitanion velutinum* Piper; subsp. *elymoides*: *Elymus sitanion* Schult., *Sitanion elymoides* Raf., *Sitanion hystrix* (Nutt.) J. G. Sm. **Vernacular Names:** Squirreltail, bottlebrush. **Life Span:** Perennial. **Origin:** Native. **Season:** C_3, cool season. **FLORAL CHARACTERISTICS: Inflorescence:** compact to open spike, 2–15 cm long, occasionally partially enclosed in uppermost leaf sheath; rachis disarticulates readily at maturity, and often only a partial spike remains; rachis internodes 3–15 mm long. **Spikelets:** usually 2, occasionally 3 spikelets per node, solitary spikelets often found in upper portion of spike, generally 1- to 6-flowered, 10–15 mm long, some florets may be reduced and glume-like below. **Glumes:** equal, 35–85 mm long, subulate, tapering into scabrous awns, entire to bifid. **Lemmas:** glabrous to puberulent, 8–10 mm long, obscurely 3- to 5-nerved; midnerve extended into awn, while 2 lateral nerves extending into bristles to 10 mm long. **Paleas:** subequal to lemma, 2-nerved, nerves often prolonged beyond paleas as bristles. **Awns:** glume awns slender, spreading, scabrous, 2–10 cm long, entire or occasionally divided into 2 to rarely 3 short, unequal segments; lemma awns 5–15 mm long, slender to stout, straight to divergent to flexuous. **Anthers:** 0.9–2.2 mm long. **VEGETATIVE CHARACTERISTICS: Growth Habit:** loosely to densely tufted. **Culms:** erect to spreading, subglabrous to white villous overall, 1–6 dm tall. **Sheaths:** open, glabrous to pubescent. **Auricles:** inconspicuous, to 1 mm long, often purplish. **Ligules:** truncate to obtuse, ciliate membrane to erose, 0.6–1 mm long. **Blades:** flat to folded to involute, 5–20 cm long, 1–5 mm wide, ascending, glabrous to hirsute to villous. **HABITAT:** Commonly found in dry areas from low elevations to subalpine. **COMMENTS:** This and *Achnatherum hymenoides* are the only 2 species that are reported from every county in the state. Two subspecies are found in Colorado:

subsp. *brevifolius*

subsp. *elymoides*

Figure 335. *Elymus elymoides.*

573

subsp. *elymoides* and subsp. *brevifolius* (J. G. Sm.) Barkworth. They can be separated by these characteristics:

1. Lowermost floret of one or both spikelets at each node sterile and reduced to a glume-like structure; palea nerves usually extended as bristles . . . . subsp. *elymoides*
1. Lowermost floret fertile, not at all reduced; palea nerves rarely extended as bristles . . . . .
. . . . subsp. *brevifolius*

314. *Elymus glaucus* Buckley (Fig. 336).

Synonym: *Elymus glaucus* Buckley var. *breviaristatus* Burtt-Davy. **Vernacular Names:** Common western wildrye, blue wildrye. **Life Span:** Perennial. **Origin:** Native. **Season:** C$_3$, cool season. **FLORAL CHAR-ACTERISTICS: Inflorescence:** erect to slightly nodding spike, 5–21 cm long, 0.5–2 cm wide, dense to loose, and interrupted below and overlapping above. **Spikelets:** elliptic to oblong, laterally compressed, generally 2 per node, 8–25 mm long, 2- to 6-flowered; disarticulation above the glumes and below each fertile floret. **Glumes:** subequal, 9–14 mm long, 0.6–1.5 mm wide, three-quarters to nearly equaling adjacent lemmas, linear-lanceolate, bases generally flat and thin, 3- to 7-nerved, broadest above base, glabrous to scabrous, concealing base of florets; margins hyaline, 0.1–0.2 mm wide; apices unawned or awned. **Lemmas:** elliptic to oblong, rounded on back, 8–15 mm long, 5- to 7-nerved, glabrous to scabrous on nerves, awn-tipped to awned. **Awns:** glumes often tapering into a short, straight awn, 1–9 mm long; lemmas awn-tipped to awned, awns slender and generally straight to flexuous, although occasionally spreading; 1–30 mm long. **Anthers:** 1.5–3.5 mm long. **VEGETATIVE CHARACTERISTICS: Growth Habit:** loose to densely tufted, occasionally with short rhizomes. **Culms:** erect, 3–14 dm tall; nodes usually exposed. **Sheaths:** glabrous to retrorsely puberulent to long-hairy; collar often purplish. **Auricles:** well developed, 2.5 mm long, purplish. **Ligules:** erose-ciliolate to entire, truncate, 0.3–1 mm long. **Blades:** flat to slightly involute, lax, 4–19 mm wide, glaucous, scabrous to sparsely pilose, occasionally glabrous below. **HABITAT:** Moist or dry mountain meadows, open forests, or shaded aspen groves (Weber and Wittmann 2001b; Barkworth et al. 2007). **COMMENTS:** The only subspecies found in Colorado is the typical one (subsp. *glaucus*).

315. *Elymus lanceolatus* (Scribn. & J. G. Sm.) Gould (Fig. 337).

Synonyms: *Agropyron dasystachyum* (Hook.) Scribn. & J. G. Sm., *Agropyron dasystachyum* (Hook.) Scribn. & J. G. Sm. var. *riparium* (Scribn. & J. G. Sm.) Bowden, *Agropyron dasystachyum* (Hook.) Scribn. & J. G. Sm. var. *riparum* (Scribn. & J. G. Sm.) Bowden, *Agropyron elmeri* Scribn., *Agropyron lanceolatum* Scribn. & J. G. Sm., *Agropyron riparium* Scribn. & J. G. Sm., *Agropyron riparum* Scribn. & J. G. Sm., *Elymus lanceolatus* (Scribn. & J. G. Sm.) Gould var. *riparius* (Scribn. & J. G. Sm.)

JRJ

Figure 336. *Elymus glaucus.*

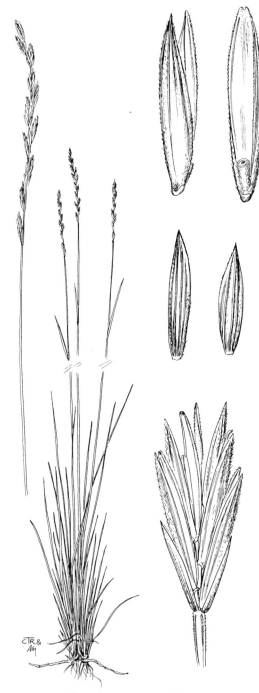

Figure 337. *Elymus lanceolatus.*

Dorn, *Elymus subvillosus* (Hook.) Gould, *Elytrigia dasystachya* (Hook.) Á. & D. Löve, *Elytrigia ripara* (Scribn. & J. G. Sm.) Beetle, *Elytrigia riparia* (Scribn. & J. G. Sm.) Beetle. **Vernacular Name:** Streambank wheatgrass. **Life Span:** Perennial. **Origin:** Native. **Season:** C$_3$, cool season. **FLORAL CHARACTERISTICS: Inflorescence:** slender spike, 4–22 cm long, erect to slightly nodding. **Spikelets:** 1 to rarely 2 spikelets per node, 11–18 mm long, 3- to 12-flowered, generally closely overlapping except low on spike; disarticulation is above the glumes and below each floret. **Glumes:** subequal, one-half to three-quaters as long as adjacent lemmas, oblong-lanceolate, scarious margined, generally scabrous to pubescent over most of surface, 3- to 5-nerved, acute to acuminate, broadest at or above midpoint, tapering to an acute or acuminate apices, apices sometimes mucronate or very short-awned; lower 4–8.5 mm long; upper 5.5–10 mm long. **Lemmas:** glabrous to villous, 7–10 mm long, 5-nerved, although obscure below; apex acute, occasionally awn-tipped. **Awns:** lemmas occasionally awn-tipped, to 2 mm long. **Anthers:** (2.5) 3–6 mm long. **VEGETATIVE CHARACTERISTICS: Growth Habit:** solitary or tufted, strongly rhizomatous. **Culms:** erect to slightly decumbent, 3–13 dm tall, glaucous. **Sheaths:** open, glabrous to puberulent, margins rarely ciliate. **Auricles:** to 1.5 mm long. **Ligules:** erose-ciliate, to 1 mm long. **Blades:** involute to flat, 5–20 cm long, 1–3.5 mm wide, stiff, glabrous to scabrous, occasionally pilose. **HABITAT:** Found on rocky slopes and other open areas, preferring gravelly or sandy soil, from the plains to montane (Weber and Wittmann 2001a, 2001b; Wingate 1994). **COMMENTS:** The only subspecies found in Colorado is the typical one (subsp. *lanceolatus*).

316. *Elymus multisetus* (J. G. Sm.) Burtt Davy (Fig. 338).

Synonyms: *Elymus sitanion* Schult. var. *jubatum* J. G. Sm., *Sitanion jubatum* J. G. Sm., *Sitanion multisetum* J. G. Sm. **Vernacular Name:** Big squirreltail. **Life Span:** Perennial. **Origin:** Native. **Season:** C$_3$, cool season. **FLORAL CHARACTERISTICS: Inflorescence:** dense spike, erect, 3–10 cm long, appears bristly, exserted to somewhat included in uppermost leaf sheath; rachis internodes 3–5 mm long; rachis disarticulates readily at maturity and then below each floret. **Spikelets:**  10–15 mm long, divergent, 2- to 4-flowered, generally 2 and rarely 3–4 per node, lower florets of 1 or both spikelets sterile and reduced to glume-like structures. **Glumes:** subequal, 10–100 mm long including awns, 3- to 9-cleft, divisions becoming divergent, scabrous awns of unequal lengths. **Lemmas:** 8–10 mm long, obscurely 3- to 5-nerved, midnerve tapering into straight, scabrous awn; 2 lateral nerves sometimes extending into bristly awns. **Paleas:** subequal to lemmas, 2-nerved, nerves exserted beyond paleas as bristles to 2 mm long. **Awns:** reddish to purplish; glumes with 3–9 awns of unequal lengths up to 70 mm long; lemma midnerve awn to 100 mm long, lateral nerves extending into bristly awns to 20 mm long. **Anthers:** 1–2 mm long. **VEGETATIVE CHARACTERISTICS: Growth Habit:** densely cespitose. **Culms:** erect, 2–6 dm tall, puberulent, hollow. **Sheaths:** glabrous to pilose, open. **Auricles:** well developed, 0.5–1.5 mm long. **Ligules:** membranous, generally entire, truncate, 0.2–0.7 mm long. **Blades:** flat to involute, 1.5–3.5 mm wide, scabrous to pilose. **HABITAT:** Found in sagebrush or grasslands (Weber and Wittmann 2001b). **COMMENTS:** Closely related to *E. elymoides* but tends to occur in slightly more mesic areas (Barkworth et al. 2007).

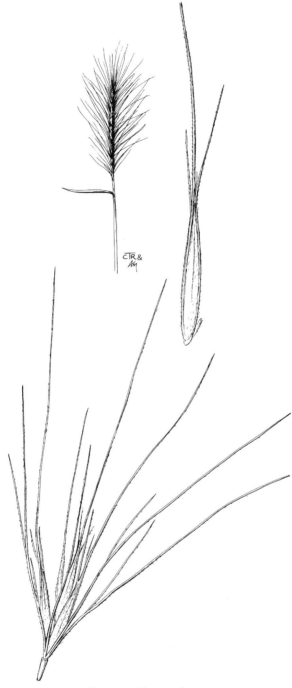

Figure 338. *Elymus multisetus.*

317. *Elymus repens* (L.) Gould (Fig. 339).

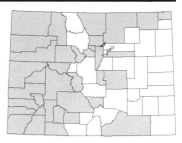

Synonyms: *Agropyron repens* (L.) P. Beauv., *Agropyron repens* (L.) P. Beauv. var. *subulatum* (Schreb.) Roem. & Schult., *Elytrigia repens* (L.) Desv. *ex* Nevski, *Elytrigia repens* (L.) Desv. *ex* Nevski var. *vaillantiana* (Wulfen & Schreb.) Prokudin, *Elytrigia vaillantiana* (Wulfen & Schreb.) Beetle, *Triticum repens* L., *Triticum vaillantianum* Wulfen & Schreb. **Vernacular Names:** Quackgrass, couchgrass. **Life Span:** Perennial. **Origin:** Introduced. **Season:** C₃, cool season. **FLORAL CHARACTERISTICS: Inflorescence:** erect spike, 10–15 cm long, 0.5–1.5 cm wide. **Spikelets:** solitary or occasionally 2 at each node, 1–27 mm long, 3- to 7-flowered, appressed or ascending; disarticulation above the glumes and below each floret. **Glumes:** subequal, keeled distally, scabrous and conspicuous, proximally inconspicuous and smooth, rigid, 5- to 7-nerved, midnerve usually scabrid, apices awn-tipped to awned; lower 9–11 mm long; upper 7–12 mm long. **Lemmas:** glabrous to scabrous, scaberulous above, 7–12 mm long, 5-nerved, acute to awned. **Awns:** glumes sometimes awned, awns 0.5–4 mm long; lemmas sometimes awned, 0.5–4 mm long, straight. **Anthers:** 4–7 mm long. **VEGETATIVE CHARACTERISTICS: Growth Habit:** strongly rhizomatous. **Culms:** decumbent to erect, 5–10 dm tall, green, occasionally glaucous. **Sheaths:** glabrous to soft-pilose below. **Auricles:** often over 1 mm long. **Ligules:** erose-ciliate membrane, to 1 mm long. **Blades:** flat, 10–30 cm long, 4–10 mm wide, smooth below, upper surface scabrous, margins scabrid. **HABITAT:** Found in ditches and gardens as a very common and persistent weed from low elevations up to subalpine. **COMMENTS:** A common weed in flower beds and ornamental juniper bushes. It is very difficult to eradicate because of the extensive rhizomes. Although often considered a noxious weed, where abundant it does provide good forage for livestock.

318. *Elymus scribneri* (Vasey) M. E. Jones (Fig. 340).

Synonym: *Agropyron scribneri* Vasey. **Vernacular Name:** Spreading wheatgrass. **Life Span:** Perennial. **Origin:** Native. **Season:** C₃, cool season. **FLORAL CHARACTERISTICS: Inflorescence:** crowded spike, 3.5–10 cm long, 3–6 mm wide including awns, generally nodding to flexuous; rachis disarticulates at maturity and then below each floret. **Spikelets:** spikelets solitary at each node, rarely 2, generally purplish, 9–14 mm long, 3- to 6-flowered. **Glumes:** equal to subequal, 4–9 mm long without awns, 2- to 3- and rarely 5-nerved, narrow-lanceolate to linear, midnerve prolonged into 1 (2) scabrous awn(s). **Lemmas:** glabrous to scaberulous, 7–10 mm long, obscurely 5-nerved, midnerve exserted as awn. **Awns:** glume awns scabrous, divergent, sometimes split into 2, 8–30 mm long; lemma awns divergent to recurved, scabrous, 15–25 mm long. **Anthers:** 1–1.6 mm long. **VEGETATIVE CHARACTERISTICS: Growth Habit:** tufted, leaves generally basal. **Culms:** decumbent to spreading, 2–5 dm tall. **Sheaths:** glabrous to short-pubescent, open. **Auricles:** small or absent. **Ligules:** ciliolate, entire to

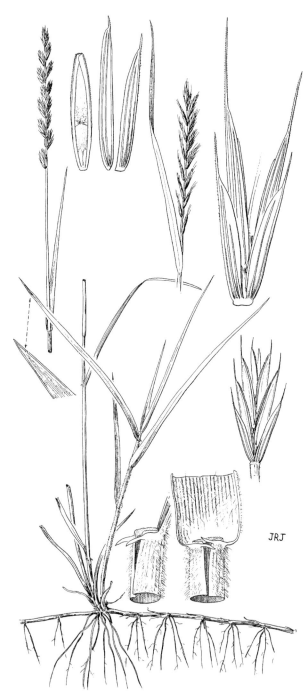

Figure 339. *Elymus repens.*

Figure 340. *Elymus scribneri*.

erose, to 0.5 mm long. **Blades:** flat to involute, rigid, 3–8 cm long, 2–5 mm wide, glabrous to puberulent on both surfaces. **HABITAT:** Found in upper subalpine to alpine communities, commonly on dry tundra (Weber and Wittmann 2001a, 2001b). **COMMENTS:** One of the more common grass species above timberline. Closely related to *E. bakeri*. *E. scribneri* can be recognized by its shorter spike in which the rachis is obscured (Allred 2005).

319. *Elymus trachycaulus* (Link) Gould *ex* Shinners (Fig. 341).

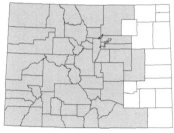

Synonyms: Subsp. *trachycaulus* Agropyron brevifolium Scribn., *Agropyron caninum* (L.) P. Beauv. subsp. *majus* (Vasey) C. L. Hitchc., *Agropyron caninum* (L.) P. Beauv. var. *hornemannii* (Koch) Pease & Moore, *Agropyron caninum* (L.) P. Beauv. var. *mitchellii* Welsh, *Agropyron pauciflorum* (Schwein.) Hitchc. *ex* Silveus, *Agropyron pauciflorum* (Schwein.) Hitchc. *ex* Silveus subsp. *majus* (Vasey) Melderis, *Agropyron pauciflorum* (Schwein.) Hitchc. *ex* Silveus var. *novae-angliae* (Scribn.) Taylor & MacBryde, *Agropyron tenerum* Vasey, *Agropyron teslinense* Porsild & Senn, *Agropyron trachycaulum* (Link) Malte *ex* H. F. Lewis, *Agropyron trachycaulum* (Link) Malte *ex* H. F. Lewis var. *majus* (Vasey) Fern., *Agropyron trachycaulum* (Link) Malte *ex* H. F. Lewis var. *novae-angliae* (Scribn.) Fern., *Elymus pauciflorus* (Schwein.) Gould, *non* Lam., *Elymus trachycaulus* (Link) Gould *ex* Shinners subsp. *novae-angliae* (Scribn.) Tzvelev, *Elymus trachycaulus* (Link) Gould *ex* Shinners var. *majus* (Vasey) Beetle, *Roegneria pauciflora* (Schwein.) Hyl., *Roegneria trachycaula* (Link) Nevski, *Triticum trachycaulum* Link. **Vernacular Names:** Slender wheatgrass, bearded wheatgrass. **Life Span:** Perennial. **Origin:** Native. **Season:** C$_3$, cool season. **FLORAL CHARACTERISTICS: Inflorescence:** slender spike, 4–25 cm long, 0.4–1 cm wide, erect. **Spikelets:** solitary at each node, 3- to 9-flowered, 10–20 mm long, strongly imbricate; disarticulation above the glumes and below each floret. **Glumes:** subequal, from 3/4 to as long as lemma, 6–15 mm long, 1.8–2.3 mm wide, oblong-elliptic to lanceolate, 5- to 7-nerved; margins scarious, 0.2–0.5 mm wide; apices acute to short-awned. **Lemmas:** glabrous to scabrous above, occasionally margins short-hirsute, 7.5–13 mm long, acute to awn-tipped to awned. **Awns:** glumes may have short awn to 11 mm long; lemma awns, when present, 1–40 mm long, generally straight, although ones less than 10 mm may be slightly curved. **Anthers:** (0.8) 1.2–2.5 mm long. **VEGETATIVE CHARACTERISTICS: Growth Habit:** cespitose, sometimes weakly rhizomatous. **Culms:** erect to decumbent at base, 3–15 dm tall. **Sheaths:** open, glabrous to pilose. **Auricles:** small to absent, 0.3–1 mm long, 1 often reduced. **Ligules:** ciliate membrane to erose, 0.2–0.8 mm long. **Blades:** flat to folded, 5–25 cm long, 2–8 mm wide, scabrous to pilose, margins with narrow whitish band. **HABITAT:** Extremely variable; found from low elevations to subalpine in open meadows and rocky slopes to closed forests. **COMMENTS:** Fairly common and considered an excellent forage species (Stubbendieck, Hatch, and Landholt 2003). Extremely variable species (as evidenced by the extensive synonymy) with numerous subspecies; the typical one (subsp. *trachycaulus*) is most frequent in Colorado.

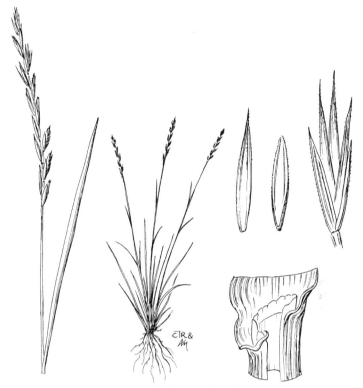

Figure 341. *Elymus trachycaulus.*

320. *Elymus violaceus* (E. Nels.) Á. Löve (Fig. 342).

Synonyms: *Agropyron latiglume* (Scribn. & J. G. Sm.) Rydb., *Agropyron trachycaulum* (Link) Malte *ex* H. F. Lewis var. *latiglume* (Scribn. & J. G. Sm.) Beetle, *Agropyron violaceum* (Hornem.) Lange var. *alboviride* (Hultén) Melderis, *Elymus alaskanus* (Scribn. & Merr.) Á. Löve subsp. *latiglumis* (Scribn. & J. G. Sm.) Á. Löve, *Elymus trachycaulus* (Link) Gould *ex* Shinners subsp. *latiglumis* (Scribn. & J. G. Sm.) Barkworth & D. R. Dewey, *Elymus trachycaulus* (Link) Gould *ex* Shinners var. *latiglumis* (Scribn. & J. G. Sm.) Beetle, *Roegneria violacea* (Hornem.) Melderis. **Vernacular Name:** Arctic wheatgrass. **Life Span:** Perennial. **Origin:** Native. **Season:** C_3, cool season. **FLORAL CHARACTERISTICS: Inflorescence:** erect spike, 5–12 cm long, 0.4–0.7 mm wide without awns; internode edges ciliate. **Spikelets:** 1 per node, appressed, 11–19 mm long, 4- to 5-flowered; disarticulation above the glumes and below each floret. **Glumes:** 8–12 mm long, 1.2–2 mm wide, subequal, from three-quarters to as long as lemmas, generally awned. **Lemmas:** glabrous to pubescent, hairs of equal length over outer surface, 9–12 mm long,

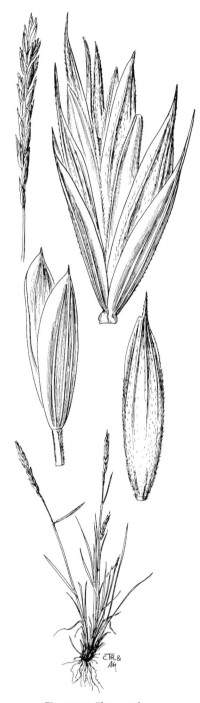

Figure 342. *Elymus violaceus.*

generally awned. **Awns:** glumes, when awned, to 2 mm long; lemmas generally awned, awn 0.5–3 mm long, straight. **VEGETATIVE CHARACTERISTICS: Growth Habit:** cespitose, non-rhizomatous. **Culms:** generally decumbent to geniculate, 1.8–7.5 dm tall, glabrous. **Sheaths:** glabrous. **Auricles:** to 0.5 mm long. **Ligules:** 0.5–1 mm long, truncate. **Blades:** flat, 10–30 cm long, 3–4 mm wide, both surfaces glabrous to hairy, lower surfaces generally less hairy and with shorter hairs. **HABITAT:** Found in alpine and subalpine communities on dolomitic to calcareous rocks. **COMMENTS:** Relatively uncommon species, not mentioned in Weber and Wittmann (2001a, 2001b) or Wingate (1994).

321. *Elymus virginicus* L. (Fig. 343).

Synonyms: *Elymus hordeiformis* Desf., *Elymus striatus* Schreb ex Muhl., *Elymus virginicus* L. var. *micromeris* Schmoll, *Elymus virginicus* L. var. *minor* Vasey *ex* L. H. Dewey, *Hordeum virginicum* (L.) Schenck, *Leptothrix virginica* (L.) Dumort; var. *virginicus*: *Elymus jejunus* (Ramaley) Rydb., *Elymus striatus* Willd., *Elymus virginicus* L. var. *jejunus* (Ramaley) Bush, *Hordeum striatum* (Willd.) Schenck. **Vernacular Name:** Virginia wild-rye. **Life Span:** Perennial. **Origin:** Native. **Season:** C_3, cool season. **FLORAL CHARACTERISTICS: Inflorescence:** erect spike, 4–17 cm long, base of inflorescence often slightly included in uppermost leaf sheath; spikelets strongly imbricate. **Spikelets:** generally 2 per node, 3- to 5-flowered, 10–16 mm long; disarticulation below the glumes and beneath each floret. **Glumes:** subequal, linear-lanceolate to subulate, firm, 7–15 mm long, broadest above base, base indurate, yellowish, curved outward, 2- to 5-nerved above base, apices awn-tipped to awned. **Lemmas:** glabrous to minutely hairy, 6–10 mm long, 5-nerved above, acute to awn-tipped to long-awned. **Awns:** glumes awn-tipped to awned, 4–20 mm long; lemma awns 0.5–25 mm long. **VEGETATIVE CHARACTERISTICS: Growth Habit:** cespitose. **Culms:** erect, 3–15 dm tall, glaucous. **Sheaths:** glabrous to pubescent below, occasionally glaucous. **Auricles:** well developed to occasionally lacking, to 1 mm long. **Ligules:** ciliolate, 0.2–0.7 mm long. **Blades:** flat, lax, 10–30 cm long, 4–15 mm wide, scabrous, especially distally. **HABITAT:** Found infrequently on the plains to foothills of eastern Colorado. **COMMENTS:** Closely related to *E. canadensis* L. and hybridizes with it where they are sympatric. Fairly good forage for livestock. The only variety present in Colorado is the typical one (var. *virginicus*).

105. *Eremopyrum* (Ledeb.) Jaub & Spach

Annuals. Inflorescence a bilateral spike with a single spikelet per node, rachis internodes flat. Inflorescence disarticulates as a whole in our species. Spikelets more than 3 times the length of internodes, divergent or pectinate, laterally compressed, with 2–5 florets. Glumes equal, becoming indurate; bases slightly connate, sharp-pointed, or short-awned. Lemma coriaceous, 5-nerved, unawned or short-awned. Palea shorter than lemmas, 2-keeled, keels sometimes prolonged into 2 tooth-like appendages. Basic chromosome number, $x = 7$. Photosynthetic pathway, C_3.

Represented in Colorado by a single species.

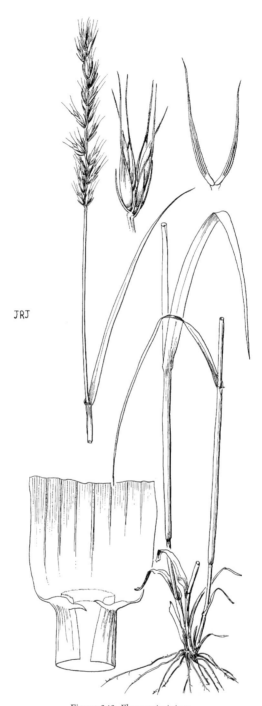

JRJ

Figure 343. *Elymus virginicus*.

322. *Eremopyrum triceum* (Gaertn.) Nevski (Fig. 344).

Synonyms: *Agropyron prostratum* (Pall.) P. Beauv., *Agropyron triceum* Gaertn., *Eremopyrum prostratum* (Pall.) P. Candargy, *Secale prostratum* Pall., *Secale pumilum* (L. f.) Pers., *Triticum prostratum* (Pall.) L. f. **Vernacular Name:** Annual wheatgrass. **Life Span:** Annual. **Origin:** Introduced. **Season:** C$_3$, cool season. **FLORAL CHARACTERISTICS: Inflorescence:** dense spike, 0.8–2.5 cm long, to 2 cm wide, ovate, long-exserted from sheaths at maturity; usually disarticulates at base of spike, occasionally above the glumes and beneath each floret. **Spikelets:** solitary at each node, 5–10 mm long, 3- to 6-flowered. **Glumes:** subequal, 5–7.5 mm long, becoming hardened but margins remaining membranous, 1-nerved, base slightly gibbous,

Figure 344. *Eremopyrum triticeum.*

acute to awned. **Lemmas:** lanceolate, 6–7.5 mm long, obscurely 3- to 5-nerved, glabrous to pubescent below, acuminate to awn-tipped to awned. **Paleas:** shorter than lemma, apices bifid. **Awns:** glumes with small awn, 0.8–1.7 mm long; lemmas, when awned, 1–7 mm long. **VEGETATIVE CHARACTERISTICS: Growth Habit:** solitary to tufted. **Culms:** erect to genicu-late at some nodes, 1–4 dm tall, hollow, retrorsely pubescent below inflorescence. **Sheaths:** open, glabrous to puberulent, upper sheath inflated. **Auricles:** small or absent. **Ligules:** membranous, 0.2–1 mm long, erose to lacerate. **Blades:** flat to loosely involute, 4–8 cm long, 1–4 mm wide, scabrous. **HABITAT:** Found in disturbed areas in western Colorado throughout the Colorado River Valley and Four Corners area and in Arapahoe County on the Front Range. **COMMENTS:** Previously referred to as *Agropyron triticeum* Gaertn., but its annual habit along with distinctive morphology separates it from the perennial *Agropyron*.

106. *Hordeum* L.

Annual. Inflorescence a spicate raceme with 3 spikelets at a node (triad). All 3 spikelets fertile and sessile on a strong rachis. Spikelets with 1 perfect floret. Glumes side by side, nar-rowly lanceolate to awnlike. Lemma rounded on back, conspicuously awned. Palea shorter than lemma and adnate to caryopsis. Basic chromosome number, $x = 7$. Photosynthetic pathway, C_3.

Represented in Colorado by a single species.

323. *Hordeum vulgare* L. (Fig. 345).

Synonyms: Subsp. *vulgare*: *Hordeum aegiceras* Nees ex Royle, *Hordeum distichon* L., *Hordeum hexastichon* L., *Hordeum hexastichum* L., *Hordeum irregulare* Aberg & Wiebe, *Hordeum sativum* Pers., *Hordeum vulgare* L. subsp. *hexastichon* (L.) Bonnier & Layens, *Hordeum vulgare* L. var. *trifurcatum* (Schlecht.) Alef. **Vernacular Names:** Barley, common barley. **Life Span:** Annual. **Origin:** Introduced. **Season:** C_3, cool season. **FLO-RAL CHARACTERISTICS: Inflorescence:** stout, erect spikes, dense, 5–10 cm long excluding length of awns; rachis continuous, generally not dis-articulating at maturity. **Spikelets:** 3 spikelets per node, central and lateral spikelets fertile and sessile. **Glumes:** subequal, linear, 6.5–20 mm long, 3-nerved, broadened below, glabrous to scabrous to ascending-pilose, tapering into scabrous awns. **Lemmas:** all 3 lemmas fertile, subequal, 6.5–12 mm long, 5-nerved, frequently obscure, generally glabrous, occasionally scabrous above, awned. **Awns:** glumes tapering into scabrous awns 4–20 mm long; lemmas tapering into stout, flattened awn 60–160 mm long with scabrous margins. **VEGETATIVE CHARACTERISTICS: Growth Habit:** single culms. **Culms:** erect, 6–12 dm tall, stout, gla-brous. **Sheaths:** smooth. **Auricles:** well developed, to 6 mm long, whitish. **Ligules:** ciliolate, erose-lacerate, 0.5–1.2 mm long. **Blades:** flat, 10–30 cm long, 5–18 mm wide, glabrous to scabrous. **HABITAT:** Cultivated and sometimes volunteering along roadsides and in fallow fields. **COMMENTS:** Other taxa commonly included in *Hordeum* have been moved to *Crite-sion*. Much more widely distributed than the map indicates, but infrequently collected. The subspecies found in Colorado is the typical one (subsp. *vulgare*).

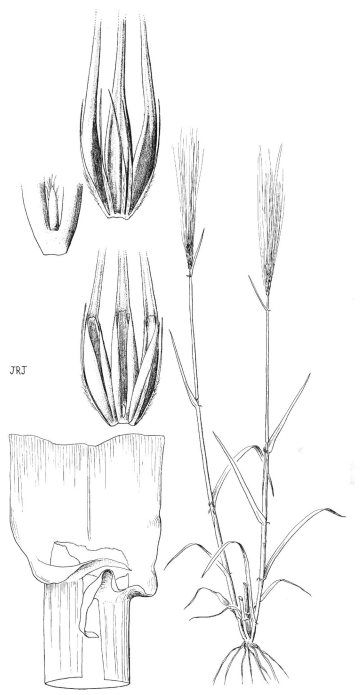

JRJ

Figure 345. *Hordeum vulgare.*

107. *Leymus* Hochst.

Perennials, cespitose or occasionally with rhizomes. Inflorescence a bilateral spike or spicate raceme. Disarticulation above the glumes and between florets. Spikelets 2 or more per node, sessile, with 2–7 florets. Glumes shorter than lemmas, narrowly lanceolate to awl-shaped, stiff. Lemma acute to short-awned. Palea shorter than lemma. Basic chromosome number, $x = 7$. Photosynthetic pathway, C_3.

Represented in Colorado by 5 species. A segregate of *Elymus*.

1. Plants strongly rhizomatous, definitely not a bunchgrass
    2. Culms 1.8–3 mm thick; glumes 5–16 mm long; spikes 5–20 cm long . . . . 328. *L. triticoides*
    2. Culms 2–12 mm thick; glumes 12–25 mm long; spikes 15–35 cm long . . . . 326. *L. racemosus*
1. Plants definitely a bunchgrass, very short rhizomes occasionally present
    3. 2–7 spikelets at each rachis node; plants 1–2 m tall . . . . 325. *L. cinereus*
    3. 1–2 spikelets at each rachis node; plants generally less than 1.2 m tall
        4. Lemma awns 0.2–2.5 mm long; nerves on adaxial surface of blade, 5–9; leaves involute; spikelets usually single at each rachis node . . . . 327. *L. salina*
        4. Lemma awns 1.3–7 mm long; nerves on adaxial surface of blade, 11–17; leaves flat or folded; spikelets usually paired at each rachis node . . . . 324. *L. ambiguus*

324. *Leymus ambiguus* (Vasey & Scribn.) D. R. Dewey (Fig. 346).

Synonyms: *Elymus ambiguus* Vasey & Scribn., *Elymus villiflorus* Rydb., *Leymus innovatus* (Beal) Pilger subsp. *ambiguus* (Vasey & Scribn.) Á. Löve. **Vernacular Name:** Colorado wildrye. **Life Span:** Perennial. **Origin:** Native. **Season:** C_3, cool season. **FLORAL CHARACTERISTICS: Inflorescence:** erect, slender spike, 6–17 cm long, 2–10 mm wide; spikelets dense. **Spikelets:** solitary at base and apex of rachis, otherwise 2 per node; 10–23 mm long, 2- to 7-flowered; disarticulation above the glumes and beneath florets. **Glumes:** subequal, subulate, scabrous above, generally 1-nerved; lower 2–10 mm long; upper 6–14 mm long. **Lemmas:** scabrous to puberulent, 6–15 mm long, at least above; 5-nerved above, below obscure to absent, apices awnless to awned. **Awns:** lemmas, when awned, with an awn 1.3–7 mm long. **VEGETATIVE CHARACTERISTICS: Growth Habit:** cespitose, occasionally with short rhizomes. **Culms:** erect, soft-pubescent near base, 3–11 dm tall. **Sheaths:** glabrous to puberulent. **Auricles:** to 1 mm long or sometimes absent. **Ligules:** 0.2–1.5 mm long, truncate. **Blades:** flat to folded, 10–30 cm long, 2–6 mm wide, (9) 11- to 17-nerved on adaxial surface, spreading, softly pilose. **HABITAT:** Found on rocky mountainsides in foothills to montane communities on the Front Range and western plateaus. **COMMENTS:** Soft-pubescence on sheath is a good vegetative character.

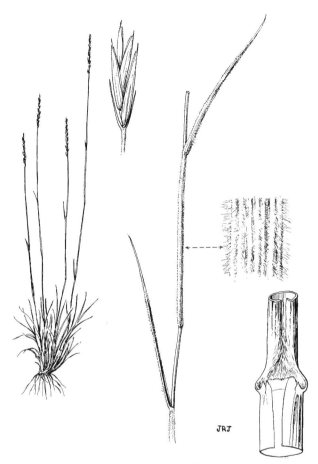

Figure 346. *Leymus ambiguus.*

325. *Leymus cinereus* (Scribn. & Merr.) Á. Löve (Fig. 347).

Synonyms: *Aneurolepidium piperi* (Bowden) Baum, *Elymus cinereus* Scribn. & Merr., *Elymus cinereus* Scribn. & Merr. var. *pubens* (Piper) C. L. Hitchc., *Elymus condensatus* J. Presl var. *pubens* Piper, *Elymus piperi* Bowden. **Vernacular Name:** Great Basin wildrye. **Life Span:** Perennial. **Origin:** Native. **Season:** C₃, cool season. **FLORAL CHARACTERISTICS: Inflorescence:** erect spike, 10–30 cm long, 7–17 mm wide, thick, dense, often interrupted below; spikelets

strongly imbricate. **Spikelets:** 2–7 per node, 9–25 mm long, 3- to 7-flowered; disarticulation above the glumes and beneath each floret. **Glumes:** subequal, subulate, 7–18 mm long,

Figure 347. *Leymus cinereus.*

0.5–2.5 mm wide, scabrous, 1-nerved to nerveless, awn pointed. **Lemmas:** glabrous to strigose, 6–12 mm long, nerveless below, 5- to 7-nerved above. **Awns:** two glumes, each with pointed awn; lemmas awnless to awned, 1–5 mm long. **VEGETATIVE CHARACTERISTICS: Growth Habit:** cespitose, forming large clumps; occasionally with short, thick rhizomes. **Culms:** erect, 7–20 dm tall, 2–5 mm thick, robust, glabrous to puberulent below nodes. **Sheaths:** glabrous to scabrous to soft-hairy with appressed to spreading hairs. **Auricles:** well developed at times, to 2 mm long or absent. **Ligules:** entire to erose membrane, 2–8 mm long, obtuse to acute. **Blades:** flat, firm, 15–40 cm long, 8–20 mm wide, upper surface scabrous, strongly nerved. **HABITAT:** Found on talus slopes and in canyon bottoms at low elevations to foothills. **COMMENTS:** It is one of the tallest native species one will encounter in the mountains. Ergot does infect the spikelets and can cause abortions or death in adult animals (Allred 2005).

326. *Leymus racemosus* (Lam.) Tzvelev (Fig. 348).

Synonyms: *Elymus arenarius* L. var. *giganteus* (Vahl) Schmalh., *Elymus giganteus* Vahl, *Elymus racemosus* Lam., *Leymus giganteus* (Vahl) Pilger. **Vernacular Names:** Mammoth wildrye, lyme-grass. **Life Span:** Perennial. **Origin:** Introduced. **Season:** C₃, cool season. **FLORAL CHARACTERISTICS: Inflorescence:** erect, dense spike, stout, 15–35 cm long, 1–2 cm wide. **Spikelets:** sessile, coarse, 15–25 mm long, 4- to 6-flowered, compressed, generally grouped 3–8 at nodes; disarticulation above the glumes and below florets. **Glumes:** linear-lanceolate, stiff, generally as long as spikelet, 12–25 mm long, 2–4 mm wide, strongly 3- to 5-nerved, usually exceeding lemmas, glabrous to inconspicuously hairy. **Lemmas:** hairy, 10–20 mm long, 7-nerved; softly hairy basally, becoming glabrous toward apices, tapering to a slender tip, awned or not. **Awns:** lemmas, if awned, to 2.5 mm long. **VEGETATIVE CHARACTERISTICS: Growth Habit:** strongly rhizomatous with long, creeping rhizomes. **Culms:** erect, generally glabrous, glaucous, stout, 2–12 mm thick basally, rarely more than 1 m tall. **Sheaths:** crowded, overlapping at base. **Ligules:** to 2.5 mm long. **Blades:** 6–15 mm wide, upper surface scabrid on nerves and margins, lower surface glabrous. **HABITAT:** Found west of Craig in Moffat County. **COMMENTS:** Native of southwestern Europe. Typically, this species is used as an ornamental, and perhaps it has escaped from cultivation.

327. *Leymus salina* (M. E. Jones) Á. Löve (Fig. 349).

Synonyms: *Elymus ambiguus* Vasey & Scribn. var. *salinus* (M. E. Jones) C. L. Hitchc., *Elymus salinus* M. E. Jones; subsp. *salina: Elymus ambiguus* Vasey & Scribn. var. *strigosus* (Rydb.) Hitchc., *Elymus strigosus* Rydb. **Vernacular Name:** Saline wildrye. **Life Span:** Perennial. **Origin:** Native. **Season:** C₃, cool season. **FLORAL CHARACTERISTICS: Inflorescence:** erect, slender spike, 5–15 cm long, 4–11 mm wide. **Spikelets:** solitary spikelet per node, occasionally paired, 9–21 mm long, 2- to 9-flowered; disarticulation above the glumes and below florets. **Glumes:** subequal, 4–12 mm long, subulate to reduced, often falcate, glabrous to scabrous, 1-nerved to nerveless, in some instances a reduced floret consisting of a single glume occurs. **Lemmas:** glabrous to scaberulous, 8–12 mm long, 5-nerved above, apices awn-tipped to short-awned. **Awns:** lemma awns, when present, 0.2–2.5 mm long. **VEGETATIVE CHARACTERISTICS: Growth Habit:** densely tufted, often short-rhizomatous. **Culms:** erect, glabrous, 3–14 dm tall, leaves generally basal, scabrous below inflorescence. **Sheaths:** glabrous to scabrous to infrequently pubescent. **Auricles:** well developed, often clasping culm, to 1 mm long. **Ligules:** ciliate, 0.2–1 mm long, truncate. **Blades:** usually involute, 10–15 cm long, 2–3 mm wide, 5- to 9-nerved on adaxial surface, glabrous to scabrous to puberulent. **HABITAT:** Found in western Colorado on dry rocky areas and desert clay hills as one of the most common and dominant species. **COMMENTS:** The specific epithet *salina* was named after Salina

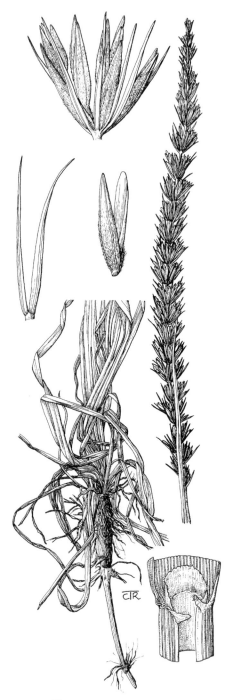

Figure 348. *Leymus racemosus.*

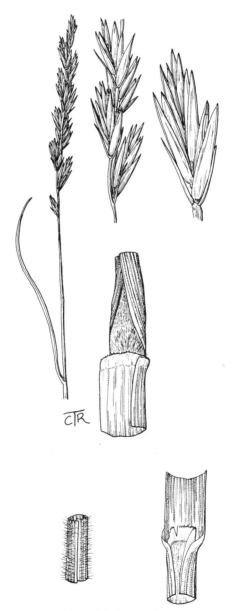

Figure 349. *Leymus salina*.

Pass, where it was first collected by Jones (Weber and Wittmann 2001b). Only 1 subspecies is found in Colorado, the typical one (subsp. *salina*).

328. *Leymus triticoides* (Buckley) Pilg. (Fig. 350).

Synonyms: *Elymus condensatus* J. Presl var. *triticoides* (Buckley) Thurb., *Elymus orcuttianus* Vasey, *Elymus triticoides* Buckley, *Elymus triticoides* Buckley var. *pubescens* Hitchc. **Vernacular Names:** Beardless wild-rye, creeping wildrye. **Life Span:** Perennial. **Origin:** Native. **Season:** C$_3$, cool season. **FLORAL CHARAC-TERISTICS: Inflorescence:** erect spike, 5–20 cm long, 5–15 mm wide, loose, slightly imbricate. **Spikelets:** brownish to purplish, 1–2 (3) per node, 12–20 mm long, 3- to 7-flowered; disarticulation above the glumes and below florets. **Glumes:** sub-equal, subulate to lanceolate, 6–16 mm long, 1- to 3-nerved, scabrous, apices awn pointed. **Lemmas:** glabrous to puberulent, occasionally only at apex, 5–16 mm long, obscurely 5- to 7-nerved, back rounded, awnless to short-awned. **Awns:** lemmas, when awned, 0.5–3 mm long. **VEGETATIVE CHARACTERISTICS: Growth Habit:** rhizomatous with creeping, scaly rhizomes. **Culms:** erect, 5–12 dm tall, slender, 1.8–3 mm thick, glaucous. **Sheaths:** glabrous to scabrous to puberulent. **Auricles:** well developed, to 1 mm long, and frequently clasping stem. **Ligules:** erose-ciliolate membrane, 0.2–1 mm long. **Blades:** flat to involute distally, 10–35 cm long, 2–5 mm wide, stiff, scaberulous to hairy. **HABITAT:** Found over Colorado on benches, clay flats, and saline meadows in some of the lower valleys (Weber and Wittmann 2001a, 2001b; Wingate 1994). **COMMENTS:** This taxon often fails to set seed because of its need for cross-pollination (Barkworth et al. 2007).

108. *Pascopyrum* Á. Löve

Perennial, rhizomatous. Inflorescence a bilateral spike, spikelets 1 at a node or occasionally paired at lower nodes. Disarticulation is above the glumes and beneath each floret. Spikelets with 2–12 florets, ascending but not appressed. Glumes half to two-thirds the length of spikelet, narrowly lanceolate, stiff, curved. Lemmas lanceolate, rounded on back; acute, mucronate, or short-awned. Paleas slightly shorter than lemmas. Basic chromosome number, $x = 7$. Photosynthetic pathway, C$_3$.

Represented in Colorado by the only species in the genus.

329. *Pascopyrum smithii* (Rydb.) Barkworth & D. R. Dewey (Fig. 351).

Synonyms: *Agropyron glaucum* Roem. & Schult. var. *occidentale* Scribn., *Agropyron molle* (Scribn. & J. G. Sm.) Rydb., *Agropyron occidentale* (Scribn.) Scribn., *Agropyron smithii* Rydb., *Agropyron smithii* Rydb. var. *molle* (Scribn. & J. G. Sm.) M. E. Jones, *Agropyron smithii* Rydb. var. *palmeri* (Scribn. & J. G. Sm.) Heller, *Elymus smithii* (Rydb.) Gould, *Elytrigia smithii* (Rydb.) Nevski, *Elytrigia smithii* (Rydb.) Nevski var. *mollis* (Scribn. & J. G. Sm.) Beetle. **Vernacular Name:** Western wheatgrass. **Life Span:** Perennial. **Origin:** Native. **Season:** C$_3$, cool season. **FLORAL CHARACTERISTICS: Inflorescence:** erect spike, 5–17 cm long; spikelets usually

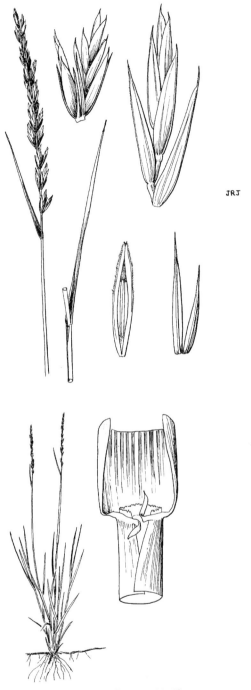

Figure 350. *Leymus triticoides*.

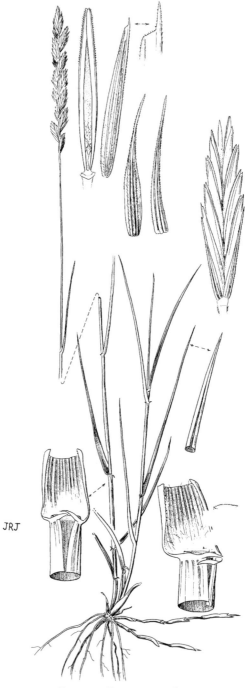

JRJ

Figure 351. *Pascopyrum smithii.*

closely imbricate, usually 1 but occasionally 2 per node. **Spikelets:** glaucous, 12–26 mm long, 2- to 12-flowered; disarticulation above the glumes and below each floret. **Glumes:** subequal, rigid, linear-lanceolate to lanceolate, obscurely 3- to 5-nerved, asymmetrical, glabrous to scabrous, tapering from midlength or below; lower 6–12 mm long; upper 7–15 mm long. **Lemmas:** lanceolate, acute, 6–14 mm long, 5-nerved, sometimes obscurely, rounded on back, awn-tipped

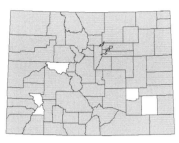

to awned. **Awns:** glumes sometimes awn-tipped; lemmas awn-tipped to awned, awns 0.5–5 mm long. **VEGETATIVE CHARACTERISTICS: Growth Habit:** strongly rhizomatous, rhizomes creeping. **Culms:** erect, 2–10 dm tall, glabrous to glaucous. **Sheaths:** open, glabrous to scabrous. **Auricles:** absent or, when present, small; 0.2–1 mm long, purple. **Ligules:** membranous, to 0.1 mm, truncate. **Blades:** flat to involute, rigid, 2–26 cm long, 1–4.5 mm wide, upper surface strongly nerved. **HABITAT:** Common plant of dry flats and grasslands and generally found from low elevations up into the foothills, but it has been found at elevations up to 9,900 ft (3,000 m) (Weber and Wittmann 2001b). **COMMENTS:** *P. smithii* was originally described as *Agropyron smithii* Rydb. It also has been placed in *Elymus* as *E. smithii* (Rydb.) Gould and *Elytrigia* as *E. smithii* (Rydb.) Nevski. It is a good forage species and cures well. A common and often dominant species of the Northern Mixed Prairie.

109. *Psathyrostachys* Nevski

Perennial, cespitose, or occasionally rhizomatous or stoloniferous. Basal sheaths closed, upper sheaths open. Inflorescence a bilateral spike. Disarticulation at rachis nodes. Spikelets 2–3 per node, sessile, with 1–3 florets, often with reduced florets above. Glumes equal to subequal, side by side, subulate, free to base, weakly 1-nerved. Lemma narrowly elliptic, rounded, 5- to 7-nerved, nerves usually prominent, apices acute to awned. Palea equal to or slightly longer than lemma, membranous, bifid. Basic chromosome number, x = 7. Photosynthetic pathway, C_3.

Represented in Colorado by a single species. *Psathyrostachys* is similar to *Leymus*, but the major difference is that *Psathyrostachys* has disarticulating rachises.

330. *Psathyrostachys juncea* (Fisch.) Nevski (Fig. 352).

Synonyms: *Elymus desertorum* Kar. & Kir., *Elymus junceus* Fisch., *Triticum juncellum* F. Herm. **Vernacular Name:** Russian wildrye. **Life Span:** Perennial. **Origin:** Introduced. **Season:** C_3, cool season. **FLORAL CHARACTERISTICS: Inflorescence:** erect spike, 6–11 cm long, 5–17 cm wide; at maturity rachis disarticulates at nodes; occasionally 2 and usually 3 spikelets per node. **Spikelets:** strongly overlapping, 7–10 mm long, 1- to 2-flowered, occasionally with reduced florets above. **Glumes:** subequal, subulate, 4–9 mm long, scabrous to short-hairy, may be glabrous at base, faintly 1-nerved; apices acute to awn-tipped. **Lemmas:** lanceolate, glabrous to hairy, 5–8 mm long, 5- to 7-nerved; apices acute to awned. **Awns:** glumes acute to awn-

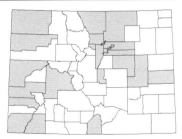

Figure 352. *Psathyrostachys juncea.*

tipped; lemmas acute to awned, awns 0.8–3.5 mm long. **VEGETATIVE CHARACTERISTICS: Growth Habit:** densely cespitose. **Culms:** erect to decumbent at base, 3–12 dm tall, generally glabrous to pubescent below inflorescence, leaves generally basal. **Sheaths:** glabrous, persistent. **Auricles:** present, 0.2–1.5 mm long. **Ligules:** ciliate, truncate, 0.2–0.3 mm long. **Blades:** flat to involute, 2.5–18 cm long, 5–20 mm wide, smooth to scabrous, generally glabrous. **HABITAT:** Introduced and used as a pasture grass or in revegetation and stabilization projects. It is tolerant of saline soils and is drought-resistant (Barkworth et al. 2007). **COMMENTS:** This species is not the same as *Thinopyrum junceum* (L.) Á. Löve.

110. *Pseudoroegneria* (Nevski) Á. Löve

Plants cespitose, occasionally with short rhizomes. Sheaths open, auricle well developed, ligules membranous. Inflorescence a bilateral spike with widely spaced, solitary spikelets; rachis internodes concave adjacent to spikelets. Spikelets with 3–9 florets, appressed; disarticulation above the glumes and beneath florets. Glumes unequal, lanceolate or narrowly lanceolate, acuminate, glabrous, 3–7-nerved. Lemma 5-nerved, usually glabrous, lanceolate, unawned or awned, awns strongly divergent at maturity; palea equal to lemma. Basic chromosome number, $x = 7$. Photosynthetic pathway, C_3.

Represented in Colorado by a single species.

331. *Pseudoroegneria spicata* (Pursh) Á. Löve (Fig. 353).

Synonyms: *Agropyron inerme* (Scribn. & J. G. Sm.) Rydb., *Agropyron spicatum* Pursh var. *inerme* (Scribn. & J. G. Sm.) Heller, *Agropyron spicatum* Pursh, *Agropyron spicatum* Pursh var. *pubescens* Elmer, *Agropyron vaseyi* Scribn. & J. G. Sm., *Elymus spicatus* (Pursh) Gould, *Elytrigia spicata* (Pursh) D. R. Dewey, *Roegneria spicata* (Pursh) Beetle. **Vernacular Name:** Bluebunch wheatgrass. **Life Span:** Perennial. **Origin:** Native. **Season:** C_3, cool season. **FLORAL CHARACTERISTICS: Inflorescence:** slender spike, 8–15 cm long, 3–10 mm wide excluding awns, lax, spikelets not imbricate. **Spikelets:** generally 1 per node, 8–22 mm long, 4- to 9-flowered; disarticulates above the glumes and beneath florets. **Glumes:** subequal, narrow, 4–13 mm long, 1/2 the length of spikelet; upper longest, 4- to 5-nerved, scabrid on nerves or glabrous; apices acute to obtuse. **Lemmas:** 7–14 mm long, glabrous, obscurely 5-nerved, acute to awned. **Awns:** glumes rarely awn-tipped; lemmas with or without awns; at anthesis, when present awns diverge strongly, 10–20 mm long. **VEGETATIVE CHARACTERISTICS: Growth Habit:** tufted, rarely with short rhizomes. **Culms:** erect, glabrous to puberulent below nodes, 3–10 dm tall, leaves generally cauline. **Sheaths:** glabrous to retrorse-puberulent to soft-hairy, persistent old sheaths in clumps. **Auricles:** well developed, 0.5–1 mm long. **Ligules:** erose-ciliolate, 0.5–2 mm long, truncate. **Blades:** flat to loosely involute, 5–25 cm long, 2–4.5 mm wide, glabrous to pubescent above. **HABITAT:** Common in the foothills to montane over much of Colorado. **COMMENTS:** It is considered an excellent forage species and makes good winter feed (Stubbendieck, Hatch, and Butterfield 1992).

Figure 353. *Pseudoroegneria spicata*.

111. *Secale* L.

Annual (in ours). Inflorescence a dense bilateral spike with spikelets solitary at nodes. Disarticulation above the glumes and between florets. Spikelets usually 2-flowered, rachilla extending behind upper floret as a minute bristle. Glumes narrow, rigid, and subulate. Lemma broad, 5-nerved, sharply keeled, ciliate on keel, tapering to a stout awn. Basic chromosome number, $x = 7$. Photosynthetic pathway, C_3.

Represented in Colorado by a single species.

332. *Secale cereale* L. (Fig. 354).

Synonyms: *Secale turkestanicum* Bensin, *Triticum cereale* (L.) Salisb., *Triticum secale* Link. **Vernacular Names:** Cereal rye, rye. **Life Span:** Annual or biennial. **Origin:** Introduced. **Season:** C_3, cool season. **FLORAL CHARACTERISTICS: Inflorescence:** dense distichous spike, laterally compressed, 4.5–12 cm long, often nodding after anthesis; spikelets solitary at each node. **Spikelets:** ascending, 10–18 mm long, 2-flowered, occasionally 3; disarticulation above the

glumes or at nodes, sometimes belatedly. **Glumes:** equal, 8–18 mm long, linear-subulate, 1-nerved, keeled, scabrous; awned. **Lemmas:** smooth to scabrous, asymmetrical, 10–19 mm long, 5-nerved, keeled, keel ciliate to pectinate-ciliate, strongly laterally compressed; awned. **Awns:** glume awns 1–3 mm long; lemma keels transforming into a scabrous awn 7–50 mm long. **VEGETATIVE CHARACTERISTICS: Growth Habit:** cespitose. **Culms:** erect, glabrous to below inflorescence where some pubescence may occur, 5–12 dm tall. **Sheaths:** open, glabrous. **Auricles:** to 1 mm long. **Ligules:** membranous, truncate, erose, 0.5–1.5 mm long. **Blades:** flat to involute, generally glabrous, 10–25 cm long, 4–12 mm wide. **HABITAT:** Originally cultivated and used as a revegetation plant in some areas; now found on disturbed areas and roadsides at low elevations. **COMMENTS:** This species was an important food source in ancient times. Frequently used as an annual species for immediate road stabilization.

112. *Thinopyrum* Á. Löve

Cespitose or rhizomatous perennials. Inflorescence a terminal spike that persists at maturity. Sheaths open; auricles present or absent; ligules membranous. Disarticulation typically above the glumes and beneath florets, sometimes in rachis. Spikelets 1–3 times the length of middle internodes, solitary, often arching outward at maturity, 3–10 florets. Glumes stiff, coriaceous to indurate, apices truncate to acute, sometimes mucronate. Lemma 5-nerved, coriaceous, truncate, obtuse or acute, sometimes mucronate or awned (to 3 cm). Basic chromosome number, $x = 7$. Photosynthetic pathway, C_3.

Represented in Colorado by 2 species and 2 subspecies. Members of this genus, at various times, have been included in *Agropyron, Elymus, Elytrigia, Lophopyrum,* or *Triticum.*

1. Glumes truncate without mucro; usually cespitose . . . . 334. *T. ponticum*
1. Glumes blunt but with a small, off-center mucro; usually rhizomatous . . . .
. . . . 333. *T. intermedium*

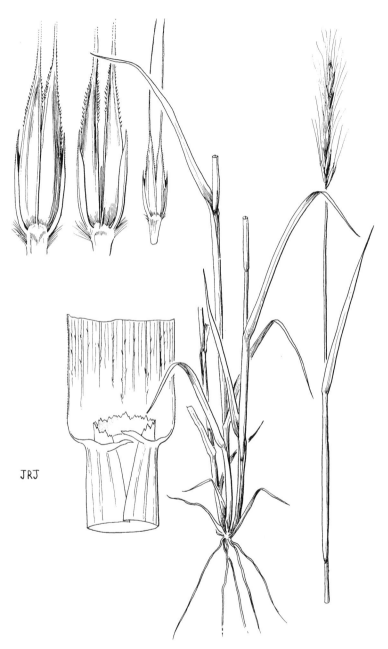

JRJ

Figure 354. *Secale cereale.*

333. *Thinopyrum intermedium* (Host) Barkworth & D. R. Dewey (Fig. 355).

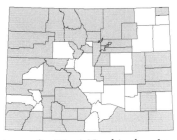

Synonyms: *Agropyron glaucum* (Desf. *ex* DC.) Roem. & Schult., *Agropyron intermedium* (Host) P. Beauv., *Agropyron intermedium* (Host) P. Beauv. var. *trichophorum* (Link) Halac., *Agropyron podperae* Nabelek, *Agropyron pulcherrimum* Grossh., *Agropyron trichophorum* (Link) Richter, *Elymus hispidus* (Opiz) Melderis, *Elymus hispidus* (Opiz) Melderis subsp. *barbulatus* (Schur) Melderis, *Elymus hispidus* (Opiz) Melderis var. *ruthenicus* (Griseb.) Dorn, *Elymus intermedius* (Host) P. Beauv., *Elytrigia intermedia* (Host) Nevski, *Elytrigia intermedia* (Host) Nevski subsp. *barbulata* (Schur) Á. Löve, *Elytrigia intermedia* (Host) Nevski subsp. *intermedia* (Host) Nevski, *Elytrigia intermedia* (Host) Nevski subsp. *trichophora* (Link) Tvzel, *Thinopyrum intermedium* (Host) Barkworth & D. R. Dewey subsp. *barbulatum* (Schur) Barkworth & D. R. Dewey, *Triticum intermedium* Host. **Vernacular Name:** Intermediate wheatgrass. **Life Span:** Perennial. **Origin:** Introduced. **Season:** C$_3$, cool season. **FLORAL CHARACTERISTICS: Inflorescence:** spike, 8–21 cm long, slender; spikelets solitary at each node, not imbricate. **Spikelets:** glabrous to hairy, compressed, 11–18 mm long, 3- to 10-flowered; disarticulation above the glumes and below florets; sometimes arching outward at maturity. **Glumes:** subequal, oblong to elliptic-lanceolate, thickened, apices obliquely truncate to acute, usually with short mucro, glabrous to hirsute; lower 4.5–7.5 mm long, 1.5–2.5 mm wide, 5- to 6-nerved; upper 5.5–8 mm long, 2–3 mm wide, 5- to 7-nerved. **Lemmas:** lanceolate, glabrous to hairy, 7.5–10 mm long, decreasing in size as they become more distal to glumes, apices blunt to awned. **Awns:** lemma occasionally awn-tipped to awned, to 5 mm long. **VEGETATIVE CHARACTERISTICS: Growth Habit:** rhizomatous, rhizomes occasionally poorly developed. **Culms:** erect, glaucous, 5–11 dm tall, glabrous to hairy, at times only nodes hairy. **Sheaths:** generally glabrous, occasionally short-hairy, margins often ciliate. **Auricles:** well developed, acute, 0.5–1.3 mm long. **Ligules:** ciliolate membrane, 0.1–0.8 mm long, truncate. **Blades:** flat to involute, 10–40 cm long, 2–8 mm wide, stiff, green to glaucous; upper surface generally sparsely strigose, lower surface glabrous, margins whitish. **HABITAT:** Introduced, and found along roadsides and in fields from low elevations up to montane. Tolerates alkaline soils well. **COMMENTS:** A segregate of *Agropyron*. The species has been successfully introduced into rangelands of the West for erosion control, revegetation, forage, and hay (Barkworth et al. 2007). Also very common on roadsides in the mountains. Two subspecies are found in Colorado: subsp. *intermedium* and subsp. *barbulatum* (Schur) Barkworth & D. R. Dewey. They differ in this characteristic:

1. Lemmas with strigose hairs over much of body . . . . subsp. *barbulatum*
1. Lemmas glabrous . . . . subsp. *intermedium*

334. *Thinopyrum ponticum* Barkworth & D. R. Dewey (Fig. 356).

Synonyms: *Elymus elongatus* (Host) Runemark, *Elymus elongatus* (Host) Runemark var. *ponticus* (Podp.) Dorn, *Elytrigia pontica* (Podp.) Holub, *Elytrigia pontica* (Podp.) Holub subsp. *pontica* (Podp.) Holub, *Lophopyrum ponticum* (Podp.) Á. Löve. **Vernacular Names:** Rush wheatgrass, tall wheatgrass. **Life Span:** Perennial. **Origin:** Introduced. **Season:** C$_3$, cool

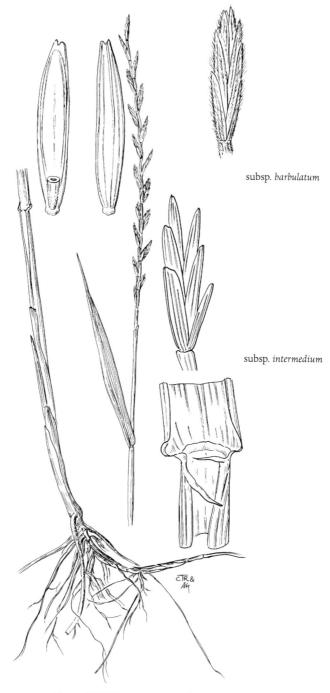

subsp. *barbulatum*

subsp. *intermedium*

Figure 355. *Thinopyrum intermedium.*

Figure 356. *Thinopyrum ponticum*.

season. **FLORAL CHARACTERISTICS: Inflorescence:** erect to lax spike, 10–45 cm long; spikelets solitary at each node; rachis not disarticulating at maturity. **Spikelets:** 13–30 mm long, 6- to 12-flowered, glabrous to hairy; disarticulation above the glumes and below florets. **Glumes:** subequal, oblong, thickened, glabrous, truncate, margins hyaline, to 0.5 mm wide, 5- to 7- rarely 9-nerved; lower 6.5–10 mm long, keeled; upper 7–10 mm long. **Lemmas:** glabrous,  lanceolate to oblong, 9–12 mm long, 5-nerved, truncate. **Paleas:** 7–11 mm long, keeled. **Awns:** none. **VEGETATIVE CHARACTERISTICS: Growth Habit:** generally cespitose. **Culms:** erect, glabrous, 5–20 cm tall, stout. **Sheaths:** glabrous, margins often ciliate. **Auricles:** not clasping culm, 0.2–1.5 mm long. **Ligules:** membranous, 0.3–1.5 mm long, truncate. **Blades:** flat to loosely involute, 10–40 cm long, 2–6 mm wide, upper surface prominently nerved, glabrous to scabrous, occasionally sparsely pilose on both surfaces. **HABITAT:** Introduced, and found along roadsides and in fields from low elevations up to montane. **COMMENTS:** Planted along roadsides for soil stabilization. It is very tolerant of arid climates and saline soils. Its tolerance to road salting during winter is enabling it to move northerly (Barkworth et al. 2007).

113. *Triticum* L.

Annuals with broad, flat blades. Inflorescence terminal, a thick bilateral spike. Spikelets solitary at nodes on a continuous rachis; does not disarticulate in our species or does so only under pressure. Spikelets with 2–5 florets, laterally flattened and oriented flatwise to rachis. Glumes thick and firm, 3- to several-nerved, toothed, mucronate or with a short awn at apex. Lemma similar to the glumes in texture, keeled, asymmetric, many-nerved, awnless or with a single awn. Basic chromosome number, $x = 7$. Photosynthetic pathway, C_3.

Represented in Colorado by a single species.

335. *Triticum aestivum* L. (Fig. 357).

Synonyms: *Triticum hybernum* L., *Triticum macha* Dekap. & Menab., *Triticum sativum* Lam., *Triticum sphaerococcum* Percival, *Triticum vulgare* Vill. **Vernacular Names:** Wheat, bread wheat, common wheat, soft wheat. **Life Span:** Annual. **Origin:** Introduced.  **Season:** C_3, cool season. **FLORAL CHARACTERISTICS: Inflorescence:** spike, erect to lax, 4–18 cm long, dense; rachis continuous, flattened; spikelets solitary at each node. **Spikelets:** subterete to flattened, 9–12 mm long, 2- to 9-flowered. **Glumes:** subequal, indurate, lanceolate to ovate, 6–12 mm long, 5- to 7-nerved, glabrous to scabrid to densely hispid; somewhat asymmetrically keeled, at least near apices; truncate and mucronate to awned. **Lemmas:** 10–15 mm long, broad; similar to glumes, keeled as with glumes, indurate with scarious margins, acute to awntipped to awned. **Awns:** glumes with mucro or awned to 4 cm; lemma awns from short

(2–20 mm) to long (47–120 mm). **VEGETATIVE CHARACTERISTICS: Growth Habit:** tufted. **Culms:** erect, robust, 5–15 dm tall, freely branching below. **Sheaths:** glabrous. **Auricles:** prominent with scattered long hairs. **Ligules:** entire, 0.5–1.5 mm long. **Blades:** flat, 2–16 mm wide, upper surface glabrous to scabrous. **HABITAT:** Found in Colorado in cultivation at lower elevations (Weber and Wittmann 2001a, 2001b; Wingate 1994; Harrington 1954). Commonly found growing on roadsides outside of cultivation. **COMMENTS:** There are many cultivars of *T. aestivum* that range from awnless to long-awned forms, and treatments based on awn length have delineated various varieties and species. Wheat and rice (*Oryza sativa* L.) are the major energy source for the peoples of the world. Much more widespread than the distribution map indicates, but infrequently collected.

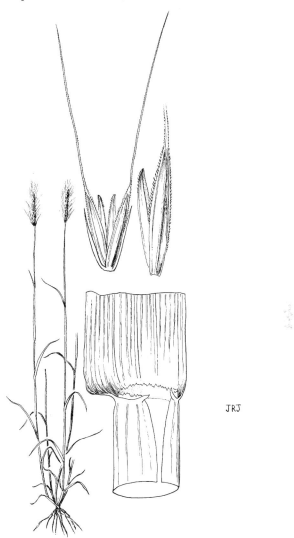

JRJ

Figure 357. *Triticum aestivum*

GLOSSARY

Abaxial. Located on the side away from the axis.

Achene. A small, dry, hard, indehiscent, one-locular, one-seeded fruit. See *caryopsis*.

Acicular. Needlelike.

Acuminate. Gradually tapering to a point.

Acute. Sharp-pointed, not abruptly or long-extended but making an angle of less than 90°, less tapered than acuminate.

Adaxial. Located on the side toward the axis.

Adnate. Fusion or attachment of unlike structure, such as a palea fused with the lemma (contrast with connate, which is the fusion of like structures).

Adventitious root. A root that arises from any organ other than the primary root or branches thereof.

Adventive. Introduced by chance or accidental seedlings and imperfectly naturalized.

Agrostology. The branch of systematic botany that deals with grasses.

Allopatric. Not growing together, occupying different geographic areas. Opposite of sympatric.

Alpine. The area above timberline.

Alveolate. Honeycombed, with angular depressions separated by thin partitions.

Androecium. The stamens of the flower referred to collectively.

Androgynous. Staminate and pistillate spikelets in the same inflorescence, staminate above the pistillate.

Annual. Of one season's or year's duration, from seed to maturity and death.

Anther. The pollen-bearing part of a stamen.

Anthesis. The period during which the flower is open and functional.

Antrorse. Directed forward or toward the apex; the opposite of retrorse.

Apical meristem. Terminal growing point.

Apiculate. Terminating abruptly in a small, short point.

Apomictic. A plant that reproduces by apomixis, embryo is formed by structures commonly concerned in sexual reproduction but there is no actual fusion of male and female gametes.

Appressed. Lying flat or close against something.

Arborescent. Treelike in size and habit.

Arcuate. Arched or bowed, like an arc.

Aristate. Awned; with a stiff bristle at the tip or terminating nerves at the back or margins of organs such as the glumes, lemmas, and paleas.

Articulation. A joint or node. See *disarticulation*.

Ascending. Growing obliquely upward, like a culm.

Asperity. Short macrohairs, also termed *prickle-hairs*.

Asperous. Rough or harsh to the touch.

Asymmetrical. Without symmetry.

Attenuate. Tapering gradually to a tip or base.

Auricle. An ear-shaped appendage; a structure that occurs in pairs laterally at the base of the leaf blade in some grasses and laterally at the sheath apex in others.

Awn. A bristle or stiff, hairlike projection; in the grass spikelet, usually the prolongation of the midnerve or lateral nerves of the glumes, lemma, or palea.

Axil. The upper angle formed between two structures such as the leaf or spikelet and stem axis.

Axillary. In an axil.

Axis (of the culm, inflorescence, or similar feature). The central stem or branch on which the parts or organs are arranged.

Basic chromosome number. The lowest, actual or theoretical, haploid chromosome number, designated by x, as $x = 7, 9$, or 10; usually applied to a group of species.

Beaked. Ending in a firm, prolonged, slender tip.

Bearded. Bearing stiff, usually long hairs.

Bidentate. Having two teeth.

Bifid. Deeply two-cleft.

Bilobed. Having two lobes.

Blade. The expanded portion of a flattened structure such as a leaf or flower petal. The blade of the grass leaf is the usually flattened, expanded portion above the sheath.

Bract. A modified leaf subtending a flower or belonging to an inflorescence; the term is sometimes used in reference to the scales of a vegetative bud or other shoot structure.

Bracteal. Of or pertaining to bracts.

Bracteate. Having bracts.

Bracteole. Small bract.

Branch. A division or subdivision of an axis.

Bristle. A stiff hair or hairlike projection.

Bulb. An underground or partially underground bud with swollen, fleshy scales, like an onion.

Burr. A structure containing spikelets and covered with spines or prickles (as in *Cenchrus*).

Callus. The hard, usually pointed base of the spikelet (as in *Heteropogon, Andropogon,* and related genera) or of the floret (as in *Aristida* and *Stipa*), just above the point of disarticulation. In the former case, the callus is a portion of the rachis; in the latter, it is a portion of the rachilla. In *Eriochloa* the callus is the thickened node and remnant lower glume; in *Chrysopogon* it is part of the pedicel.

Canescent. With very short white or gray hairs, giving the surface a gray or white appearance.

Capillary. As applied to hairs, very slender and fine.

Capitate. Head-shaped; collected into a head or dense cluster.

Capsule. A dry, more or less dehiscent fruit composed of more than one carpel.

Carinate. Keeled, with one or more longitudinal ridges.

Carpel. A unit of the pistil; a simple pistil is formed from a single carpel and a compound one from two or more carpels.

Cartilaginous. Firm and tough but flexible; like cartilage.

Caryopsis. A dry, hard, indehiscent, one-seeded fruit with the thin pericarp adnate to the seed coat; the characteristic grass fruit. This differs from the achene only in the fusion of the pericarp and seed coat.

Caudeus. Like wood; referring to the woody or thicken base of perennial herbaceous plants.

Cauline. Of or pertaining to the culm (stem).

Cespitose (or **caespitose**). In tufts or dense clumps.

Chartaceous. With the texture of stiff writing paper, like parchment used to make charts.

Chasmogamy. Pollination and fertilization of open flowers or florets; the opposite of *cleistogamy.*

Ciliate. With a marginal fringe of hairs.

Ciliolate. Ciliate but the hairs minute.

Clavate. Club-shaped, thickened or enlarged at the apex from a slender base.

Cleistogamy. Pollination and fertilization within closed flowers or florets; the opposite of *chasmogamy.*

Coleoptile. The sheath of the shoot of the embryo; protects the developing shoot.

Coleorhiza. The sheath of the primary root of the embryo; protects the developing radicle.

Collar. The outer side of a grass leaf at the junction of the blade and sheath.

Column. The modified base of an awn (see *Aristida, Andropogon, Bothriochloa*).

Compressed. Flattened, usually laterally.

Compressed-keeled. Flattened laterally with a ridge or keel.

Conduplicate. Folded lengthwise down the middle.

Connate. Fusion or attachment of like structures, such as the margins of the leaf sheath (contrast with adnate, which is the fusion of unlike structures).

Continuous. An axis that does not disarticulate at the nodes at maturity; usually applied to a rachis or inflorescence branch.

Contracted. An inflorescence that is narrow and dense with, typically, short, appressed branches or pedicels.

Convolute. Rolled up longitudinally; used primarily for leaf blades; same as involute.

Cordate. Heart-shaped, with a broad, notched base and a pointed tip.

Coriaceous. Leathery in texture.

Corm. The enlarged, fleshy base of a culm (stem).

Cormous. With a corm.

Cotyledon. The leaf in a grass embryo.

Crenate. Tooth with teeth rounded at apex.

Crown. The persistent base of a perennial grass.

Culm. The stem of a grass.

Cuneate. Wedge-shaped; narrowly triangular.

Cuspidate. Tipped with an abrupt, sharp, short, firm point.

Cyme. A usually broad and flattish determinate inflorescence, with the central or upper-most flower maturing first; grass inflorescences are cymose in that they are determinate, with the terminal spikelet maturing first.

Deciduous. Falling naturally, like the awns in *Sorghum halepense*.

Decumbent. Lying on the surface but ascending at the tip.

Decurrent. Extending downward from the point of insertion.

Dense. Inflorescences in which the spikelets are crowded.

Depauperate. Impoverished, stunted.

Dichotomous. Divided in two equal parts, like the inflorescence branches in *Achnatherum hymenoides* and *Argillochloa dasyclada*.

Diffuse. Widely spreading.

Digitate. Like the fingers of a hand; usually referring to an inflorescence in which all the branches originate from a common point (*Chloris, Digitaria*).

Dioecious. Unisexual, with staminate and pistillate flowers on separate plants.

Diploid. With respect to polyploid chromosome series, having twice the basic (x) number of chromosomes.

Disarticulation. Separation at the joints or nodes at maturity.

Distal. The end opposite the point of attachment.

Distichous. Distinctly two-ranked, in two vertical rows.

Distinct. Separate, as in parts not united.

Diverging. Widely spreading, distinct.

Dorsal. The back side or surface; the surface turned away from the central stalk or axis; abaxial.

Dorsally compressed. Flattened from the front and back; said of spikelets, florets, glumes, lemmas, and panicles. Dorsal compression is common in the Paniceae.

Drooping. Inclined downward.

Elliptic. In the form of a flattened circle, more than twice as long as broad, widest in the middle, and the two ends equal.

Elongate. Narrow, the length many times the width.

Emarginate. With a small, shallow notch at the apex.

Embryo. The undeveloped plant within a seed.

Endemic. Indigenous or native in a restricted locality.

Endosperm. Nutritive tissue arising in the embryo sac of most angiosperms following fertilization of the two fused polar nuclei (primary endosperm nuclei) by one of the two male gametes. In most organisms the cells of the endosperm have a *3n* chromosome number.

Entire. Undivided, the margin continuous, without teeth or lobes.

Erose. Irregular and uneven, as if gnawed or worn away.

Excurrent. Extending out or beyond.

Exotic. Not native, introduced.

Extravaginal. A type of branching in which the tip breaks through the enclosing sheath; the opposite of intravaginal.

Falcate. Flat and curved to one side, tapering toward the apex.

Fascicle. A cluster or close bunch; usually used in reference to culms, leaves, or branches of the inflorescence.

Fertile floret. A floret capable of producing a seed; a fertile floret may be perfect or pistillate.

Fibrous. Composed of fibers, fiber-like.

Filiform. Long, slender, and terete.

Flabellate. Fan-shaped, broadly wedge-shaped.

Flexuous. Bent alternately in opposite directions.

Floret. As applied to grasses, the lemma and palea with the enclosed flower. The floret may be perfect, pistillate, staminate, or sterile.

Flower. A plant reproductive structure composed of an ovary, stamens, and enveloping bracts (lemmas and paleas). The flower can be perfect, pistillate, staminate, or sterile.

Foliaceous. Leaflike, or with leaves.

Fruit. As applied to grasses, the ripened ovary; the typical grass fruit is a caryopsis.

Geniculate. Abruptly bent, as at the elbow or knee joint; frequently used to describe an awn or a culm.

Gibbous. Swollen on one side, like a pouchlike swelling.

Glabrate. Becoming glabrous with age.

Glabrous. Without hairs; about the same as smooth.

Gland. A protuberance, depression, or appendage that secretes or appears to secrete a fluid.

Glandular. Glandlike or bearing glands.

Glaucous. Covered with a whitened or bluish waxy bloom, as a cabbage leaf or a plum.

Globose. Spherical or rounded; globelike.

Glomerate. In densely contracted, headlike clusters.

Glume. One of a pair of bracts usually present at the base of the grass spikelet.

Grain. In respect to grasses, the threshed or unhusked fruit, usually a caryopsis; used to refer to the mature ovary alone or to the ovary enclosed in persistent bracts (palea, lemma, or glumes).

Gross morphology. The external structure and form.

Gynaecandrous. With pistillate and staminate spikelets on the same inflorescence, pistillate above.

Gynoecium. The female portion of a flower as a whole; the pistil of the grass flower.

Haploid. With respect to polyploid chromosome series, the basic (x) number of chromosomes. With respect to sporophyte-gametophyte generations, the gametic (ln) chromosome number; half the number of chromosomes of the somatic (diploid) cells.

Hastate. Spear- or arrowhead-shaped.

Herb. A plant without persistent woody stems, at least above the ground.

Herbaceous. Without persistent woody stems; dying to the ground each year.

Herbarium. A collection of dried and pressed plant specimens, usually mounted for permanent preservation; the room, building, or institution in which such a collection is kept or to which it belongs.

Hermaphroditic. A spikelet with both male and female reproductive parts. Same as perfect and bisexual.

Hilum. Scar where the seed was attached to the ovary wall.

Hirsute. With rather coarse and stiff hairs that are long, straight, and erect or ascending.

Hirsutulous. Somewhat hirsute.

Hirtellous. Minutely hirsute.

Hispid. With erect, rigid, stiff bristles or bristle-like hairs.

Hispidulous. Minutely hispid.

Hooked. Curved at the tip.

Hyaline. Thin in texture and transparent or translucent.

Hydrophytic. Water-loving, wet; in respect to moisture, habitats are classified as *hydrophytic* (wet), *mesic* (moist), or *xeric* (dry).

Hygroscopic. Altering form or position because of changes in humidity; like the awn in *Achnatherum* and *Hesperostipa*.

Imbricate. Partly overlapping, as the shingles of a roof.

Imperfect. Unisexual flowers or florets; with either male or female reproductive structures but not both.

Included. Not protruding; like the awn in some *Agrostis*.

Indehiscent. Closed, not opening; like a caryopsis.

Indigenous. Native.

Indurate. Hard.

Inflated. Bladder-like; blown up like a balloon.

Inflorescence. The flowering portion of a shoot; in grasses, the spikelets and the axis or branch system that supports them, the inflorescence delimited at the base by the uppermost leafy node of the shoot.

Intercalary meristem. An actively growing primary tissue region somewhat removed from the apical meristem; intercalary meristems are present at the base of the internodes in young grass shoots.

Intercostal. Area between the nerves.

Internode. Portion of the stem or other structure between two nodes.

Interrupted. With the order broken; used to describe inflorescences with major gaps in spikelets or branches.

Intravaginal. Growth of a shoot is intravaginal when the tip does not break through the enclosing sheath but emerges at the top; the opposite of extravaginal and the more common condition in grasses.

Introduced. A plant intentionally brought in for a specific purpose; it may escape and persist.

Involucre. A circle or cluster of bracts or reduced branchlets that surround a flower or floret, or a group of flowers or florets.

Involute. Rolled inward from the edges.

Joint. The node of a grass culm; an articulation.

Keel. A prominent dorsal ridge, like the keel of a boat. Glumes and lemmas of laterally compressed spikelets are often sharply keeled; the palea of some florets is two-keeled.

Lacerate. Irregularly cleft or torn.

Laciniate. Margins incised in to narrow and pointed lobes.

Lamina. The blade or expanded part of a leaf.

Lanate. With long, tangled, woolly hairs.

Lanceolate. Lance-shaped; relatively narrow, tapering to both ends from a point below the middle.

Lateral. Borne on the sides of a structure.

Laterally compressed. Flattened from the sides; said of spikelets, florets, glumes, lemmas, and/or paleas; most common type of spikelet compression in the grasses.

Lax. Loose or drooping.

Lemma. The lowermost of the two bracts enclosing the flower in the grass floret. The most important structure for grass identification.

Lignified. Woody; converted into lignin.

Ligule. A membranous or hairy appendage on the adaxial surface of the grass leaf at the junction of sheath and blade. Often one of the few vegetative characters used in identification.

Linear. Long and narrow and with parallel margins, like the majority of grass leaf blades.

Lobed. With lobes or segments of an organ cut less than halfway to the middle, apex of segments rounded.

Lodicule. Small, scalelike processes, usually two or three in number, at the base of the stamens in grass flowers. Lodicules are generally interpreted as rudimentary perianth segments.

Loose. Open, like a panicle with spreading branches.

Membranous. With the character of a membrane; thin, soft, and pliable.

Meristem. Embryonic or undifferentiated tissue, the cells of which are capable of active division.

Mesic. A moist habitat; in respect to moisture, habitats are classified as *hydrophytic* (wet), *mesic* (moist), or *xeric* (dry).

Midnerve. The central nerve of an oddly nerved structure. Same as midrib.

Midrib. The central rib of a structure.

Monoecious. Flowers unisexual, with male and female flowers borne on the same plant; like corn, the ears are female and the tassels are male.

Monotypic. Having a single type or representative, as a genus with only one species.

Morphology. The science that deals with the form and structure of plants (or animals). Broadly interpreted, morphology includes anatomy, histology, and the nonphysiological aspects of cytology and embryology. In plant taxonomy, morphology is concerned mainly with external form.

Mucro. A short, small, abrupt tip of an organ, as the projection of a nerve of the leaf.

Mucronate. With a mucro.

Muriculate. Finely covered with short, sharp points.

Naked. Lacking some structure.

Nerve. A simple vein or slender rib of a leaf or bract (glumes, lemma, and/or palea).

Neuter. Without functional stamens or pistils; same as sterile.

Node. The joint of a stem; the region of attachment of the leaves.

Nodding. Inclined from an upright position; used primarily for panicles, as in several species of *Bromopsis*.

Oblanceolate. Inversely lanceolate; lance-shaped and attached at the pointed end.

Oblong. Two to three times longer than broad and with nearly parallel sides.

Obovate. Inverted ovate.

Obovoid. Egg-shaped, attached at the narrow end, three-dimensional.

Obtuse. Blunt or rounded at the apex.

Open. Loose, like a panicle with spreading branches.

Opposite. Two structures at a node and across from each other.

Ovary. Part of the pistil that contains the ovules.

Ovate. Egg-shaped, with the broadest end toward the base.

Ovule. The structure that develops into the caryopsis.

Palea. The uppermost of the two bracts enclosing the grass flower in the floret; the palea is usually two-nerved and two-keeled and often enclosed by the lemma.

Panicle. As applied to grasses, all inflorescences in which the spikelets are not sessile or individually pediceled on the main axis.

Paniculate. Borne in a panicle.

Papillate, papillose. Bearing minute nipple-shaped projections (papillae).

Pappus. Tuft of hairs or bristles at the basal tip of a structure.

Pectinate. With narrow, closely set, and divergent segments or units, like the teeth of a comb.

Pedicel. The stalk of a single flower; in grasses, applied to the stalk of a single spikelet.

Peduncle. A primary flower stalk, usually applied to the stalk of a flower cluster or an inflorescence.

Pedunculate. With a peduncle.

Pendulous. Suspended or hanging.

Perennial. Lasting year after year.

Perfect. A flower (or spikelet, or floret) with both functional male and female reproductive structures.

Perianth. The floral envelope, undifferentiated or of calyx and corolla.

Pericarp. The fruit wall developed from the ovary wall; the ripened outer wall of the mature fruit.

Persistent. Remaining attached.

Petiole. A leaf stalk.

Phenotype. The aggregate of visible characteristics of an organism.

Phytomere. The basic unit of structure of the grass shoot, an internode together with the leaf and portion of the node at the upper end and a bud and portion of the node at the lower end.

Pilose. With soft, straight hairs; about the same as villous.

Pistil. The female (seed-bearing) structures of the flower, ordinarily consisting of the ovary, stigma, and style.

Pistillate. Having a pistil but not stamens.

Plano-convex. Flat on one side and rounded on the other.

Plicate. Folded into pleats, usually lengthwise, as in a fan.

Plumose. Feathery, having fine, elongate hairs on either side.

Plumule. Embryonic culm- and leaf-producing structure.

Pollen grain. Microgametophyte at the stage when it is shed from the anther; usually two- or three-nucleate.

Polyploid. A plant with three or more basic sets of chromosomes; any ploidy level above the diploid.

Prickle-hairs. Short macrohairs, also termed *asperities.*

Primary shoot. The shoot developing from the plumule of the embryo.

Procumbent. Culms lying on the ground but not rooting at the nodes.

Prophyll (prophyllum). The first leaf of a lateral shoot or vegetative culm branch. The prophyll is a sheath, usually with two strong lateral nerves and numerous fine intermediate nerves; a blade is never developed.

Prostrate. Lying flat on the ground.

Proximal. End of a structure by which it is attached.

Puberulent. Minutely pubescent.

Pubescent. Hairy, with short, soft hairs.

Pulvinus. A swelling at the base of a leaf or of a branch of the inflorescence.

Pungent. Terminating in a rigid, sharp point.

Pustulate. With irregular, blister-like swellings or pustules.

Pyramidal. Like a pyramid; a cone-shaped inflorescence.

Raceme. As applied to grasses, an inflorescence in which all the spikelets are borne on pedicels inserted directly on the main (undivided) inflorescence axis, or some are sessile and some pediceled on the main axis.

Racemose. Bearing or like a raceme.

Rachilla. Axis of a grass spikelet.

Rachis. Main axis of an inflorescence.

Radicle. Primary root of the grass embryo.

Rame. Disarticulating inflorescence branches; the basic unit of the typical Andropogoneae inflorescence. An inflorescence branch composed of spikelet units (usually a fertile, sessile spikelet; a reduced pedicellate spikelet; and the internode that extends from the sessile spikelet to the next distal sessile spikelet).

Reduced floret. A staminate or neuter floret; if highly reduced, then termed a *rudimentary floret.*

Reflexed. Curved outward or backward.

Reniform. Kidney-shaped.

Retrorse. Directed backward or downward.

Revolute. Rolled backward from the margin.

Rhizomatous. With rhizomes.

Rhizome. An underground stem, usually with scale leaves and adventitious roots borne at regularly spaced nodes.

Root. The descending axis of the grass, without nodes and internodes, and absorbing water and nutrients from the soil.

Rosette. A dense basal cluster of leaves arranged in a circle; like the winter rosette of *Dichanthelium*.

Rudiment. An imperfectly developed organ or part, as the rudimentary floret or florets of some spikelets.

Rugose. Wrinkled.

Saccate. Sac- or pouch-shaped.

Scaberulent. Slightly scabrous.

Scaberulous. Slightly scabrous.

Scabrid. With a rough surface, roughened, somewhat scabrous.

Scabridulous. Minutely roughened or scabrid, slightly scabrous.

Scabrous. Rough to the touch, usually because of the presence of minute prickle-hairs in the epidermis.

Scarious. Thin, dry, and membranous; not green.

Scutellum. A band of tissue between the embryonic shoot and the endosperm.

Secund. Borne on one side of the axis.

Seed. A ripened ovule.

Seminal root. Literally translated, "seed root"; the primary root and all other roots that arise from embryonic tissue below the scutellar node.

Sericeous. Covered with long, straight, soft appressed hairs.

Serrate. Saw-toothed, with sharp teeth pointing forward.

Serrulate. Finely serrate.

Sessile. Attached directly at the base, without a stalk.

Seta. A bristle or a rigid, sharp-pointed, bristle-like organ.

Setaceous. Bristly or bristle-like.

Sheath. The tubular basal portion of a leaf that encloses the stem, as in grasses and sedges.

Sinuous (sinuate). With a strong, wavy margin.

Spathe. A large bract enclosing an inflorescence, as in some genera of the Andropogoneae.

Spicate. Spikelike.

Spicule. Short, stout, pointed projection of the leaf epidermis; spicules often grade into prickle-hairs.

Spike. An inflorescence with flowers or spikelets sessile on an elongated, unbranched rachis.

Spikelet. The basic unit of the grass inflorescence, typically consisting of a short axis, the rachilla, bearing two "empty" bracts, the glumes, at the basal nodes and one or more florets above. Each floret usually consists of two bracts, the lemma (lower) and the palea (upper), which enclose a flower. The flower usually includes two lodicules (vestigial perianth segments), three stamens, and a pistil.

Spikelike. Similar to a spike inflorescence.

Sprawling. Spread out in an irregular fashion, viney.

Stamen. The male organ of the flower, consisting of a pollen-bearing anther on a filiform filament. Collectively, the stamens of a flower are referred to as the androecium.

Staminate. Having stamens but not pistils.

Stigma. The part of the ovary or style that receives the pollen for effective fertilization.

Stipe. A stalk.

Stipitate. Having a stalk or stipe, as an elevated gland.

Stolon. A modified horizontal stem that loops or runs along the surface of the ground and spreads the plant by rooting at the nodes.

Stoloniferous. Bearing stolons.

Stramineous. Of or resembling straw.

Strigillose. Like strigose, but hairs shorter.

Strigose. With appressed, stiff, short hairs.

Style. The contracted portion of the pistil between the ovary and the stigma.

Sub-. Latin prefix meaning "almost," "somewhat," "of inferior rank," "beneath."

Subtending. Below or beneath something, often enclosing it.

Subulate. Awl-shaped.

Succulent. Fleshy or juicy.

Sucker. A vegetative shoot of subterranean origin.

Sympatric. Growing together, occupying the same geographic area. Opposite of allopatric.

Synoecious. Having male and female flowers in the same inflorescence.

Taxon. Any taxonomic unit, as species, genus, or tribe. Taxa is plural.

Teeth. Pointed divisions of a structure.

Terete. Cylindrical, round in cross-section.

Tiller. A subterranean or ground-level lateral shoot, usually erect, as contrasted with horizontally spreading shoots referred to as either stolons or rhizomes.

Translucent. Allowing the passage of light rays but not transparent.

Transverse. Extending across a structure, such as the transverse rugose lemma of some panicoid genera (some *Eriochloa*, *Setaria*, and others).

Trifid. Divided into three parts.

Truncate. Terminating abruptly, as if cut off transversely.

Tuberculate. Having small, pimple-like structures, or covered with tubercules.

Turgid. Swollen; tightly drawn by pressure from within.

Umbonate. Bearing a stout projection in the center.

Uncinate. Hooked at the tip.

Unilateral. One-sided; developed or hanging on one side.

Unisexual. With either male or female sex structures but not both.

Urceolate. Pitcher-shaped.

Utricle. A small, bladdery, one-seeded fruit.

Ventral. Inner side of an organ; upper surface of a leaf; adaxial.

Verticil. A whorl; with three or more members or parts attached at the same node of the supporting axis.

Vestigial. Rudimentary and almost completely reduced; with only a vestige remaining.

Villous. Bearing long and soft, not matted hairs; shaggy; same as pilose.

Viscid. Sticky, glutinous.

Viviparous. Plant that reproduces by vivipary; apomictic reproduction in which the progeny are produced by means other than seed. In grasses the term generally refers to vegetative reproduction involving spikelet structures (glumes, lemmas, and/or paleas forming leaves), while vegetative reproduction refers to rhizomes, stolons, etc.).

Weed. Any plant growing where it is not wanted.

Wing. Any membranous expansion bordering a structure.

Winter annual. A plant that germinates in the fall and completes its life cycle the following spring or summer.

Woolly. With long, soft, interwoven hairs.

Xeric. Characterized by, or pertaining to, conditions of scant moisture supply, very arid region or environment; in respect to moisture, habitats are classified as *hydrophytic* (wet), *mesic* (moist), or *xeric* (dry).

LITERATURE CITED

Allred, K. W. 2005. *A Field Guide to the Grasses of New Mexico*, 3rd ed. Department of Agricultural Communications, College of Agriculture and Home Economics, Agricultural Experiment Station, New Mexico State University, Las Cruces.

Bailey, R. G. 1995. *Description of the Ecoregions of the United States*, 2nd ed. USDA Forest Service Misc. Publ. no. 1391 (rev.). Government Printing Office, Washington, DC.

Barkworth, M. E., K. M. Capels, S. Long, L. K. Anderton, and M. B. Piep, eds. 2007. *Magnoliophyta: Commelinidae* (in part): *Poaceae, Part 1. Flora of North America North of Mexico*, vol. 24. Oxford University Press, New York.

Barkworth, M. E., K. M. Capels, S. Long, and M. B. Piep, eds. 2003. *Magnoliophyta: Commelinidae* (in part): *Poaceae, Part 2. Flora of North America North of Mexico*, vol. 25. Oxford University Press, New York.

Barkworth, M. E., D. R. Dewey, and R. J. Atkins. 1983. New Generic Concepts in the Triticeae of the Intermountain Region: Key and Comments. *Great Basin Naturalist* 43:561–572.

Chaplick, G. P., and K. Clay. 1988. Acquired Chemical Defenses in Grasses: The Role of Fungal Endophytes. *OIKOS* 52:309–318.

Chapman, G. P. 1996. *The Biology of Grasses*. CAB International, Oxon, UK.

Chronic, J., and H. Chronic. 1972. Prairie, Peak, and Plateau: A Guide to the Geology of Colorado. *Colorado Geological Survey Bulletin* 32, Denver.

Colorado Agricultural Statistics. 2005. http://www.nass.usda.gov/co/pub/prof_t.pdf [accessed January 2007].

Colorado Geography. http://www.netstate.com/states/geography/co_geography.htm [accessed June 7, 2006].

Columbus, J. T. 1999. An Expanded Circumscription of *Bouteloua* (Gramineae: Chlorideae): New Combinations and Names. *Aliso* 18:61–65.

Cronquist, A., A. H. Holmgren, N. H. Holmgren, J. L. Reveal, and P. K. Holmgren. 1977. *Intermountain Flora, Vascular Plants of the Intermountain West, U.S.A.,* vol. 6: *The Monocotyledons.* Columbia University Press, New York.

Finot, V. L., P. M. Peterson, R. J. Soreng, and F. O. Zuloaga. 2005. A Revision of *Trisetum* and *Graphephorum* (Poaceae: Pooideae: Aveneae) in North America North of Mexico. *Sida* 21:1419–1453.

Good, R. 1953. *The Geography of Flowering Plants,* 2nd ed. Longmans, Green, London.

Gould, F. W. 1951. *Grasses of the Southwestern United States.* University of Arizona Press, Tucson.

———. 1975. *The Grasses of Texas.* Published for the Texas Agricultural Experiment Station by Texas A&M University Press, College Station.

Gould, F. W., and R. B. Shaw. 1983. *Grass Systematics,* 2nd ed. Texas A&M Press, College Station.

Grass Phylogeny Working Group. 2001. Phylogeny and Subfamilial Classification of the Grasses (Poaceae). *Ann. Missouri Bot. Gard.* 88:373–457.

Griffiths, M., and L. Rubright. 1983. *Colorado: A Geography.* Westview, Boulder.

Harrington, H. D. 1946. *Grasses of Colorado (Including Cultivated Species).* Colorado A&M College, Fort Collins.

———. 1954. *Manual of the Plants of Colorado.* Swallow, Ohio University, Athens.

———. 1977. *How to Identify Grasses and Grasslike Plants.* Swallow, Ohio University, Athens.

Hartley, J. R. 1954. The Agrostological Index. *Australian J. Bot.* 2:1–21.

Hartman, R. L., and B. E. Nelson. *A Checklist of the Flora of Colorado.* Rocky Mountain Herbarium, University of Wyoming, Laramie. Posted at www.rmh.uwyo.edu/ [accessed May 2007].

Hitchcock, A. S. 1935. *Manual of the Grasses of the United States.* USDA Miscellaneous Publication no. 200. U.S. Government Printing Office, Washington, DC.

———. 1951. *Manual of the Grasses of the United States,* 2nd ed., revised A. Chase. USDA Miscellaneous Publication no. 200. U.S. Government Printing Office, Washington, DC.

Hitchcock, C. L., A. Cronquist, M. Ownbey, and J. W. Thompson. 1969. *Vascular Plants of the Pacific Northwest. Part 1: Vascular Cryptogams, Gymnosperms, and Monocots.* University of Washington Press, Seattle.

Krishnamachari, K. A., and R. V. Bhat. 1978. Poisoning by Ergoty Bajra (Pearl Millet) in Man. *Indian J. Med. Res.* 64:1624–1628.

Lorenz, K. 1979. Ergot of Cereal Grains. *CRC Crit. Rev. Food Sci. Nutr.* 11:311–354.

National Research Council. 1993. *Vetiver Grass: A Thin Green Line Against Erosion.* National Academy Press, Washington, DC.

Rubright, L. 2000. *Atlas of the Grasses of Colorado.* Lynnell Rubright, Denver.

Schantz, H. L. 1954. The Place of Grasslands in the Earth's Cover of Vegetation. *Ecology* 35:143–145.

Shaw, R. B., S. L. Anderson, K. A. Schulz, and V. E. Diersing. 1989a. Floral Inventory for the U.S. Army Piñon Canyon Maneuver Site, Colorado. *Phytologia* 67(1):1–42.

————. 1989b. Plant Communities, Ecological Checklist, and Species List for the U.S. Army Piñon Canyon Maneuver Site, Colorado. Range Science Series no. 37. Colorado State University, Fort Collins.

Stubbendieck, J. S., S. L. Hatch, and C. H. Butterfield. 1992. *North American Range Plants,* 4th ed. University of Nebraska Press, Lincoln.

Stubbendieck, J. S., S. L. Hatch, and L. M. Landholt. 2003. *North American Wildland Plants.* University of Nebraska Press, Lincoln.

Trewartha, G. T. 1968. *An Introduction to Weather and Climate.* McGraw-Hill, New York.

Vicari, M., and D. R. Bazely. 1993. Do Grasses Fight Back? The Case for Antiherbivore Defenses. *Trends in Ecol. and Evol.* 8:137–141.

Watson, L., and M. J. Dallwitz. 1992. *The Grass Genera of the World.* CCAB International, Oxon, UK.

Weber, W. A., and R. C. Wittmann. 1992. *Catalog of the Colorado Flora: A Biodiversity Baseline.* University Press of Colorado, Niwot.

————. 2001a. *Colorado Flora: Eastern Slope,* 3rd ed. University Press of Colorado, Niwot.

————. 2001b. *Colorado Flora: Western Slope,* 3rd ed. University Press of Colorado, Niwot.

Welsh, S. L., N. D. Atwood, S. Goodrich, and L. C. Higgins. 1993. *A Utah Flora,* 2nd ed., revised. Brigham Young University Press, Provo.

Wingate, J. L. 1994. *Illustrated Keys to the Grasses of Colorado.* Wingate Consulting, Denver.

Zuloaga, F. O., L. M. Guissani, and O. Morrone. 2007. *Hopia,* a New Monotypic Genus Segregated from *Panicum. Taxon* 56:145–156.

627

INDEX

Page numbers in italics indicate illustration; page numbers in bold indicate description of family, tribe, subtribe, genera, or species.

629

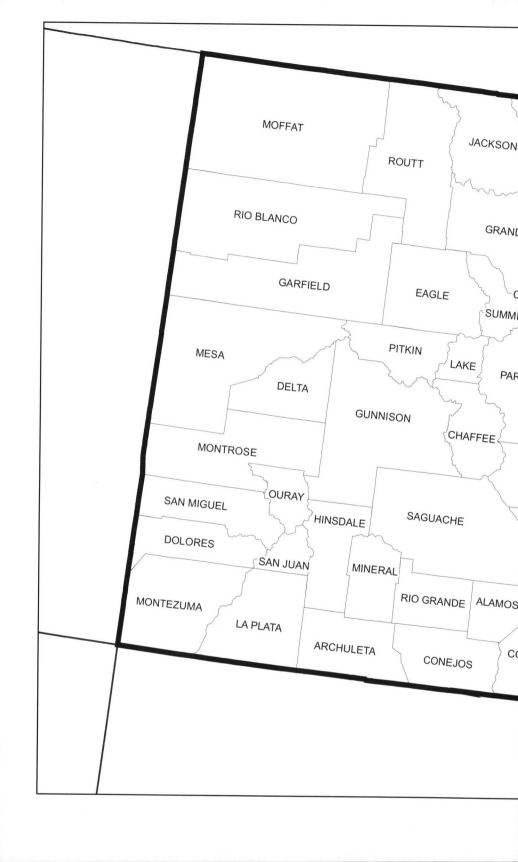

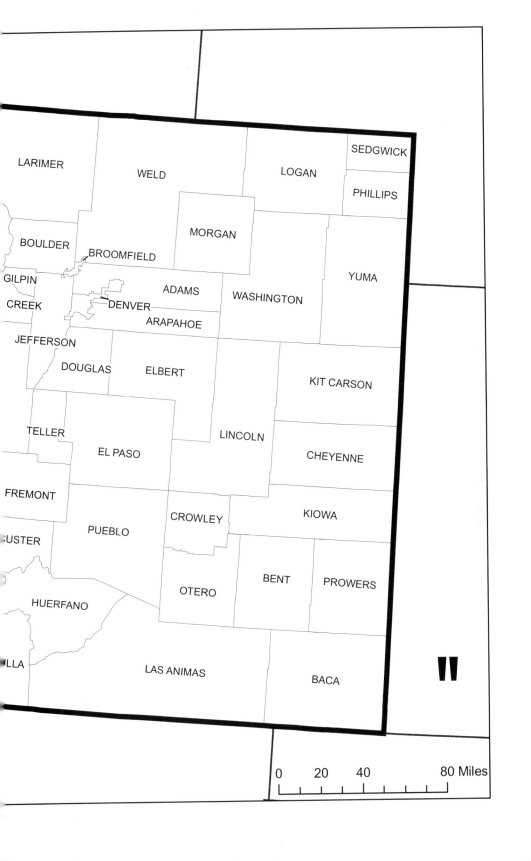

LARIMER

WELD

LOGAN

SEDGWICK

PHILLIPS

BOULDER

BROOMFIELD

GILPIN

CREEK

MORGAN

JEFFERSON

ADAMS

DENVER

ARAPAHOE

WASHINGTON

YUMA

DOUGLAS

ELBERT

KIT CARSON

TELLER

EL PASO

LINCOLN

CHEYENNE

FREMONT

CROWLEY

KIOWA

CUSTER

PUEBLO

OTERO

BENT

PROWERS

HUERFANO

LLA

LAS ANIMAS

BACA

0   20   40   80 Miles

INDEX OF FAMILIES, SUBFAMILIES, TRIBES, AND GENERA

The list below is in alphabetical order and shows the family, subfamilies, tribes, and genera followed by the page number where the treatment is found.